FACILITY PLANNING:
Principles, Technology, Guidelines

STEPHANIE HENAH
814-450-
+107

Jeffrey E. Clark
Harrington College of Design, Chicago

PEARSON
Prentice
Hall

Upper Saddle River, New Jersey 07458

For my students

The essence of good teaching begins with respect for students. °

Library of Congress Cataloging-in-Publication Data
Clark, Jeffrey E., 1942-
    Facility planning : principles, technology, guidelines/Jeffrey E.
    Clark Pearson Prent.
        p. cm.
    ISBN 0-13-114936-9
1. Facility management. 2. Production engineering—Planning—Data
processing. I. Title.
    TS177.C58 2007
    658.2—dc22   2006026282

Editor-in-Chief: Vernon R. Anthony
Editorial Assistant: ReeAnn Davies
Design Coordinator: Miguel Ortiz
Managing Editor: Mary Carnis
Production Liaison: Janice Stangel
Manufacturing Buyer: Cathleen Petersen
Manufacturing Manager: Ilene Sanford
Marketing Assistant: Les Roberts

Pearson Education Ltd.   Pearson Education North Asia Ltd.
Pearson Education Singapore, Pte. Ltd.   Pearson Educación de Mexico, S.A. de C.V.
Pearson Education Canada, Ltd.   Pearson Education Malaysia, Pte. Ltd.
Pearson Education—Japan   Pearson Education, Upper Saddle River, New Jersey
Pearson Education Australia PTY, Limited

10 9 8 7 6 5 4 3 2
ISBN-10: 0-13-114936-9
ISBN-13: 978-0-13-114936-6

BRIEF CONTENTS

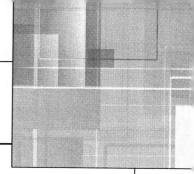

PREFACE   XV

CHAPTER 1   Classical Principles   1
Time Capsule—The Ramesseum   11

CHAPTER 2   Geometry and Planning   13
Time Capsule—The Parthenon   30

CHAPTER 3   Place and Modernism   33
Time Capsule—A Residence at Delos   50

CHAPTER 4   Dwellings   51
Case Study—Operations Centers   63

CHAPTER 5   Management and Networks   67
Time Capsule—The Pantheon   90

CHAPTER 6   Workplace Programming   93
Time Capsule—Cathedrals and the Medieval Program   119

CHAPTER 7   Space Allocation   125
Case Study—Three Facility Programs   142

CHAPTER 8   Space Planning   153
Time Capsule—Cathedrals of Today   176

CHAPTER 9   Area-Measurement Standards   183
Case Study—Global-Technology Company   206

CHAPTER 10   Relational Databases   209
Case Study—International Chemical Company   229

CHAPTER 11   Computer-Aided Facilities Management   233
Case Study—BP America   264

CHAPTER 12   Sustainable Design   269

CHAPTER 13   Building Codes   293

CHAPTER 14   Egress   323

CHAPTER 15   Accessibility   349

CHAPTER 16   ADA Accessibility Guidelines   361

APPENDIX A   Facility-Planning Checklist   403
APPENDIX B   Space-Requirements Worksheet   411
APPENDIX C   Loading AutoCAD Macros   419
APPENDIX D   Archibus/FM Space-Management Module   425

REFERENCE NOTES   439

INDEX   461

CONTENTS

PREFACE **XV**

 Top-Down or Bottom-Up? **xv**
 Facility Planning **xvi**
  What Is a Facility? xvi
  Building Blocks xvii
  Evolution xvii
  Motivation xviii
 Historical Links **xix**
 What Else Is in the Book? **xx**
 Acknowledgments **xxvii**

CHAPTER 1 **Classical Principles**   **1**

 Marcus Vitruvius Pollio **1**
  Fundamental Principles 3
  Vitruvian Man 6
 Omphalos **6**
  Anastomosis of Navelcords 8
  Man and Place 9
 Closing Words **10**

 Time Capsule - The Ramesseum 11

CHAPTER 2 **Geometry and Planning**   **13**

 Ancient Measures **14**
 Earth and Man **16**
 Φ **17**
  Leonardo Fibonacci 17
  Adolf Zeising 18
  Le Corbusier 19
  Anthropometry and Ergonomics 22
 Nineteenth-Century Influences on the Twentieth **23**
  Friedrich Froebel 23
  Owen Jones 25
  Victor Hugo 27
 Renaissance **28**
 Closing Words **29**

 Time Capsule - The Parthenon 30

 Order in the Plan 30

CHAPTER 3 **Place and Modernism**   **33**

 The Potency of Place **33**
  Scientific Method 33
  Temperament and Objectivity 36
  Pre-modernism 36
  Modernism 37
 Barcelona to Bilbao **38**
  Relationships in Play 39
  Utility and Delight in Planning 47

Closing Words   49

Time Capsule - A Residence at Delos   50

CHAPTER 4   Dwellings   **51**
Site Conditions   51
Project Conditions   51
    Module and Ideogram   52
Program Requirements   53
    Space Allocation   56
    Block Planning   56
Schematic Layout   58
    Visualization   59
Details: The *Operations* Center   60
    Appliances   60
    Countertops   60
    Cabinetry   61
    Flooring   61
    Walls, Ceiling, and Lighting   61
Closing Words   61

Case Studies - Operations Centers   63

Control Room for a Large Computer Center   63

CHAPTER 5   Management and Networks   **67**
Scientific Management   67
    Scheduling and Quality Control   68
    Theory Y   69
    Understanding Management's Agenda   70
Influences on Organization and Culture   71
    Product and Service Lifecycles   71
    Technological Change   72
    Globalization   73
    Inflection Points   74
    Communication and Vision   75
Defining the Project Elements   76
    Programming   77
    Concept Development   78
    Implementation   79
Project Management   79
    Project Networks   79
    Project Definition   80
    Dependencies, Durations, and Resources   81
    Monitoring the Project   86
Quality and Facility Management   86
    Determine Goals and Targets   87
    Determine Methods of Reaching the Goals   87
    Engage in Education and Training   88
    Implement Work   88
    Check the Effects of Implementation   88
    Take the Appropriate Action   88
Closing Words   88

Time Capsule - The Pantheon   90

CHAPTER 6   Workplace Programming   93

Beyond Theory X and Theory Y   93
The Facility Program and the Balance Sheet   95
Types of Organizations   96
The Programming Process   99
Generating Facility-Requirements Data   100
Analyzing the Data Worksheet   102
Structuring the Data   102
Organizational Mechanisms   105
Codes   107
Space-Allocation Guidelines   109
Synthesis   113
Adjacency Relationships   114
The Sandpaper and Army Blanket Technique   115
Program Case Study   115
Closing Words   118

Time Capsule - Cathedrals and the Medieval Program   119
Order in the Universe   120
Earthly Order   121
Salisbury   121
Ad Quadratum   123

CHAPTER 7   Space Allocation   125

From Programming to Facility Management   125
Data Pipelining   127
Programmed Area Requirements   127
Populating the Database   128
Management and Planning Reports   129
Area-Allocation Tools   135
Closing Words   140

Case Study - Three Facility Programs   142
Blue Cross and Blue Shield   142
Phase I   142
Phase II   142
Phase III   143
Phase IV   144
An International Pharmaceutical Company   144
Analysis of Existing Conditions   145
Spatial Planning Models   145
Implementation and Phasing   145
Causal Factors   148
A State Government Agency   148
The Building Envelope   148
Blocking and Stacking   150

CHAPTER 8   Space Planning   153

Wayfinding   153
Primary Circulation   153
Corridors and Aisles   154
Building Character   158
The Space Layout—Partition Plan   160
Area-Allocation Analysis   162
Drawing Area Polylines   162

Area-Allocation Layers   164
Allocation Process   167
Facility-Management Data   168
Room Categories and Types   171
Data Pipelining   174
Closing Words   174

Time Capsule - Cathedrals of Today   176
Sears Tower   178
Bank of China   179
Torres Petronas   180
Torre Espacio   180
Summing Up   180

CHAPTER 9   Area-Measurement Standards   183
ANSI/BOMA   183
BOMA Usable Area   184
BOMA Rentable Area   184
Area Measurement   185
ASTM-E1836 (IFMA)   192
Facility Area Relationships   192
Facility Rentable Area   193
Facility Usable Area   194
GSA/PBS   195
GSA Area Measurement   195
Space Categories and Types   198
PEFIC   199
Nonassignable Categories   200
Room-Use Codes   200
Other PEFIC Classifications   203
Comparison   204
ASTM/IFMA   204
PEFIC   204
Closing Words   205

Case Study - Global-Technology Company   206
Scope of the Project   206
Project Management and Consulting   206
Assembling Drawings and Data   206
Drawing-Conversion Management   207
FAST Web Interface   207

CHAPTER 10   Relational Databases   209
Origins of the Relational Model   209
Hierarchies   210
Networks   211
Relational Databases   211
Codd's Twelve Rules   211
Rule 1: Information   211
Rule 2: Logical Accessibility   212
Rule 3: Null Values   212
Rule 4: Self-Descriptive   213
Rule 5: Comprehensive Data Sublanguage   213
Rule 6: Views   214
Rule 7: Set-Based Relational Operations   214
Rule 8: Physical-Data Independence   215

Rule 9: Logical-Data Independence   215
Rule 10: Integrity Independence   215
Rule 11: Distribution Independence   216
Rule 12: Nonsubversion   217
Relational Properties and Normalization   217
First Normal Form   217
Second Normal Form   218
Third Normal Form   218
Boyce-Codd Normal Form   219
Fourth Normal Form   220
Fifth Normal Form   221
The SQL Language   222
SELECT   222
INSERT   223
Program Requirements   224
Database Administration   224
Protection and Security   224
User Interface   226
Closing Words   227

Case Study - International Chemical Company   229
Data-Network Hosting   229
Infrastructure Architecture   229
Hosting Methods   230
Implementation   230

CHAPTER 11   Computer-Aided Facilities Management   233
CAFM Applications   234
TRIRIGA   234
FAMIS   235
Service Orientation   237
Archibus/FM   238
Real Property and Lease Management (1.0)   239
Strategic Master Planning (2.0)   240
Space Management (3.0)   242
Design Management (4.0)   245
Furniture and Equipment Management (5.0)   248
Telecommunications and Cable Management (6.0)   250
Building Operations Management (7.0)   253
Task Categories   256
System (8.0)   259
Closing Words   263

Case Study - BP America   264
Drawing, Database, and Application Issues   264
Customized Archibus/AutoCAD Procedures   264
Merger-Integration Metrics   264
Collaborative Space   265
Key Performance Indicators for the New Work Environment   265
Managing Change   267

CHAPTER 12   Sustainable Design   269
Environmental Awareness in the United States   270
Unforeseen Consequences   270
Global Environmentalism   271
Principles of Sustainability   273

Six Major Areas of Sustainability Awareness   274
    Site Considerations   274
    Water Supply   274
    Energy Management   275
    Materials Utilization   276
    Interior Environment   276
    Waste Management   277
Leadership in Energy and Environmental Design—LEED   278
    LEED Green Building Rating System   278
    LEED for Commercial Interiors—LEED-CI   279
    LEED for Existing Buildings—LEED-EB   287
    LEED for New Construction—LEED-NC   290
Closing Words   291

CHAPTER 13   Building Codes   293
Use and Occupancy Classification   294
    Assembly   295
    Business   297
    Educational   297
    Factory and Industrial   297
    High Hazard   297
    Institutional   298
    Mercantile   298
    Residential   298
    Storage   299
    Utility and Miscellaneous   299
    Special Building Types   299
Types of Construction (600)   304
    Use Classifications and Construction Types   305
    Building Height and Area Limitations (500)   307
Fire-Resistance-Rated Construction (700)   308
    Fire-Resistance Ratings and Fire Tests (703)   309
    Exterior Walls (704)   310
    Fire Walls (705)   310
    Fire Barriers (706)   311
    Shaft Enclosures (707)   312
    Fire Partitions (708)   312
    Smoke Barriers and Smoke Partitions (709–710)   312
    Horizontal Assemblies (711)   312
    Penetrations (712)   313
    Fire-Resistant Joint Systems (713)   313
    Fire-Resistance Rating of Structural Members (714)   313
    Opening Protectives (715)   313
    Ducts and Air-Transfer Openings (716)   314
    Other Fire-Resistance Requirements (717–721)   315
Fire-Protection Systems (900)   315
    Automatic Sprinkler Systems (903)   315
    Alternative Automatic Fire-Extinguishing Systems (904)   317
    Standpipe Systems (905)   317
    Portable Fire Extinguishers (906)   318
    Fire Alarm and Detection Systems (907)   318
    Emergency Alarm Systems (908)   321
    Smoke-Control Systems (909)   321
    Smoke and Heat Vents (910)   321
    Fire Command Center (911)   322
Closing Words   322

CHAPTER 14   Egress   **323**
  Fire Has an Attitude 324
  Means of Egress (1000) 327
      General Requirements 327
      Occupant Load and Egress Width 328
      Accessible Means of Egress (1007) 330
      Doors, Gates, and Turnstiles (1008) 331
      Changes in Level or Floor 334
      Means of Egress Illumination and Signage 336
      Aisles, Corridors, and Exit Access (1013) 337
      Exits and Exit Discharge 340
      Special Requirements 344
  Closing Words 347

CHAPTER 15   Accessibility   **349**
  ICC/ANSI A117.1-2003 350
      Scoping Requirements (1103) 350
      Accessible Route and Entrances (1104–1105) 350
      Parking and Passenger Loading (1106) 353
      Dwelling and Sleeping Units (1107) 354
      Special Occupancies (1108) 356
      Other Features and Facilities (1109) 357
      Signage (1110) 358
  Closing Words 359

CHAPTER 16   ADA Accessibility Guidelines   **361**
  Terminology and Documentation 362
      What Does "Accessible" Mean? 363
  Scope 364
      Sites and Exterior Facilities: New Construction 364
      Buildings: New Construction 364
      Building Additions and Alterations 365
      Historic Preservation 365
  Building Blocks 367
      Floor or Ground Surfaces (302/4.5) 368
      Changes in Level (303/4.8) 368
      Turning Space (304/4.2) 368
      Clear Floor or Ground Space (305/4.2) 369
      Knee and Toe Clearance (306) 370
      Protruding Objects (307/4.4) 370
      Reach Ranges (308/4.2) 370
      Operable Parts (309/4.27.4) 372
  Accessible Routes 373
      Walking Surfaces (403) 373
      Doors, Doorways, and Gates (404/4.13 and 14) 374
      Ramps and Curb Ramps (405–6/4.7 and 8) 377
      Elevators (407/4.10) 377
      Limited-Use and Private-Residence Elevators (408–409) 378
      Platform Lifts (410/4.11) 379
  General Site and Building Elements 379
      Parking Spaces and Passenger-Loading Zones (502–3/4.6) 379
      Stairways (504/4.9) 379
      Handrails (505) 379
  Plumbing Elements and Facilities 382
      Drinking Fountains (602/4.15) 382
      Bathrooms, Water Closets, and Toilet Compartments (603–4) 382

Urinals (605/4.18)   385
Lavatories and Sinks (606/4.19)   385
Bathtubs (607/4.20)   386
Shower Compartments (608/4.21)   387
Grab Bars (609/4.26)   390
Seats (610/4.27)   391
Other Plumbing-Related Elements (611–12)   392
Communication Elements and Features   393
Fire Alarm Systems (702/4.28)   393
Signs (703/4.30)   393
Telephones (704/4.31)   394
Detectable Warnings (705/4.29)   394
Assistive-Listening Systems (706/4.33)   396
Automated Teller and Fare Machines (707/4.34)   396
Two-Way Communication Systems (708)   396
Special Rooms, Spaces, and Elements   396
Wheelchair Spaces, Companion Seats, and Designated
Aisle Seats (802/4.33)   397
Dressing, Fitting, and Locker Rooms (803/4.35–37)   397
Kitchens and Kitchenettes (804)   397
Medical Care and Long-Term Facilities (805/6.0/*IBC* I-2)   397
Transient Lodging Guest Rooms (806/9.0/*IBC* R-1))   398
Holding Cells and Housing Cells (807/12.0/*IBC* I-3)   398
Courtrooms (808/11.0/*IBC* A-3)   399
Residential Dwelling Units (809)   399
Transportation Facilities (810/10.0/*IBC* A-3)   399
Built-In Elements   400
Dining Surfaces and Work Surfaces (902/5.0/*IBC* A-2)   400
Benches (903/4.37)   400
Checkout Aisles, Sales and Service
Counters (904/7.0/*IBC* A-3, B, M)   401
Recreation Facilities (1000/15.0/*IBC* A-3)   401
Closing Words   401

Appendix A   Facility-Planning Checklist   403
1. Space Allocation and Use
2. Interior Systems
3. Building Systems
4. Building Enclosure
5. Architectural Concepts
6. Structural System
7. Site Conditions
8. Building Services
9. Safety and Environmental Issues

Appendix B   Space-Requirements Worksheet   411
Worksheet Instructions
1—Groups
2—Positions
3—Support Facilities
4—Projections
5—Spaces
6—Furnishings
7—Storage
8—Machines

9—Relationships
10—Departmental Adjacencies
11—Comments

APPENDIX C   Loading AutoCAD Macros   419
Fpdata.ini

APPENDIX D   Archibus/FM Space-Management Module   425

REFERENCE NOTES   **439**

INDEX   **461**

PREFACE

In the world of investing, a *top-down* strategy means looking at the different industries or business sectors to start with and then narrowing down the choices to the best-performing companies within those sectors or industries. A top-down strategy looks first at the overall economy, then groups of similar companies, and then specific companies that are expected to show exceptional performance. The final step is to analyze the fundamentals of a given security.

A *bottom-up* strategy begins with the individual company: its management, business model, history, prospects for growth, and other distinctive qualities. The company is evaluated on its own merit based on the belief that it is superior to its industry peers and can be expected to outperform even in the face of adverse industry or economic trends. The final step may be to compare the company with others in its business sector.

Two other terms that are polar opposites in the investment industry are *technical* versus *fundamental* analysis. Technical analysis relies largely on perceived trends within business sectors or industries within the context of the overall business climate, relying heavily on mathematical models and statistics. Fundamentalists believe that a company's potential can best be determined from its internal financial data and management performance.

TOP-DOWN OR BOTTOM-UP?

One of the buzzwords in the information technology sector, as business systems continue their migration to the Web, is *service-oriented architecture* or *SOA*. An *SOA* is essentially a collection of services on a network that communicate with each other. (A *service* is a self-contained routine or process that performs a well-defined function supporting other programs. It does not depend on the context in which it is called or the state of other services.) The communication can simply involve passing data back and forth or it can involve several services coordinating some activity. All that is required is some method of connecting the services to one another.

Service-oriented architecture is not a new idea. Microsoft's *Distributed Component Object Model* (*DCOM*) was among the first to be widely supported across networks. If you are running a Windows PC, you can observe upwards of forty local services running on your machine.°

Most developers agree that an SOA should support the needs of the business, but there is a wide divergence of opinion on how this should be brought about, which leads back to the top-down versus bottom-up debate that often portends a clash of cultures.

Top-down here means senior management vigorously driving business process management so that system implementations reflect strategic process definitions. Bottom-up can mean that existing systems are simply repackaged within Web services to create a service layer. This approach is prone to creating mismatches between the new Web service interaction style and obsolete constraints of the legacy system. The solution is usually to meet somewhere in the middle.

SOA development is not purely top-down nor bottom-up. On the one side, the developer must make a continuing effort to understand and respond to the needs of management and business processes. On the other, he must integrate the applications and technologies that implement those processes. But the developer must always be aware that no amount of technology can save the effort if the overall developmental approach is getting in the way.

FACILITY PLANNING

Why do we begin a book on facility planning by talking about investment strategy and software development? The answer has several dimensions.

One element is that the polarity of top-down versus bottom-up is common to all three activities. The first set of choices facing the planner faced with developing a facility program is how (and where) to begin. Discussions with upper management can give strategic direction to the effort but cannot offer much detail. Interviews with managers and surveys of workers can lead to massive amounts of data being accumulated, but finding a coherent structure can be challenging. Somehow the views from the top and the bottom have to be reconciled.

A second aspect is that some understanding of investing and computer systems development, along with a multitude of other business processes, is an essential part of the facility planner's knowledge base. The articulation of space within and among assemblies of different scales is the planner's product: *utility*, *strength*, and *delight* are its attributes. The planner must learn to listen to and be able to understand the client's language in order to effectively translate a dynamic organization's immediate and long-term goals into a physical environment that has these attributes.

This is not a *design* book in the conventional sense. There are few color photos of finished design because our focus is not upon specific design solutions, furniture types, or systems, nor upon colors and materials except from a functional standpoint. Recognizing that most office environments of today are movable and changeable, many in fact have no hard-walled offices at all. This is in keeping with the dynamic of the postmodern business organization. One of our main case studies talks about "collaborative space" in relation to "key performance indicators."

Our primary interest is process. The success of programming depends to a large degree on successfully communicating with managers and workers who do not (in most cases) immediately relate to the profusion of shapes and configurations (panels, work surfaces, storage components, wire management systems, etc.). So we speak traditionally in terms of private, semiprivate, and open office areas and varied combinations of representative furniture components. It is the task of the planner to assimilate the mass of detail working from the bottom up while finding a coherent schema that will unite all the elements from the top down.

What Is a Facility?

A conventional view of space planning (and design) is that it involves buildings, offices, and interior environments. The term *facility* is most commonly used to refer to commercial office buildings and the physical environments. Some organizations such as banks, for example, use the word in a specialized sense when referring to a branch office.

The basic principles underlying the planning process, however, are common to all types of buildings including dwellings. For the home, Le Corbusier used the term *machine a habiter*: a "machine for living in." Deceptively straightforward looking and appealing on the exterior, his Villa Savoye is a meticulously planned and highly complex structure. Differences are largely a matter of scale and in the mechanics of developing the program.

Much of this book is devoted to methods of planning appropriate for commercial projects of fifty thousand, to five-hundred thousand, square feet and up. Automated database applications for facility management are appropriate for the large organization or business enterprise. But we begin our discussion of planning methodology with a small family dwelling because it allows us to focus on some of the very basic, scalable concepts. The approach to developing a set of program requirements varies as a function of scale at least as much as they are a consequence of project type.

Building Blocks

If we look at the facility-planning process as a whole, including its ongoing maintenance after completion, we can identify three distinct classes of activities:

Facility planning = programming + space planning + facility management

In a large-scale enterprise, each class has many distinct parts that must be considered individually and holistically while the project is underway.

Programming = management consulting + strategic planning + organizational structure + physical space and equipment + relationships + project management

The programming process begins and ends with management and in that sense is top-down, but it also depends for its success on a deep understanding of low-level requirements such as furniture, equipment, and physical relationships among the various elements. Often, the development of the program begins at the middle, so that the organization can be understood in cross section. Then a look upward to senior management gives strategic direction while a thorough analysis of detailed functional requirements begins to shape the plan.

Space planning = building geometry + measurement standards + interior environment + ergonomics + office layout + building codes + accessibility + sustainability

Space planning and design cover a great deal of territory, and the challenge is to integrate the requirements from the top (management) with the requirements from the bottom (physical/functional) to create an environment that is safe, comfortable, and flexible; has clarity; and performs effectively.

Facility management = data management (space, telecom, real estate, etc.) + integration with CAD + operations + maintenance

Facility management, in a sense, encompasses the other two activity classes in that it is an ongoing process, even during the implementation of a major project. It can involve all aspects of the operations and management of a facility including real estate, furniture and equipment inventory, telecom management, and maintenance.

Evolution

Over the course of some thirty-odd years, I have managed numerous projects ranging in size from a few thousand to several million square feet, each averaging slightly over three-hundred thousand square feet. They all fall into the commercial or *contract* as opposed to residential design category, and Table 1 breaks them down by organization type. The majority of them involved substantial formalized programming. Nearly twenty of the projects listed as *corporate* were for *Forbes 500* companies. Most of the governmental projects were federal, although several were for state and local agencies.

There were other projects to be sure. Over the years, I did my share of tenant planning, representing buildings and producing layouts for prospective tenants ranging in size from a few hundred to a few thousand square feet. Virtually all of these projects involved preparing and maintaining records of building areas and rent factors.

Although these projects had in common the fact that there was a good deal of number crunching involved, none of them could be called *typical* in the sense that it provided a prototype applicable to the others. I once tried to compare projects in similar businesses to see if common denominators could be found. For instance, did banks have a predictable utilization rate (area per person) just by virtue of the fact that they were banks? There were no such commonalities to be found. Even within such groupings, the organization

Table 1 Programming, Planning, and Facility-Management Projects

Type of Organization	Companies	Rentable Area	CEO Directly Involved	
Corporate	34	14,300,000	11	30%
Financial	16	3,000,000	6	35%
Governmental	7	1,600,000	—	—
Insurance	6	2,700,000	3	50%
Publishing	3	500,000	—	—
Retail	5	400,000	1	20%
TOTALS	71	22,500,000	21	30–40%

structures were different. The product mixes were different, the management philosophies were different, and space allocations were different.

The notion of a corporation being treated as a person for accounting and tax purposes is functionally accurate. Each has its own personality and character and must be treated as such. Organizations have certain common characteristics to be sure. I have attempted to categorize some of them in Chapter 4, but such comparisons are rarely quantifiable and are useful only up to a point.

Having to tabulate personnel and square footage figures involving upwards of a thousand people and several hundred thousand square feet prompted me to devise some early computer applications when none were commercially available. When the PC appeared in the early 1980s, it was a great liberation for architects and design firms for whom the cost of computer access had been largely prohibitive up to that point. PC-based, computer-aided design programs such as AutoCAD came into their adolescence in the latter part of that same decade, as did many varieties of spreadsheet and data-management software.°

There were Luddites among the revolutionaries, however. One of the designer-principals of the firm where I started out in the mid-1960s insisted throughout most of the 1970s that you couldnot show a banker a computer printout! A "management" report had to be typewritten with a proportionally spaced font and bound in imitation leather to have credibility with a senior executive. Thirty years later, an architect who knows as much about putting together high-rise buildings as any now practicing refused to commit projects to CAD because he could not walk around the office and see work in process on the drafting tables. He was steadfast in his refusal to buy a plotter yet could not deal with the fact that the work was "in the machine"!

But the PC, the World Wide Web, office furniture systems, accessibility, and "green" design principles have all been integral components of workplace transformations over the past three decades. Architects such as Frank Gehry, who proudly admitted on an *ABC Nightline* episode that he didnot know how to turn the "darn" thing on, nonetheless professes complete faith in computers as indispensable design tools. His architectural tour de force in Bilbao was designed using software originally developed for aircraft engineering long before the humble PC came on the scene.

Motivation

Another thing that the projects listed in Table 1 had in common was that virtually all of them had the attention of senior management. Nearly one-third benefited from the direct involvement of the chief executive officer, president, or chairman. This tended to be the case when there was substantial programming involved, either prior to a major building program or when space needed to be extensively reconfigured following a major organizational change.

One case in point was a major oil company that was completely reorganized following the first energy crisis in the early 1970s. The company was realigned from a traditional organizational structure into three functionally independent business units: chemicals, petroleum products, and natural resources. It was the major employer in a small town at a time when nearly five thousand people had been laid off. Over the course of several years, several million square feet of office and laboratory space were reconfigured to satisfy a new business model with new adjacency requirements and space-allocation parameters.

Much of this change, some of it painful indeed, was brought about by the CEO's conviction that the slimmed-down company might be obliged to divest itself of one or more of its businesses and that the three new business units needed to be capable of operating autonomously. The company survived intact but was eventually acquired by another, larger enterprise.

Senior management is nearly always vitally interested in a major building program, simply because of the magnitude of the capital resources involved. When the chief executive is personally invested, there are often other factors in play as well. Building, expanding, or extensively reconfiguring a corporate headquarters is an activity that is outside the domain in which the business executive usually feels comfortable. His or her expertise is, after all, in running the business. This is why the planner-programmer often fills a role similar to that of management consultant. He or she brings to the table up-to-date, specialized knowledge and technology that can aid in realizing business objectives, perhaps also helping to identify new ones.

Another often overlooked factor is that the finished project may have significance as a physical symbol of the senior executive's leadership or the company itself. The planner's ability to listen to his client, understand her strategic goals (and motivations), and offer creative and practical solutions during the earliest phases of a project is often the critical factor in building a successful long-term relationship. This is the time at which important issues that the end user may not have even considered such as accessibility, sustainability, and new building technologies can be most easily integrated into the planning process.

HISTORICAL LINKS

The driving force behind the ascent of man has been his intellect and his use of the tools he has created for building. The earthbound temples of the Egyptians glorified their god kings. The Greeks glorified their civilization and their gods with temples of the finest materials and highly refined geometry. The Romans built an empire that spanned three continents and half a millennium along with temples dedicated to a pantheon of gods. Romanesque churches gave way to towering Gothic cathedrals in the Christianized West, using technology learned from the Islamic East.

From the Renaissance forward, there was a shift in focus from soaring masonry cathedrals built for the glory of God to secular palaces culminating in the towering skyscrapers of today. The Renaissance inspired a renewed interest in the Classical forms, which, when they were applied as mere decoration, led to the excesses of the eighteenth and nineteenth centuries. There was ultimately a disconnect between the experience of history and its architecture, about which Victor Hugo had much to say in his novel of Notre Dame.

The fundamental relationships of geometry were well known to the Greeks, and the theory and practice of engineering were raised to an art form during the first century of Roman dominance. Many of the only roads that existed across Europe through the Middle Ages were those built by the Romans. The elemental geometric principles have a universality that places them at the core of rational planning. That is why they are the focus of the first part of this book.

Some may feel that the historical references are superfluous in a textbook on facility planning. I have included them to illustrate the direct application of geometry in giving order and significance to a plan. The "Time Capsules" illustrate several architecturally significant buildings as products of complex programs within their respective cultural frameworks. The execution is the result of functional requirements completely foreign to our own, yet the expressions of order, arrangement, and balance resonate with our contemporary ways of seeing. The "Time Capsules" provide a counterpoint to the "Case Studies" representative of contemporary programming and facility-management projects.

WHAT ELSE IS IN THE BOOK?

The following pages give a wide view of the book's overall plan and the material that is covered in each chapter

Chapter 1: Classical Principles

Marcus Vitruvius Pollio was a Roman engineer and architect in the service of Augustus Caesar. He wrote the first and one of the most influential treatises on planning and construction, *De Architectura*, around 28 BCE, the same year Agrippa's original Pantheon was built.

Most of the ten books comprising *De Architectura* deal with the planning of cities and the construction of war machines, but the first and third set forth requirements for an architect's education and six fundamental principles of design. As they have not been much improved upon in two thousand years, we have taken these two enumerations as our starting place. We propose that *technology* be added to the eleven fields in which Vitruvius says the architect/planner should be educated.

In the initial chapter of his first book, Vitruvius describes architecture as a science arising out of many other sciences, concluding that the architect must be generally well informed in many different disciplines. It is only mathematicians he holds in higher regard. There is also the Vitruvian Triad, added almost as an afterthought in the final paragraph of the third chapter: three attributes resulting from his principles being properly applied.

It is in this third book that Vitruvius compares the symmetries of the temple to those of the human form. Thus was written the original specification for da Vinci's image of man as universal measure. Vitruvius was the classical authority for Michelangelo, Alberti, Bramante, Serlio, Palladio, and others who revitalized his instructions during the Renaissance.

Chapter 2: Geometry and Planning

The Vitruvian principles, if not so much the classical orders, remain useful tools for planning. In Chapter 2, we follow the symbolism of the omphalos, the navel, the center, along a golden cord from Vitruvius to da Vinci to Le Corbusier, making the connection between ancient Greece and modern science, technology, and architecture. The architect and planner of today must, just as in early Rome, maintain a broad worldview. Our systems of measurement are connected to our comprehension of the earth and its place in the cosmos. The application of a simple mathematical series such as that named for Fibonacci, and the golden section from which it originates, provides a rational schema for an ever-expanding geometric system of proportion.

The second chapter examines four figures of the nineteenth century who influenced the development of several great architects of the twentieth: Friedrich Froebel, Victor Hugo, Owen Jones, and Adolph Zeising. Our principal interest here is to underscore the fact that many important influences on planning and design come from seemingly unrelated fields. It is up to the educated person to make the connections, for, in Vitruvius' words: "it is by his/her judgment that all work done by the other arts is put to the test."

The last of Vitruvius' six principles, economy, is echoed in Ockham's Razor: "It is useless to do with more what can be done with less," restated with Mies van der Rohe's characteristic verbal parsimony as "Less is more!"

Chapter 3: Place and Modernism

In Chapter 3, we take a comparative view of the design principles of three foremost architects of the twentieth century: Ludwig Mies van der Rohe, Le Corbusier, and Frank Lloyd Wright.

Victor Hugo prophesied in 1832 that the "great accident of an architect of genius may happen in the twentieth century, like that of Dante in the thirteenth." It was no accident that the result of advances in science and technology made possible new materials demanding a new expression. There are differences but also many commonalities in the philosophies of the triad of master builders who fulfilled Hugo's prediction.

We continue the development of some earlier themes, among them the omphalos, the center of our sense of place. We touch upon some of the *faux pas* that were made in the name of the Beaux Arts tradition and during the transition to the modern era. It took a special vision and some experimentation before new principles were discovered that were not simply distortions of the past. It was largely new materials, strong in tension, that allowed the separation of the enclosure of a building from its structure and interior space. The omission of load-bearing walls in the "free plan" permitted space to flow so that space could be defined rather than confined. Space planning, the articulation of the voids, became a primary focus.

The late-twentieth-century work of Frank Gehry stands in stark contrast to the modernism of earlier decades, but is entirely rational in its execution. We look at his Guggenheim Museum in Bilbao, one of the last great structures of the century. The monumental display spaces within admirably fulfill their function while the form of the building transcends sculpture.

This chapter explores an additional pertinent idea: the principle of tolerance. Vitruvius said that the architect must have natural gifts but must also be amenable to instruction. In the last century, we saw all too clearly the results of intolerance and dogma that continue to affect our world culture. An integral part of the planning process is considering the possibility that you are on the wrong track. In this regard, we also propose that *humility* be among our guiding principles.

> The ear that hears the reproof of life stands among the wise. He that refuses instruction despises his own soul but he that hears criticism gains understanding. °

Chapter 4: Planning a Dwelling

We have discussed the development of systems from the top-down versus bottom-up. When it comes to facility planning, we might use as well the construct of inside-out as opposed to outside-in. Even with an assembly such as the *machine a habiter*, we have the internal functional requirements (the program) to consider on the one hand and the external envelope (the site) on the other.

Here we begin to examine the space programming and planning process. Most of our discussion will center on offices and larger-scale projects that lend themselves to systematized facility management, but we choose to begin with the dwelling for two reasons. First, the basic activities are similar. Sitting down with the owners to document their program for a home is a highly personalized activity. It is in keeping with the theme of building to satisfy our sense of place.

Planning a five-hundred thousand-square-foot office facility may be an equally personal experience for the owner or manager of the small business. Projects of every different order of magnitude require a different approach. Twenty five thousand to fifty thousand square feet is usually the threshold for a formalized programming analysis, and five-hundred thousand square feet certainly requires all the resources of a major project. Still, planning a new corporate headquarters may be an equally significant undertaking for its chief executive.

Second, looking at a simple dwelling allows us to introduce some of the basic techniques for information gathering, documentation, and visualization of the program. We can build upon these in developing effective approaches for larger-scale projects. At all levels of detail, we must always keep in mind the character and constraints of the building or site as well as the personalities and culture of the people with whom we are working.

Chapter 5: Management and Networks

In order to appreciate the functional requirements of an organization, we must have some knowledge of its mission and how it operates. From the beginning of the Industrial Revolution, managers attempted to measure the factors that led to greater productivity. Management in the first two industrialized ages was, by and large, as dehumanizing in its methods and practices as those of the monarchies and cottage economies that preceded them. This led Karl Marx, writing in 1844, to conclude that money is the *only* need produced by an economic system. Gradually, however, but at an accelerating rate in the age of the computer and an economy increasingly based on service and professional skill, more humane and effective management systems have come to be applied.

An understanding of management issues and concerns is at the root of effective programming and planning. Organizational theory, statistical quality control, product lifecycles, and globalization all need to be added to the traditional areas of study. The facility planner must have some perspective on how business practices have changed over the past century in order to realize how a particular enterprise may be evolving in today's competitive environment.

The organization structure and culture, management's plans for future growth and change, business systems in place or being put in place are all forces that will shape the planned environment. The possible effects of changing any of these variables must be adequately foreseen in determining the facility's form. Communications and the support of networks are critical elements, person to person as well as machine to machine.

Systematic management of the project at hand is also a necessary component. Toward the end of this chapter, we discuss how to set up a project plan and manage the planning process stage by stage. Finally, we apply some of the principles of quality management to the process and show how programming usually offers a natural point of entry into the quality cycle.

Chapter 6: Workplace Programming

The modern business enterprise is a complex organism made up of people, equipment, and machines. Some spaces must be designed to serve specialized functions while many others are planned for multiple uses and productive interaction. To a certain extent, the space layout will reflect management's hierarchy. The actual plan may take many different forms depending on the organization's character and formal structure. Respecting the individual's sense of place should be a conscious concern so that the work environment fosters a sense of involvement rather than depersonalization.

Chapter 6 focuses on the mechanics of developing the facility program and understanding the subtleties of each project. Earlier, we observed that each organization is unique and must be analyzed on its own terms, but there are certain identifiable physical and cultural features by which some classes of company can be identified. We begin the sixth chapter with a look at some of the types of organization one is likely to encounter in planning for business ventures.

Next, we discuss the kind of information that it is necessary to collect and suggested methods for going about it. Distinguishing between two principal kinds of functional space, we identify ten essential data categories that can be used to describe virtually all work spaces and support areas. A significant part of the development of facility-management databases requires the precise classification of all kinds of spaces for allocation and accounting purposes. We distinguish between spaces that are *occupiable* and *assignable* versus those that may not be, although they may be functional areas in various respects. Certain kinds of areas are not programmable except as quantities that must be estimated and factored in to the general space inventory.

Giving the facility program a structure requires that the management hierarchy be represented. Locations of the numerous organizational components must also be charted in ways that will allow different layout options to be analyzed and evaluated. The open network, a tree structure, is among the most useful organizational mechanisms for developing the facility-management database. Other aspects of the process are covered: space allocation, developing space-assignment guidelines, use of the building-planning module, and analysis of adjacency relationships.

The chapter concludes with a short "Case Study" illustrating the use of some basic graphic tools to visualize organizational and adjacency relationships in a medium-size commercial space, that of a credit-card-processing company.

Chapter 7: Space Allocation

Chapter 7 concentrates on some specific, computer-based tools for documenting facility-requirements data, focusing on the use of a general-purpose Microsoft Access database application, a spreadsheet, and a computer-aided design (*CAD*) program. With an emphasis on ways in which these applications can be used together, passing data back and forth while converting them from numeric to graphic form, we will see how the elements of the PROGRAM → IMPLEMENT → EVALUATE → REFINE cycle develop, from the facility program to building a computer-aided facility management (*CAFM*) database.

The Access database summarizes program requirements and produces detailed reports for use in planning, some of the data from which can be imported into Excel for further analysis. Automation services are used to convert the spreadsheet data into graphic form in CAD, which can then be fashioned into adjacency schemes (bubble and stacking diagrams) and plan documents (block diagrams and detailed space layouts). Space-allocation plans are linked to a facility database using AutoCAD polylines having an *area* property that accurately delineates each space. The polylines can be directly linked or pipelined into various CAFM databases from which accurate allocation and utilization reports can be produced.

Three Facility Programs

Three strategic master occupancy plans are presented in conjunction with this chapter as case studies to show how these tools may be applied. The bubble diagrams, block layouts, and stacking configurations illustrate ways in which different planning options can be represented. The first master plan involves an extensive reorganization and consolidation of executive and operating departments within a new headquarters, developed on a phased basis over a period of nearly ten years.

The second, contrasting plan focuses on deferring the need for new construction for several years through the efficient reconfiguration of existing space, examining options offered by several different arrangements of a matrix organization.

In the final case, the program called for the relocation and consolidation of a state government agency. This study shows the application of space-allocation standards and stacking diagrams as the basis for developing a new headquarters building.

Chapter 8: Space Planning

Once the program has been established, the location of every required space must be determined by a detailed layout within a specific building envelope. Any given building may offer several alternative configuration possibilities, and it is at this point that alternative approaches can be studied. They are evaluated both in terms of the efficiency with which the occupiable space can be utilized as well as the organization of the circulation system. If a new building is being planned, both of these elements must be considered in arriving at proposed architectural options.

The implementation of the program that determines the arrangement of physical space becomes the network through which, at any given moment, a sense of place can be experienced. The perception of an orderly flow when traveling from point *A* to point *B* is often referred to as *wayfinding*. Exit signage, graphics, and special illumination can reinforce an occupant's sense of orientation where life safety is an issue; but visual, tactile, and even aural stimuli contribute to every individual's sense of comfort and well-being on a continuing basis. These elements together comprise people's mental images of the space and the quality of the environment while helping them to gain their bearings in both new and familiar settings. °

This chapter begins with a discussion of wayfinding considerations in relation to principal corridors and aisles and looks at ways in which circulation can be handled in several different types of building. Each building's character is an important factor in determining how space can be configured within its boundaries; therefore, circulation possibilities should be examined before detailed space planning begins.

Continuing our discussion of the final "Case Study" from Chapter 7, we examine the stages in the development of a detailed space layout for one typical floor. AutoCAD polylines are used to illustrate how accurate area measurements can be made using CAD, and a set of layer-naming guidelines is defined for the purpose of consistently allocating and accounting for space. CAD objects such as polylines provide a continuing link between the facility layouts and the CAFM database. The remainder of this chapter is concerned with ways in which data can be extracted from space plans while they are in development, so that they can be checked for accuracy and evaluated in terms of utilization efficiency.

Chapter 9: Area Measurement Standards

There is a large and confusing array of nomenclature associated with measuring areas in buildings. There are also numerous sets of guidelines as to how this activity should be carried out, several of which are discussed in Chapter 9.

ANSI/BOMA is the leading standard for calculating space in office and other commercial buildings. It focuses on determining what is termed *rentable* area, which is the figure most often used in commercial leasing. ASTM/IFMA takes this process one step farther, providing terminology for subcategorizing rentable space according to its functional use. These are both national standards, and their terminology is commonly in use across the country. In this chapter, we examine both sets of definitions, followed by a detailed explanation of how area measurements are made according to the BOMA method with AutoCAD polylines.

Chapter 9 also considers the way in which the federal government, through its General Services Administration, applies the BOMA methodology in classifying the space it manages. Finally we examine a classification system widely used by postsecondary-educational institutions.

Chapter 10: Databases

Once again, we return to the subject of application software. The relational database is one of the most advanced and elegant means of information management in use today. If one is involved in the implementation of CAFM systems, or even working with small Access databases, it is useful to have a basic understanding of the underlying theory.

The idea of the relational database stems from algebraic number theory and relies on the simple concept of linked tables, or *relations*, based on certain key sets of data known as *domains*. Such data structures provide a means of storing information without redundancy so that it may be easily retrieved and updated. The relational model for information management was first proposed in 1970 by a mathematician named E. F. Codd working at IBM. There are numerous large-scale commercial database systems available today from companies such as Oracle, Microsoft, and Sybase.

Chapter 10 begins with a brief discussion of the history and development of the relational model by Codd and his collaborators. In 1985, Codd published a list of rules defining twelve characteristics of the ideal database, which together describe the essence of the relational concept. A section follows that describes how tables must be designed to fit together unambiguously, allowing data to be input, queried, and extracted without the user having to be concerned with the underlying structure.

We enter and extract data from a database by making queries using a language known as *SQL*. Some examples of how structured queries operate are explained, and a final section on database administration covers important issues such as security, recovery, and the user interface.

Chapter 11: Computer-Aided Facility Management

CAFM applications began to appear in the 1980s, first using flat files and hierarchal databases. As relational database systems became available, they gained almost universal favor as the engines driving these specialized applications. A typical CAFM system integrates the database with a CAD program and provides methods of linking drawing objects with the many kinds of information stored in the database.

In this chapter, we examine several CAFM applications belonging to the subclass of programs that manage the allocation of an organization's resources. All the physical components of the typical company's infrastructure are deployable assets having a measurable useful life. Simply looking at the main components of one leading application should make this clear:

- Space
- Building Operations
- Real Property and Leases
- Furniture and Equipment
- Telecommunications and Cabling
- Strategic Master Planning

With the exception of the final category, each of these modules makes use of a specialized inventory to track and control asset allocation. The sixth uses information from the various inventories to aid in developing forecasts and budgets and the administration of standards. Having a physical asset-management system in place can greatly simplify the programming process when a relocation or building project is in the offing. Day-to-day changes such as moving furniture and equipment, as well as routine maintenance, can also be expeditiously coordinated.

Chapter 12: Sustainability

The five final chapters address several aspects of the framework in which all space planning must be undertaken: the many codes and guidelines that exist. Modern building, zoning, and fire codes have been developed over the course of the last century, primarily for the purpose of ensuring public safety and establishing local authority over real estate usage. *Accessibility* has always been a focus of fire codes, but in more recent years the word has taken on a new level of meaning, emphasizing the usability of facilities by people with disabilities under normal circumstances.

The *Americans with Disabilities Act* of 1990 (*ADA*) is not a building code but is actually a civil rights law. Similarly, the movement toward sustainable or "green" design has its roots in the period of rising environmental consciousness during the 1960s and 1970s that coincided with unprecedented initiatives to end other kinds of institutionalized discrimination.

Chapter 12 begins by tracing some of the significant events in the rise of global environmentalism during this period. We believe it is not coincidental that the same cyclical model that describes *total quality management* (see Figure 5-12) underlies the practice of sustainable design and construction. Operating on a global scale, the International Standards Organization established defined quality standards (*ISO 9000*) in 1987, followed by a set of environmental standards (*ISO 14000*) in 1996.

Following a discussion of the principles of sustainability, this chapter considers the LEED guidelines (Leadership in Energy and Environmental Design). LEED is an initiative by the U.S. Green Building Council, currently being implemented to bring about a "transformation" of the construction industry through consensus-based standards, market incentives, and competition on a basis of best practices. Compliance with LEED falls squarely into the interest and responsibility area of architects, space planners, and facility managers.

Chapter 13: Building Codes

The question of building codes covers a vast amount of territory. There have been many varieties of codes throughout history dating back nearly fourthousand years to the *Code of Hammurabi*, which placed a rather stringent penalty for structural failure upon the builder.

After the great fire destroyed the City of London in 1666, Christopher Wren developed a draft master plan that, although never fully implemented, earned Wren considerable acclaim as the developer of many modern principles of city planning. Not the least of these was the consideration of making principal streets wide enough to prevent the spread of fire from one side to the other, and zoning to the extent that smoke and odor-producing trades would have been separated from the main part of the city and kept away from the riverfront. Chicago, especially since the great fire of 1871, has had particularly strict fire codes in spite of which there have been some serious events.

It would be impossible to cover, even in a volume entirely devoted to the subject, all the important building and fire codes currently in effect. We have chosen to use the recently published *International Building Code* (2003) as a model. The *IBC* is a consolidation of three major codes (described in the chapter) that establishes, on a "prescriptive and performance-related" basis, a uniform set of building standards intended to be adopted and used by local jurisdictions on a worldwide basis.

Chapters 13 through 15 address those sections of the *IBC* that are most applicable to the interests of facility-planning professionals and facility managers. Beginning with a section briefly describing the overall structure of the model code, this chapter covers the following areas:

- Use and Occupancy Classification
- Types of Construction
- Fire-Resistance-Rated Construction
- Fire-Protection Systems

All current codes are periodically updated; therefore, the current documents adopted by state and local authorities must be consulted regarding specific code issues.

Chapter 14: Egress

Chapter 14 explores the *IBC* standards as they relate to means of egress. Here will be found general requirements covering occupancy loads and the necessary widths of all passageways used for exiting. Areas of refuge for people who may be temporarily unable to reach an exit are discussed. Aisles, corridors, doors, changes in level, illumination, and signage are addressed in conjunction with exit access, exits, and exit discharge. These factors are all integral parts of the planning process known as *wayfinding*.

Chapter 15: Accessibility

The *IBC* is produced by an organization now known as the International Code Council, Inc. (formerly the Council of American Building Officials or *CABO*). The American National Standard titled *Accessible and Usable Buildings and Facilities* is also published by the ICC.

Also known as *ICC/ANSI A117.1*, this second document not only is largely the basis of the accessibility portion of the *IBC* but is closely coordinated with the *ADA Accessibility Guidelines* (ADAAG) and the *Fair Housing Accessibility Guidelines* (FHAG) as well. In the 2003 edition of the *IBC*, paragraph 2 of the Accessibility section (1100) incorporates *ANSI A117.1* by reference. Chapter 15 covers the following subjects:

- Accessible Route and Entrances
- Parking and Passenger Loading
- Dwelling and Sleeping Units
- Special Occupancies and Other Features
- Signage

Chapter 16: ADA Accessibility Guidelines

The *Architectural Barriers Act* (ABA) was enacted in 1968 to ensure "access to facilities designed, built, altered, or leased with federal funds." The ADA of 1990 extended the right of accessibility to facilities in the private as well as the public sector: "places of public accommodation and commercial facilities."°

The U.S. Access Board that administers the guidelines (otherwise known as the U.S. Architectural and Transportation Barriers Compliance Board) issued a revised and updated document in July 2004. Its stated goals were fourfold, as follows:

- bringing the specific guidelines up to date to ensure that the needs of persons with disabilities are adequately met;
- refining the format of the guidelines for easier use in order to promote compliance;
- ensuring consistency among the requirements of both ABA and ADA; and
- "harmonizing" the guidelines with current industry standards and model building codes such as the *IBC*.

The final goal listed here specifically refers to the *ANSI A117.1* standard, which is at the heart of the *IBC* as we have seen. The new document contains scoping provisions for ABA and ADA in two parts that specify what has to be accessible. The third part contains a common set of technical requirements to which both scope documents refer.

Supplements to the guidelines cover requirements that are specific to certain types of facilities and elements, such as state and local government facilities including courthouses and prisons, children's use play areas, and other recreation facilities (1998–2002).

Chapter 16 is directed primarily to the technical requirements of Part 3. These consist largely of interior planning criteria such as sizes and clearances for basic building elements referred to as "building blocks." Specific dimensional standards for things such as changes in level, unobstructed floor space, knee and toe clearances, and reach ranges are shown. Horizontal surfaces, changes in level, doors, entrances, and plumbing-related and communication utilities are among the elements considered.

Appendixes

There are four appendixes that relate to the material covered in specific chapters. The first is a checklist of items to be considered in developing a project plan (Chapter 5). Appendix B contains a space-requirements worksheet that can be used during the first stages of data collection for a facility program (Chapter 4). Appendix C covers the installation of the area-allocation tools (AutoCAD macros) described in Chapter 7. The final appendix outlines some of the procedures and reports offered by one of the leading CAFM applications treated in Chapter 11, as they relate to space.

ACKNOWLEDGMENTS

I wish it were possible to properly thank all the people who have contributed to this book, many of whom were clients over forty years of professional practice. Most of what I have learned, especially about computer applications, was learned in the field solving the problems that they presented to me.

Many people at Harrington College of Design where I have taught for the past five years gave me ideas and support for this project, particularly my students. They are too numerous to name, but some individuals do deserve special credit for their help. Many thanks to Mary Beth Janssen for helping me develop the residential programming material in Chapter 4, and to Pam Vanek, Mary Ellen Steele, and Chris Birkentall, who contributed their unique insights into the specialty of kitchen design.

I would like to honor the memory of a good friend and long-time member of the Harrington faculty, Lenore Levy, who passed away in November of 2006. Lenore taught at the Harrington Institute of Interior Design for over three decades, and ended her design career as its Director of Education. We worked on a number of important projects together at the architectural firm of C.F. Murphy Associates in the early 1980s, and Lenore was directly instrumental in my coming to teach at Harrington in the Fall of 2001.

Special thanks are due to the students in several of my classes at Harrington, first for putting up with the rough drafts of my chapters in process that I tested on them, and also for asking good questions.

Thanks also to the folks at Archibus, Inc., who, in making their software and support available for my facilities management class through their educational program, enabled me to write a detailed overview of one of the leading CAFM products available in the marketplace.

I would like to thank Bob Verdun and John Pabon at Computerized Facility Integration, LLC, one of the foremost consultants and systems integrators specializing in CAFM, where I spent a transitional four years between my planning and design career and my present one. They graciously provided the material for the two case studies that bracket Chapter 10. For the insightful "Case Study" of BP America, thanks to Ernest P. Pierz, General Manager of Global Property Management and Services. It was my pleasure to work with Ernie and his staff during most of my time at CFI.

Cyndi Crimmins was my first contact at Prentice Hall for this project, and I am grateful to both of the editorial teams who have followed its progress over the past three years. (It was supposed to take less than two!) Vernon Anthony was helpful with the early structuring of the project. Janice Stangel, Senior Production Editor at Prentice Hall, and Judy Ludowitz, Project Editor at Carlisle Publishing Services crossed the finish line with me, and I appreciate their help and patience.

Finally, thanks to the members of my small family who have been a constant source of support and inspiration over the years: my mom, dad, and brother, and especially my daughters Elizabeth and Samantha and granddaughters Meridian and Avalon. I have seen less of my progeny than I would have liked while working on this project.

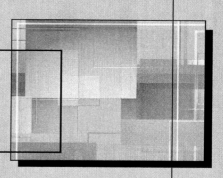

CHAPTER **1**

Classical Principles

Frustra fit per plura quod potest fieri per pauciora.
—William of Ockham

There is no constructive endeavor as centered on the needs of man as architecture. No art says more about the human race, which builds.

Due in no small part to specialization and the growth of technology, the traditional meaning of the word *architect* has been diluted significantly in current usage. Changes in the profession over the course of the past fifty years have lessened the stature of the *architect*—the *master builder*—as a professional title. The architect today is most often but one of several members of a project team that may include a construction manager, a team of engineers, interior designers, a real estate broker, a facility manager, and any number of other design and construction professionals. Within all but the smallest individual practices, architectural offices recognize a hierarchy of functional titles such as *project architect, design architect,* and *associate architect;* and, on large projects in which a number of firms are involved, one is usually given the legal distinction of *architect of record.*

The title has also been appropriated by other professions. We now recognize the title of *software architect*—and even such obscenely contradictory approbations as *architect of destruction.*

In his seminal work on architectural theory titled *Le Modulor,* published in 1954, Charles-Edouard Jeanneret, known as *Le Corbusier,* provided the following definition of *architecture*:

> The art of building houses, palaces and temples; ships, cars, railway trucks and aeroplanes . . . Domestic and industrial equipment and the equipment of trade . . . The art of typography as it is used in the making of newspapers, periodicals and books.

It is in this general sense that we use the term *architect,* sometimes interchangeably with *designer,* to describe those to whom this book is directed. We are interested in *process* and include in our prospective audience all those who plan, build, and manage houses, palaces, and temples, including facilities for commerce, public administration, and scientific investigation. This is consistent with what is, to our knowledge, the first written job description, by Marcus Vitruvius Pollio, known as *Vitruvius.* Le Corbusier's inclusion of the machine, writing, and typography are highly relevant to the conceptual underpinnings of design and planning, which is the focus of our discussion in this chapter and part of the next.

MARCUS VITRUVIUS POLLIO

Vitruvius, writing about the necessary qualifications of the architect, observes that the architect must be a generalist, first and foremost.° The architect's education must consist both of theory and practice. The *practice* consists of the regular production of drawings and other documents that determine the form and substance of her work. The *theory* allows the architect to articulate the rationale behind her solutions. Those who acquire the necessary manual skills without scholarship seldom attain the position of authority necessary to implement their plans. Similarly, the scholar theorist, as Vitruvius puts it, hunts the shadow and not the substance.

Raising the bar even higher, Vitruvius asserts that the architect should be both "naturally gifted and amenable to instruction," which implies that talent and authority alone do not produce great works. On another level, it might be taken to mean that the ego must be tempered with humility. It is the architect's *judgment* by which the work done by the other arts (one might read *contractors* in today's vernacular), not to mention the members of the project team, is put to the test.

Vitruvius goes on to list the various fields of study in which the architect should be educated, recognizing that it is not within the architect's power to be an all-around *expert* in all of them because he cannot hope to understand the theory underlying all. It is remarkable that this inventory of job qualifications, written over two thousand years ago, still has such currency.

- *Drawing* is the means by which the architect shows the appearance of the work he proposes.
- *Geometry* is necessary for a basic understanding of form, as well as the use of drafting tools, the "rule and compasses . . . the square, the level, and the plummet."
- *Arithmetic* is the means by which measurements are computed and the costs of projects calculated. Difficult issues involving symmetry and proportion often require the application of both geometry and arithmetic.
- *Optics* is integral to questions of lighting, both interior and exterior.
- *History* offers an understanding of the context in which the architect's work exists on a symbolic and spiritual level. Vitruvius gives two examples relating to temple architecture being enriched by ornament depicting significant events of the times. In similar ways, more recent practitioners have seized upon different symbolic expressions deriving from nature (Louis Sullivan, the cotyledon); the materials used in the structure itself (both Frank Lloyd Wright and Ludwig Mies van der Rohe); and, once again, historical and physical context (postmodernism, contextualism).
- *Philosophy* engenders an attitude of high-mindness without being self-assuming. The architect's mind must be clear, without being preoccupied with what we now call *perks*. She must maintain her dignity and honor by cherishing her good reputation because "no good work can be rightly done without honesty and incorruptibility."
- *Physics* is essential to the management of structure. Here, Vitruvius asserts, a higher level of knowledge is required because the laws of physics are immutable. The behaviors of water, wind, fire, and earthquakes (the four classical elements) are all direct concerns. The builders of the Gothic cathedral at Beauvais learned that in 1282 when its vault and upper clerestory fell flat. Although it remains the tallest cathedral in France at 157 feet, a second disaster befell the structure when its tower collapsed in 1573.
- *Music* contributes to an understanding of *canonical* and mathematical theory. From our modern perspective, though we seldom judge the adjustment of a tension member by the tone it produces, we have a rich variety of work from which to gain inspiration. The principles that govern the harmonic intervals of third, fifth, and octave have a direct relationship with geometry and mathematics.
- *Medicine* bears upon the relationship of building types to climate and the relative healthiness of sites, not to mention hazardous materials and the diagnosis of what we refer to as *sick building syndrome*. [Later in this chapter, we briefly compare the physical stature of Vitruvian man to modern, anthropometric man and woman.]
- *Law*, of course, is necessary in developing contractual relationships between client, architect, and contractor. Vitruvius enumerates some of the ancient issues that now fall under the headings of *building codes* and *errors and omissions*: drainage, the water supply, party walls, water dripping from the eaves, and windows. Before beginning

a project, he cautions us, "Be careful not to leave disputed points for the house-holders to settle after the works are finished."

- *Astronomy* gives us the four points of the compass—hence, the coordinate system we use in geometry, together with the solstice, the equinox, and the movement of the stars. Although we are not so concerned with sundials as were Vitruvius' clients, a recent theorist in city and regional planning, *Ludwig Hilberseimer,* placed a great deal of importance on the winter solstice in determining the height, spacing, and orientation of buildings.

Vitruvius makes quite a point of his distinction between practice and theory, return-ing to it following his listing of the essential arts. The actual production of the work is done by those trained in the individual subject. Theory, on the other hand, provides the com-mon ground whereby certain principles are shared among several of the arts. Musicians and physicians have in common "the rhythmical beat of the pulse and its metrical move-ment," Vitruvius says, but it is the physician to whom we turn when we are ill and the musician we call when we need our piano tuned. It would be difficult to expand this exhaustive list, except to add a twelfth discipline: *technology.*

Many significant inflection points in history have resulted from the application of new technology. The *Domesday Book* of 1086° shows over 5,600 water mills in use in England, which supplanted much of the slave labor used throughout the ancient world. By 1300, there were over eighty cathedrals and five hundred abbey churches in France alone, con-structed in a spacious and elegant style that came to be known as *Gothic.* The efficient dis-tribution of weight through the use of such architectural techniques as the Gothic arch resulted from knowledge gained through contacts with the Arabic world.

The Renaissance, which began with the invention of printing, reached its apex between 1450 and 1550. The three revolutions of the late eighteenth and early nineteenth centuries were also major technological watersheds. In the last twenty years, computer science has be-come all but indispensable to the designer and architect. A number of software applications developed to aid in facility planning are discussed in the course of this book.

Fundamental Principles

In the second chapter of his first book of architecture, Vitruvius defines six fundamental principles: *order, arrangement, eurythmy, symmetry, propriety,* and *economy.*

- *Order* and arrangement are planning essentials. Order encompasses the elements of a work separately and as a whole. The basis of order is the program, which clearly outlines the quantities required and the "selection of modules" from the elements of the work itself.
- *Arrangement* is the process of placing all the defined elements in correct functional relationships, never losing sight of the overall "elegance of effect." Vitruvius identi-fies the *ground plan, elevation,* and *perspective* as the means by which these relation-ships are expressed. The first of these is created through "the proper successive use of compasses and rule" and shows the outlines and interior divisions of the build-ing. The elevation portrays the front of the building, picturing all the elements in their correct proportions. Finally, the perspective shows the front and sides, with the sides converging to a point.

Today, we would add the *section* and *detail,* for a total of five basic types of drawings depicting arrangement (plan, elevation, perspective, section, and detail). The section cuts through the building vertically, showing the internal construction of the exterior walls and interior divisions combined with interior elevations of whatever is seen in the direction of view chosen. The detail may consist of any of the first four views, enlarged as much as required to show how the element pictured is to be constructed.

With characteristic economy, Vitruvius reduces his description of the creative process to two words: reflection and invention. He defines *reflection* as "careful and laborious thought, and watchful attention directed to the agreeable effect of one's plan." *Invention* is described as the "solving of intricate problems and the discovery of new principles by means of brilliancy and versatility." These are grandiloquent words in English translated from Latin written fifty years before the beginning of our common era, but the meaning is clear.

- *Eurythmy,* a word not in our everyday usage, refers to maintaining visually pleasing relationships among the dimensions of individual elements of the work: length to width to height. In other words, *eurythmy* refers to the harmonious proportionality of the discrete members.
- *Symmetry* is the achievement of a successful balance among the individual elements of the work and the relationships between the different parts and the overall scheme, "in accordance with a certain part selected as standard."

Vitruvius cites the human body as an example, demonstrating in his third book the "symmetrical harmony between forearm, foot, finger and palm." We speak more of these relationships presently. He also mentions temples where symmetries can be found in the thickness of columns, triglyphs, and modules. This fact is exemplified by the Parthenon of Athens, illustrated in Figure 1-1, in which its elevation is subdivided into a series of golden section rectangles.

These principles, symmetry and eurythmy, complement each other and are shared by all the arts. Exploring new concepts and solutions while maintaining harmony among the elements is always the challenge. In his *Musikalisches Opfer* (Musical Offering), composed in 1747, Johann Sebastian Bach achieved absolute symmetry, backward and forward, in the canon pictured in Figure 1-2.

The last two principles, propriety and economy, are also sisters.

- *Propriety* is the perfection of form that is attained when a building is constructed on a basis of sound principles. For Vitruvius, these were symbolic as well as practical. In classifying temples, he states that temples honoring Jupiter, the sun, moon, or heavens should be open to the sky. The Doric order should be used for temples of Mars and Hercules because of its masculine expression of strength. The Corinthian order, with its slender figures and leaves, should be used to honor the "delicate divinities" such as Venus, Proserpina, and the nymphs. The Roman gods—such as Bacchus and Diana—in the middle position are properly represented by the Ionic order, which constitutes "an appropriate combination of the severity of the Doric and the delicacy of the Corinthian." (In Figure 1-3, the three orders are dimensioned as multiples of a common diameter.°)

Figure 1-1   The Parthenon.

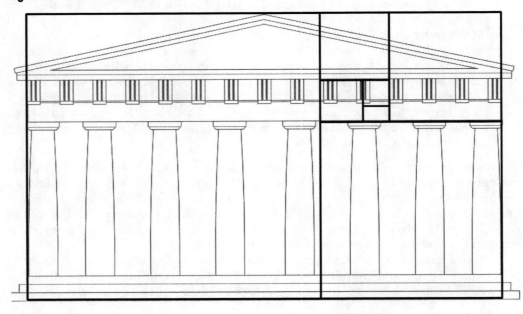

Figure 1-2   A Bach "Mirror" Canon.

Figure 1-3   The Doric, Ionic, and Corinthian Orders, According to Vitruvius.

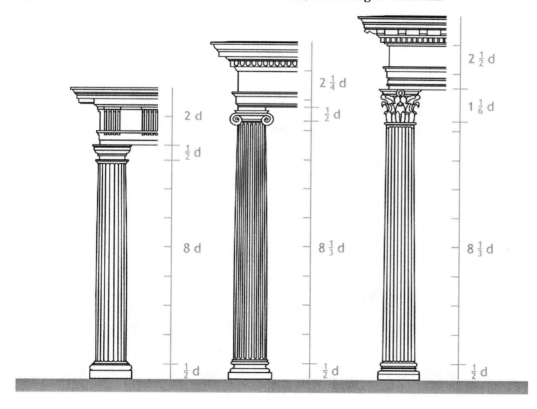

In practical terms, propriety means making sure that all the various elements of a work are in sync with one another. A monumental building or space should have an entrance in keeping with its scale; a minimal, three-foot-wide door would not be appropriate. There is a "natural" propriety in placing bedrooms and libraries where they will receive eastern light, western light being preferred in winter for baths and winter apartments. Northern light should be provided for art galleries and other uses where steady light is desirable throughout the day.

One of Professor Hilberseimer's program requirements was that all principal rooms in a building (that is, living and sleeping spaces) must be oriented so as to receive at least four hours of sunlight on December 21, the shortest day of the year. Keeping buildings of differing heights out of each other's shadows imposes a significant discipline on the spacing between them.° This simple requirement limits density, helps maintain a human scale, and establishes a balance among elements in the landscape.

Vitruvius' observation on the last of his six principles could be taken from a contemporary guide to project management:

- *Economy* requires the effective management of materials and the site, maintaining a balance between cost and common sense during the construction process. The architect should not specify components that are not available or cannot be obtained at reasonable expense. If proper housing is to be widely available, it must be made ready at an affordable cost.

Vitruvius is unapologetic for the fact that the wealthy and powerful may have special requirements that result in more "opulent and luxurious" dwellings, declaring that "the proper form of economy must be observed in building houses for each and every class."

Vitruvian Man

Vitruvius begins his third book of architecture with a reiteration of his definition of symmetry, comparing the temple to the human form. Nature designed the parts of the human body with symmetries and precise relationships comparable to those in the well-shaped temple, he says, listing some of these relationships in narrative form. For the entire body, the proportions of parts to the whole are as follows:

- 1/10—face, chin to top of forehead
- 1/10—open hand, wrist to tip of middle finger
- 1/8—head, chin to crown
- 1/6—neck and shoulder, top of breast to hairline
- 1/6—length of the foot
- 1/4—middle of breast to crown
- 1/4—breadth of the breast
- 1/4—length of the forearm

For the face itself, the following relationships apply:

- 1/3—bottom of chin to underside of nostrils
- 1/3—underside of nostrils to eyebrows
- 1/3—eyebrows to hairline

The center of the human body is the navel. If we draw the figure of a man as Leonardo da Vinci did, with his arms and legs outstretched, and set a compass point at his navel, its circle will define the extent of his fingers and toes. Likewise, the distance from the soles of his feet to the crown of his head is equal to the total length of his outstretched arms, the limits defined by a square (see Figure 1-4). With these relationships in mind, Vitruvius reasoned that the perfect building must, as a consequence, have its individual members in an exact relationship with the overall scheme.

Vitruvius does not specifically identify the proportion found in the dimensions of crown to navel and navel to sole, but they are approximately equal to the golden ratio, later made precise in Le Corbusier's *Modulor*. If one measures the height of the navel in the da Vinci drawing, it is found to be within 2 percent of the golden section values of 0.618 and 1.618. This suggests a greater set of symmetries than that of Vitruvius' anthropocentric figure: one rooted in pure mathematics.

OMPHALOS

The oracle of Apollo at Delphi (derived from *delphos*, the womb) is pictured as a tripod sitting atop a fissure in the earth. The legs of the tripod symbolize the periods of time

Figure 1-4   Leonardo da Vinci's Drawing of Vitruvian Man.

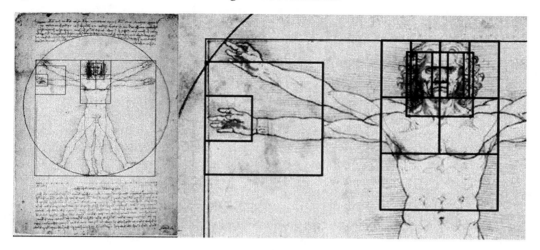

controlled by Apollo: past, present, and future. At the apex of the tripod, which defines in space a sacred Pythagorean tetrahedron, the priestess is seated above the abyss of the oracle. The spiritual nature of man is thus represented, suspended above the "abyss of oblivion by three golden threads of divine power." The Greeks believed the site of the oracle to be the "umbilicus of the earth," suggesting that they regarded our planet as a living being.°

One of the significant elements in the structure of James Joyce's *Ulysses*, the labyrinthine twentieth-century novel based on Homer's epic, is the navel, or *omphalos*. Early in Homer's *Odyssey*, Odysseus is stalled on Calypso's island *Ogygia*, described as the "navel of the sea." The ancients believed that the soul of man, his consciousness, resided in the area of the heart and stomach, which has been variously regarded as the seat of internal divine light and the source of poetic and prophetic inspiration. Given the navel's purely physiological function, it is not difficult to imagine why this portion of the body acquired many symbolic associations.

In the opening scene of *Ulysses*, Stephen Dedalus (Joyce's Telemachus) fantasizes an association between the tower in which he lives and the seat of the Delphic oracle. Leopold Bloom as Odysseus later begins his journey from an omphalos as well, his home at No. 7 Eccles Street in Dublin corresponding to Calypso's island of Ogygia. Much later in the novel, Bloom and Dedalus meet in the "nighttown" district where Stephen hallucinates a scene in the National Maternity Hospital where a birth is taking place:

> A chasm opens with a noiseless yawn. Tom Rochford, winner in athlete's singlet and breeches, arrives at the head of the national hurdle handicap and leaps into the void. He is followed by a race of runners and leapers. In wild attitude they spring from the brink. Their bodies plunge. . . . It rains dragons' teeth. Armed heroes spring up from furrows. They exchange in amity the pass of knights of the red cross and fight duels with cavalry sabers. . . . On an eminence, the centre of the earth, rises the field alter of Saint Barbara. From the high barbicans of the tower two shafts of light fall on the smokepalled altarstone. On the altarstone Mrs. Mina Purefoy, goddess of unreason, lies naked, fettered, a chalice resting on her swollen belly.°

The associations in this comical phantasmagoria are plentiful. The first allusion is to the Greek myth of Cadmus who, commanded by Athena, killed a dragon and sowed its teeth in a field, giving birth to armed men who immediately began to kill each other. Cadmus and the last five of these soldiers, whom Joyce compares to the Knights Templar, founded Thebes. The "centre of the earth" refers to the ninth and lowest circle of Hell, as pictured in Dante's *Inferno*. Saint Barbara is the patron saint of soldiers, gunsmiths, firefighters, and architects, whose name is invoked as protection against lightning, explosives, and fire. The altar of the *omphalos* is thus a powder magazine: "womb of dynamic outburst."°

In a historical parallel, the Parthenon survived virtually intact for two millennia, until its midsection was destroyed in 1687 during a siege by the Venetians. At the time, the structure was being used by the occupying Turks as a powder magazine!

Anastomosis of Navelcords

So far, we have touched upon several of the fields of study that Vitruvius included in his list: music, physiology, and geometry. In ancient and contemporary literature, we encounter repeated references to the centers of our being and our origins.

Our natural inclination toward order and symbolic connections is an intrinsic part of human consciousness. It is part of the creative urge that forces us to try to give form to shapeless matter. In the experience of space, it is the determination of position that lends it its character. "Each space is something special to us as a space of a particular kind of life. Each space imparts a certain attitude to us."°

Virtually the entire history of philosophy, physics, and astronomy has been driven by man's need to understand his place in the universe while constructing shelter in which to seek refuge from it. Only in the last two hundred years has architecture, for the *common man*, become more than merely shelter. Only recently has Frank Lloyd Wright's eighth principle of the "inherent human factor" entered our design vocabulary, that cover and shade and changing light are all desirable and attainable elements of our environment.

Writing some three hundred years before Vitruvius, Aristotle articulated the notion that *place* has *potency*. He believed that places are *qualitatively* different from one another and that one must have a sense of place, a center, before attempting to define space. The coordinate system named for René Descartes is a mathematical expression of this idea, its *omphalos* being the *0, 0, 0* point at which the infinite *x, y,* and *z* axes meet.

But, of course, Aristotle was talking about space in the largest possible sense:

As to [the earth's] position there is some difference of opinion. Most people— all, in fact, who regard the whole heaven as finite—say it lies at the center. But the Italian philosophers known as the Pythagoreans take the contrary view. At the center, they say, is fire, and the earth is one of the stars, creating night and day by its circular motion about the center. . . . Their view is that the most precious place befits the most precious thing: but fire, they say, is more precious than earth. . . . Reasoning on this basis they take the view that it is not earth that lies at the center of the sphere, but rather fire.°

It is a modern myth that, prior to Columbus' voyage to the Americas in 1492, most people believed the earth to be flat. Aristotle knew the earth to be spherical and also of no great size because traveling small distances could be observed to produce large changes in the appearance of the horizon.

Eratosthenes, about one hundred years after Aristotle, actually made a fairly accurate measurement of the earth's circumference. The city of Syene in Egypt (now Aswan) lies nearly due south of Alexandria on the Tropic of Cancer. Therefore, the sun is directly overhead at noon on the summer solstice and produces no shadow. By measuring the sun's shadow in Alexandria on that day, Eratosthenes was able to calculate that the angle of the sun was approximately one-fiftieth of a circle. Multiplying the five-hundred-mile distance between the two cities by fifty gave him a circumference of twenty-five thousand miles.

Man's early view of place relative to the heavens, however, was anthropocentric: "I am the focal point of the universe." Now that we know this belief to be incorrect, we might even characterize it as *egocentric*. It was, in any case geocentric.

Claudius Ptolemy, who followed Vitruvius by about two hundred years, gave his name to this view of reality. The Ptolemaic model of the universe persisted as the dominant world view (at least in Western civilization) for nearly a millennium and a half. Ptolemy attempted to use physical observation to reconstruct the process by which man arrived at his geocentric view, but it was virtually impossible for him to conceive of Earth being in constant motion about a different center.

It was not until the sixteenth century, when Nicholas Copernicus revived the ancient doctrine of the Pythagoreans with the notion that the paths of the heavenly bodies are uniform, continuous, and circular, that the idea of a heliocentric universe became plausible. Less than a hundred years later, Galileo confirmed the Copernican model, in that "the fact that we observe the planets sometimes nearer the earth and sometimes farther away is logical proof that the center of the earth is not the center of their orbits."°

Galileo's telescopic observations showed without rational doubt that not only does the earth revolve around the sun, our moon revolves about the earth. Beyond even that, what Galileo regarded as his most important discovery, made in 1610, was that Jupiter had four moons that orbited that planet. As Copernicus himself had put it, "There is more than one center." In fact, mankind ultimately had to accept the idea that the universe is a veritable clockwork of wheels within wheels.

Acceptance did not come easily. Galileo was ultimately obliged to recant his observations, being acutely aware of the fate of his peer, Giordano Bruno of Nola, who had been burned at the stake ten years earlier by the Inquisition for espousing similar views.

> The problem of the heavens, stripped . . . of metaphysical obscurities, was laid bare to the reason as one of pure mechanics: The planets came to be treated as ordinary projectiles, and distinct reasoning about the nature of their paths was rendered possible. [Sir Isaac] Newton's great task was thus prepared and defined by Galileo.°

It was only in the last three hundred years that Newtonian mechanics became a commonly accepted model of physical reality. While Copernicus and Galileo established that the planets in our solar system orbit the sun rather than the earth, Newton discovered the laws that govern their motion.

Newton favored the idea that light is composed of particles; and it took less than a hundred years (1783) for another Cambridge don, John Mitchell, to theorize about stars so massive that they were invisible because light could not escape. In 1900, Max Planck suggested that light is composed of "quanta" that are wave packets of energy in proportion to their frequency.

Albert Einstein published his general relativity theory in 1915. Werner Heisenberg formulated his *principle of uncertainty* and Erwin Schrödinger his quantum mechanical model of the atom in the 1920s. These events led to our current model of light as a wave-particle duality, the possibility of stars collapsing into black hole singularities, and the notion that time may have had a finite beginning and could equally well have an end at some point in the future.°

The poet William Butler Yeats wrote in 1920:

> Turning and turning in the widening gyre
> The falcon cannot hear the falconer;
> Things fall apart; the centre cannot hold;
> Mere anarchy is loosed upon the world, . . .

In the context of the early twentieth century, between the two world wars, Yeats voiced his concern about the impending third millennium. If we take the falconer as referring to the soul and the falcon as the intellect—"the life gyre . . . sweeping outward," as Yeats himself put it: "All our scientific, democratic, fact-accumulating, heterogeneous civilization belongs to the outward gyre and prepares not the continuance of itself but the revelation as in a lightning flash."° Yeats was voicing his sense of the world about him and the loss of the qualities of life that, to him, were most valuable under the dying social tradition. This is the central part of the poet's magical belief in *magnus annus*, the great year at the end of two thousand years that brings extinction and replacement in the Second Coming.

Man and Place

We have so far survived the transition from the second millennium, but maintaining our sense of place, the centers by which we define space and our place in it, requires a delicate balance. While we have had to reluctantly accept that we are not the center of the universe, architecture remains intrinsically anthropocentric.

The three revolutions of the late eighteenth and early nineteenth centuries set the stage for unprecedented changes in the two centuries to follow. The American and French revolutions dominated the half century from 1765 to 1815. It was the *Industrial* Revolution, however, that produced one of the most important inflection points in the history of mankind.

> The machines changed the organization of society, and shifted the centre of a
> man's life from his cottage home to the daily factory. In that shift, the man
> ceased to be a member of his family and his village, and in the long run became
> simply himself: a person. Because the machine in the factory changed the order
> in his life, it slowly changed the status of the worker who served it. It regimented
> and brutalized and starved him, it exploited him and (for a long time) his
> family, and it robbed him of everything but his skill. And yet, by these acts in
> the end it made him a man—a man alone.°

With the perspective of five hundred years, we accept the fact that our society was first remade during the Renaissance, which broke religious unity. Foreshadowing the Industrial Revolution, Johannes Gutenberg printed the Bible from movable metal type in 1456. It took just sixty-one years for a thirty-four-year-old Augustinian friar named Martin Luther to instigate the reformation of the Christian Church in 1517. Whether or not Luther actually nailed his theses to the church door at Mainz is arguable; but after circulating several copies, he reportedly had an uneasy surprise when he received them back from South Germany, *printed*.°

CLOSING WORDS

It is not possible to overstate the extent to which the Industrial Revolution upset patterns of society that had been evolving for hundreds of years, changing the flow of thought and action throughout Western civilization. Even more than the first, the second Industrial Revolution that extended from 1850 to the beginning of the First World War was technology-based. New technologies led to the creation of the electrical, chemical, and petrochemical industries, not to mention the internal combustion engine.

From the time of Descartes through the period called the Enlightenment, people attempted to come to terms with these radical changes in the social structure. At the same time, classical principles continued to be used as the basis for planning. Proportional systems, such as those articulated by Vitruvius and later visualized by da Vinci, were rationalized in the mid-nineteenth century using the Fibonacci number series discovered in the twelfth century. In the next chapter, we look at this development along with the work of several nineteenth-century figures who had a significant influence on the best-known architects of the twentieth century.

Another of the consequences of industrialization was the perceived need to standardize systems of measurement in order to facilitate trade between regions. This ultimately led to the development and adoption of the metric system around the globe, with the conspicuous exception of the United States. As measurement and geometry are central to our discussion of planning, we take a comparative look at the relationships between measurement units in common use today.

TIME CAPSULE   The Ramesseum

The Ramesseum was constructed around 1300 BCE, in Thebes, for the pharaoh Ramses II. Its plan (Figure TC1-1) is rigorously linear, consisting literally of chambers leading to other chambers. One enters, unless one is coming from the royal pavilion to the south of the first court, through a monumental set of pylons 220 feet wide, into a grand hall containing a colossal statue of the pharaoh. Another name by which Ramses was known is contained in the inscription that once appeared on the base of the statue: "King of Kings I am, Ozymandias. If anyone would know how great I am and where I lie let him surpass my works."°

The main entrance is only ten feet wide, opening spectacularly into a first court enclosing some twenty thousand square feet. From the first court, one passes into a second of equal size, containing double rows of columns along the sides and pillars depicting the god Osiris at the front and rear. Toward the rear of the second court are located three ramps with stairs leading to an equal number of openings. Two smaller colossi flank the central stairway, between the portals leading into the ten-thousand-square-foot grand hall beyond. This grand hypostyle hall contains forty-eight columns supporting an elevated roof and clerestory. The twelve columns along each side of the central passageway are crowned with characteristic bell capitals.

Beyond the grand hall, one passes through three smaller halls of 1,600 square feet each, before finally reaching the sanctuary buried deep at the back of the structure. One has traveled more than five hundred feet in a straight line upon reaching this point, where the

image of the god was housed. According to Banister Fletcher,° there are no provisions for the circulation of processions around the sanctuaries of mortuary temples, but one can easily imagine such ceremonial processions in the outer courts and grand hall.

The Ramesseum is a mortuary temple in which the pharaoh was worshipped and offerings were made, but it was not only a place for religious expression. Temples in Egypt were built as residences for the gods embodied in the person of the pharaoh and were therefore integral to the administration of the state. They were not accessible to the general public, except for the outer chambers where people would come to call upon the help of the gods. These outer chambers were furnished with statues of the gods and kings for this purpose, and the walls were covered in hieroglyphs that facilitated the rituals. None of the populace, however, would likely have been permitted to gaze upon the statue of the god king within the inner sanctuary. The structure itself turns its back on the outer world, exhibiting an image of royal power and sacerdotal mystery.

Among the many carvings on the walls of the Ramesseum was the text of a treaty of alliance between Egypt and the Hittites, concluded by Ramses in his twenty-first year as pharaoh.° Although the treaty is presented here as an overwhelming victory, Hittite sources indicate that the Egyptians were fortunate to escape with their skins from the final confrontation. According to historical sources, the campaign was actually a stalemate. The presence of these writings testifies to the use of the temple as a political instrument.

Figure TC1-1   The Ramesseum, Thebes, circa 1300 BCE.

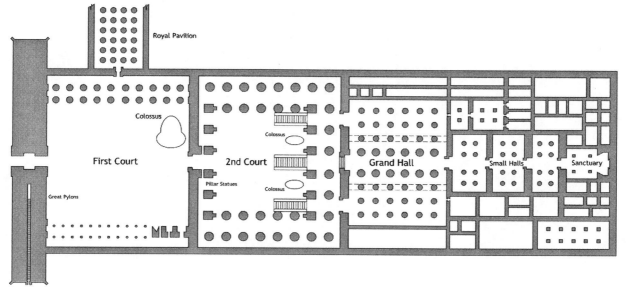

Royal Pavilion

Colossus

First Court

Great Pylons

Colossus

Pillar Statues

Colossus

2nd Court

Grand Hall

Small Halls

Sanctuary

(continued)

(continued)

Documents such as these, both on the walls of the temple and on the many papyrus scrolls of historical and local records stored within, indicate that the structure functioned as a library and archive as well. In fact, the temple served many more purposes than that of a religious institution. Its massive walls and highly secure compartmentalization made it a safe and defensible place for treasure. It was also a learning center for the privileged boys who would advance to priestly or administrative stations. The high priest was often a former member of the court or military leader who had managerial responsibility for public works such as temple construction and maintenance of the granaries. The granaries at the Ramesseum were of such magnitude as to be capable of feeding nearly 3,500 families for a year, and another temple at Karnak reportedly employed a labor force of over eighty thousand.°

The Ramesseum was, in effect, a governmental center with an extremely complex functional program. The unbroken linearity of its plan, combined with its sheer magnitude, demonstrates the way in which the authority of a god king can be given symbolic expression. At the same time, its uses facilitated the exercise of his officials' power.

Looking at the remains of this great structure today, one cannot help but be reminded of the concluding lines of the poet Shelly's famous sonnet, which incorporates the inscription quoted earlier:°

> *. . . Nothing besides remains. Round the decay*
> *Of that colossal wreck, boundless and bare*
> *The lone and level sands stretch far away.*

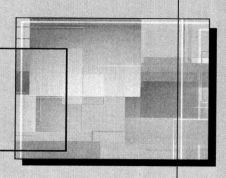

CHAPTER 2

Geometry and Planning

Now let us, by a flight of imagination, suppose that Rome is not a human habitation but a psychical entity with a similarly long and copious past—an entity . . . in which nothing that has once come into existence will have passed away . . .
—*Sigmund Freud°*

In March of 1790, Talleyrand proposed to the French legislature that a measure derived from nature be adopted as the standard to replace that derived from history or the fiat of kings. He further proposed that the fundamental length be part of a system in relation to which all other units would be defined. The legislature added a third feature, that the system would be decimal. Two months later, the name *meter* was proposed for the basic unit of length; and, in 1793, the idea of using Greek and Latin prefixes was proposed by the Commission on Weights and Measures.

Talleyrand's initial proposal also was that the fundamental unit of length be based upon the length of a pendulum oscillating once per second. Galileo had demonstrated that the period of a pendulum's swing was wholly determined by its length. However, Galileo had not accounted for the fact, since discovered, that the period of a pendulum's swing varies with gravity, which also varies with latitude. The Royal Society of England also proposed a standard based on nature: a new *yard* based on a pendulum swing at one-second intervals in the Tower of London (39.2 inches).

So the question of where the measurement would be made became an international political issue. Jean-Charles de Borda, chairman of the Commission on Weights and Measures and inventor of the Borda circle, the most accurate surveying instrument developed to date, argued that it was inappropriate that one unit of measure (length) should be based on another (time). The matter was settled in March of 1791, when the legislature recommended that the meter be set equal to one ten-millionth of the distance from the North Pole to the equator. This would require that a sufficient portion of an Earth meridian be measured to permit a sufficiently accurate extrapolation of its total length.

Borda proposed four criteria for choosing the segment of meridian to be measured:

- Its length must exceed at least ten degrees of latitude to permit accurate extrapolation.
- It must straddle the forty-fifth parallel, midway between pole and equator, in order to minimize the effect of Earth's oblate shape.
- Its endpoints must be located at sea level, the natural outline of Earth's shape.
- It must run through an area already surveyed to minimize the time required to accomplish the task.

The only meridian that met all these requirements ran from Dunkirk, on the northern coast of France, through Paris to Barcelona. Two savants belonging to the Royal Academy of Sciences set out to accomplish the task in June of 1792. Jean-Baptiste-Joseph Delambre set out to begin from the north end while Pierre-Francois-Andre Méchain began in the south. The task was to take a year.

The French Revolution was in full swing at this point, the Bastille having fallen on July 14, 1789. The new French Republic was founded on September 22, 1792. The surveying mission was actually completed in November of 1798; and on June 22, 1799, a platinum bar the length of the *definitive* meter was presented to the French legislative assembly.

A law passed in 1793 had implemented a *provisional* meter based on an earlier survey in 1740. Curiously, due to a 1794 error that had been made in the survey at the Barcelona end of the meridian, it turned out that the definitive meter deviated twice as much as the provisional one from what we now know to be the precise circumference of the earth.

Over the next 135 years, European acceptance of the metric system progressed to the point at which multiple standard meters were required for each nation and the 1875 *Convention of the Meter* became the permanent standards organization, responsible for all international metric standards. The standard is known today as the *International System of Units*, or SI (*Le Systéme International d'Unités*).

In 1889, new platinum-iridium meter bars replaced the original Archive Meter. The standard was redefined most recently in 1983 as the distance light travels in a vacuum in 1/299,792,458 second. Time ultimately did become the fundamental unit as defined by an atomic clock, yet each of these redefinitions meticulously preserves the 1799 meter. Current satellite surveys set the actual pole-to-equator distance as 10,002,290 meters.°

ANCIENT MEASURES

Eratosthenes calculated the circumference of the earth within a tolerance of 0.5 percent circa 230 BCE. (The one-fiftieth wedge of the earth from Alexandria to Aswan in Egypt that he extrapolated is shown on the right side of Figure 2-1.)

The Great Pyramid at Giza predated Eratosthenes by over two thousand years. At a latitude of 29°58'51.06", it is located precisely one-sixth of Earth's circumference from the North Pole. It is also aligned with the four points of the compass. The variation in length from one side of its base to another does not exceed eight inches, giving it an average base dimension of 755.8 feet. Its height is 481.4 feet, approximately the height of a forty-story building.

From these dimensions, we can determine that the distance from the midpoint of one side to the structure's apex is exactly 612 feet. If we set the length of this side equal to one, as in Figure 2-1, then the base of this triangle has a proportionate value of 0.6175, which is within one-tenth of 1 percent of the golden section. The height of the Great Pyramid, proportionately, is 0.7866, which is the square root of this same value. But the subtlety of the spatial relationships does not stop here. A circle drawn through the apex of the

Figure 2-1   The Great Pyramid at Giza.

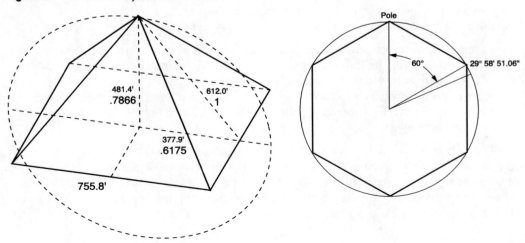

pyramid, centered at the midpoint of the base, has nearly the same circumference as its perimeter of 3,023 feet, within 0.05 percent.

Of course, it is highly doubtful that the ancient Egyptians made their measurements in English feet! Their unit was the Egyptian *royal cubit*. The exact length of this unit is a matter of some debate, but a good nominal value is equivalent to 20.6 inches. Within a tolerance of 0.06 percent of the length of each side, this allows the structure's key dimensions to be expressed as integers:

Base = 440 cubits, and height = 280 cubits

This means that the perimeter of the base of the Great Pyramid, as well as the circumference of a circle having the pyramid's height as its radius, is equal to 1,760 cubits. If we use the ancient approximation of π (pi) to make the calculation (22/7, which was the value calculated by Archimedes, a contemporary of Eratosthenes, and not materially improved upon for nearly two thousand years), the results are absolutely precise.

The *rod* is equal to 16.5 feet, or 198 inches. Otherwise referred to as a *pole* or *perch*, the rod is part of the English measurement system, still listed in the National Institute of Standards and Technology (NIST) handbook° and most commonly used today by surveyors. If we choose an exact value of 20.625 inches for the royal cubit, we can demonstrate precise relationships, not only between the cubit and the rod, but with the mile as well.

Because we are so used to thinking in terms of 5,280 *feet* in a mile, it is easy to become confused as to what kind of units we are looking at in Table 2-1. In Chicago, Illinois, 660 *feet* is the length of one city block, the entire city having been laid out in one-eighth-mile squares according to Daniel Burnham's master plan; and 660 feet squared equals 435,600 square feet, which is exactly ten acres. There are 640 acres in a square mile.

What is of particular interest, however (other than the fact that the units are inches), is that certain key multiples of the Egyptian royal cubit can be expressed as powers of 2, with, in the case of the mile, the number 3 as a coefficient. The rod has a coefficient that is exactly one-fifth of the number 3. We include the rod here for reasons that will become apparent shortly.

Let us take a moment to point out that English architectural scales are also based on identical relationships. One of the reasons that the English (Egyptian?) measurement system offers such flexibility from an architectural planning standpoint is that it ultimately derives from a *binary* base. Thus, measurements given in inches, feet, or miles can be easily divided by any number of integer factors. The mile, for example, expressed in inches, can be factored into the numbers 2, 3, 5, and 11:

$$2^7 * 3^2 * 5 * 11 = 63,360$$

Table 2-2 illustrates the dimensional equivalencies that are shown on a typical architectural scale. They are accompanied by the scale factors most often used in computer-aided design (CAD) applications. In the third row are the common factors based on powers of 2. All of the common factors are shown up to 2^{10}, and those scales that are seldom used are shaded in gray. The scales that have the coefficient of 3 are emboldened.

The compatibility between the ancient and English systems is so perfect that if forty-nine Great Pyramids were placed edge-to-edge (Figure 2-2) they would overshoot the edge of a one-mile square by 10.6 feet, or 0.2 percent. How rich these measurement systems are in the ways distances can be subdivided.

Table 2-1 Equivalence of Egyptian Royal Cubit, English Rod, and Mile (Inches)

Royal Cubit	Rod	1/8 Mile	Mile	
20.625 in.	198 in.	660 ft.	5,280 ft.	63,360 in.
1	$0.6 * 2^4$	$3 * 2^7$	$3 * 2^{10}$	

Table 2-2   English Architectural Scales

Scale	12	6	4	3	2	$1\frac{1}{2}$	1	3/4	1/2	3/8
Scale Factor	1	2	3	4	6	8	12	16	24	32
Common Factors	2^0	2^1	$3*2^0$	2^2	$3*2^1$	2^3	$3*2^2$	2^4	$3*2^3$	2^5

1/4	3/16	1/8	3/32	1/16	3/64	1/32	3/128	1/64	3/256
48	64	96	128	192	256	384	512	768	1,024
$3*2^4$	2^6	$3*2^5$	2^7	$3*2^6$	2^8	$3*2^7$	2^9	$3*2^8$	2^{10}

Figure 2-2   Forty-nine Pyramids Fit (Almost) Exactly into One Square Mile.

EARTH AND MAN

Do these measurement systems relate to the dimensions of the earth? This question can be answered in a couple of ways, both of which open the door to speculation regarding the mathematical and conceptual sophistication of the ancients, as well as the mystical significance attached to their numbers.

If we start with a circle having a circumference of one rod, or 198 inches, and we divide it by the Archimedean π of 22/7, we get a radius of 63 inches. Multiplying 63 by 25,000,000 and converting the units yields 24,858 miles. This figure deviates from current satellite measurements of the earth's circumference by less than ten miles, or approximately 0.01 percent. The pole-to-equator distance is 10,001,270 meters by this method, roughly halving the error that was made in defining the meter!

The second method, which brings the element of time into the equation, uses the equatorial circumference, which is slightly larger than the polar distance due to the earth's rotation and gravity. (One might assume that the *omphalos* in this model would properly be the intersection of the equator and the prime meridian.) This circumferential distance is 24,907 miles, or 131,508,960 inches. Consider the following equation:

$$365.256 * 360 * 1,000 = 131,492,160$$

The deviation is slightly more than three miles, again about 0.01 percent, but here the relationship is quite different. There are 365.256 days in a sidereal (celestial) year, and 360 is the number of degrees (by our convention) in a circle. Each degree is thus divided into as many segments as there are days in a millennium. One foot may thus be said to be equivalent to the duration of a single day.

Throughout history, man has attempted to create or define his sense of *place* in the way that Aristotle uses the word. Nowhere is this fact more directly manifested than in our cities and in our architecture. Man continues to attach divine significance to many of the numbers and relationships we have been discussing. For example, the integer 63 is exactly equal to the square of three times the square root of seven. The triangle and the square root of seven were both thought to be symbolic of spiritual wisdom.°

Popular myth has it that the English yard was defined as the distance from Henry I's nose to his outstretched thumb. The Egyptian royal cubit is defined in legend as the distance from the pharaoh's elbow to the tip of his middle figure. Moreover, this value was said to change with each pharaoh, which is in complete agreement with one of the axioms of modern quality management: "If standards and regulations are not revised in six months, it is proof that no one is seriously using them."°

We humans tend to take our sense of place very seriously, however, as evidenced by the 6 and 1/2 years of effort that Delambre and Méchain put into trying to refine the definition of the meter. The egos of monarchs and politicians notwithstanding, it seems reasonable to believe that some considerably deeper thought was applied to the development of measures both ancient and modern. What should be clear, however, is that ancient measurement units are part of a much more unified system than is generally perceived and that they are intrinsically harmonious with the measure of man.

In the gyre of the last ten pages or so, we have touched on most of the fields of study that Vitruvius listed. Keeping some of these connections (and of course, the *omphalos*) in the back of our minds, let us expand upon our discussion of mathematics and the golden section (symbolized by the Greek letter *phi*: φ).

The use of φ is a recent convention, adopted at the beginning of the twentieth century in the name of Phidias who was the creator of the Parthenon sculptures. Previously, the golden ratio had been designated in mathematics by the letter *tau*: τ, from the word τομη (to-mi), which simply meant *the section*.

The German astronomer Johannes Kepler was an exact contemporary of Galileo, and he maintained that in geometry there are two treasures: the Pythagorean theorem and the division of a line segment in extreme and mean ratio. "The former," he said, "can be compared to the value of gold, the latter can be named as a gemstone." Pythagoras was born around 570 BCE on the island of Samos in the Aegean Sea, and it is widely believed that the Pythagoreans discovered not only the theorem that bears his name but the concept of incommensurability and the golden ratio as well.°

Leonardo Fibonacci

Fibonacci was a Pisan, born around 1170 CE. One of his singular accomplishments, other than discovering the subject number series, was the introduction of the Hindu-Arabic decimal, positional number system into Europe with the publication of the *Liber Abaci* in 1202. The Fibonacci series, at first glance, is simplicity itself:

$$0 \ 1 \ 1 \ 2 \ 3 \ 5 \ 8 \ 13 \ 21 \ 34 \ 55 \ 89 \ 144 \ 233 \ 377 \ 610 \ 987 \ldots$$

Each successive term is the sum of the two terms immediately preceding it. The series was given its name by the French mathematician Edouard Lucas (1842–1891). It was Kepler who, in 1611, had discovered the link between the Fibonacci series and the golden ratio, which is that any two successive terms in the series, when divided by the other, converge toward the golden section value as the numbers grow larger (see Figure 2-5).

Both the Pythagorean theorem and the golden section can be handily demonstrated by the same construction, pictured along with their formulas in Figure 2-3. If we construct a rectangle of one by two units, its diagonal is the square root of five *a*. We can see this in *b* where the squares have been constructed on each of the sides of the right triangle formed

Figure 2-3   Constructing the Golden Section.

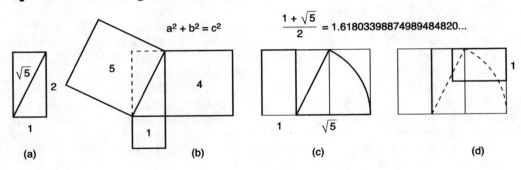

Figure 2-4   Infinite Golden Spiral.

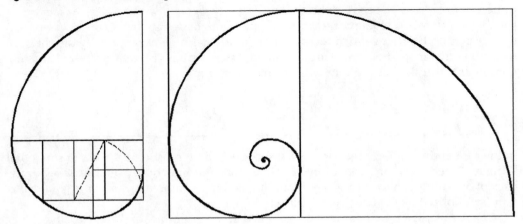

by half of the one by two rectangle. In *c*, we have mirrored the rectangle to the left and used the diagonal to set the right boundary of a golden rectangle whose base is equal to one plus the square root of five. Finally, in *d*, we bisect this rectangle in both directions, which gives us a unit rectangle in the golden ratio of 1 to 1.61803398. . . .

From the unit rectangle, we can construct the golden spiral pictured in Figure 2-4. Infinite in both directions, the Fibonacci series and its derivations are manifested throughout history and throughout nature in the florets of sunflowers, the scales on the surface of pineapples, and in the arms of the Milky Way.

Adolf Zeising

In the quarter century from the early 1830s to 1856, the work of four individuals had a significant influence on one or more of the three most influential architects of the twentieth century: Le Corbusier, Ludwig Mies van der Rohe, and Frank Lloyd Wright. One of them was Adolf Zeising. At the time of great scientific discoveries and theories that began in the Renaissance and accelerated in the Industrial Revolution, Zeising was among the first to apply an analytical approach to understanding the methodology of design. His work brought the golden section into the realm of architectural theory.

Zeising retired in 1853 as a secondary-school professor and a year later published his *New Theory of the Proportions of the Human Body*. In it, Zeising developed his thesis that in the golden section is contained the fundamental principle of all formation striving to beauty and totality in the realm of nature and in the field of the pictorial arts, and that it from the very beginning was the highest aim and ideal of all figurations and formal relations, whether cosmic or individualizing, organic or inorganic, acoustic or optical, which had found its most perfect realization, however, only in the human figure.°

Zeising's model is pictured at left in Figure 2-5, adjacent to the Fibonacci series from which its basic units are drawn. The upper and lower parts are divided at the navel on a

Figure 2-5   Zeising's Man, the Fibonacci Series, and the *Omphalos*.

Zeising		Fibonacci Series	(x+1)/x	Inches		Millimeters	
21		0		1.5		39	
34		1		2.5		63	
34		1	1.000000	2.5		63	
34		2	2.000000	2.5		63	
21		3	1.500000	1.5		39	
34		5	1.666667	2.5		63	
55		8	1.600000	4.0		102	
55		13	1.625000	4.0		102	
55		21	1.615385	4.0		102	
34	377	34	1.619048	2.5	27.5	63	699
89		55	1.617647	6.5		165	
55		89	1.618182	4.0		102	
89		144	1.617978	6.5		165	
55		233	1.618056	4.0		102	
89		377	1.618026	6.5		165	
55		610	1.618037	4.0		102	
34		987	1.618033	2.5		63	
55		1597	1.618034	4.0		102	
34		2584	1.618034	2.5		63	
55	610	4181	1.618034	4.0	44.5	102	1131
987	0.61803	6765	1.618034	72.0	0.61732	1830	0.61804
		10946	1.618034				
		⋮	⋮				

basis of two successive numbers: 377 and 610. The figure is then subdivided into Fibonacci ratios without regard to actual height. In the right-hand columns, these ratios are proportionately adjusted to the six-foot height of Vitruvian man in da Vinci's drawing, rounded to the nearest half inch and whole millimeter. The deviation of the total height from 0.618, minuscule in both cases, is shown next to the totals.

Le Corbusier

Beginning in 1943, Le Corbusier took a similar approach in the development of his *Modulor*, pictured in Figure 2-7 adjacent to Zeising's skeleton (dimensions are in centimeters). It began with an exercise to determine the harmonious placement of a square aligned with two other equal squares set side by side. The solution was to align the square at the "place of the right angle" as determined by the golden section. The intersection of the square and the right angle thus determined two points through which an oblique line could be constructed. As pictured in Figure 2-6, the oblique line defines a diminishing series on the left and an ascending series on the right. (It must be pointed out, as it was to Le Corbusier, that the "right angle" is actually 89.6 degrees. It is not this angle, however, that determines the Modulor; it is the golden ratio.)

Two such squares stacked vertically, 113 centimeters on a side, will accommodate a person six feet in height with her arm upraised. The key dimensions are shown in Figure 2-7. The value of 113 furnishes a golden mean of 70, and their sum is 183, which in centimeters is equivalent to six feet within 0.05 centimeter. This number starts off what Le Corbusier called the red series: . . . 4, 6, 10, 16, 27, 43, 70, *113*, 183, 296. . . .

The second (blue) series is based on the navel: . . . 8, 13, 20, 33, 53, 86, 140, *226*, 366, 592. . . centimeter. The base number 226 furnishes the golden mean values of 86 and 140.

The *Modulor*, in Le Corbusier's words, "is a measure based on mathematics and the human scale . . . they are the effect of a choice made from an infinity of values. [They] possess the properties of numbers but the manufactured objects whose dimensions these numbers are to determine are either *containers* of man or *extensions* of man."°

In his "preamble" to *Le Modulor*, Le Corbusier addressed some of the issues that are the central subjects of this chapter:

> Do we understand clearly enough what it meant when, one fine day, the zero—
> key to the decimal system—was created? Calculation is a practical impossibility

Figure 2-6 The Construction of the Modulor.

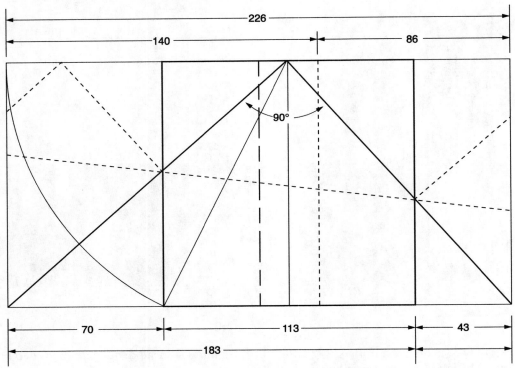

Figure 2-7 Zeising's Man Compared with Le Corbusier's Modulor.

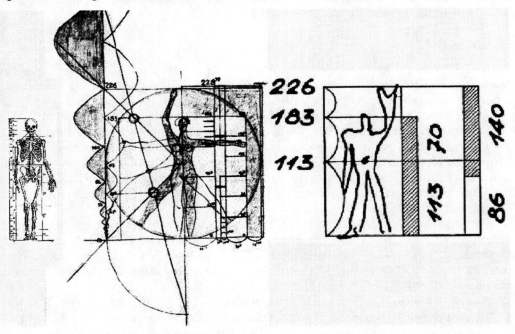

without the zero. The French Revolution did away with the foot-and-inch system with all its slow and complicated processes. That being done, a new system had to be invented. The savants of the Convention adopted a concrete measure so devoid of personality that it became an abstraction, a symbol: the metre, forty-millionth part of the meridian of the earth. The metre was adopted by a society

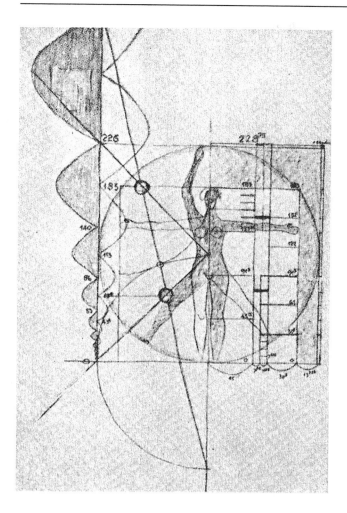

Figure 2-8   Serralta and Maissier's Drawing of *Modulor Woman.*

steeped in innovation. [Over two] centuries later, when factory-made goods are circulating all over the globe, the world is divided into two halves: the foot-and-inch camp and the metre camp. The foot-and-inch, steadfast in its attachment to the human body, but atrociously difficult to handle: the metre, indifferent to the stature of man, divisible into half metres and quarter metres, decimeters, centimeters, millimeters, any number of measures, but all indifferent to the stature of man, for there is no such thing as a one-metre or a two-metre man.°

The middle image in Figure 2-7 (and shown individually in Figure 2-8) was originally prepared by Justin Serralta and Maissier, two of Le Corbusier's associates, for the *Divina Proportione* Exhibition at the 1951 Milan Triennial. Exhibited alongside of manuscripts or first editions of the works of Vitruvius, Dürer, da Vinci, Alberti, and others, it elicited a singular comment from a noted mathematics professor: "How beautiful this drawing is!" According to Le Corbusier: "[S]ince Serralta has a soft spot for the ladies, his man is a woman 1.83 metres tall: brrrrh!"°

Before and since Vitruvius, philosophers, physicists, and mathematicians have sought the answers to the questions of the universe. Architects seek to balance Vitruvius' six fundamental principals—order, arrangement, eurythmy, symmetry, proportion, and economy—in order to achieve harmony between man and his environment. The Modulor offers a scalable system of proportion that attempts to unite metric and man.

In Figure 2-9 is pictured a "Modulor" grid at three different scales. Each shows Le Corbusier's *red* and *blue* series superimposed (the lighter lines indicate the *blue* series). The series overlap in a manner similar to the common factors of the English architectural scales shown in Table 2-1. In the grid on the left, the *red* series covers nearly eleven miles while

Figure 2-9   Modular Grid Based on Le Corbusier's Modulor.

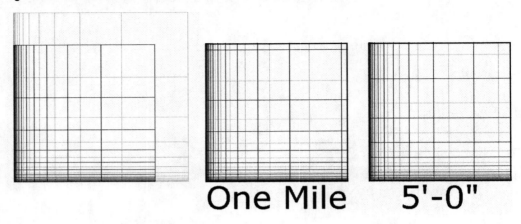

the *blue* series extends to slightly more than thirteen. As with the infinite golden spiral shown earlier, we would see the same pattern in a grid one inch on a side, although the Modulor grid never gives us exactly one inch.

In the spring of 1946, Le Corbusier visited Albert Einstein at Princeton. Einstein's assessment of the *Modulor* was this: "It is a scale of proportions which makes the bad difficult and the good easy."° We shall return in the next chapter to the discussion of the Modulor and the relationship of the golden section to two significant examples of twentieth-century dwellings.

Anthropometry and Ergonomics

Since the Industrial Revolution, the amount of attention given to designing for individuals having widely varying characteristics has steadily increased. This has given rise to the science of *anthropometry*, which is the compilation of statistical information on which to base design decisions. The word results from a conjoining of the Greek words *anthropos*, meaning *human*, and *metrikos*, meaning *pertaining to measurement*.

A second, related compound derives from the words *ergos*, meaning *work*, and *nomos*, meaning *natural laws*. *Ergonomics* is the application of anthropometric information to the design of objects, systems, and environments for human use.

The term *human factors* covers the elements affecting human performance and well-being within man-made environments and in the use of tools. The term encompasses both physiology and psychology and is the foundation of modern industrial design. The Second World War saw the development of detailed human engineering standards for the branches of the military. Man-machine relationships became essential as military tactics evolved to the point that machines, as opposed to personnel, were perceived to be decisive.

During the last four decades of the twentieth century, human factors standards have become progressively more refined, extending to all segments of our population.

- 1960s—The United States Department of Health, Education and Welfare published data pertaining to the height, weight, and bodily dimensions of adults.
- 1970s—The automotive industry added a detailed survey of children between the ages of two months and eighteen years.
- 1980s—The elderly were added to the collective database when they began to constitute a significant segment of the population.
- 1990s—Many of these data were enacted into law with the passage of the Americans with Disabilities Act (ADA), which established rules of accessibility and acceptable levels of protection for persons in wheelchairs together with the visually and hearing impaired.

Figure 2-10   Vitruvian Man Compared to Anthropometric (U.S.) Man and Woman.

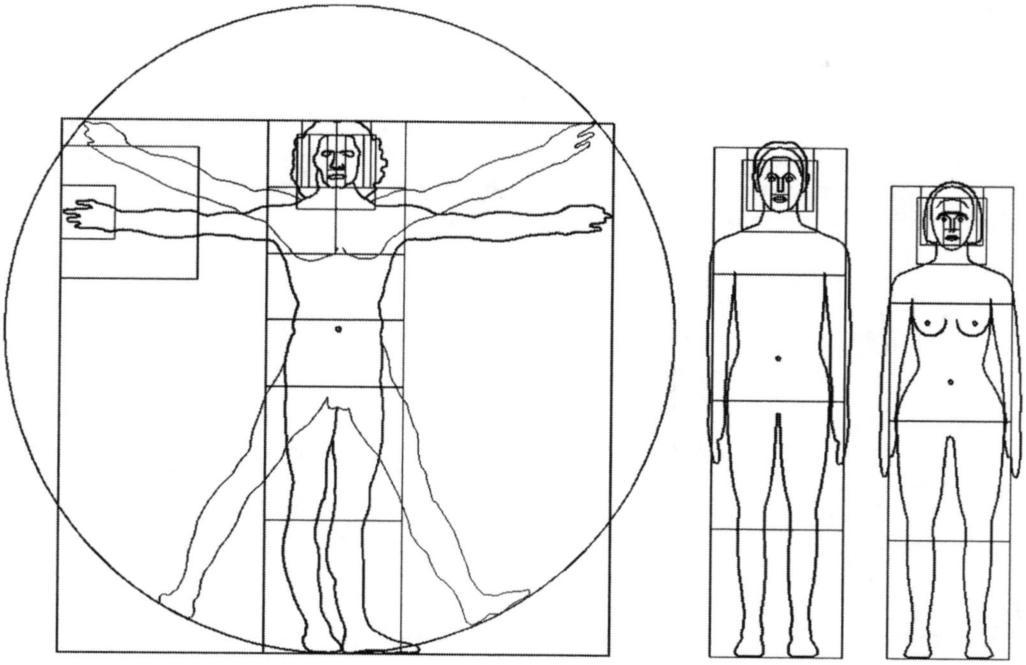

Table 2-3   Comparative Statures of Man and Woman°

Percentile	99	50	1
Man	75.6	69.1	62.6
Woman	69.8	64.0	58.1

Cognitive science now addresses related functional issues such as awareness and decision making. In a later chapter, we discuss how building codes and the ADA govern planning and design as we enter the twenty-first century.

In order to validate the assumptions we have been making regarding the scale of modern man, Figure 2-10 shows Vitruvian man at a comparable scale with the fiftieth percentile, the average U.S. man and woman. Table 2-3 covers all but the 2 percent of the population that do not fall within the documented range of variance.

Architects and designers must now accommodate the whole of mankind in all of its diversity. The United States is at this point the only major holdout in adopting the metric system, but it is at the forefront of development in the human factors area.

NINETEENTH-CENTURY INFLUENCES ON THE TWENTIETH

We said earlier that there were three Europeans besides Zeising who influenced the development of design and architectural theory in the mid-nineteenth century. This is, of course, not an exclusive list but includes three people whose theories and criticism appear repeatedly linked to whom it may concern: Wright, Le Corbusier, and, through the Bauhaus, Mies van der Rohe. The three were Friedrich Froebel, Owen Jones, and Victor Hugo.

Friedrich Froebel

Like Zeising, Froebel was an educator. He was something of a mystic in that he believed that a child's early development was linked through playthings to the nature and life of the cosmos. Through observing and handling geometric forms, Froebel held that the child gained an understanding of certain key relationships on a symbolic level through meaningful play.

In Froebel's vocabulary, knowledge, beauty, and life were represented by the sphere and cube. The sphere was thought to correspond primarily to feelings and the heart, while the cube symbolized rational thought and the intellect. Froebel characterized this process as a *metaphorical dance*, through which the child gained a hands-on familiarity with lines, edges, sides, surfaces, and volumes that Froebel referred to as *dance forms*. The area in which these activities took place was described as a "garden for children," from which we have the word *kindergarten*.

One of Froebel's principal activities involved the use of wooden building blocks, packaged in increasing levels of complexity, which Froebel termed "gifts" as they were given to the children. The gifts were developed in the 1830s and are still used to teach the basic elements of geometric form, mathematical relationships, and creative design.

Frank Lloyd Wright
As a child, Frank Lloyd Wright played for several years with a set of Froebel blocks given him by his mother. The smooth maple blocks measured two inches on the short side. Wright recalled working with them on a kindergarten table with *unit lines* painted four inches apart, saying that the blocks remained *in his fingers* throughout his life.

Wright developed, through playing with these blocks, a symbolic sense of geometry. For him, the square signified integrity, the circle infinity, and the triangle aspiration, all "with which to 'design' significant new forms." When Wright spoke of the "nature" of an object or condition, it was its "innate sense of origin" to which he referred. He compared the human heart as the "interior activating impulse" to the machine in terms of its "dependable mechanism" to the essence of a brick lying in its "brickness."

Johannes Itten and the Bauhaus
Johannes Itten, who developed the basic course at the Bauhaus, was originally an elementary school teacher trained by Froebel. Itten was invited to join the Bauhaus by Walter Gropius, who had just formed the design school in Weimar in 1919.

In his book titled *Design and Form*, Itten ascribes his own set of attributes to geometric forms:

> The three basic forms, the square, the triangle, and the circle, are characterized by the four different spatial directions. The character of the square is horizontal and vertical; the character of the triangle is diagonal; the character of the circle is circular. . . . Thus the circle was experienced as an evenly curved line in continuous motion. . . . Experiencing the square requires an angular, tense form of movement because the right angle always recurs. . . . In the triangle the whole diversity of angles appears.°

Itten believed that the primary goal of all teaching should be to develop genuine seeing, feeling, and thinking. He went so far as to define three basic classifications of the creative spirit:

- *Spiritual-expressive*—Guided by intuitive feelings, sometimes neglecting the constructive forms and patterns of order
- *Naturalistic-impressive*—Observes and recognizes natural varieties and represents them realistically without adding expression
- *Intellectual-constructive*—Begins with the construction of the object, attempting to grasp and geometrize its aspects clearly

One of Itten's exercises in life drawing was to have students interpret a subject expressively, naturalistically, and constructively, then to synthesize the essence into a *generally valid representation*. Most often, according to Itten, the student will achieve the most success with the approach that conforms to her own temperament, but the exercise helps to bring out a rational consciousness of the less-instinctive ways of seeing and interpreting. Itten's methodology is built on the belief that

[t]hrough his reason, man is able to recognize the impersonal principle and to use it objectively. In a deeper sense, all measuring and constructing is a method to overcome personal limitations and shortcomings and to arrive at an objective and generally valid statement. . . . When a man is genuine; everything he does becomes a reflection of his own formative powers.

Earlier, we noted, quoting from Bauhaus literature, that it is part of the creative urge that forces us to try to give form to shapeless matter. The Bauhaus was one of the first major teaching institutions where design psychology was seriously considered.

The painter Oskar Schlemmer headed the sculpture and theater workshops at the Bauhaus from 1923 to 1929. He taught a third-semester course, simply titled *Man*, which was divided into biological, formal, and philosophical parts. The philosophical segment considered man as "a 'thinking and feeling being' in his world of conceptions and ideas." Among the challenges to the students was the creation of visualizations of their theoretical investigations in the form of life drawing exercises, figural renderings, and dance forms. Schlemmer's own visualization of man from one of the Bauhaus books is pictured in Figure 2-11, another variation on one of the principal themes of this chapter.°

Owen Jones

In 1856, Owen Jones compiled a folio of over one hundred plates beautifully and encyclopedically illustrating various historical styles of ornament from around the world, which he called *The Grammar of Ornament*.° The list is truly ecumenical. Represented are Egyptian, Greek, Roman, Arabian, Persian, Indian, Chinese, Medieval European, and Renaissance geometry—only a partial list.

Figure 2-11 Man and Artistic Figure.

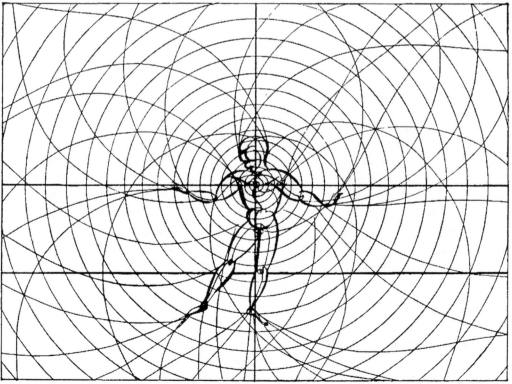

Figure 2-12 Moresque Ornament Modules (see Plate One).

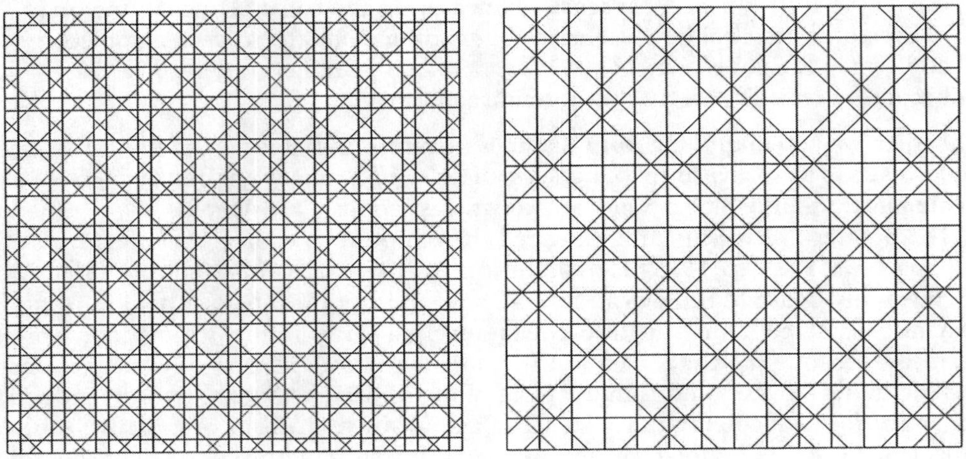

In Chapter 10 of Jones's folio are diagrams illustrating two general principles of ornament from the Alhambra Palace in Grenada (Figure 2-12). The first consists of equidistant lines crossed diagonally by perpendicular lines on each square. The second is constructed with an equidistant grid crossed by diagonals only on each alternate square. Jones suggests that an infinite number of patterns can be produced from these two systems, notwithstanding the use of color.

In his 1943 *Autobiography*, Frank Lloyd Wright admits to having been fascinated by Jones's work, particularly these two modular patterns. Wright describes being taken by Jones's visualizations in much the same manner as his involvement with his Froebel blocks, "tracing the multi-fold design . . . until . . . I needed exercise to straighten up from this application."° Wright instinctively looked at these forms not just as mere ornament but as the basis for an ordering system that is seen in the 30-, 45-, and 60-degree angles of not only his buildings but in Wright's own ornament.

Jones expressed some misgivings in the preface to his book and would certainly have been gratified to know the manner in which at least one great architect built upon his work:

> I have ventured to hope that . . . in bringing into immediate juxtaposition the many forms of beauty which every style of ornament presents, I might aid in arresting that unfortunate tendency of our time to be content with copying, whilst the fashion lasts, the forms peculiar to any bygone age. . . . Ignoring the peculiar circumstances which rendered an ornament beautiful, *because it was appropriate* . . . entirely fails as expressive of other wants when . . . transplanted.

> It is more than probable that the first result of sending forth to the world this collection will be seriously to increase this dangerous tendency, and that many will be content to borrow from the past those forms of beauty which have not already been used up *ad nauseam*. It has been my desire to arrest this tendency, and to awaken a higher ambition.

> If the student will but endeavor to search out the thoughts which have been expressed in so many different languages, he may assuredly hope to find an ever-gushing fountain in place of a half-filled stagnant reservoir.

> In the following chapters I have endeavoured to establish these main facts—

> *First*. That whenever any style of ornament commands universal admiration, it will always be found to be in accordance with the laws which regulate the distribution of form in nature.

> *Secondly*. That however varied the manifestations in accordance with these laws, the leading ideas on which they are based are very few.

Thirdly. That the modifications and developments which have taken place from one style to another have been caused by a sudden throwing off of some fixed trammel, which set thought free for a time, till the new idea, like the old, became again fixed, to give birth in its turn to fresh inventions.

Lastly. I have endeavored to show . . . that the future progress of Ornamental Art may be best secured by engrafting on the experience of the past the knowledge we may obtain by a return to Nature for fresh inspiration. To attempt to build up theories of art, or to form a style, independently of the past, would be an act of supreme folly. It would be at once to reject the experiences and accumulated knowledge of thousands of years. On the contrary, we should regard as our inheritance all the successful labours of the past, not blindly following them, but employing them simply as guides to find the true path.°

Between the preface and the plates to *The Grammar of Ornament,* Jones sets forth thirty-seven propositions relating to the proper application of ornament and color. Many of them are equally germane to planning. Proposition 6, for example, states: "Beauty of form is produced by lines growing out, one from another in gradual undulations. There are no excrescences; nothing could be removed and leave the design equally good or better." One might call this the *principle of parsimony.*

Victor Hugo

The building of great cathedrals largely came to an end before the Renaissance. Before Luther distributed his *95 Theses,* printed in German, architecture had *been* the world's mass medium.

In 1832, still younger than Martin Luther was when he upset the established Church's apple cart, Victor Hugo wrote *Notre-Dame de Paris* or, as it is better known, *The Hunchback of Notre Dame.* In Chapter 2 of Book 5, which he called *"This Will Kill That,"°* Hugo characterizes the development of architecture as paralleling the development of language from hieroglyph to alphabet to word, proper name, phrase, and finally complete sentences. "In fact," he said, "from the origin of things down to the fifteenth century of the Christian era, inclusive, architecture is the great book of humanity, the principal expression of man in his different stages of development, either as a force or as an intelligence. . . . During the first six thousand years of the world, from the most immemorial pagoda of Hindustan, to the cathedral of Cologne, architecture was the great handwriting of the human race. And this is so true, that not only every religious symbol, but every human thought, has its page and its monument in that immense book."

In the epigraph at the beginning of his *Testament,* Frank Lloyd Wright quotes Hugo describing the Renaissance as "the setting sun all Europe mistook for dawn." Hugo foreshadowed Jones in describing three "lesions" in the architecture of the Middle Ages as time, revolution, and *fashion,* which he describes as having been caused by the "anarchical and splendid deviations of the Renaissance [that] followed each other in the necessary decadence of architecture." However, Hugo also prophesied that the "great accident of an architect of genius may happen in the twentieth century, like that of Dante in the thirteenth."

In an earlier chapter titled "A Bird's-Eye View of Paris," there is a wonderful paragraph describing the Paris Stock Exchange:

As for the Palace of the Bourse, which is Greek as to its colonnade, Roman in the round arches of its doors and windows, of the Renaissance by virtue of its flattened vault, it is indubitably a very correct and very pure monument; the proof is that it is crowned with an attic, such as was never seen in Athens, a beautiful, straight line, gracefully broken here and there by stovepipes. Let us add that if it is according to rule that the architecture of a building should be adapted to its purpose in such a manner that this purpose shall be immediately apparent from the mere aspect of the building, one cannot be too much amazed at a structure which might be indifferently—the palace of a king, a chamber of communes, a town-hall, a college, a riding-school, an academy, a warehouse,

a court-house, a museum, a barracks, a sepulchre, a temple, or a theatre.
However, it is an Exchange.

Fast-forward to Chicago, 1997, and the opening of the largest trading floor in the world. It is the antithesis of Hugo's Bourse, a technological marvel, bomb resistant, a static cube without ornament and devoid of grace.

RENAISSANCE

We are halfway through the first decade of the twenty-first century, and the Industrial Revolution continues to accelerate. Technology and terrorism dominate our worldview, and the cathedrals of today are largely the buildings that house our corporations and governmental and cultural institutions. Some say this is an age of decadence; but, if it is, it shares the dualist nature of the Renaissance.

Decadence is a loaded word; *ambivalence* might be more fitting. A historian wrote in the year 2000 that decadence

> . . . implies in those who live in such a time no loss of energy or talent or moral
> sense. . . . On the contrary, it is a very active time, full of deep concerns, but
> particularly restless, for it sees no clear lines of advance. . . . The forms of art as
> of life seem exhausted, the stages of development have been run through. . . . °

Vitruvius was unapologetic for the fact that the wealthy and powerful had special requirements in their dwellings. Habitat for Humanity was an idea whose time would not come for another two thousand years. Consider the proposition that the Roman Empire might have enacted into law a code mandating accessibility for the disabled to all its public buildings! History is full of failures and contradictions, but the principles stand.

The great architects of the twentieth century did not spring forth fully formed like Cadmus' soldiers. They were threads in the historical fabric, subject to the influences of those who mentored them, the ideas of their time, and their own creative spirits. Given the relatively few historical influences that we have chosen to mention here, the extent to which they resonate together is remarkable.

The work of the three we mentioned specifically exhibit these principles, consciously or out of their intuitive spirits, in their buildings. Zeising laid the groundwork upon which Le Corbusier built a system of proportion that has mathematical precision. Froebel directly influenced Itten (and through him the Bauhaus) as well as Wright. Jones echoed, from a Victorian perspective, the principles first articulated by Vitruvius (Figure 2-13) and, with Hugo, voiced concerns about the greatest hazards that befall the architect: dishonesty and lack of original thought.

Throughout the remainder of this book, we focus primarily on the functional aspects of planning; but we must keep in mind the overall form and image, so that our work forms a harmonious whole. We have dwelt on some of the significant threads of history, marking some connections to literature in order to point to the idea that no single art exists in a vacuum. We weave space into the fabric of our time.

We need not be apologetic for responding to the needs of our clients, be they wealthy and powerful or not. But we must look beneath the surface for the essential principles, those of Vitruvius, the *principle of uncertainty*—as applied to all knowledge, which Jacob Bronowski said might better be called the *principle of tolerance*°—and the *principle of parsimony*.

The principle of parsimony is often colloquially translated as "Keep it simple, Stupid!" But Ockham never said that. What he said was: *Frustra fit per plura quod potest fieri per pauciora*, which translates as "It is useless to do with more what can be done with less."° So, from this fourteenth-century heretic and political activist echoes Mies van der Rohe's paradoxical dictum, "Less is More!"

Figure 2-13   Vitruvius' Six Fundamental Principles.

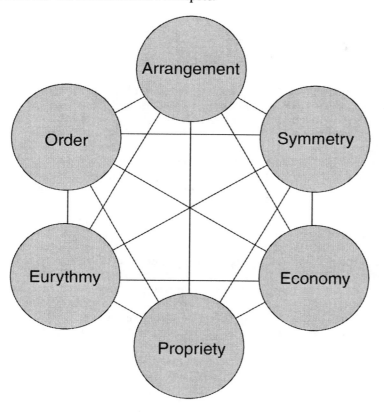

CLOSING WORDS

In this chapter, we have looked at four nineteenth-century personages whose ideas influenced three of the foremost twentieth-century architects: Frank Lloyd Wright, Ludwig Mies van der Rohe, and Le Corbusier. This triad fulfilled Victor Hugo's prophecy, in part by finding appropriate ways to use modern materials in need of a new expression.

Next, we continue to develop some earlier themes, among them the *omphalos*, the symbolic center of our sense of place. New construction materials and ways of building permitted the separation of building enclosures from their interior space and structure, but it took a modern vision to advance architecture and planning beyond the habits of the past. Vitruvius said that the architect must have natural gifts but also be amenable to instruction, which implies the need to look forward and outward with an open mind.

TIME CAPSULE The Parthenon

Between 447 and 432 BCE, the architects Ictinius and Callicrates designed the temple dedicated to the goddess Athena. In stark contrast to the Ramesseum, the Parthenon stood open to the city within the Acropolis. Its columns and much of its rich relief sculpture were on the *exterior*. The pediments at each end were filled with the finest sculpture of the master Pheidias. As Banister Fletcher describes, the east end depicted the birth of Athena, and the contest between Athena and Poseidon for the Attic land was represented on the west end.° Below the pediments, extending around both sides of the temple, were metopes sculptured in high relief alternating with marble triglyphs. These sculptures depict gods and giants on the eastern end, Centaurs and Lapiths on the south, Greeks and Amazons on the west, and scenes from the siege of Troy on the north.°

Sculptured friezes ornamented both the inside and outside of the temple. Illuminated by the natural light reflected upward from the white marble pavement, the celebrated outer frieze depicting the Panathenaic procession encircled the entire building outside the naos, crossing both ends above the six columns of the pronaos (at the eastern end) and opisthodomos (at the western end). (Its location is indicated by the dashed outer rectangle in Figure TC2-2*b*.) The procession culminated at the eastern end above the temple's main entrance where the forty-two-foot-high, ivory and gold statue of the goddess could be seen through the open door to the naos. Bright colors were used to highlight the sculptures on the pediments, metopes, and friezes.

The ancient city-state up to this time, even in Greece, was characteristically totalitarian by nature. Individual liberty among citizens was limited; even more so was that of foreigners who were generally regarded with suspicion as potential enemies. Although there were a sizable number of slaves to be sure, Athens under Pericles was liberalized to an unprecedented degree. In addition to forty thousand citizens, there were half again that number of resident foreigners known as "metics," meaning *those who had changed homes*. Perhaps 40 percent of the craftsmen who worked on the Parthenon were metics.° Although they could not become citizens or own land, the metics were an important part of the city's culture and economy. Metics often engaged in retail and wholesale trade, import-export, banking, and the industrial crafts in which they enjoyed a virtual monopoly.° Thucydides quoted Pericles as saying, "Our city is open to all men: we have no law that requires the expulsion of the stranger in our midst, or debars him from such education or entertainment as he may find among us."

The great annual festival depicted on the outer frieze lasted for two days, and every fourth year it lasted at least twice as long. The Panathenaea began with athletic contests and was brought to a close by the procession through the center of the city toward the Acropolis. Priests, civic leaders, and the general populace (including metics) participated in the event, bearing an embroidered robe to be placed upon the shoulders of the statue of the patron goddess within her temple. Upon reaching the Parthenon, they made a mass sacrifice of oxen, sheep, and cows sufficient to feed the entire city.°

The refinement of the architecture and the openness of the rituals surrounding it reflected the character of Athenian democracy that had developed by this time. This was not representative democracy as we know it today, however. The center of power was vested in an Assembly, membership in which was the right of all adult male citizens. Meetings of the Assembly on a hill outside the city often included upward of six thousand people, about one-fourth of those eligible to attend. A talent for oratory was highly valued, heated debates were the order of the day, and propositions were decided by majority vote. Continuity of government was provided by the Council of Five Hundred, elected officials who met throughout the year. Issues to be discussed in the Assembly were determined by the Council.°

It is significant that verbal communication and oratory were highly valued in Periclean Athens. This was an era of extraordinary trust and cohesiveness within Greek society (at least among the citizenry), which spawned the great philosophers Socrates, Plato, and Aristotle. Only a century later, written contracts began to be used much more widely. A fourth-century orator is reported to have observed, "We make written contracts with one another through distrust, so that the man who sticks to the terms may get satisfaction from the one who disregards them."°

ORDER IN THE PLAN

Charles Freeman observes that "the moving spirit behind the building of the Parthenon was certainly Pericles but there was never any doubt that this was a city enterprise inspired by democratic pride."° He identifies three features that make us see the Parthenon as one of the great buildings:

1. It was constructed throughout using the finest materials, principally twenty-two thousand tons of marble that were transported ten miles from Mount Pentelicus.
2. The proportions give an overall sense of lightness: The corner columns are slightly thicker and closer together than the others and lean inward slightly. It is said that the columns extended would meet one mile above the structure.

Figure TC2-1   Parthenon Elevation Symmetries, circa 440 BCE.

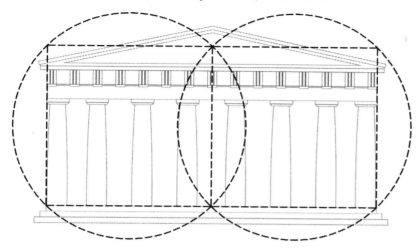

3. It is the most richly ornamented temple ever constructed, and its sculptures "represent the climax of the Classical revolution in art."

In plan, the Parthenon has seventeen columns along the sides and eight across each end forming an open colonnade around the entire structure, nine feet wide on the sides and eleven feet wide across the front and rear. A second colonnade stands within the naos, having ten columns on each side and five across the west end. The entire structure stands on a twenty-three thousand-square-foot crepidoma, or pedestal of three steps, raising it five feet above its site. Inside the naos, in front of the colonnade at the west end, stood Pheidias' monumental statue of Athena Parthenos.°

We looked briefly at some of these relationships in Chapter 1 when discussing Vitruvius' description of symmetries to be found in the elements of temples in terms of the golden section (Figure 1-1). Figure TC2-1 shows another set of proportions in the ordering of the Parthenon elevation. A double-square rectangle can be placed symmetrically upon the Parthenon's end elevation with one long side resting upon the crepidoma. From the top of the column capitals (which is the same line that defines the bottom edge of the frieze) to the dashed line at the top of each square is one-third the height of the squares. This dashed line bisects the height of the pediment horizontally, making the height of the columns three-fifths the height of the building's elevation at its centerline. This diagram is based on Banister Fletcher's comparison of the principles of proportions used in several building types.°

In Chapter 2, we spoke of regulating lines, planes, and masses that define boundaries and edges. As we move through a space, we perceive volumes, masses, and enclosing surfaces in different ways depending upon the kind of materials used, their degree of opacity or transparency, and such focal points as may occur along our path or immediately beyond it. If the points of visual and physical access do not correspond, we may lose our

bearings and become disoriented or, at the very least, uncomfortable. It is the dynamic relationship among the five continuous quantities defined by Aristotle that contribute to our experience of geometric organization.

In Figure TC2-2, we have analyzed several more of the proportional relationships that shape the Parthenon's plan. Such regulating lines are not always consciously apparent to a person within a space, but they are felt, nonetheless, and contribute to her sense of place within the orderly flow of the spatial environment.

If two golden section rectangles the width of the crepidoma are placed, one at each end, the region where they overlap defines a square. (Solid diagonals identify two such rectangles in Figure TC2-2*a*.) This square can then be used to mark off congruent areas at the ends of the plan, shown in the figure by the diagonals with *dashed* lines.° The shaded rectangle between the two end squares, exactly in the center of the plan, determines the alignment of the five columns at the west end of the inner colonnade. This ambulatory is formed by a double tier of Doric columns. The eastern boundary of this region determines the placement of the statue of Athena facing the door into the naos. The inner colonnade is also, as we mentioned earlier, five by ten columns, giving it the form of a double square.

The four *Ionic* columns in the virgin's chamber opening from the west end of the Parthenon are centered on the western boundary of the central square. It is noteworthy that Ionic columns were used within this predominantly Doric structure for their slender grace. Vitruvius later speaks of the Ionic order as being "less severe" than the Doric. This was an early instance of the orders being mixed, which became increasingly common afterward.

The shaded rectangle central to the plan has precisely the same proportions as the two that define the perimeters of the pronaos and opisthodomos, and it is very nearly equal to the combined areas of the other two.

The golden section appears yet again in the perimeter of the enclosure surrounding the naos and virgin's chamber. The dashed lines in Figure TC2-2*b* show a

(*continued*)

(continued)

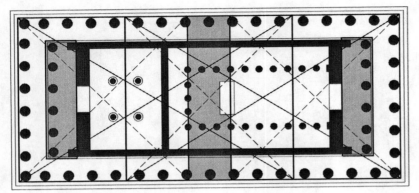

Figure TC2-2 Exterior and Interior Symmetries in the Parthenon's Plan.

(a)

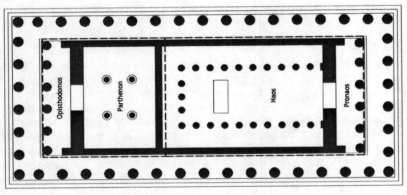

(b)

golden rectangle plus a square enclosing the two inner chambers and the porticos at each end. This proportion has extraordinary mathematical significance because the golden section (φ) is the only number that, added to the number 1, gives the same result as squaring its value:

$$\varphi + 1 = \varphi^2.$$

In the Time Capsule that follows Chapter 6, we analyze the plan of the medieval cathedral at Salisbury, England, which uses another classical system of visual organization (Figure TC5-4).

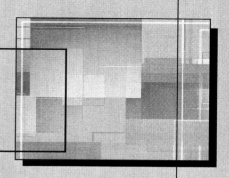

CHAPTER 3

Place and Modernism

> . . . [W]hatever may be the future of architecture,
> in whatever manner our young architects may one day solve the question of their art,
> let us, while waiting for new monuments, preserve the ancient monuments.
> —Victor Hugo°

"Howard Roark laughed." While we do not recommend getting expelled from architectural school as a motivator, we quote here the opening paragraph of Ayn Rand's architectural fiction° to remind ourselves of the essential role of exuberance in design and planning. But the book did not start the modern revolution.

At the time Rand's novel was written, Frank Lloyd Wright (on whom the character of Roark was largely based) was an architect of stature at the age of seventy-four. Le Corbusier was fifty-six. The Bauhaus had closed ten years earlier due to World War II, and Ludwig Mies van der Rohe, at fifty-seven, was in the process of establishing his architectural office in Chicago. When he had become head of the architectural department at the Illinois Institute of Technology (IIT) five years before, a number of the former Bauhaus faculty and students formed the nucleus of that school as well as what later became known as the Institute of Design. Mies van der Rohe had been the last Bauhaus director from 1930 until its demise.

The Fountainhead has attained something of a cult status among students of design and architecture since its appearance in 1943. In the late 1940s, a motion picture was made of the novel in which Gary Cooper plays (without laughter) the iconoclastic twentieth-century architect who, upon learning that his antagonists have frosted one of his projects with Beaux-Arts icing, dynamites the buildings—not a recommended solution for the twenty-first-century master builder!

THE POTENCY OF PLACE

In the last chapter, we discussed some of the basic principles of architecture and planning, beginning with Vitruvius, who was a moderately successful engineer and architect in the Roman Empire at the time of the emperor Augustus. Vitruvius' early manifesto has been referred to and often misinterpreted by numerous theorists beginning with the resurgence of interest in classical architecture in the Renaissance. Leonardo da Vinci was one among many who followed Vitruvius in relating the basis of proportion in architecture to the proportion of man.

We enumerated the six fundamental principles of architecture according to Vitruvius. These are the starting points. In the *third* chapter of his first book, Vitruvius declared that there are three qualities all architecture must possess: *firmitatis*, *utilitatis*, and *venustatis*. These are the qualities of the end product, the three golden threads referred to as the *Vitruvian triad*: strength, utility, and delight. While Vitruvius spoke of these almost as an afterthought, Renaissance writers such as Leon Battista Alberti, Andrea Palladio, and

Sebastiano Serlio seized upon them as the guiding principles in their books of architecture.°
(Alberti's *De Re Aedificatoria* was the first architectural book printed with movable type in
1485. Vitruvius' own work was first printed in Rome a year later.)

Scientific method, which Descartes formalized in the seventeenth century, led to a *reductionist* approach to problem solving. The *analysis/synthesis* method is to divide a problem
into as many individual elements as possible, then develop a solution by dealing with the
simplest tasks first, proceeding to the more complex in some logically determined fashion.
Descartes used the term *difficulties* to describe the individual tasks into which a complex
problem could be broken.

Descartes' mathematical tool, analytic geometry, has a natural affinity with the architectural and planning process because it expresses numerical relations visually. Conversely,
visual relationships can be turned into numbers. Since Descartes, analytical reasoning is
often viewed as the exclusive corridor to truth, but the compulsion to reason comes from
the creative spirit. This may explain why rationalists occasionally become fanatics. It is
worth remembering that the inspiration behind Descartes' system was not arrived at systematically but was the result of three dreams.°

Our current usage describes the elements of the construction process as *tasks* or
activities, and *difficulties* encountered are referred to as *issues* to be resolved. The engineering and construction disciplines now are often separated from architectural design, and
project progress is measured in terms of activities, or *milestones*, completed. Progress monitoring is formalized using a *critical-path* project-management algorithm. All the tasks are
identified and linked into a network according to precedence, defining which activity must
be successfully completed before another can be undertaken.

We said at the outset that the architect or designer no longer has total control over the
building process. The inevitable result of specialization and division of labor is that engineering and construction have become separate disciplines. The design professional is now
primarily responsible for the form and planning layout but must rely upon and coordinate
with other professionals to get the project put together. New construction materials, new
construction technology, and the fragmentation of the design and construction process
due to specialization add up to a new triad that has shaped architecture since the end of
the nineteenth century.

When we enumerated the fields of study in which Vitruvius believed the architect
should be educated, we added a twelfth, technology, with which the modern designer must
contend. The impact of that single factor relative to the original eleven cannot be underestimated, however, and a modern architect's major *difficulty* is to maintain a harmonious
balance between the elements of today's triad: *function, aesthetics,* and *technology*.

Because this book is primarily focused on *planning*, we direct most of our attention to
the *functional* thread from here forward. But it is important to keep in mind that the other
two components of the modern triad keep the design process inexorably in tension. Delight, beauty, the *aesthetic,* whichever word we choose, is that ineffable quality that distinguishes architecture from basic construction, and it is not easily subordinated or
compromised. Without it, architecture reverts to mere building.

Since the beginning of the Industrial Revolution, *technology*, the third modern thread,
has been the single factor that has had the greatest effect on the shape of architecture. All
ancient building materials were limited to use in compression by their physical properties.
The development of steel, used alone and as concrete reinforcement, opened up a new
world of possibilities in need of appropriate expression. We still seek *propriety*, the perfection of form that is attained when a building is constructed on a basis of sound principles.
The proper expression of these new materials and structure is the modern equivalent of
Vitruvius' "appropriate combination of the severity of the Doric and the delicacy of the
Corinthian."

The temptation that must constantly be overcome is to derive designs from previous
solutions without regard to the principles upon which those solutions were based.
All too often, answers that may have worked in the past are adopted or modified without regard to their origins, chosen according to current fashion or the designer's personal
style.

Scientific Method

One of the original premises of scientific method was that the order of our world is knowable, which implies that we can gain control over it through the acquisition and application of knowledge. This is all well and good in dealing with problems of structure: *firmitatis*. Even though we now know that physics is ultimately governed by the scientific principle of indeterminacy, we are well within the boundaries of Newtonian mechanics while we engineer structures here on Earth.

The difficulties are multiplied when we deal with the problems of function: *utilitatis*. Depending upon the building type, *function* is often a moving target. The dwelling or pavilion can be programmed relatively easily; and the small building presents opportunities for a simple, clear expression. Synthesis can focus on the structure and the form; but simple, clear expressions do not scale easily. Cathedrals may have been God's skyscrapers; but our skyscrapers, like it or not, are the cathedrals of our time. Facilities measuring in the millions of square feet, even if limited in height, whether residential, business, institutional, assembly, or a mixture of occupancy types, present functional difficulties that need to be *managed* on a continuing basis. Moreover, the option taken remains in the landscape for a long time.

Function plus technology does not equal beauty. During the twentieth century, a large amount of formalistic and brutally inhumane building was produced, in some cases at a monumental social cost. Le Corbusier's plan for a city of three million was one of many prototypes for the city of the future consisting of large apartment structures isolated on superblocks. Many of the same theories may be seen in the work of Ludwig Hilberseimer: greater density in the urban centers, separation of mechanized from pedestrian transport, and provision for common open spaces.

These theories were tested by many of the developers of public housing in the United States at midcentury. The top portion of Figure 3-1 shows an imagined view of Le Corbusier's 1922 plan. Below it is the actualization in the Robert Taylor Homes project in Chicago, which was recently demolished. Regimented apartment blocks of reinforced concrete cannot be synthesized into gold. Or, as T. S. Eliot put it in his 1925 poem, "The Hollow Men," "Between the idea and the reality . . . Falls the Shadow."°

The problem of delight, *venustatis*, is the most subtle and difficult of all. Here the scientific method fails us because it demands that we strive for knowledge with certainty, which requires complete objectivity. The Cartesian method allows for no preconceptions—and no personality—and requires the total elimination of the subject. But man is the builder and man's sense of place is at the very center of his identity. Thus we return to stand face-to-face with the center, the *omphalos*.

Figure 3-1   The Idea and the Reality.

The relationship of man to the golden section expressed in many ways throughout history is more than just a convenient determinant of scale. On the objective level, it does that; but on a symbolic level, it also expresses the qualitative meaning of space. Establishing a sense of place requires that space be defined in a way that makes the necessary connections: physical, psychological, and spiritual. Le Corbusier said in 1937 that ugliness "comes from incongruity, from incoherence, from the separation that occurs between the idea and its realization."° He blamed the distressing state of architecture on the schools of the time, but thoughtlessly inflating essentially sound ideas led to much the same result.

Temperament and Objectivity

If we apply Johannes Itten's classifications of the creative spirit to the three twentieth-century masters, they would be as follows; Wright was innately the *spiritual-expressive*—guided by intuitive feelings; Mies van der Rohe and Le Corbusier were predisposed to the *intellectual-constructive* approach—attempting to grasp and geometrize the construction of the object. We might even go so far as to equate Itten's classifications with the symbolic meanings given by Vitruvius to the three orders: Corinthian, Doric, and Ionic. The work of each clearly uses his individual expression to address the problem of propriety, tempered with the objective application of the impersonal principles.

Itten's method of teaching the artist how to reach beyond her natural preconceptions implies that the problem of delight must be approached on its own terms. Attempting to look at one's own expression from other points of view is an effective means of overcoming one's limitations and shortcomings. The individual expression is neither eliminated nor objectified. The process is an application of the principle of tolerance. It demands the willingness to admit there may be a better answer and to search it out or, in the stronger words of a contemporary of Descartes, Oliver Cromwell, "I beseech you, in the bowels of Christ, think it possible you may be mistaken."°

Premodernism

Part of the design methodology of the École des Beaux-Arts was the *concours,* typically a twelve-hour surprise design problem. At the end of a limited time period, the student submitted a conceptual sketch incorporating a floor plan and principal elevation, which was called the *esquisse.* The student then had two months to develop the esquisse into detailed plans and elevations. However, if the final scheme differed too greatly from the initial idea, the student was disqualified. This approach fixed the functional layout *and the image* at the beginning of the design process.

This approach has a certain validity as a teaching method. Once you are locked in to your initial mistakes, you must live with them, which almost guarantees something for the professor to critique at the end of the term. On the other hand, it only allows for refinement of the first concept, however misconceived that may be.

The image of the brainstorm quickly sketched on the napkin is a popular one, but often it is necessary to throw the napkin away and start over. As valid as the original concept may be, most often it is necessary to rationally evaluate numerous possibilities until a workable solution is settled upon.

During the Renaissance, there was a great reawakening of interest in the Greek and Roman classical forms, which were often used in contexts radically different from those in which they evolved. By the nineteenth century, the so-called Beaux-Arts method was often manifested by a flowery façade hung on a steel skeleton, completely dissociating the form from the decoration.

This was what concerned Owen Jones when he made a dictionary of ornamental forms widely available, and this was what motivated late nineteenth-century architects such as Louis Sullivan and Wright to reject the method in its entirety. Jones expressed his concern in terms of an "unfortunate tendency" to copy the current *fashion.* Alfred Caldwell, lecturing on art history at IIT in the 1960s, managed to vocalize the word *style* with the expression of an obscenity.°

The Beaux-Arts movement had its final apotheosis in Chicago's Columbian Exposition in 1893. In his *Testament,* Wright quotes Daniel H. Burnham as saying, "The Chicago

World's Fair became the occasion of modern architecture's grand relapse." This was not an unprecedented judgment, however. Victor Hugo had prophesied, some sixty years earlier, "Our fathers had a Paris of stone; our sons will have one of plaster." He went on to describe one modern monument as "certainly the finest Savoy cake that has ever been made in stone."

Modernism

Modernism is a blanket term used to describe the initial attempts to deal with the new building materials and to seek their honest expressions. The imitative aspects of the Beaux-Arts method were rejected by the best of the design thinkers at the opening of the last century, but the historic principles of order and arrangement were not ignored. Thoughtful designers still look to classical forms and the experience of the Renaissance for inspiration and guidance, recognizing that modern building is shaped by current technology in the context of today's culture. It took the entire century to make a reasonable start toward learning the appropriate uses and expression of the technology.

In the final two decades of the twentieth century, names taken from literary movements were appropriated to describe architects' approach to the problem of delight. Some are classified as *postmodernist* or *contextualist*, which has more recently given way to the term *deconstructionist*. Postmodernists tend to look upon what they view as the bland forms and spaces of the modernists with some disdain, responding with bold—in some cases, discordant—forms and elements borrowed from classical styles.

Robert Venturi's inversion of Mies van der Rohe's famous dictum "Less is a bore!" is often quoted by the postmodernist school. If we remember William of Ockham's version, however, how much is appropriate depends on the complexity of the problem. Oversimplification of intrinsically complex problems with contradictory requirements produces artificially simple solutions. Living in a pavilion requires certain compromises in one's lifestyle; living in a Robert Taylor Homes housing unit virtually destroys it.

Venturi argues that two critical factors, complexity and contradiction, have been recognized everywhere except in architecture.° Following on the heels of Werner Heisenberg and his uncertainty principle, the mathematician Kurt Gödel formulated his *incompleteness theorem* in 1931. In practical terms, Gödel's theorem states that however solid or flexible a system may be, it is impossible to prove absolutely *from within* that there are no contradictions in the system. Venturi argues that these principles apply to architecture and that it is necessary to exploit the uncertainties and embrace the contradictions in order to deal effectively with the complexities.

Borrowing terminology from noted literary critic Cleanth Brooks,° Venturi said that he preferred the idea of *both-and* to *either-or*. Either-or does not admit much tolerance: things are black *or* white. On the other hand, it is possible to have "black *and* white and sometimes gray," and maybe even a bit of color besides.

Alongside the aesthetic considerations of order, arrangement, eurythmy, and symmetry, though, remain the functional considerations of propriety and economy. The functional requirements of many of today's buildings are uniquely complex, if only on a technical level and as a consequence of scale. Even the dwelling, as Venturi puts it, may be simple in scope but "is complex in purpose if the ambiguities of contemporary experience are expressed." The underlying meaning of "less is a bore" is found in an earlier statement:

> An architecture of complexity and contradiction has a special obligation toward the whole: its truth must be in its totality or its implications of totality. It must embody the difficult unity of inclusion rather than the easy unity of exclusion. More is not less.°

Postmodernism, though in some quarters viewed as reactionary, is not an entirely irrational extension of modernism. *Deconstruction* stems from an approach to philosophical and literary criticism whereby words and ideas are broken down and analyzed in terms of polarities. This process, the word being a compound of *destruct* and *construct*, is not inconsistent with the idea of analysis/synthesis. Architecturally, the effects have so far been manifested in terms of form, with some truly stunning results.

BARCELONA TO BILBAO

One of the last new monuments of the twentieth century is also one of the most exuberant: Frank Gehry's Guggenheim Museum in Bilbao, Spain. Arguably a masterpiece of deconstructionism, it specifically subverts the principles of linear rationality. A mathematical parallel might be the leap from Euclidean to non-Euclidean geometry. Gehry has stepped outside of the system, introducing elements of unpredictability and even chaos into this building, and the result is spectacular.

The obvious comparison to make is with Frank Lloyd Wright's original Guggenheim Museum in New York City. Its simple, circular form in concrete made a strong ideogram in the landscape. It satisfies the exhibition function by providing the visitor with a single, continuous, spiral ramp, down which she may perambulate from top to bottom. This was the midcentury expression and evoked no small amount of critical comment when it was built.

The first Guggenheim's end-of-century counterpart is immensely more complex, reflecting a fundamental change in the very concept of *museum*. Museums today, partly in order to survive, must appeal to a mass audience in the same way the Columbian Exposition did a hundred years ago. The positive side is that the mass audience is there to visit and appreciate the traveling loan exhibitions that move from one to the next. The Bilbao project, completed in 1997, is one of several in the expanding scale of operations of the Guggenheim in New York. We can only hope that we will not see it replicated like so many franchise restaurants.

The Bilbao Guggenheim goes beyond sculpture. It is theater. Its flowing interior and lighting effects appear choreographed. In fact, a theater consultant was an integral part of the project team. The building's ground area is equivalent to three football fields, and its interior provides a uniquely complex spatial environment.

Upon close inspection, though, the underlying plan is entirely rational. Figure 3-2 shows an ideographic floor plan at the third level and a section through the museum looking south. Beneath the fluid contours of the Bilbao museum's exterior are several regular structural grids with columns typically spaced approximately ten feet (three meters) apart, which define the rectilinear shapes of two main galleries and most of the building's support functions.

Figure 3-2   The Guggenheim Museum in Bilbao, Spain.

Two-column grids intersect at a seventy-degree angle on the third level and are integrally woven into the curvilinear gallery and circulation spaces. A central atrium connects all the galleries horizontally and vertically with the main entrance and theater on the lowest of the floors.

The building's sweeping curves, which are clad in titanium panels applied like shingles, were engineered using advanced computer technology originally developed for aircraft design. The application, known as *CATIA*, belongs to a software class referred to as *PLM*: Product Lifecycle Management. These applications are employed in product synthesis, shape design, and machining to manage the tremendous amount of detail involved in responding to change throughout the life span of large development programs. Only in this way could an architectural project of such scope have been brought to completion at a manageable cost.

Relationships in Play

In the next several pages, we attempt to encapsulate some of the main guiding principles of the three masters, Wright, Mies van der Rohe, and Le Corbusier, any one of which could be called a fountainhead of architecture in the twentieth century. They were all products of a common era, although their works exhibit very different sensibilities and character. Viewing them from the historical perspective of *this* century, we can see that all three strove to reach honest and thoughtful solutions to the universal principles and that they have common denominators.

Le Corbusier—Modulor as Game

Perhaps Le Corbusier's most significant contribution to architectural theory is the use of the Modulor as a eurythmic regulating grid. The mean and extreme ratios of Fibonacci numbers enable the golden section to provide a vocabulary from which elements can be used selectively. Even in the 1920s, Le Corbusier had written of the capacity of regulating lines to ensure against "capriciousness."

One of several games that are suggested in *Modulor I* is the "panel exercise," in which a square is divided up in various ways using Modulor elements.° Corbusier emphasized in his writing the "play" aspect of examining the infinite number of harmonious combinations possible within the bounds of choice, requirements, and the means of execution. In the book, he describes an associate playing the game at a figurative kindergarten table in Manhattan: "I opened your file at nine o'clock in the morning. I started to calculate, to make drawings. . . . It was six in the evening before I realized how time had flown. . . ."°

The example in Figure 3-3 uses elements from the *red* series to create several different square panels. The panels are then subdivided using elements of the *blue* series. The effect is not dissimilar from Owen Jones's ornament modules in Figure 2-12, except that the patterns are based on the golden ratio rather than on the square.

The proportion of the golden section rectangle derives from the square root of 5. Likewise, the square is intrinsically bound to the square root of 2. Either can be the basis of

Figure 3-3   Panels from the Red and Blue Series.

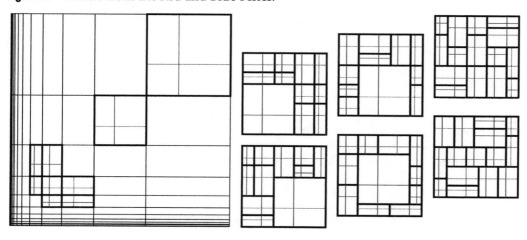

order and *arrangement,* one based on a progression of mathematically related, proportionate forms and the other deriving from the expansion and elaboration of a regular grid or module. The module is a regulator of *symmetry,* the Modulor of *eurythmy.*

Le Corbusier considered the plan as fundamental to architecture. Secondarily, the façade should have a precise organization based on regulating lines. Finally, the site should be regarded as an integral part of the overall plan.

Relating specifically to the nature of a house, Le Corbusier said that a house must serve two functions. It should first be a "machine for living," meaning that it should be efficiently planned and equipped to make necessary daily and weekly activities as simple and effortless as possible. Second, but of equal importance, the home must provide "surroundings where meditation can take place, and a place in which beauty brings the repose of spirit that is so indispensable."°

Almost from the beginning of his career, Le Corbusier promoted the idea of mass-produced houses. His house (*domus*), designed on the Dom-ino principle (*domus* + *innovation*), expressed his idea that simplified, rationalized designs could be mass produced, whether individually or in large-scale apartment blocks. The basic Dom-ino unit consisted of three horizontal floor/roof slabs, six columns, and a stair on the periphery. The column supports were to join the horizontal surfaces smoothly and be set back from the edge of the slab.

Le Corbusier systematized the principles of the Dom-ino house into his *five points of the new architecture,* published in 1927:°

1. *Pilotis* (columns)—provide for an independent structural system.
2. Free plan—Bearing walls are omitted, allowing space to flow freely. Different levels are organized independently of one another.
3. Free façade—Freestanding curtain walls are attached to the structure, incorporating glass.
4. Horizontal windows should be used in the façade.
5. Roof terrace—provides access to sunlight and air.

These principles are all incorporated in the *Villa Savoye,* considered to be Le Corbusier's architectural manifesto, built in 1931. Not a mass-produced dwelling by any stretch of the imagination, the Villa Savoye is very much the *machine for living.* The open ground level is virtually given over to the entrance and parking spaces for three automobiles. Raised on narrow supporting columns, the living space on the second floor is reached by a ramp as well as a stairs. Horizontal windows ring the façade at the second level. The roof doubles as a terrace and solarium, in keeping with what Le Corbusier considered the "essential joys": sun, space, and trees.

Ludwig Mies van der Rohe

The architecture program at IIT espoused Mies van der Rohe's vision, which of course included many of the design principles of the school from which he had come. The emphasis was on learning by doing, which, on several levels, was not dissimilar to Wright's kindergarten exercises with the Froebel blocks. To gain a *feel* for the tools, for example, the student spent his first full year executing rather highly structured drafting exercises with pencil and a ruling pen. Only when these basic tools had been mastered was it time to move on to materials and construction methods.

In the second year, there were courses in wood and brick construction. Steel and glass came later. The brick exercises at the *kindergarten* level actually involved the use of wooden blocks to teach, using Wright's descriptive "the essence of 'brickness.'" Mies, who was characteristically given to making sentences using as few words as possible, reportedly had his own way of contemplating *brickness,* observing, "A brick . . . now that's something!"

Mies van der Rohe's process did not allow for jumping to conclusions or solutions quickly. His responses to an idea were often, "Ja, ve kann do dat," and that all the ideas should be put down on paper. All the ideas would at some point be set alongside and rationally sorted out, narrowing them down to one or two promising ideas to be taken to the next step.

Figure 3-4   Wall Intersection Studies in Brick.

The basic brick exercise was presented to the student in much the same manner as Professor Hilberseimer's sun-penetration problem—as a simple rule. The objective of the "game" was to construct masonry walls and corners of varying thicknesses, the only constraint being that there must be no vertical mortar joints extending through any two adjacent courses of the wall. Three-quarter bricks were permitted in addition to full bricks. These exercises are sometimes pictured with the project for a brick villa that Mies van der Rohe designed in 1923.

Several variations, pictured in Figure 3-4, which are only found through much trial-and-error experimentation with the $1\frac{1}{2}$ inch scale bricks, is the masonry configuration known as *English bond*. This is among the most stable of masonry bonds because it does not rely on ties or other reinforcing to hold the bricks together; nor are there any cut or broken bricks other than the three-quarter unit necessary to finish the ends of the wall. Though demanding of the mason's expertise, the English bond is highly resistant to water damage because the staggered joints do not permit water to run down through the wall and freeze. An elegant variation can be found in the English cross bond, in which alternate stretcher courses are shifted one-half brick so that the vertical mortar joints march in orderly diagonals across the wall's surface.

Learning to construct a wall this way leaves a basic understanding of brick construction *in the fingers*, second only to laying full-size brick with mortar and a trowel. (Wright's method put his students even closer to the ground, engaging them in construction projects at Taliesin.) The essential principle here is that the tranquil, *playful* approach to problem solving allows the designer, by looking at multiple options, to objectively choose a generally valid solution. The functional program, together with the materials, largely determines the rational aspects. The spiritual features are incorporated through an intuitive grasp *in the fingers* of the principles of form.

The design exercises at IIT were based on simple criteria but were highly disciplined. At one point, as it is in the nature of students to do, several asked when there was going to be a "free problem." This came, no doubt, from their burning desire to design a building. The answer given (not by Mies) was, "Free from what?"

The brick exercise encapsulates Mies van der Rohe's view of the importance of understanding function, the character and use of building materials, and their effects. In his inaugural address at IIT in 1938, Mies put this succinctly:

Each material has its specific characteristics which we must understand . . . We must remember that everything depends on how we use a material, not on the

Figure 3-5   Mies van der Rohe's Study for a Brick Villa.

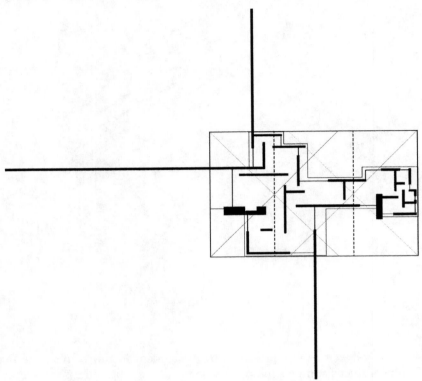

material itself . . . New materials are not necessarily superior. Each material is only what we make of it . . . We must be as familiar with the functions of our buildings as with our materials. We must learn what a building can be, what it should be, and also what it must not be . . . Just as we acquaint ourselves with materials, just as we must understand functions, so we must become familiar with the psychological and spiritual factors of our day. No cultural activity is possible otherwise, for we are dependent on the spirit of our time.

A schematic plan of Mies van der Rohe's brick villa of 1924 is pictured in Figure 3-5. The roof line can be seen following the outline of the carefully composed, intersecting brick walls that define the areas both in the landscape and within the building. The individual rooms are not the compositional elements of the plan. Rather, space flows continuously among the masonry elements. Take away the roof and glass separating interior from exterior, and the masonry alone becomes a cromlech, as strong a figure in the landscape as Stonehenge.

In Figure 3-5, a golden section rectangle has been extrapolated from the extreme edges of the roof. Although the masonry elements of the plan are completely asymmetrical, there is a formality to the central longitudinal gallery connecting the living spaces at either end. A second such rectangle in the 1.618 ratio (dashed line) has been placed in the center of the plan at right angles to the first. It is symmetrical with respect to the gallery space and is bisected by one of three long walls extending outward into the landscape. The two remaining rectangles at either end are each composed of two exact squares, as indicated by the diagonal gridlines within the large rectangle.

Mies van der Rohe's method was the antithesis of the Beaux Arts. Many of his associates have observed that one of the big problems in working with Mies was getting him to say what he wanted. He would defer making an important design decision until the last possible moment. In an interview, Alfred Caldwell quoted Mies as saying:

"I hate to make the decision because I think that maybe there is another wonderful possibility that has not yet occurred to me. When I make the decision

I have killed every possibility save the one I have chosen." Mies explained himself very, very well. Wright always said his buildings were thought-built, which was true. But, maybe to an even greater degree Mies' buildings were thought-built. He really thought it through. . . .°

Consistent with his taciturnity, the principles of Mies van der Rohe's architecture can be seen in terms of four major insights that he spent the better part of a lifetime articulating.°

1. The interplay of reflection and refraction of light is of primary importance in glass-enclosed buildings.

Traditional considerations of mass, light, and shadow no longer apply because a building is no longer viewed as a solid mass and the landscape no longer vanishes once inside. This principle was discovered in 1921–22 while Mies van der Rohe was working on early glass skyscraper projects. Outside the high-rise, one sees the reflection of the sky and surroundings during the day and the transparency of the envelope at night. Experiencing this transparency from inside, one cannot help but be aware of the constantly changing landscape both day and night.

2. The structure and its enclosure can be completely independent of one another.

Both of these principles reached a high degree of refinement in the Tugendhat residence of 1928 and pavilions such as Barcelona (1929). A column grid, once established, defines the roof and podium as the horizontal enclosing planes. The vertical elements, enclosing as well as dividing, can be freely arranged so that the interaction of opaque, translucent, transparent, and reflective elements (water, columns) constantly changes with the occupant's point of view and conditions of light.

The second insight embodies the same concept of separation that is stated in Le Corbusier's first two principles: *pilotis* and the free plan. It is specifically enabled by a structural framework capable of having elements in tension.

3. Space should be defined, not confined.

There is an integral relationship between a building and its site and, in the case of a pavilion, its platform. Often the platform is used to make a transition between the building and the indeterminate landscape beyond and to establish relationships among multiple buildings. This conviction becomes a common thread beginning with the brick villa and is particularly evident in the small city museum and concert hall studies of 1942. Wright articulates the principle, insofar as it relates to building and site, as the "kinship of building to ground."

Mies van der Rohe and Wright both applied this approach to interior spaces as well, once they recognized the extent to which the free plan allowed it. Spatial interpenetration can be seen in Wright's early work such as Unity Temple (1905) and the Robie House of 1908. The de-emphasis of the corner in his constant striving to "break out of the box" contributed to the dynamic character of Wright's work. The confinement of rooms was relieved by overlapping spaces and dividing elements that established less formal spatial zones.

In Mies van der Rohe's pavilions, most notably in the one in Barcelona, he emphasized the layering of both interior and exterior elements. Crown Hall, his architecture building at IIT, is a black steel and glass hall measuring 120 by 220 feet. The columns are on the exterior, and the roof is suspended from four exposed trusses, resulting in a completely unobstructed interior. The main entrance on the south is approached via a monumental travertine terrace. The dark-gray terrazzo floor is six feet above the ground, and a white acoustic-tile ceiling is eighteen feet above the floor.

Crown Hall is a pavilion set in a cityscape. Its outer skin is glass, transparent above and at the entrances and sandblasted below. The only division of the interior is by means of freestanding oak partitions that separate the learning areas around the perimeter from a central exhibition space. By tradition, the incoming freshman architecture class occupied the southwest corner (the hottest), eventually migrating to the northeast corner by the time they reached graduate school.

4. Colors should be neutral because the environment constantly changes.

An aphorism often heard around Crown Hall in the 1960s was, "Follow the buff brick road to the flat black city!" It has been speculated that for most of his career Mies van der Rohe saw the landscape as a monochrome photograph, but one can certainly see the subtlety of his use of color in materials as early as the pavilion in Barcelona. His minimal preference, at least so far as painted steel was concerned, was black in the city and white in the country.

Mies van der Rohe's last realized dwelling, the Farnsworth House, stands on a seven-acre site on the west bank of the Fox River near Plano, Illinois. Completed in 1951, it is a 1,600-square-foot, single-story, glass-enclosed, white pavilion. The second owner writes:

> With its glass walls suspended on steel *piloti* almost two meters above the flood plain of its meadow, life inside the house is very much a balance with . . . and an extension of nature. . . . With an electric storm of Wagnerian proportions illuminating the night sky and shaking the foundations . . . to their very core, it is possible to remain quite dry! The overriding quality . . . is one of serenity.°

Mies van der Rohe is quoted as having remarked on a number of occasions, regarding the effect of color and light in the surfaces of both Crown Hall and the Farnsworth House, "There is your frieze."

Frank Lloyd Wright

Wright also shared with Mies van der Rohe a propensity for seeming to procrastinate. Where Mies advocated putting all the ideas down on paper, then putting drawings up on a wall and sorting out the most workable schemes, Wright did much of his conceptualizing in his head. This was perhaps an instinctive reactionary behavior against the Beaux-Arts method described earlier. Wright said on a number of occasions that he had to have a clear mental picture of every detail of the inside of a structure before picking up a pencil, and he cautioned his disciples against committing to paper prematurely. On occasion, he was also given to changing details when a building was already under construction.

For Wright, the grid, or "unit system" as he preferred to call it, was most often a four-foot planning module. The forty-eight-inch unit is evenly divided by all but one of the English architectural scale factors less than itself; and that one, thirty-two inches, goes evenly into two such modules (see Table 2-2). The alternate scale factors above forty-eight are divisible by that value with integer results. It is not coincidental that the nominal sizes of American wood and masonry construction units are multiples of eight, sixteen, and forty-eight inches, making four feet a convenient increment for residential construction. But a larger pattern of order is apparent in Wright's buildings; nowhere is it more apparent than in his Kaufman House at Bear Run, Pennsylvania, otherwise known as Fallingwater.

There is an archaic Beaux-Arts term, *parti*, that is used to describe the essential visual essence of a structure. The word was derived from the French phrase *parti pris*, which translates as "option taken." It refers to the immediate, abstract visual image projected by the building, which, in conjunction with the functional layout, constitutes its *ideogram*. The strength of the ideogram, the unspoken expression of the idea of a thing, is a measure of the structure's visual clarity and impact. A pure example of abstract form as ideogram is Eero Saarinen's Gateway Arch in St. Louis.

The gestation of Wright's Fallingwater took him the better part of nine months, but the ideogram is reported to have come to him rather quickly. At the entrance level, which has a staircase descending to the stream below, the main portion of the structure defines two intersecting golden section rectangles that overlap at the south corner (Figure 3-6*a* and Figure 3-7*a*). The rectangle, with its long dimension oriented in the east-west direction, follows the perimeter of the lowest balcony.

On the level above (Figure 3-6*b* and Figure 3-7*b*), two golden section rectangles are again evident, this time with the long axis of one centered upon the cross-axis of the other. The north-south-oriented rectangle outlines the cantilevered balcony crisscrossing the one below. The two rectangles at the first level, together with the northernmost one at the second, form a nearly perfect square when viewed in plan. The fourth rectangle (all are of

Figure 3-6   Ideographic Floor Plans of Fallingwater.

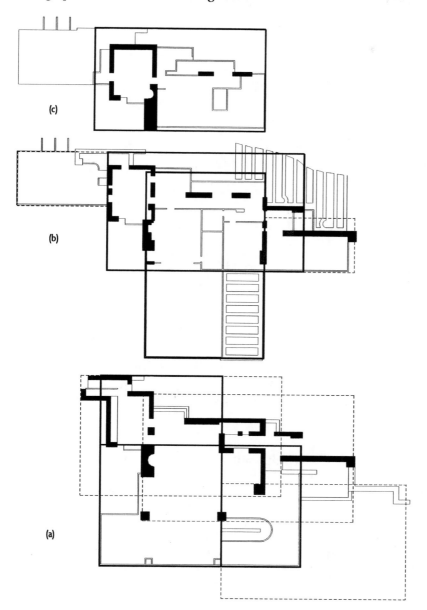

(c)

(b)

(a)

nearly equal size: thirty-seven by sixty feet) is centered in the square in one direction but offset toward the south, overhanging the stream (Figure 3-7*d*).

Two additional rectangles of similar size are defined in the ground floor plan, maintaining close alignments with the principal structural and visual elements of the house. A fifth rectangle extends to the southeast, from the staircase down to the stream at one end of its long axis and to the entry bridge at the other. Its short axis straddles the banks of the stream encompassing the retaining wall intersecting the north end of the bridge. These three rectangles are outlined using dashed lines in Figure 3-6*a*.

These seven, equal-size rectangles, all in the golden ratio, form a linkage of what we may call *regulating planes* that define the precise organization of Fallingwater's spatial weave. In addition, there are two small rectangles balancing each other at the east and west extremes of the second level and a third encircling the third floor.

One begins to experience this complex interweaving of spaces as soon as one approaches the structure, which is effectively united with its site. After crossing the entrance bridge, the drive circles around behind the house, under a horizontal trellis, ultimately

Figure 3-7 Principal Golden Sections and Composite Ideogram.

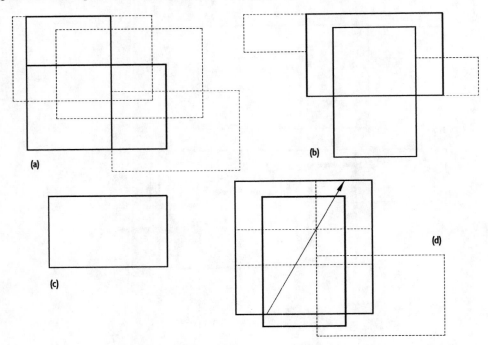

reaching the upper-level garage and guesthouse beyond. The entrance to Fallingwater is via a *porte-cochere* just before the trellis.

Frank Lloyd Wright summed up nine principles of his architecture in his *A Testament*, published in 1957.°

1. Appropriate Character

Materials should be used rationally, particularly newly developed technology (in Wright's case, steel and glass), to create new forms and methods of putting together the modern structure.

2. Materials

Materials are each capable of characteristic effects. They provide the grammar through which new concepts and methods of construction make the building an organism in accord with its nature and purpose, inside and out. Wood, plaster, and concrete should not aspire to resemble cut stone or marble.

3. Use of Tension and the Cantilever

Modern materials allow us to develop space from within, not being limited to rigid box shapes. Both push and pull are now possible.

4. Space

Space must be developed with a sense of serenity and exuberance. Wright speaks of the house being a work of art rather than simply a dwelling.

5. The Third Dimension

The third dimension is seen as *depth* rather than simply thickness or weight. When space is woven, walls define and differentiate space, becoming "humanized screens," but the exterior and interior spaces are allowed to interpenetrate.

6. Form

Form should have a sense of spiritual integrity. Modern materials allow a sense of lightness to be achieved and geometrical forms employed in ways that were not possible throughout history. Walls require a minimal amount of floor area, and the flexible use of interior space presents infinite possibilities.

7. The Inherent Human Factor

The building is more than just shelter. Architecture must be a function of the human scale. Cover and shade and changing light are all a part of a gracious, expanding environment.

8. Kinship of Building to Ground

The term *organic* refers to the use of modern machine methods and structural materials to serve the needs of man. The unit system is used in three dimensions. Wright likens the vertical module of the elevation to "laying warp on the loom." The substance (fabric) of the design is thus woven into and becomes the visual expression of the space.

9. Decentralization

Concentration in urban centers is antithetical to the concept of breaking away from the rigid box shapes that characterized nineteenth-century architecture. Technology should not be used to further increase congestion in cities to the detriment of "human profit and delight in living."

Utility and Delight in Planning

It is a natural tendency, as we have seen, to emphasize one attribute of the Vitruvian triad over another according to one's temperament. Itten's drawing exercises began with naturalistic, constructive, and expressive interpretations, attempting finally to synthesize the essence, incorporating the "impersonal principles" to attain a balance transcending any of the individual aspects. In art, the whole is more than the sum of the parts.

The individual expressions of the architects whose guiding principles we have touched upon are of course very different, but it is not too hard to discern common threads among them. Most of the guidelines of Corbusier, Mies van der Rohe, and Wright either add texture to or make specific one or several of Vitruvius' six original principles. These are the starting points, the qualities that architecture must possess to satisfy the triad. The triad of attributes, then, is a basis for evaluating the results (Figure 3-8).

Strength is usually taken to mean structural stability. This is what Vitruvius actually said: beginning with a solid foundation upon a stable base, using the proper materials without cutting corners. But it is more than that. It is the strength of the image that is projected by the end product—its force or impact.

Utility derives from the program statement, the list of room requirements, the bubble diagram, the adjacency matrix, the facility database, and so forth. It arises from a judicious distribution of all the parts so that their functions are provided for and each is properly situated in relation to the whole.

The property of *delight* is largely determined by the principles of eurythmy, symmetry, and proportion applied to all the parts, including the plan. Delight as a daily experience, however, is a matter of the inhabitants' lifestyle and temperament that can only be enhanced, not created, by the environment.

There are few sites that offer the challenges of a Bear Run. Nothing could be more unique than the thickly wooded cliffs and waterfall with its overhanging balconies. Wright went to some lengths to create the effects of falling water and cantilever at his own home and studio at Taliesin, but he found this environment ready-made in Pennsylvania. It is noteworthy that he chose to make no changes whatever to Fallingwater's physical environment. The building and the nature that surrounds it are delicately and flawlessly integrated.

By contrast, the Farnsworth House sits in a midwestern meadow, less than a hundred feet from a riverbank surrounded by deciduous trees. When the original owner lived there, it was possible to view the house only by renting a canoe and, at some physical risk, paddling upstream. The fifty-five-by-twenty-nine-foot great room, porch, and entrance terrace float above the green meadow in summer, a flawlessly detailed jewel box of industrial technology.

Figure 3-8   Vitruvian Principles and Attributes (see Plate Two).

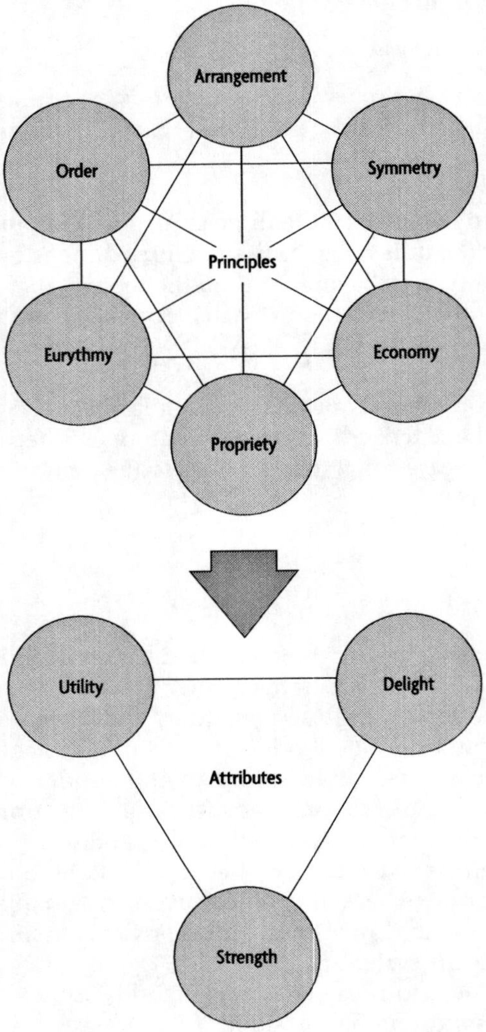

For all the beauty of their surroundings, it is ironic—sad, really—that the occupants of these two homes were not happier in them. For reasons having as much to do with the owner's failed relationship with the architect as with the building itself, Edith Farnsworth lived in her house angrily for twenty years before selling it. Lord Palumbo had a far more salutary experience with the home designed for a single person, writing:

> At Farnsworth, the dawn can be seen or sensed from the only bed in the house, which is placed in the northeast corner. . . . Shortly after sunrise the early morning light, filtering through the branches of the linden tree, first dapples and then etches the silhouette of the leaves in sharp relief upon the curtain. It is a scene no Japanese print could capture to greater effect.°

Another recent author, writing about Fallingwater, observed that it "would be hard to find a house plan that better charted the dynamics of a dysfunctional family."° Keeping in mind the fact that Wright made the hearth the center of most of his residential plans from the very beginning, it is curious that

> [w]hen we put Fallingwater on the [psychiatrist's] couch, we see E. J. and Liliane locked in their solitary bedrooms, Junior isolated in the penthouse, each one of them corralled into a separate terrace. . . . Even while isolating the three Kaufmanns in non-communicating bedrooms, [Wright] married their fireplaces

into a chimney stack. The three-in-one fireplaces . . . seem to have reflected Wright's sad take on a tortured family.

CLOSING WORDS

In the Preface, we briefly discussed the development of information systems from the top down. Facility programming implies a structured approach to developing functional requirements from the inside out, within the constraints of an external envelope and location; in this chapter, we have considered some of the design factors in a residential context.

As we look at the details and the tools for carrying out the programming and space-planning process, we begin in the next chapter with a simple dwelling. This allows us to start with some of the most basic techniques for data gathering, documentation, and visualization. There are no building types more intimately linked to our sense of place than those in which we live, so sitting down with prospective owners to document their program needs to be a highly personalized activity. Planning a small office facility may be an equally personal undertaking on the part of a small business owner or manager, so the activities are similar as a matter of scale.

In subsequent chapters, we build upon this foundation to develop effective approaches for larger-scale projects. There is no question that the tools needed to plan a half-million-square-foot office facility differ from the approach to a residence two orders of magnitude smaller, but the character and constraints of the building and its environment, as well as the personalities and cultures of the people involved, are essential considerations in successfully accomplishing either.

TIME CAPSULE   A Residence at Delos

The *Maison de la Colline* at Delos is by no means typical of the *average* home in ancient Greece, falling rather into the category Vitruvius described as "opulent and luxurious." It has much of the formality of a temple, in plan forming nearly a perfect square, and houses of its type influenced the forms of later Roman houses. Banister Fletcher describes it as follows:

> The court was fully colonnaded, with a water cistern centrally below it, its north side lighting the *pastas* [an interior porch] which extended the full width of the house. From the latter opened a large room, occupying half the available width, and two other principal apartments of differing sizes. The entrance was on the west, with a kitchen adjoining, and in the south-west corner was a wooden staircase to bedrooms opening from a gallery on a second floor.°

The first floor of this carefully planned and detailed structure is shown in Figure TC3-1. The placement of the colonnaded court and cistern is determined by a series of inscribed squares rotated forty-five degrees with respect to the one before. The area of each square is one-half that of the previous one, a method diagrammed in the *Meno* of Plato and later described by Vitruvius, which has come to be known as *Ad Quadratum*, meaning *of the square*.° The subtlety in the plan of Maison de la Colline is that the second inscribed square is offset by the swing of an arc based upon its diagonal. Shifted toward the south, the continuing series of squares determines the placement of the court, colonnade, and cistern as well. Thus, the areas of these features constitute $\frac{1}{4}$, $\frac{1}{16}$, and $\frac{1}{64}$ of that of the house, respectively.

The interior court and cistern, surrounded by eight columns in the Doric style, were open to the sky, and each of the two floors occupied nearly four thousand square feet. This residential structure, dating from the second century BCE, manifests all the geometric refinement of a Villa Savoye or a Mies van der Rohe pavilion.

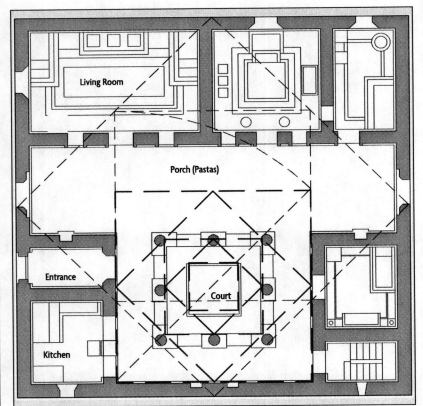

Figure TC3-1   Maison de la Colline at Delos.

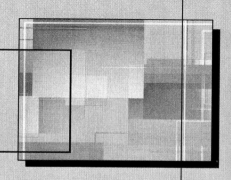

CHAPTER 4

Dwellings

*In preparing for battle, I have always found that plans are
useless, but planning is indispensable.*
—Dwight D. Eisenhower°

Let us now turn our attention to the mechanics of planning a family dwelling. The process can be simple or complex in direct proportion to the scope of work and the personal involvement of all the interested parties. It usually begins with a simultaneous analysis of the functional and physical requirements as embodied in

1. the *site*, and
2. the *program*.

For the purposes of this chapter, we have abstracted the site requirements in order to concentrate on the layout of the interior spaces, their relationship to one another, and the connections or flow between interior and exterior.

SITE CONDITIONS

The ideogram of our site begins with a deck measuring sixty-two by one hundred feet. On it, we plan to develop a one-story residence occupying approximately one-half of the deck area, or three thousand square feet. There are two masonry elements at right angles within the deck area, one of which is suitable for a fireplace. These elements have been used in conjunction with the deck perimeter to define the boundaries of the enclosure. The defining elements are pictured in Figure 4-1.

The terrain immediately surrounding our deck has a moderate slope toward a lake. The northeast corner of the structure is about twenty feet from the water's edge, so the corner of the deck in that direction provides direct access to the water. The rough floor elevation of the entire deck is three feet above the water level so the east end can serve as a boat dock. The exterior deck at the west end must provide for principal and secondary entrances. We will assume that the site can provide for vehicular access and parking in reasonable proximity to the west end of the platform.

PROJECT CONDITIONS

This home is intended to be a relaxing country retreat for a successful professional couple with one child. The existing substructure is sound, allowing some main elements to be retained. The foundation, the two masonry walls constructed of local stone (one of which incorporates a large fireplace), and the deck platform itself are intact. Existing exterior walls and roof may be modified or replaced as required. All interior elements may be laid

Figure 4-1   Ideographic Plan of Deck and Defining Elements.

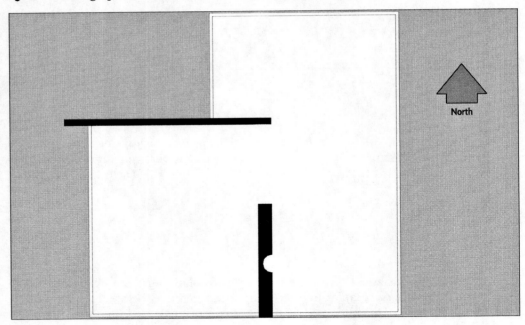

Figure 4-2   Four-Foot Modular Grid.

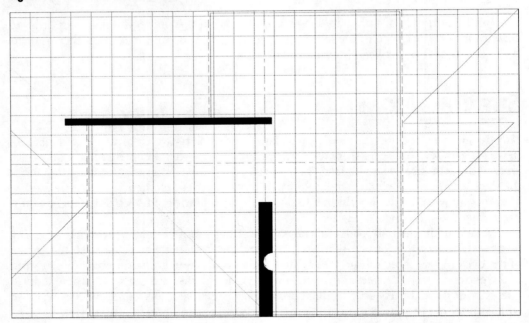

out in accordance with the family's functional requirements, with comfort and privacy being prime considerations.

The owners do a substantial amount of entertaining and often have houseguests, some of whom are disabled, so the main traffic areas of the home and guest rooms need to be accessible to persons with disabilities.

Module and Ideogram

As an aid to orderly planning, a regular four-foot grid has been established across the deck, as shown in Figure 4-2. As the north-south dimension of the platform is not evenly

Figure 4-3   Ideographic Studies of the Building Envelope.

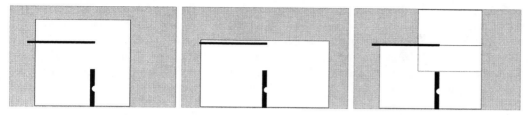

divisible by 4, the grid is centered on the platform's longitudinal axis. Centerlines are added as an additional visual guide to subdividing the platform area, together with several diagonals as potential *regulating lines*. The locations of the diagonals indicate possible alignments of edges and endpoints with other planning elements.

Figure 4-3 illustrates several studies of ways in which the building envelope may be situated upon the deck. They all follow the modular grid.

The necessary enclosed area can be defined by a square perimeter. At left, a square, anchored upon the masonry elements, encloses the desired three thousand square feet, using nearly the full width of the platform. This approach provides exterior deck around three sides of the house with a sizable area adjacent to the lake.

The middle plan shows a rectangular enclosure symmetrically placed with respect to the deck and the stone walls. Most of the open deck area is on the north side, extending from west to east. The stone wall at the northwest corner of the house appears to make placement of the main entrance potentially awkward.

At right in Figure 4-3 is the ideographic plan that was chosen for further development. The enclosed area is formed by an overlapping square and rectangle that have a clear and balanced visual relationship with the stone elements. The setback at the northwest corner suggests a logical location for the front entrance adjacent to the east-west stone wall. Allowing the house to occupy the full width of the deck from north to south ensures maximum privacy of the exterior platform area on the lake side. Finally, the overlap, situated between the two pieces of masonry, may be thought of as the symbolic *omphalos* of the home. This is an optimal location for the kitchen and dining areas, the center where most social interaction may be expected to occur.

PROGRAM REQUIREMENTS

The summary of functional requirements for the project can be divided into three parts:

1. *A narrative description of the owner/occupants' usual routine, lifestyle, and specifics as to the character of the environment they want to achieve.* This should be a written statement based on initial conversations and visits to the site, a comprehensive owner profile, and statement of goals. In addition to addressing the utilitarian issues, the owner profile should capture the temperament and preferences of the future occupants as they relate to aesthetic considerations. Special design objectives and features to be provided should be described here, including technologies for communication and entertainment that may be of particular interest.

2. *Supporting the owner profile, a room-by-room inventory of required individual spaces must support the general planning goals* (Table 4-1). By defining in detail the sizes and functional requirements of every area, including all the furnishings and equipment that need to be accommodated, oversights can be avoided from the outset. Equipment descriptions for every space must be specific enough to allow for the proper planning of electrical and mechanical support systems. Attention to detail here can avoid potentially costly additions later on.

   In preparation for the next stage of plan development, physical-adjacency relationships to be established and maintained among individual spaces must be

OWNER PROFILE

The lady of the house is a Web designer and writer, and her husband is a consulting engineer who travels extensively. Both are avid amateur chefs and boating enthusiasts. They have a child who has just reached driving age and attends a nearby high school. In addition to their boating activities, all the family members enjoy bicycling and hiking on the nearby trail system, so storage facilities for their various outdoor equipment are necessary.

When the husband is at home, the family tries to make a point of having an early breakfast together, usually around 7 a.m. He usually leaves for work at 7:45 and drops their child off at school on the way. When the husband is traveling, his wife usually takes their child to school. Occasionally, since learning to drive, their child uses the second vehicle, a sport utility vehicle (SUV), to get to school.

The wife is self-employed and maintains an office in the home. All of her computer-oriented activities—which include research, writing, Web page development, and keeping in touch with her clients by e-mail and telephone—are carried out in her office. Many of her projects, and her husband's as well, involve digital photography and computer graphics, so all of this equipment is usually stored in her office. There is also a computer workstation in the kitchen that is used for recipe storage and household management.

When the husband works at home (occasionally in the evenings and on weekends), he usually sets up his laptop on a small desk in their bedroom or in the dining area. Network connections are desirable so that the office printers and file server may be accessed from any of these locations.

Entertaining is usually done on weekends between Friday evening and Sunday afternoon. As well as being the physical center of the home, the kitchen and dining areas are often the center of the couple's indoor social activities. Meals are usually informal and usually involve anywhere from two to eight guests. Weather permitting in their midwestern location, meals are often taken outdoors on the lakeside deck.

Because both spouses take their cooking quite seriously, their kitchen must be well equipped. Professional-grade appliances are required, including refrigerator and freezer, range top, and at least two ovens, as well as an effective exhaust hood. Adequate, organized pantry storage is a must.

As some of the owners' friends, relatives, and business associates who are occasional house guests are disabled, they have decided that their home and guest facilities should meet the Americans with Disability Act (ADA) standards for accessibility. This includes access to the guest bedroom and bath as well as to the boat mooring that is a part of the lakeside deck. Given the thirty-six-inch grade differential between the floor and water levels, an accessible ramp is required.

determined and prioritized. Even if some *rooms* are not going to be fully enclosed, the character and function of each space should be defined in detail. In addition to identifying the areas that must be planned in logical proximity to one another, this part of the program statement addresses physical proximity and privacy criteria to be established, while identifying opportunities for overlapping spaces or *interweaving* them for functional or aesthetic effect.

3. *A chart in the form of an adjacency matrix is useful in helping visualize the layout relationships to be maintained among the planning elements.* While it may seem like overkill for a small project, the simple matrix can help to prioritize important functional affinities.

   In Figure 4-4, we have listed all the spaces in the functional-requirements inventory. If desired, links to exterior and other functions might be indicated as well. Each pair of elements in the matrix can show a symbol or number signifying the importance of the relationship to be established.

Table 4-1   Room-by-Room Functional-Requirements Inventory

Room	Area	Functional Requirements	Equipment	Adjacencies
Entrance Hall	100	The entrance hall is the main formal entry to the home. It should serve as a gracious welcoming area for both guests in the home and short-term visitors who will proceed no farther than the office.	Quarry Tile Floor Guest Closet	Office Kitchen Main Living Area
Dining Area	300	Most indoor meals are taken in the dining area, breakfast included.	Seating for 4 to 6, Expandable to 12 Eight Chairs Buffet and China Cabinet (72″ w × 20″ d × 80″ h)	Kitchen Main Living Area
Main Living Area	700	The main living area serves as both formal living room and family entertainment area. It must comfortably accommodate up to a dozen people. This room has a dual focus: the hearth and the home entertainment system.	Comfortable Lounge Seating Occasional Tables Flat-Screen HDTV Unit Audio System with Surround Shelf Storage for Books, CDs, Etc.	Dining Area Kitchen
Kitchen	200	The kitchen is the hub of household operations and the center of most activities. Barstool seating should be provided at center island for conversation and "snacking."	Double Sink Refrigerator (18 cu. ft.) Freezer (18 cu. ft.) Six-Burner Gas Range Top Exhaust Hood Two Full-Size Ovens Warming Oven Center Island Computer Workstation (Stand-Up) Storage	Dining Area Main Living Area Entrance Hall Office
Office	150	The office is used principally by the wife who is self-employed and works primarily from the home.	Work Counter along One Wall (30″ d) Storage Counter with Filing Beneath (24″ d) Shelving for Books (36 lin. ft.) PC and 20″ Flat Screen Monitor Laserjet Printer (Monochrome) Color Printer/Scanner /Fax/Copier Various Peripherals (UPS, etc.)	Entrance Hall
Master Bedroom	300	Acoustical privacy is required.	King-Size Bed Dresser (84″ w × 20″ d × 42″ h) Dressing Table (42″ w × 20″ d × 30″ h) 2 Night Tables (20″ Square)	Master Bathroom Master Closet
Master Closet	75		15 lin. ft. of Hanging Storage	Master Bedroom
Master Bathroom	100		Toilet Vanity/ Sink Bathtub & Separate Shower	Master Bedroom
Bathroom	100	ADA Accessible	Toilet with Grab Bars Accessible Vanity with 2 Sinks Bathtub with Grab Bars Accessible Shower Unit	Bedroom and Guest Bedroom
Bedroom	200	Acoustical Privacy Required	Queen-Size Bed Dresser, Night Stand, Television Desk and Computer Lounge Chair and Ottoman	
Bedroom Closet	50		10 lin. ft. of Hanging Storage	Bedroom
Guest Bedroom	250	ADA Accessible, Acoustical Privacy Required	Queen-Size Bed	

(continued)

Table 4-1   Room-by-Room Functional-Requirements Inventory (*continued*)

Room	Area	Functional Requirements	Equipment	Adjacencies
Guest Closet	50	ADA Accessible	10 lin. or linear ft. of Hanging Storage	Guest Bedroom
Laundry Room	100		Laundry Sink Washing Machine Gas Dryer Ironing Area Storage for Cleaning Supplies and Vacuum Cleaner Linen Storage	Utility/Storage
Utility/Storage	100	Serves as "Mud Room"	Furnace Water Heater Water Softener Incidental Storage (Boots, Etc.) Three Bicycles	Exterior Entrance Laundry Room
Total Net	2575			
Circulation (12 percent)	400			
Total Gross	2975	Utilization = 87 percent		

Figure 4-4   Adjacency-Relationship Matrix.

Space Allocation

Once the program requirements data have been documented, the optimal size of each space agreed upon, and the proximity relationships understood, development of the space layout can begin. The space-allocation diagram shown in Figure 4-5, often referred to as a *bubble diagram*, shows the individual spaces in their desired relationship, roughly to scale. Any number of preliminary studies may be required until a satisfactory planning scheme can be agreed upon. More, of course, will be required if it is necessary to test multiple configurations of the building envelope.

Block Planning

Another tool for visualizing the allocated areas is the block diagram. In Figure 4-6, we have used selected area blocks from a "Modulor" grid to develop a preliminary block layout. Figure 4-6*a* shows a vocabulary of Modulor rectangles selected to match the programmed area requirements. These elements were selected from the blue-series grid and chosen on a basis of how closely they represent the areas listed in the functional-requirements inventory. The areas will seldom match exactly, but the idea is to build a tool kit of blocks that have a proportional relationship.

The inventory of blocks can usually be chosen in such a way that it contains a small, mostly contiguous cluster in one area of the Modulor grid that represents areas of similar

Figure 4-5   Space-Allocation "Bubble" Diagram.

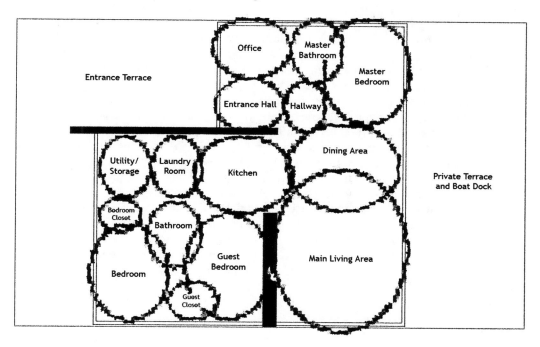

Figure 4-6   Space Allocation Using the "Modulor" Approach.

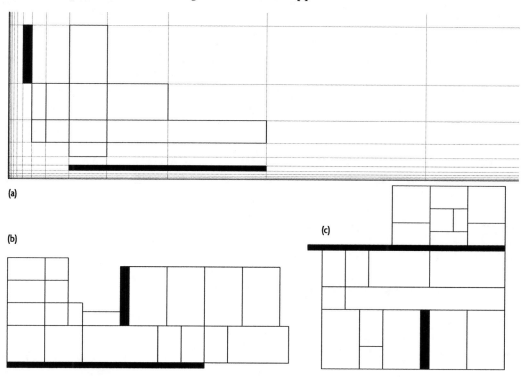

(a)

(b)

(c)

scale. In this case, the two rectangles shaded in black represent the masonry walls that act as *regulating masses*, or anchors, to our plan.

Figure 4-6b shows the selected blocks arranged according to size and orientation. Some blocks of similar size may appear more than once, depending upon the room requirements. The necessary orientation of individual elements may not be immediately apparent but can be refined as the blocks are placed into the actual floor layout. In Figure 4-6c, we see one possible arrangement of the blocks that fits within the envelope of our building while satisfying the identified adjacency relationships.

The main living area is shown in this block diagram as two equal rectangles, reflecting the dual-focus property of this space: the hearth and the home-entertainment area. The long, horizontal block balanced upon the end of the hearth structure in Figure 4-6c represents a circulation axis that unites all the central gathering areas in the home. This establishes a basic flow of space that unites the living, dining, and kitchen facilities, which can be further strengthened in the detailing of interior elements in all dimensions.

Developing in this way the special relationships of the functional elements in our plan gives us several ways in which to look at the arrangement. The bubble diagram allows us to play with the *relationships* rather precisely without being particularly concerned with physical constraints (other than area) in order to visualize the flow of space.

Using a deliberately limited vocabulary of square and rectangular elements creates a degree of repetition that helps establish a eurythmic, *Modulor* rhythm. Looking at the plan in relation to the *modular* grid (Figure 4-2) gives us another checkpoint in terms of symmetry. It is not necessary to strictly follow either the square (modular) grid or the Modulor, so long as an overall sense of order, proportion, and rhythm is maintained.

SCHEMATIC LAYOUT

A detailed schematic floor plan of our home is illustrated in Figure 4-7. All walls and doors are shown, as well as floor patterns defining the main entrance hall and hearth areas. A great room measuring twenty-five by thirty-eight feet combines the functions of hearth, entertainment, and dining areas. The visual extension of the corridor leading to the bedrooms, bathroom, laundry room, and utility room suggests the functional division between living and dining areas.

There is no dividing partition between the kitchen—the *operations center* of the house—and the great room. Whoever is working there is not isolated from the living space; it is the social center [dare we use the term *omphalos* one more time?] of the home.

The master bedroom is a self-contained suite of rooms on the opposite side of the entry hall from the common spaces. It is easily accessible to the rest of the house, but private,

Figure 4-7   Schematic Floor Plan.

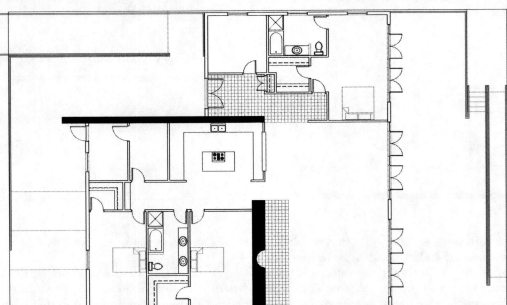

and at the diagonally opposite corner of the structure from the other two bedrooms. The master bedroom, dining, and main living areas have direct access to the private terrace and boat dock.

The home office is immediately adjacent to the main entry so that the occasional business visitor may be taken there without having to pass through any other areas within the building envelope.

The utility/storage room on the opposite side of the stone wall from the main entrance hall has a secondary door to the entrance terrace. This facilitates such use as bicycle storage and provides direct access to the laundry facilities and kitchen without having to pass through the living spaces.

The central living areas and guest rooms are planned to facilitate access by people with disabilities, which was a special requirement stated in the program. Consistent with this requirement, both front and rear terraces are provided with ramps meeting ADA requirements.

Visualization

Figure 4-8 shows a three-dimensional (3-D), wire-frame view of the schematic plan. This kind of graphic is a useful aid to visualizing spatial and volumetric relationships and can be invaluable in presenting space-planning concepts to a client. Such views can be quickly and easily generated using computer-aided-design (CAD) applications such as AutoCAD. With a little foresight in setting up the two-dimensional plan drawing, the conversion to a 3-D plan can be largely automated.

Multiple views of a virtual 3-D model can provide an effective and flexible means of analyzing the plan from several points of view. The details that follow and the Case Study that follows this chapter show how a professional kitchen designer approaches the "operations center" of our small house. This process is compared with the development of the control room for the operations center of a large bank, leading us into the discussion of facility planning at the corporate level. In a later chapter, we discuss several more uses of 3-D models in developing facility plans, along with some tools for creating and working with them.

Figure 4-8 Three-Dimensional, Wire-Frame Layout.

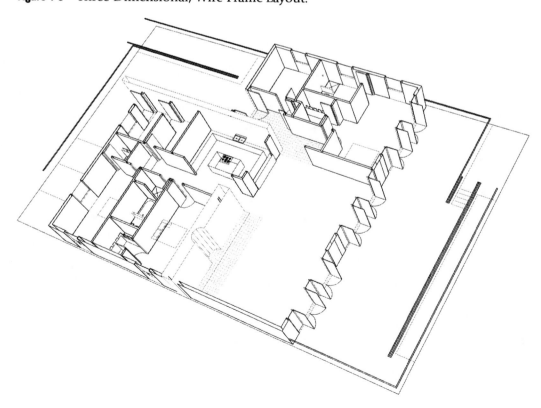

DETAILS: THE *OPERATIONS* CENTER

Once the overall plan configuration has been established, the next step is to develop refinements and details in the individual spaces. To illustrate this process, we focus on the components of the kitchen and the areas closely related to it, probably the most complex in the home.

The role of the kitchen has changed as technology has advanced and lifestyles have accelerated, but the basic functionality of a kitchen remains its most important feature. One of the rules of thumb is the formation of a *work triangle*. The three main elements—sink, refrigerator, and a cooktop or range—should form a triangle defined by a logical work pattern. Beyond the essential functions, aesthetics are the principal consideration in how a kitchen is planned.

As a place for social interaction, the kitchen has largely supplanted the living room and has come to function as the home's operations center. In conjunction with the breakfast nook, the kitchen is often one focus of a central *great room*, the other focus being the family entertainment center. In addition, minikitchens are often found in other areas of the home such as wet bars that are part of a media room or outdoors as permanent adjuncts to a terrace. The work area is a center of family communication while creating a meal in a simple, swift, and flexible manner to accommodate a variety of schedules. Whether the family gathers daily for an evening meal or entertains for special occasions, the kitchen is usually the hub of activity.

Kitchen designs often combine the professional look of a high-end restaurant with the personal style of the home owner. The restaurant atmosphere is achieved by specifying professional-grade appliances. Appliances are the first of seven elements to consider when designing a kitchen:

- Appliances
- Countertops
- Cabinetry
- Flooring
- Walls
- Ceiling
- Lighting

Appliances

Middle- to high-end kitchens display an assortment of appliances such as double ovens, a cooktop, multiple-burner ranges, indoor grills, microwaves, deep fryers, smokers—the options are virtually endless. Stainless-steel finishes have become the new neutral. The finish becomes an extension of colors within the room. Metals with a satin finish are generally easier to clean and give the professional look that is much sought after.

For those who prefer complete coordination of finishes, the fronts of refrigerators, dishwashers, and microwaves are often covered with matching door panels. Everything is neatly tucked away, lending a more traditional, cozy feel.

Refrigerators have evolved from the single icebox to fully automated systems capable of actually ordering product. High-end kitchens usually are equipped with full-size refrigerators and freezers. To accommodate these, the kitchen must be of adequate size. Although standard refrigerator depth is usually close to twenty-four inches to align with the countertop, necessary width clearances may range up to ninety-six inches.

Countertops

Countertops are often a dominant element fabricated from stone or other stonelike material. Even plastic laminate countertops can give a convincing look of textured stone. Style and durability are important factors in the choice of countertop material. Stone, engineered solid surface, and laminate are the three major classes. Granite is an extremely hard natural stone with unique veining and texture. Several varieties of engineered stone have textures and colors resembling the natural beauty of granite yet are bacteria resistant. Engineered surfaces are repairable and provide a wide range of colors, edges, and inlays from which to choose.

Cabinetry

Kitchen cabinetry selections should be consistent with the atmosphere of the home and support the lifestyles of its occupants. Contemporary solutions often utilize glass cabinets, wall niches, or an adjacent matching hutch. Traditional kitchens may feature raised-panel doors with a distressed finish topped off with a glaze, a popular technique that offers a pre-painted or aged appearance. Contemporary kitchens emphasize smoother lines, not only in the cabinet door styles but in the choice of appliances as well. Common wood door materials include maple, cherry, oak, pecan, walnut, pine, ash, birch, red birch, and hickory, with an unlimited variety of colors and stains.

Cabinet units are manufactured using both frameless and framed construction. With framed construction, the strength is in the front; American manufacturers typically use this technique. Hinges are attached to the solid wood frame, which is visible when the door is opened. With frameless construction, strength of the cabinet is in the rear. This is characteristic of European casework; the hinge is attached directly to the interior panel. Both types of cabinets offer many extra features such as roll-out trays, plate organizers, pull-out toe-kick stepladders, a recycling center, appliance garages, and elevated dishwashers. Roll-out trays are the number one accessory choice, consisting of shelves on casters that make it easy to access the entire contents.

Flooring

Flooring is also an important element. Many types of natural wood are specified; there are engineered products, and bamboo is classified as a rapidly renewable material environmentally. Tile having the look of stone and various types of laminate flooring resembling natural materials are economical alternatives.

Walls, Ceiling, and Lighting

Wall coverings should enhance the cabinetry, countertop, and other finishes, which are the most expensive elements of the kitchen. It is important to carry through the design with a choice of complementary color and pattern. Painting the ceiling using a light tone enhances a sense of spaciousness, and light colors appear softer than just white.

The backsplash area provides a good opportunity for the use of accents. Tumbled stone tiles are very functional and can be used on the backsplash and floor. Basic ceramic tiles enhanced with glass or metal accent pieces or a hand-painted mural can be used here to complement the other materials in the space.

Lighting is the final element to be considered; today's kitchen is equipped with more than just a single hanging light over the breakfast table. Pendants may be hung above an island or bar area, creating a warm atmosphere. Lighting units under cabinetry are a must for all working or task areas. Recessed lights above the cabinet units are mood enhancers, augmented by dimmers and colored lamps. Glass-fronted cabinets and open shelving can be used to display collectibles if lit from inside or below.

CLOSING WORDS

We have considered here the *machine for living* as a means of presenting some of the basic approaches and tools for programming and planning. Using an imagined site and a hypothetical owner profile, as developed from preliminary investigations and one or more fact-gathering meetings, we formulated a program statement defining the internal elements to be planned and then charted the relationships among them using a priority matrix.

A preliminary arrangement was achieved using proportional blocks to visualize the functional relationships within the constraints of the site.

Considering the proportionality of the various spaces using a selection of *Modulor* components as a guide, we developed a schematic plan showing the key elements of the home. This more literal, detailed layout provided a basis for evaluating how well the arrangement satisfied the owner profile and functional-requirements inventory. From this point forward, the planning and design process becomes a matter of refining the details, such as the kitchen as the operations center of the home.

Having seen some of the basic techniques of information gathering, documentation, and visualization, we can now examine some of the more complex, multidimensional issues to be found in the development of a planning approach that is appropriate for a large organization with a multitiered hierarchy or matrix structure. An early, essential task involves getting to know the character and culture of the business. Many of the conceptual planning approaches and 3-D visualization techniques are equally applicable to small- and large-scale planning efforts, particularly in operational hubs, as the following Case Study illustrates.

In the next chapter, we look at some past business practices as a reference point with which to compare the way companies operate in today's competitive global environment. Management's plans for growth, changing product focus, and new business strategies, as well as environmental considerations, all influence the planning process. Project management is also a critical component, which goes hand in hand with conscious exercise of control over the quality of the work.

CASE STUDY   Operations Centers

To the extent that the necessary components can be represented in computer-accessible form, complex spaces can be analyzed and the presentation of alternatives in pictorial form largely automated. Numerous office furniture and other product manufacturers publish their entire catalog offerings in the form of 3-D CAD symbols that may be used to enhance a CAD model. Such furniture components are often intelligent symbols containing attributes that can be extracted from the model to produce bills of materials and cost estimates. Architectural software enhancements built around a kernel of a CAD program may partially or wholly automate the building of the model based on parameters specified to the system.

One such product for kitchen design created the illustrations that follow. The process is widely used in retail design settings where the visualization is literally created in front of the client or customer, making the customer an integral part of the development process. The basics—walls, cabinets, and surface textures—are given to the software, and the designer need only press the proverbial button to create instant elevations and perspectives (Figure CS1-1). The visual representation of the kitchen is the direct result, but what is appealing to designer and retailer alike is the fact that a client proposal and an order sheet are produced along with it. Different pricing structures can be evaluated with a minimal time investment and a few keystrokes.

The step-by-step process begins with the use of a kitchen floor plan taken from the overall house plan, as shown in Figure CS1-2. The step-by-process is as follows:

1. Open the drawing and specify drafting preferences (measuring units, component library, etc.).
2. Define walls and openings (doorways, windows, etc.).
3. Add corner cabinets.
4. Add appliances.
5. Add remaining cabinets, beginning with base followed by wall cabinets.
6. Add soffits, top trim, and countertops using the design tools.
7. Choose surfaces and finishes for the walls, floor, and ceiling.

At any point in the process, the designer can override the automated choices and exercise control over countertops, trim selections, toe kicks, soffits, and other details.

Once an acceptable configuration has been reached, the model can be further enhanced with lighting, switches, and outlets, up to and including window treatments and dishes in the sink. With the proper tools, this can be accomplished in a matter of minutes rather than hours.

CONTROL ROOM FOR A LARGE COMPUTER CENTER

In corporate and institutional environments, the operations center is more often a one-of-a-kind affair. Standard modular equipment is available to be sure (Figure CS1-3), but the complex relationships among the various components must be meticulously planned.

In the case of the control room pictured in Figure CS1-4, located in the operations center of one of the largest U.S. banks, the objective was to consolidate all of the control functions. The existing computer room, pictured at left in the illustration, occupied approximately twenty thousand square feet divided into two major spaces. Control functions were scattered throughout the facility, having been created as new systems were added. The consolidation was undertaken due to security considerations and in order to improve the effectiveness of the control and monitoring functions.

Several of nearly two dozen plans that were developed are pictured here, together with computer-generated perspectives of the interior space. A major consideration that required accurate perspective renderings was the establishment of sight lines to each of the fifty-plus monitors that were to occupy the room. Most of these would be arrayed in three-high cabinets around the perimeter of the room, with the primary operating consoles in a central location. With the successful layout, a small contingent of operations personnel could actively monitor the entire system on a 24-7 basis.

The floor plans were developed using 3-D AutoCAD blocks representing the computer consoles, display monitor cabinets, and other items that could be placed within the plan and moved about as required. Then interior views could be established from key locations and at typical viewing heights within the room. A single drawing of the room envelope, showing the corresponding two-foot-six-inch square ceiling grids, was used for all the studies. They were not created in real time in front of the client but were economically prepared in the space of a few days for presentation.

(continued)

(continued)

Figure CS1-1   Interior Elevation and Perspective of Finished Kitchen.

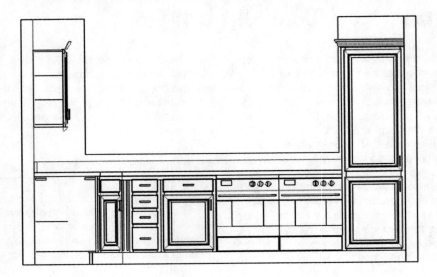

Figure CS1-2   Kitchen Floor Plan Showing Cabinetry and Applicanes (see Plate Three).

Figure CS1-3   Modular Local Area Network Server Racks.

(*continued*)

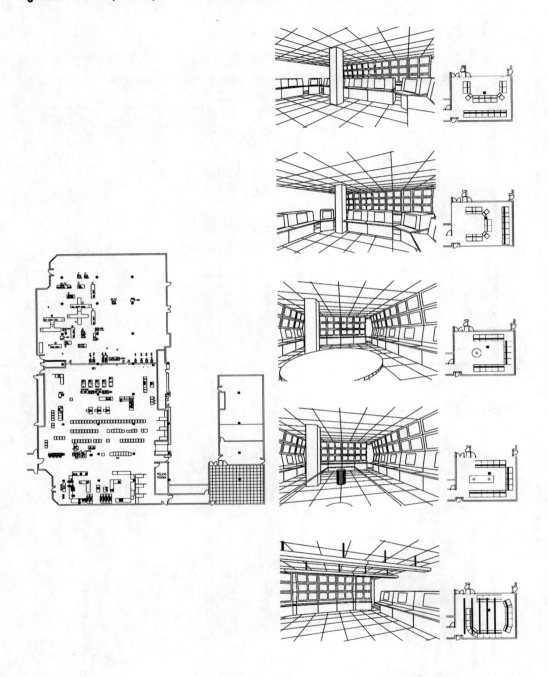

(*continued*)

Figure CS1-4   Computer Operations Center Plan and 3-D Studies.

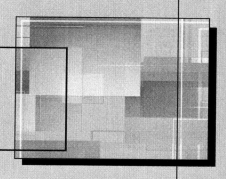

CHAPTER 5

Management and Networks

Change means movement. Movement means friction.
Only in the vacuum of a nonexistent world can movement or change occur
without the abrasive friction of conflict.
—Saul Alinsky

Because large-scale programming and facility planning are intimately bound to the management process, this chapter is devoted to the subject of *management* on several levels.

The program for a family dwelling described in Chapter 4 is a simple statement of project requirements. The requirements are summed up rather succinctly in a general narrative of the owner's needs, a descriptive inventory of the spaces and functions to be provided, and a statement or chart of the desired relationships among the various elements. Detailed specifications may be developed for specialized and potentially high-cost areas such as the kitchen. Similar approaches may lend themselves to the development of specialized technical facilities.

Managing facility projects on a large scale requires a sustained and systematic approach, and the planner must come to understand some of the inner workings of the organization. Formalized project-management procedures help to keep the facility needs analysis on track, and an understanding of modern quality-management processes will contribute to improving the planner's own procedures as well as her understanding of the client's management method and organizational culture.

SCIENTIFIC MANAGEMENT

During the first Industrial Revolution, Charles Babbage, who developed a mechanical prototype of the modern computer, was also one of the first to argue in favor of a scientific approach to management. Babbage originated the idea of mechanical principles governing manufacturing. He observed that the physical layout of the early factory needed to be based on good organizational data and saw the factory as a system that realized great economies of scale through effective management and the division of labor. Babbage foresaw what is now called lifecycle management of capital equipment by observing that machinery used to manufacture products in great demand seldom actually wears out. Technological improvements resulting in increased capacity or improved quality often make replacement desirable from a competitive standpoint long before the equipment actually fails.

Babbage made extensive tours of factories throughout England and Europe while formulating these principles, believing that collecting sound data is essential to good planning and organization. He encouraged managers to follow this example, arguing that the errors arising from the application of theory without facts are far more numerous and long lasting than those resulting from incorrect conclusions drawn from sound data.

At the beginning of the twentieth century, Frederick Winslow Taylor took the idea of scientific management to the next level, which Peter Drucker credits with lifting the lifestyle of the "working masses" well above previous levels, even for those considered wealthy. ° This was the greatest period of industrial expansion in the history of Western culture, although the aspects of inhumanity in Taylor's approach have been noted. Writing in 1933, John Dos Passos portrayed the image of Taylor dying with a watch in his hand. °

The Taylor method called for a clear division of work responsibility between management and labor, which contributed in no small measure to the polarization of those who planned and supervised the work and the laborers who carried it out. In fairness, however, one must consider the era in which Taylor worked and the fact that handling ninety-pound pigs of iron at the Bethlehem Steel Company typified the kind of job he was analyzing. °

At the core of the Taylor method was work measurement: establishing "one best way" of doing a job and establishing performance standards based on the amount of work a "first-class" worker could be expected to perform in a day. In theory, managers were to scientifically choose the person best suited to each job, who was then trained to do the job in the exact manner prescribed until he performed at the so-called first-class level.

One of the dangers of what we earlier called the objective application of impersonal principles is that it can be dehumanizing. From the beginning of the Industrial Revolution until the middle of the last century, it was often assumed that people have an inherent dislike of work and will avoid it if possible. This attitude stemmed from a conviction that the average person prefers to be directed, does not willingly take responsibility, generally has little ambition, and values security above all else. Despite the fact that one of Taylor's stated fundamentals was that "the principal object of management should be to secure the maximum prosperity for the employer, coupled with the maximum prosperity for each employee," the perception in practice was quite different. °

Scheduling and Quality Control

Along with work measurement came methods for scheduling and quality control. Henry Gantt, a mechanical engineer and associate of Taylor, originated the *Gantt Chart*, the staggered bar chart along a time line, which remains most project managers' favorite representation of a schedule.

Less known but perhaps more significant is the fact that Gantt was one of the first to recognize the need for performance incentives. Gantt observed that "whatever we do must be in accord with human nature. We cannot drive people; we must direct their development." ° To that end, he implemented bonuses to factory foremen as an inducement to act as a *teacher*.

Quality control initially consisted of the inspection of finished goods in order to identify defective products. These defective items either became rejects or were reworked in order to meet established standards. Experience taught that products that do not go straight through the manufacturing process from beginning to end tend to be more damage prone after reaching the end user. Quality control therefore evolved into *quality management*, which attempts to maximize the percentage of product that goes straight through the process.

The concept of total quality management (*TQM*) was developed by W. Edwards Deming (it was Deming who introduced statistical quality control to Japan in 1950°). The essentials of TQM are measurement, communication, and cooperation, all used to gain an understanding of the potential sources of defects and flaws so that they may be removed from the manufacturing process. ° In that way, *process* can be continually improved, possible causes of failure foreseen, and costs kept under effective control by minimizing the necessity of rework.

One of the methods of TQM in manufacturing is to minimize inventories of raw materials by closely coordinating delivery schedules with suppliers. Close working relationships may be established with a small group of preferred suppliers, or even a sole source, to ensure that deliveries are made "just-in-time." In effect, each supplier functions as an integral part of the production process, with the necessary infrastructure of measurement, coordination, and cooperation being maintained through electronic data interchange.

Figure 5-1 illustrates Deming's concept of the TQM cycle, a continuous monitoring and feedback loop whereby the manufacturing process is constantly improved and refined. We return to the discussion of both project and quality management later in this chapter.

Figure 5-1   Deming Quality Management Cycle.

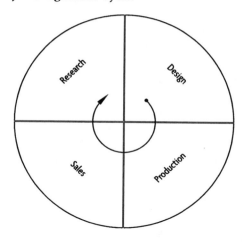

Theory Y

Around 1930, a series of productivity studies was conducted at the Western Electric Hawthorne Works near Chicago. In one of these studies, the lighting level in a relay assembly area was progressively reduced to nearly the level of moonlight, and it was found that productivity actually went up. Significantly, it was observed that the experimental groups tended to have higher overall productivity rates than other groups within the plant. A major conclusion of the Hawthorne Studies was that workers sometimes improve their productivity when they believe management pays attention to them and is concerned about their welfare.

In 1960, a social scientist named Douglas McGregor published a book titled *The Human Side of Enterprise*, in which the traditional view of employee motivations was characterized as *Theory X*. McGregor proposed, in true Cartesian fashion, a radically different view of worker motivation that he called *Theory Y*.

Theory Y takes the view that the average person does *not* inherently dislike work and, furthermore, depending on conditions that are controllable, work can be a positive source of satisfaction and self-actualization. McGregor saw the worker's commitment to meeting organizational objectives as a function of the perceived rewards associated with their attainment. The individual's capacity to use creativity and ingenuity to solve organizational problems is widely found in the general population but only partially utilized most of the time. Avoidance of responsibility, lack of ambition, and preoccupation with security are learned responses, according to Theory Y; and the properly motivated worker can be expected to exercise self-direction in the satisfaction of goals to which he is committed.

Edgar Schein, a psychologist and collaborator of McGregor's, was the first to use the term *corporate culture*.° A number of factors contribute to an organization's culture, which is often perceived as a nonverbalized climate that can be felt or sensed within the environment. Among examples he cites are the following:

- The management philosophy underlying the attitudes of senior management toward staff and customers.
- The dominant values promoted by the organization, along with its formal rules, procedures, and processes.
- Evolved patterns in working groups and observable regularities in the ways people interact.

Schein identified five basic categories of behavior that are considered *natural* within different organizations. These are often diametric opposites and establish a continuum between Theory X and Theory Y:

1. The natural disposition of some organizations to attempt to dominate the external environment while others operate in a more reactive manner.

2. The nature of reality and truth and the ways in which a particular organization arrives at "the truth".

3. The nature of human activity.

    Some organizations emphasize the completion of defined tasks as opposed to self-fulfillment and personal and professional development.

4. The nature of human relationships.

    Some organizations seek to facilitate interaction while others tend to look upon it as unnecessary distraction.

5. The nature of human nature.

    Some individuals act in the manner predicted by Theory X while others actively seek to maximize their potential in the furtherance of their own and the organization's goals.

Understanding Management's Agenda

Conducting a facility-planning analysis requires insight into the motivations that drive the enterprise—not necessarily all the specifics, but it helps to develop an understanding of the corporate culture. More often than not, the programming effort involves senior management, simply due to the size of the capital expenditures involved. Equally important, these individuals are the ones who have the greatest influence upon the shape of the culture in the first place. Many of the factors that Schein identified can provide a useful checklist of things for which to look.

The organizational structure must be thoroughly understood, both the formal one that may be graphically expressed on the organization chart and the informal groupings that emerge while surveying or interviewing department heads and workers. This information will be used directly in structuring the personnel, equipment, and space-requirements data that are obtained and in reporting to management. More important, all the discernible patterns that can be identified, including physical-adjacency relationships, will help determine the planning strategies that are formulated in the course of the analysis.

Senior management often initiates a facility-programming effort with questions such as these:

- Given my current business plan and projected growth rate, how much longer can I use our present facilities and what are my options to meet my changing needs?
- My various units are expanding at different rates resulting in, among other things, continual isolated requests for additional space. Can you help me develop a strategic plan for my projected facilities requirements, including timing and costs, so that I can adopt a corporate policy and plan to meet those needs along with my other business objectives?
- I have just reorganized my business. How best can I house my various units to accommodate changes in function and reporting while providing for growth?
- I am considering relocating my headquarters. How much and what kind of space do I need, and should I consider leasing or building?

These are all facility-related management questions, and the planning analysis needs to address them within management's specific agenda. The planning study may also be the first step toward creating an ongoing facility-management program within the company.

Successfully mapping out a long-range facility plan is a direct function of understanding the business objectives of the people who drive and manage the organization. There is usually constant pressure for rapid results because time and cost constraints often force people to launch projects with a large short-term payoff, often to the neglect of longer-term considerations. Patience, however much of a virtue it may be, is often in short supply, yet the cost of errors can be considerable. Figure 5-2 pictures the relative *stability* of a project framework, emphasizing the importance of thorough problem definition up front as a means of minimizing costly changes at the end.

Figure 5-2   Constant Cost of Documents and Construction.

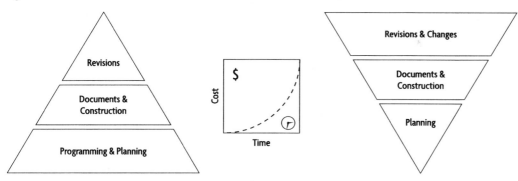

INFLUENCES ON ORGANIZATION AND CULTURE

We have talked about work measurement, which many large organizations have been doing for nearly a century. Industrial-engineering techniques are well established for quantifying basic business processes such as product assembly, inventory control, marketing, and sales.

Measurement goes hand in hand with TQM, which focuses on continuous process improvement. In the past, it was possible to make improvements in a particular area, then let the area stabilize and evaluate the result. But companies have found that continuous process and quality improvement leads to increased sales and profits, so there is always pressure to look for more savings and better performance. Pressures for change come from many different sources in today's marketplace.

Product and Service Lifecycles

The basic processes for manufacturing steel did not change for several decades. Not all that long ago, the time line from product development to production and sales was often measured in years. Now, especially in areas related to technology, product and service lifecycles are sometimes measured in months.

The four stages of the product lifecycle are shown in Figure 5-3. Both physical products and services follow this curve, anything produced that is capable of satisfying customer needs. Shapes of individual curves vary, but the stages do not. For example, communication technologies might be classed as follows:

- Teleconferencing—Introduction
- E-mail—Growth
- Fax or facsimile transmission—Maturity
- Business letters—Decline

Introduction is a high-risk stage. Substantial research and development costs may have been incurred in getting the product to this point, and the marketing effort is high. If the product is successful, it will enter its *growth* stage. This is often when companies dedicate significant resources to promoting the product in an effort to maximize market share.

The *maturity* stage traditionally is the longest, but in technology may be only a few months. Competition is usually intense in proportion to the product's degree of success, as competitors attempt to win market share. Over the long term, research and development are concentrated on modifying and improving the product to keep the company's share of the market. AutoCAD, for example, was introduced in 1982, not long after the PC itself, but product upgrades are released almost yearly.

Ultimately, all products enter the *decline* stage, in which the market begins to shrink. At this stage, management may be monitoring production costs very closely and looking for ways to reduce them, trying to sell the product into other markets, or preparing to end the product altogether.

Figure 5-3 Lifecycles of Products and Services.

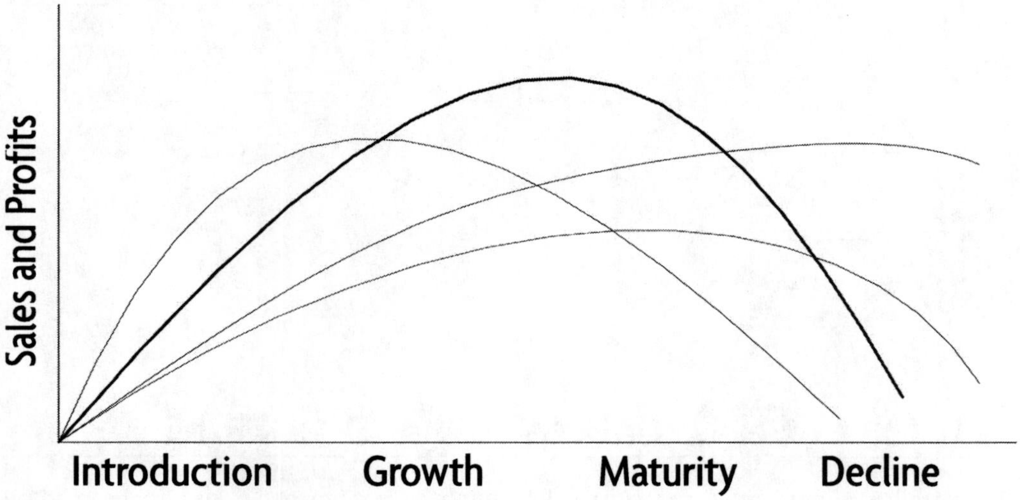

Technological Change

Although the personal computer is only twenty-five years old, the extent to which computing and communications technology has affected the global marketplace has almost become a cliché. In spite of some economic setbacks in recent years, the growth and convergence of these technologies have just begun.

Business Process Reengineering

Often an organization will analyze, either internally or using consultants, its basic processes: product manufacturing, marketing and sales, and financial systems. The motivation for process reengineering is usually to reduce costs, improve production efficiency or quality, better position the company for further growth, or some combination.

Successful reengineering crosses many departmental boundaries and involves contact among organizational units that may have had little contact in the past. Workflow analysis looks at the way in which work or product moves through the organization. Division of labor specifies the manner in which tasks are divided into separate jobs. The entire organizational structure of a company may be fundamentally changed, and the facility is often reconfigured to reflect such changes.

A traditional, functional departmental structure may begin with production, marketing, and finance divisions reporting to the chief executive. The functional and several other systems of departmentalization are shown in Table 5-1.

Any of these may be the basis for a master facility-planning strategy; in a large organization, they may be combined.

Downsizing

Modern corporations, especially those that are publicly owned, are always under pressure to appear financially stable and to be competitive in their markets. Downsizing may be the result of new technology applications or a financial downturn, or a major product may be reaching the end of its lifecycle with nothing "in the pipeline" to replace it. We may include with downsizing another major trend called *outsourcing*.

Often, in the course of reengineering the company, management finds it necessary to reduce staff or determines that it would be more profitable to have certain functions that were traditionally managed "in-house" performed by others. Outsourcing may eliminate entire departments, such as when marketing or customer-service call centers are contracted to companies overseas.

In the 1960s, a large company's payroll system required a staff dedicated to that function. A group of programmers might have been assigned to develop proprietary software to handle all the required functions on the company's mainframe computer. There might

Table 5-1   Traditional Departmentalization Structures

Function	Production	Marketing	Finance
Product	Natural Resources	Chemicals	Petroleum Products
Geographic	East	Midwest	West
Process	Receiving	Fabrication	Shipping
Customer	Consumer	Industrial	Governmental

have been remote terminals set up to handle data input. Developing technologies, together with changes in management methods, opened major windows of opportunity that drove many of the changes we have been discussing. Payroll processing is now among the most common functions to be outsourced, and there are many new companies for which payroll processing is their core business.

Any of these kinds of changes can require an immediate response from those responsible for managing the facility. Excess space can quickly become a liability, and consolidating the functions that remain may require extensive reconfiguration.

Systems Integration

The term *Systems integration* is often used to describe the process of making software applications work smoothly with one another; but the implications of systems integration are much more far-reaching, often resulting from the convergence of several, fundamentally different technologies. Table 5-2 shows most of the major technological developments of the past two centuries, all of which resulted from the synthesis of earlier concepts or products.

Globalization

Lest we leave the impression that manufacturing alone drives today's global economy, let us look at some recent statistics. In the final year of the last century, manufacturing output in the United States was 18 percent of gross domestic product (GDP). Services constituted 72 percent. Japan was only slightly more oriented toward manufacturing at 24 percent versus service output at 62 percent of GDP. Slightly more than 140,000 patent applications were filed in the United States, with nearly triple that number in Japan. The five largest *exporters* of commercial services were the United States, the United Kingdom, France, Germany, and Japan with a combined total of $600 billion. The United States and Japan were the largest exporters of intellectual property with the United States leading by a factor of over 4 to 1.°

Looking again at Table 5-2, one can see that the convergence of technologies during the last century was largely focused on the development of new products. Product development has not slowed in *absolute* terms, but the ability to capture and utilize information that can be applied to better decision making and more efficient application of effort has grown exponentially.

The work-measurement processes that Frederick Winslow Taylor and Elton Mayo (the Hawthorne Studies) pioneered have given way to continuing efforts to collect and make sense out of the massive amounts of data that organizations generate daily. The term *data mining* was coined several years ago to describe the process of capturing and storing raw data and converting it into useful information by giving it structure. Through the application of scientific analysis, hidden trends can often be discerned in the structured data that can lead to useful knowledge.

A host of specialized software applications have been developed, and entire new companies have come into being with them, with acronyms such as *ERP* (enterprise resource planning), *SCM* (supply chain management), and *CRM* (customer resource management). Another acronym, *CAFM* (computer-aided facilities management), describes a specialized class of ERP system.

One of the major sustained sources of change in today's marketplace comes from competition on a global scale. This has in part resulted from the constant change that imposes

Table 5-2 Convergence: Integration of Technologies (Dates in parentheses indicate the decade of first commercial application.)

Electric Motor	1831 (1870)	Electricity + Magnetism + Conductivity + Metallurgy
Telephone	1876 (1880)	Telegraphy + Microphone + Loudspeaker
Analog Recording	1877 (1890)	Telegraphy + Amplifier + Loudspeaker
Automobile	1885 (1900)	Coach Building + Internal Combustion Engine
Radio	1895 (1910)	Wave Theory & Detection + Wireless Induction
Airplane	1903 (1910)	Gliders + Rigid Airship + Internal Combustion Engine
Vacuum Tube	1904 (1910)	Vacuum Incandescent Lamp + Electron Flow
Air Conditioning	1911 (1910)	Refrigerator + Ice Maker
Television	1920 (1950)	Kinetoscope + Cathode Ray Tube + Photoelectricity
Antibiotics	1928 (1940)	300 Years of Experimental Science
Digital Computer	1937 (1950)	Mechanical and Electrical Relays, Binary Computation
Photocopying	1937 (1950)	Mimeograph + Electrostatics
Microwaves	1941 (1940)	Shortwave Radio + Radar + Magnetron Tube
Jet Engine	1941 (1950)	Technologies from 1920s and 1930s
Atomic Energy	1942–1945	Quantum Theory, Atomic Particles + Radioactive Disintegration
Transistor	1948 (1950)	Electron Flow + Semiconducting Materials
Digital Recording	1951 (1950)	Plastic Recording Tape + Binary Data Theory
Software	1951 (1960)	Card-Based Data Storage + Stored-Program Computing
Satellites	1957 (1960)	Multistage Missiles + Solid-State Electronics
Integrated Circuit	1959 (1960)	Transistor + Electron Microscopy + Silicon Doping
Apollo Moon Landing	1969	
Internet	1970 (1990)	Mainframe Computers + Time-Sharing + Packet-Switching Theory
Genetic Engineering	1973 (1990)	Electron Microscopy + Micromanipulation + Supercomputing
Personal Computer	1977 (1980)	

severe time pressure to get new products into the marketplace before they become obsolete. Gordon Moore, one of the cofounders of Intel Corporation in 1968, formulated what is now referred to as *Moore's law*, which states that the amount of data that can be stored on a given quantity of silicon doubles every eighteen months while the cost of that chip drops by half. Intel's first chip, which came out in 1972, was capable of processing 3,500 instructions per second. Six years later, the Intel 8088 (which became the central processor in the first IBM PC) processed twenty-nine thousand instructions in the same period, roughly an eightfold increase. Twenty-eight years later, Intel's processor is seven thousand times faster than the 8088.

Inflection Points

In 1996, the *other* cofounder of Intel, Andrew Grove, published a book titled *Only the Paranoid Survive.*° If one considers the true meaning of the word *paranoia*, remembering Henry Kissinger's apocryphal observation that it is impossible to be paranoid in Washington, one might argue that the book is mistitled. However, it opens with the following statement: "Business success contains the seeds of its own destruction. The more successful you are, the more people want a chunk of your business and then another and then another until there is nothing left."

Figure 5-4   The Inflection Curve.

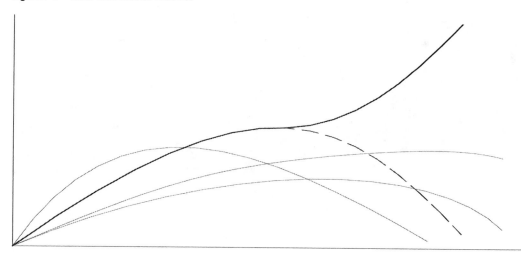

Grove used the engineering term *inflection point* to describe a potential reversal in the traditional product/services curve, not as applied to products and services but to companies and careers. An inflection point occurs when the slope of a curve stops accelerating in one direction and begins to slope in the other (mathematically, the curve's second derivative). If, at maturity, you are not able to navigate through an inflection point, you—or your business—go through a peak and then decline (the dashed curve in Figure 5-4). If you are able to navigate strategic inflection points, where the curves profoundly and permanently change, your business may be able to ascend to new heights. In Intel's case, a major inflection point came in the middle 1980s, when Grove and Moore decided to get out of the memory-chip business in the face of increasing Japanese competition. Intel's decision to become a *microprocessor* company allowed it, in Grove's words, to "break out of a plateau and catapult to a higher level of achievement."

Looking inward can be a painful process, especially in a crisis. It may require admitting that conditions are less than optimal; in fact, it may require acting on the realization that the organization is on a road to disaster. Coming to grips with real knowledge about the way an organization functions is often difficult because past practices are often deeply rooted in its culture.

Communication and Vision

Communication is the key element in coordination and cooperation, which is why communication networks have become the global circulation system of information management. Robert Metcalfe, one of the developers of the first local area network (LAN) called the *Ethernet*, formulated another law, which states that the value of a network increases exponentially as a function of the number of people connected to it. When Tim Berners-Lee coined the term *World Wide Web* to describe his unique protocol for accessing the Internet, he understated neither the power nor the delicacy of the worldwide network of networks that connects the contributing members of the global community.°

The paradigm of the network, at the *macro* level, is central to the operation of our technological society. In this chapter, we have attempted to highlight some of the ways in which scientific management has influenced organization and corporate culture. It is important to understand these principles on multiple levels because being able to adjust our perspective will help us to make planning decisions within a rational framework.

Figure 5-5 depicts an abstracted view of the Internet at the beginning of this century. If we consider it at different scales, the picture could just as well represent a highway network, a detailed corporate organization chart, or the *Bürolandschaft* approach to an office landscape (minus the live plants).

Figure 5-5 A View of the Internet around 2001.

At the very highest level, the network model represents all the connections that allow global enterprise to function. Each business establishment and societal institution has connections to the network both symbolic and concrete, but the web is a tension structure. It is the stress on the individual node that promotes the questions managers ask, such as those listed earlier. Growth presents only one set of issues. Some follow-up questions that may be used to parameterize the facility problem are: At what rate? Is it positive or negative? Which functions are likely to require more (or less) space? What are the best options for providing it? And, often not asked, What is the likelihood that my assumptions are totally *wrong* and that the situation may change?

Differential growth, some elements expanding while others contract, presents different questions and possibly other options, such as having space potentially available to multiple units as they come online. Reorganizations and relocations present yet a different set of considerations because they can be disturbing to the culture of the organization or be a consequence of a major change that has already occurred.

Listening to each client and understanding the particular organization's structure, culture, and direction are essential elements of successful planning. Programming is the first step in a space-planning project and also can serve as a beachhead toward the establishment of an ongoing facility-management effort. The objective of both is the translation of management's vision into a spatial reality, and the process connects with numerous financial and operating functions such as the following:

- *Long-range (strategic) planning*—Three-to-five-year plans, ten-year plans, and beyond.
- *Financial management*—Capital commitments and operating costs, construction management.
- *Real estate*—Building and site acquisition (purchase, build, or lease), disinvestment.
- *Architecture and Engineering*—Architectural and structural design, telecommunications, electrical, mechanical systems, building and accessibility code compliance.
- *Space management and interior planning*—Inventory and allocation, layout planning, furnishings and equipment specification.
- *Installation, maintenance and Operations*—Furniture and equipment installation; move management; energy management; preventive maintenance; interior, exterior, and landscape maintenance; housekeeping and waste disposal; and so on.

Let us now change our focus to the *micro* level and look at the network from a project-management perspective.

DEFINING THE PROJECT ELEMENTS

Following the Second World War and concurrent with the development of the first-generation computers, the application of scientific methods by corporations and governmental agencies began to increase at an exponential rate. One of the first uses of the *analysis/synthesis* approach to project management resulted in the launch of the first nuclear-powered submarine, the *Nautilus*, by the United States Navy in 1954. This application was the *critical-path method*, which utilized an approach called the *program evaluation and review technique*, or *PERT*. The PERT approach gave us an effective set of tools and methods for managing complex projects, which are at the core of most project-management software in use today.

As we have seen, in the second half of the twentieth century, technological growth continued to accelerate. Along with it came an increased recognition of the relationship between the value and velocity of money. We said at the beginning of this book that the traditional meaning of the word *architect* was diluted significantly during this same period, due in part to the increased complexity of project management and in part to a loss of nerve. It is now the construction-management firm that often assumes the responsibility (and the financial liability) of managing the project schedule and controlling budgets.

To manage a project, you need to look at scheduling and budgeting from two points of view: the client or owner's and your own. The definition of activities to be performed and the time frame are most often determined by the client's requirements, but the cost is determined by the resources needed to complete the project. A realistic understanding of your own level of involvement and your client's expectations is essential to a quality—and profitable—result. Regardless of the scale of the undertaking, responsibilities need to be defined so that all the participants' roles are clearly understood and agreed upon at the outset.

We are reminded of the injunction to "be careful not to leave disputed points for the householders to settle after the works are finished." A well-defined project plan will go a long way toward satisfying the balance between Vitruvius' principles of *propriety* and *economy*. The formulation of the plan should begin with the earliest phase of work and encompass a detailed analysis of the scope of work on two fronts:

1. The tasks to be performed and the dependencies among those tasks.
   In other words, what activities must be in process or complete before others can commence?

2. The resources that will be required to perform those tasks.

Programming

Programming is a first-class example of Descartes' reductionist approach to problem solving, described at the beginning of Chapter 4. We commonly use the word *program* to refer to a set of discrete coded instructions that control the operation of a computer. More generally, the *Oxford English Dictionary* defines a program as "a plan of future events" or a series of studies.° If the process of planning and design is problem solving, then a proper definition of *programming* is *defining the problem*. Some refer to the program definition process as "problem seeking."°

The American Institute of Architects publishes several standardized contracts for architectural and design services that specify the phases of work into which projects are often divided. The details vary depending on the scope of work and the contractual understanding between planner and client, but the following list is typical:

1. Programming
2. Planning and schematic design
3. Design development
4. Contract documents
5. Competitive bidding or contract negotiation
6. Contract administration

The term *programming* was adopted as the official name for Phase one around 1970. Prior to that, the contract terminology used the phrase *project requirements*, which suggests the problem-defining nature of the activity, albeit in a less structured fashion. A facility program is often an extension of management's long-range or strategic-planning effort, carried out by real estate, operations people, or an organization's facility manager. At some point, an architect or designer may be engaged along with (or following) a team of business consultants.

Concept Development

Progress through Phase two and Phase three is characterized by increasing levels of conceptual detail. There may be some overlap among the first three phases (two steps forward and one back as options are identified and either pursued or discarded), but contractually the end of each is defined by the "deliverables" that formally bring that phase to a close. Schematic design commences when the program requirements are reasonably well understood and possible approaches to solutions begin to be examined. There may be programmatic issues that still need to be resolved, though, and the visual conceptualization of alternatives will be general: bubble diagrams, block-planning layouts, and stacking diagrams where multifloor and multibuilding facilities are entailed. In the latter case, site studies of a similar nature are made. If relocation to new quarters in an existing building (or buildings) is a given, then these studies are likely to center on a comparative evaluation of candidate buildings.

The process of developing the program through the schematic planning stage is the focus of the next chapter, but programming is complete for project-management purposes when all the basic planning requirements have been documented, including blocking and stacking diagrams. In Phase two the project is taken to the next level of detail with the preparation of detailed plans showing walls and doors, possibly including lighting, furniture, and equipment layouts for areas with special needs. Functional areas needing further consideration are identified and design parameters established. A presentation of the final space layouts to the client or owner usually marks the formal conclusion of the second phase of work.

The transition from schematic to definitive design occurs when detailed space layouts have been completed and approved. In addition to wall and door locations, the space layouts show principal corridors and traffic aisles established in accordance with the adjacency and workflow requirements defined by the program. Simultaneously, even beginning during the two earlier phases, background information is being developed that will support the critical decision making that is encompassed within Phase three.

If we may borrow some terminology from the TQM vocabulary, design development involves the synthesis of many *causal factors*. Process is a collection of causal factors, the acquisition and organization of information, and narrowing options to plot a course of action. Once the physical layout has been substantially formalized, a number of factors are brought into play that will determine the shape of the final planning solution.

1. The corporate or organizational culture largely determines the design goals, the image or outward expression that the facilities must project. Management philosophy and organizational culture, whatever form it takes, will directly influence the choice of areas whose special importance needs an appropriate expression. Design parameters are based upon decisions ranging from the standards to be used in assigning quantities and types of space to the quality and expense of materials and finishes.

2. The collection of background information on the space to be occupied, together with furnishings and equipment to be used, should already have begun during the two initial phases. If the intent is to reuse movable assets such as furniture, a detailed inventory is needed to determine each item's remaining value and physical condition. In the same circumstance, a field investigation will verify existing physical conditions at the site including exterior wall and window conditions; utility core elements; existing ceiling, lighting, and mechanical conditions; and so on.

3. Throughout the definitive design phase, budgetary targets are established and continually checked and the design concept becomes more detailed. Statements of probable cost for construction, furnishings, and infrastructure based on the space program and approved layouts serve as guidelines and are continually updated during this phase of project development.

At the end of Phase three, detailed facility layouts are complete, the overall design concepts are agreed upon, the necessary technical background information has been assembled, and a detailed master schedule has been established for implementing the project.

Implementation

The preparation of contract documents begins the process of implementation. These largely consist of drawings and specifications that describe the work to be done in sufficient detail to allow the builders to properly assess the cost of every project element and actually do the construction. On that basis, those would-be participants must determine the resources they must bring to the project in order to complete the necessary work.

There are several ways in which the actual contracting and construction phases may be structured. Traditionally, the architect managed a design-bid-build process in which multiple prospective contractors examined the drawings and specifications for the architect's completed design. Each prospective bidder makes an independent assessment of what will be required to complete the job in a timely and profitable manner. Using this method, formal written bids are submitted and the project is awarded to the lowest qualified bidder. This is followed by the construction and contract-administration phase.

Today, particularly for large projects, a construction manager may lead the project team from the very beginning to the very end; this manager occupies the top position on the team's organization chart and even assumes the *financial* responsibility for getting the project completed on time and within budget. As the project executive, she may direct the efforts of several architects, engineering firms, and other specialized consultants.

PROJECT MANAGEMENT

We began this chapter by talking about Babbage and Taylor and how their methods were based upon quantifying the amounts of effort needed to accomplish different tasks and dividing the efforts of labor accordingly. These principles are extensions of the Cartesian process of reducing a problem to its individual, component elements in order to construct an ordered framework within which to accomplish work.

Project Networks

Communication networks, governed by Metcalfe's law, are open networks as exemplified by the Internet at a global scale as pictured in Figure 5-5 We can also picture Descartes' *analysis/synthesis* method as a network, but it is a *closed* network that has a beginning and an end. Analysis and synthesis may be represented as two trees that are mirror images of each other, as shown in Figure 5-6. Processes such as the convergence of technologies (Table 5-2) may be viewed as *synthetic*.

This expansion from a single start event to multiple paths, followed by the subsequent narrowing to a single conclusion is the basis of PERT, otherwise known as the *critical-path method*. One can establish links among all the discrete activities that make up a project on a basis of precedence by asking a single question: Which task must be completed before another can commence? Approached in this way, certain tasks will be found to have critical dependencies while others will be less dependent on activities that may take place before, after, or concurrently with them.

If we make several passes from beginning to end and then *backward*, from the end to the beginning, we can calculate the singular, shortest path through the project network. This shortest set of links is the *critical path*. The total time required to complete all the tasks along the critical path determines the shortest time in which the whole project can be completed. This assumes, of course, that sufficient resources are available to work on all the other activities, which, *in theory*, can be carried on simultaneously.

All the noncritical tasks in the network are said to have a property called *float*, which means that they each have an earliest possible starting time and a latest permitted completion depending on where they are linked into the network. These activities may begin

Figure 5-6 Analysis and Synthesis.

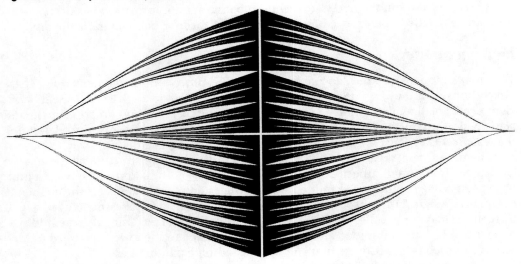

Figure 5-7 Project Network: PERT Diagram.

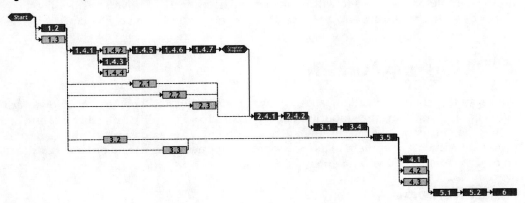

later or be completed *earlier* than those determined to be critical without affecting the overall project schedule. Being able to differentiate between critical and noncritical tasks gives the project manager a powerful tool for prioritizing where effort and resources can be most effectively expended.

Figure 5-7 is a PERT diagram of the tasks that comprise the simple planning project just described. The numbers on each activity correspond to the outline of tasks that are listed by phase in Figure 5-8 and Figure 5-9. The dark-gray network nodes with white text are all on the critical path, as determined by *Microsoft Project*. (There are many project-management software applications available, of which *Project* is one of the most accessible for general use.)

The noncritical tasks have the light-gray shading, such as 3.2 and 3.3 in Figure 5-7. The field check of site conditions and the furniture inventory are examples of the *background information* referred to earlier, which can be assembled at any time almost from the beginning of the project up to the time the design concept is finalized (Task 3.5). When these tasks are actually performed is at the project manager's discretion within these limits.

Project Definition

Within the locus of the network model lies not only the key to taking the project from beginning to end in a controlled fashion but also the mechanism for allocating resources to each activity along the way. This permits us to look at the construction of the project plan from two important perspectives:

1. **The owner/client who will be the end user of the completed facility:** This includes everyone from upper management down to users of the individual spaces. This, from the point of view of the planner doing the work, is the *external* view.

2. **The project manager, planners, designers, and other members of the project team:** Your rational self-interest requires that the amount of time put into getting the job done is balanced against the cost of the resources used. This is the *internal* view.

Mapping out the project plan begins by breaking down the major stages of work into individual activities or tasks. This begins with the main phases like the six described earlier. Each phase should be broken down into as much detail as possible using the information currently available and listed in outline format. The object is to identify the major parts of the project in terms of a general time sequence without (at least not yet) getting caught up in the detail of individual precedences. We are thinking strategically at this point: how to get through the whole enterprise from start to finish in a more or less linear fashion.

The initial project budget also needs to be identified. There are actually two project budgets: one contained within the other. First is the client's overall budget, however large or small the project may be. Within the owner's total budget is a defined portion that is going to pay for your time and effort in putting everything together. As we observed earlier, this is one of those "points," according to Vitruvius, not to be left unsettled until after the works are finished!

Comparing the outline of tasks in Figure 5-8 with the PERT chart in Figure 5-7, we can see that there is substantially more detail in the outline than there is in the network. This was done partly to simplify the example, but it also says something about the nature of the process. Our example shows the programming and space-planning phases detailed down to the third level of the outline in the PERT chart. Phases three through five are broken down only as far as the second level, and the contract-administration phase is shown as a single activity. The project plan and the network can be refined and expanded in as much detail as required as work progresses. At the outset, we are focusing primarily on the first two phases immediately ahead of us and taking a more general view of the work to follow.

Milestones should be identified as early as possible. These are major reference points marking progress and often determine the dates when certain deliverables are due. Project milestones can be thought of as activities without any time duration. Figure 5-7 shows two milestones: *Start* and *Complete Program*.

Dependencies, Durations, and Resources

With the outline of activities complete, we can begin to look at resource allocation. This is the first step in determining our costs. We need to look at each task in terms of what level of effort is required, in two dimensions:

1. *Horizontally:*   Assuming there are different levels of experience (and hourly rates) among the people on the project team, what is the mix of expertise required for each defined task?

2. *Vertically:*   What proportion of the total time to be spent by the project team do we anticipate being allocated to each task?

Using a worksheet like the "Project Manpower Allocation" table shown in Figure 5-8, we can begin working out the application of manpower in each dimension separately. Working across the sheet row by row, we can make an estimate of the proportion of time to be spent by staff in each job category: principal, project manager, senior and junior planners, draftspersons, other support staff, and so forth. Each of these estimates should add up to 100 percent in the right-hand column.

Using the column titled "Percentage Allocation by Task," we then estimate the amount of time to be spent on each activity, relative to the total project. Once again, the allocated total at the bottom of the sheet should be 100 percent.

Figure 5-8   Project Manpower Allocation.

Phase 1 Subtotal:	20.00%
Phase 2 Subtotal:	12.00%
Phase 3 Subtotal:	18.00%
Phase 4 Subtotal:	25.00%
Phase 5 Subtotal:	10.00%
Phase 6 Subtotal:	15.00%
Allocated Percent Total:	100.00%

A.B.C. Manufacturing Corporation
Worksheet Criteria (Assumptions)

PROJECT HOUR DISTRIBUTION MATRIX

Task	Percentage Allocation by Task	Principal	Project Manager	Senior Planner	Junior Planner	Drafts-Person	Support	CheckSum
1. Programming								
1.2. Project Orientation	1.5	10%	40%	40%	10%			100%
1.3. Schedule Development	0.5	20%	30%	30%	20%			100%
1.4. Space Occupancy Program								
1.4.1. Distrib. Program. Questionnaires	1.0		50%	50%				100%
1.4.2. Personnel Projections	1.5		20%	20%	20%	10%	30%	100%
1.4.3. Space Utilization Standards	3.0	10%	30%	20%	20%	20%		100%
1.4.4. Area Requirements	2.5		20%	40%	20%	20%		100%
1.4.5. Adjacency Requirements	2.5	10%	20%	30%	20%	20%		100%
1.4.6. Block/Stacking Plans	4.0	10%	30%	20%	20%	20%		100%
1.4.7. Prepare Management Report	3.5		30%	20%	20%		30%	100%
Allocated Percent Subtotal:	20.0							
2. Space Planning								
2.1. Base Sheets	3.0		10%	20%	40%	30%		100%
2.2. Building Standards	1.0		30%	40%	30%			100%
2.3. Site Visit	1.0	20%	40%	40%				100%
2.4. Space Planning								
2.4.1. Develop Initial Space Plan	3.0	10%	20%	50%	20%			100%
2.4.2. Finalize Space Plan	4.0	10%	20%	30%	20%	20%		100%
Allocated Percent Subtotal:	12.0							
3. Design Development								
3.1. Design Orientation								
3.1.1. Establish Design Concept	1.5	20%	30%	40%	10%			100%
3.1.2. Color & Material Palettes	1.0	10%	10%	40%	30%	10%		100%
3.1.3. Establish Target Budgets	0.5	20%	30%	20%			30%	100%
3.2. Furniture Inventory	3.0		10%	20%	30%	10%	30%	100%
3.3. Field Check								
3.3.1. Verify Site Dimensions	2.0		10%	20%	40%	30%		100%
3.3.2. Modify Base Sheets	1.0			20%	40%	40%		100%
3.3.3. Record/Assignment Drawings	1.0		10%	30%	30%	30%		100%
3.4. Statement of Probable Cost	2.0	10%	30%	20%	10%		30%	100%
3.5. Design Concept								
3.5.1. Select Furniture & Furnishings	1.5	10%	20%	40%	30%			100%
3.5.2. Prepare & Present Des. Concept	0.5	20%	50%	30%				100%
3.5.3. Refine Design Concept	1.5		20%	40%	30%	10%		100%
3.5.4. Finalize Technical Information	2.5		20%	30%	30%	20%		100%
Allocated Percent Subtotal:	18.0							
4. Contract Documents								
4.1. Construction Documents								
4.1.1. Construction Plan	4.0	10%	20%	40%	20%	10%		100%
4.1.2. Electrical & Communication Plan	3.0	5%	20%	40%	25%	10%		100%
4.1.3. Reflected Ceiling Plan	3.0	10%	20%	40%	20%	10%		100%
4.1.4. Room Finish Plans	2.0	5%	10%	30%	35%	20%		100%
4.1.5. Door & Hardware Schedules	2.0	5%	10%	30%	35%	20%		100%
4.1.6. Design Details	2.0	5%	15%	35%	25%	20%		100%
4.2. Engineering Drawings	1.5		40%	40%	20%			100%
4.3. Furniture Distribution								
4.3.1. Furniture Distribution Plan	4.0	5%	30%	40%	15%	10%		100%
4.3.2. Furniture Specifications	3.5		30%	20%	10%		40%	100%
Allocated Percent Subtotal:	25.0							
5. Competitive Bidding								
5.1. Construction Bidding								
5.1.1. Prepare Constr. Bidders' List	1.5	10%	30%	20%	20%		20%	100%
5.1.3. Construction Bidding	3.0	10%	50%	40%				100%
5.1.5. Issue Construction Contracts	1.0		30%	50%	20%			100%
5.2. Furniture & Furnishings								
5.2.1. Prepare Furniture Bidders' List	1.5	10%	30%	20%	20%		20%	100%
5.2.3. Furniture Bidding	2.0	10%	50%	40%				100%
5.2.5. Order Furniture	1.0		50%	50%				100%
Allocated Percent Subtotal:	10.0							
6. Contract Administration								
6.4. Field Observation	8.0	20%	30%	30%	20%			100%
6.5. Shop Drawing Review	4.0		30%	50%	20%			100%
6.8. Construction Punch List	1.5	20%	50%	30%				100%
6.9. Furniture Punch List	1.5	10%	50%	40%				100%
Allocated Percent Subtotal:	15.0							
SUMMARY								
Allocated Percent Subtotal:	100.0							

The "Project Hour Distribution" worksheet illustrated in Figure 5-9 gives the results. At the top of the sheet, a "Billing Rate Schedule" contains the hourly rates for each job category. All that remains is to enter an estimated "Total Project Budget" on the top line and the calculated results are distributed across the matrix. A schedule at the top of the sheet shows the manpower allocation in hours and weeks along with the proportion of time involvement by team members in each rate category. Estimated hours for each task are summarized by phase along with their associated costs in the right-hand column. A what-if

Figure 5-9   Project Hour Distribution.

Total Project Budget: $133,000

	Principal	Project Manager	Senior Planner	Junior Planner	Drafts Person	Support	Total Time Committed
Billing Rate Schedule:	$200	$160	$100	$75	$50	$40	
Project Total in Weeks:	1	51	11	9	6	4	37
Project Total in Hours:	48	218	432	350	245	171	1464
Percent of Total Project:	7.3%	26.2%	32.5%	19.8%	9.2%	5.2%	100.0%

PROJECT HOUR DISTRIBUTION MATRIX

A.B.C. Manufacturing Corporation

		Principal $200	Project Manager $160	Senior Planner $100	Junior Planner $75	Drafts-Person $50	Support $40	Total Hours	Total Dollars
1.	Programming								
1.2.	Project Orientation	1	5	8	3			17	1,995
1.3.	Schedule Development	1	1	2	2			6	665
1.4.	Space Occupancy Program								
1.4.1.	Distrib. Program, Questionnaires		4	7				11	1,330
1.4.2.	Personnel Projections		2	4	5	4	15	31	1,995
1.4.3.	Space Utilization Standards	2	7	8	11	16		44	3,990
1.4.4.	Area Requirements		4	13	9	13		40	3,325
1.4.5.	Adjacency Requirements	2	4	10	9	13		38	3,325
1.4.6.	Block/Stacking Plans	3	10	11	14	21		59	5,320
1.4.7.	Prepare Management Report		9	9	12		35	65	4,655
	Phase Subtotal:	8	47	72	65	68	50	310	26,600
2.	Space Planning								
2.1.	Base Sheets		2	8	21	24		56	3,990
2.2.	Building Standards		2	5	5			13	1,330
2.3.	Site Visit	1	3	5				10	1,330
2.4.	Space Planning								
2.4.1.	Develop Initial Space Plan	2	5	20	11			38	3,990
2.4.2.	Finalize Space Plan	3	7	16	14	21		61	5,320
	Phase Subtotal:	6	20	55	51	45		177	15,960
3.	Design Development								
3.1.	Design Orientation								
3.1.1.	Establish Design Concept	2	4	8	3			16	1,995
3.1.2.	Color & Material Palettes	1	1	5	5	3		15	1,330
3.1.3.	Establish Target Budgets	1	1	1			5	8	665
3.2.	Furniture Inventory		2	8	16	8	30		3,990
3.3.	Field Check								
3.3.1.	Verify Site Dimensions		2	5	14	16		37	2,660
3.3.2.	Modify Base Sheets			3	7	11		20	1,330
3.3.3.	Record/Assignment Drawings		1	4	5	8		18	1,330
3.4.	Statement of Probable Cost	1	5	5	4		20	35	2,660
3.5.	Design Concept								
3.5.1.	Select Furniture & Furnishings	1	2	8	8			19	1,995
3.5.2.	Prepare & Present Des. Concept	1	2	2				5	665
3.5.3.	Refine Design Concept		2	8	8	4		22	1,995
3.5.4.	Finalize Technical Information		4	10	13	13		41	3,325
	Phase Subtotal:	6	27	68	83	63	55	302	23,940
4.	Contract Documents								
4.1.	Construction Documents								
4.1.1.	Construction Plan	3	7	21	14	11		55	5,320
4.1.2.	Electrical & Communication Plan	1	5	16	13	8		43	3,990
4.1.3.	Reflected Ceiling Plan	2	5	16	11	8		42	3,990
4.1.4.	Room Finish Plans	1	2	8	12	11		33	2,660
4.1.5.	Door & Hardware Schedules	1	2	8	12	11		33	2,660
4.1.6.	Design Details	1	2	9	9	11		32	2,660
4.2.	Engineering Drawings		5	8	5			18	1,995
4.3.	Furniture Distribution								
4.3.1.	Furniture Distribution Plan	1	10	21	11	11		54	6,320
4.3.2.	Furniture Specifications		9	9	6		47	71	655
	Phase Subtotal:	9	46	117	94	69	47	382	33,250
5.	Competitive Bidding								
5.1.	Construction Bidding								
5.1.1.	Prepare Constr. Bidders' List	1	4	4	5		10	24	1,995
5.1.3.	Construction Bidding	2	12	16				30	3,990
5.1.5.	Issue Construction Contracts		2	7	4			13	1,330
5.2.	Furniture & Furnishings								
5.2.1.	Prepare Furniture Bidders' List	1	4	4	5		10	24	1,995
5.2.3.	Furniture Bidding	1	8	11				20	2,660
5.2.5.	Order Furniture		4	7				11	1,330
	Phase Subtotal:	5	35	48	14		20	122	13,300
6.	Contract Administration								
6.4.	Field Observation	11	20	32	28			91	10,640
6.5.	Shop Drawing Review		10	27	14			51	5,320
6.8.	Construction Punch List	2	6	6				14	1,995
6.9.	Furniture Punch List	1	6	8				15	1,995
	Phase Subtotal:	14	42	72	43		20	171	19,950
Summary									
	Project Total:	48	218	432	350	245	171	1464	133,000
	Percent of Total Project:	7%	26%	32%	20%	9%	5%	100%	

analysis of the project budget can be performed by changing any of the three parameters controlling the distribution: the total budget, the time distribution by activity, and the experience level of staff assigned.

Dependencies

With the detailed activity information in hand, we are well prepared to begin constructing the project network in an application such as Microsoft Project. The task-list information can be imported directly into Project, maintaining the hierarchy of phases, subphases, and individual activities, which can be modified simply by indenting or outdenting an item. Estimated task durations can be developed from the budgeted hours from the "Project Hour Distribution" worksheet.

Milestones are established whenever a task is given zero duration to create a reference point or mark a significant event such as the end of a phase. Milestones can also be used to represent external dependencies, events that are not actually part of your project network but that can affect the schedule. For example, you cannot take possession of a facility until the former occupant has finished removing some hazardous material.

Once the task list is complete and task durations have been assigned, the next step is to define the task dependencies. This establishes how the activities are sequentially related to one another through the network. Initially, these relationships can be specified as simply as by looking at each task and determining its predecessors. In Project, these are specified using the sequential numbers of the activities in the task list. Figure 5-10 shows the bar-chart view of our example project in the format developed by Gantt, labeled at the left with each task name, duration, and the task's predecessors. Projected start dates for each activity are generated by the application as dependencies are added, defining the network

Note that many scheduling programs allow you to create a Gantt Chart without the underlying network simply by dragging the bars along the horizontal time line; but with the bar chart alone it is difficult to measure—or even observe, in many cases—whether the schedule is attainable. With the network schedule, we can have reasonable confidence about whether the proposed schedule is realistic or not, and we can test our assumptions. Equally important, the network provides a mechanism for updating the schedule while the work is in process. Actual completion dates can be inserted at any point in time and the schedule recalculated to determine whether milestones are within acceptable limits. In this manner, slippage and other such issues may be anticipated and necessary action taken, such as applying more effort to a critical task or tasks.

Task dependencies can be of four kinds, of which *finish-to-start* is the most common and usually the default. Given two activities A and B, A must be completed before B can commence. Two other types of dependency are *finish-to-finish* and *start-to-start*. These links determine that related activities must either finish or start together. For example, if the quality inspection of certain equipment takes place while the devices are being installed, the inspection cannot be finished before the installation is complete. These kinds of links are often used to pin down the earliest or latest start/finish dates for noncritical activities that may otherwise be indeterminate. Finally, *start-to-finish* dependencies may be specified that say that activity B cannot be completed until activity A begins. This is a rarely used, inverse dependency that essentially says that activity A must start but need not be completed

Figure 5-10   Project Network: Gantt Chart.

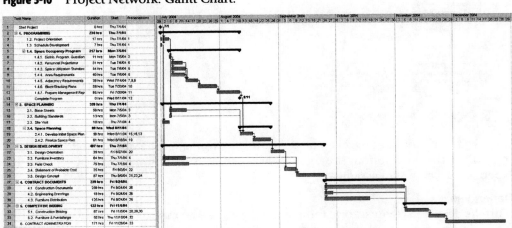

Figure 5-11   Types of Task Dependencies.

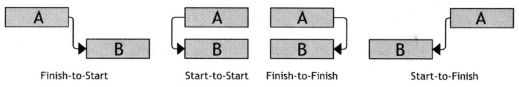

| Finish-to-Start | Start-to-Start | Finish-to-Finish | Start-to-Finish |

Table 5-3   Degrees of Scheduling Constraint

Schedule is flexible.	**As Soon as Possible** (Default When Tasks Are Scheduled from Start Date)		
	As Late as Possible (Default When Tasks Are Scheduled from Completion Date)		
Schedule is moderately constrained.	Start	**No Earlier** (Start On or After Specified Date)	
		No Later (Start On or Before Specified Date)	
	Finish	**No Earlier** (Complete On or After Specified Date)	
		No Later (Complete On or Before Specified Date)	
Schedule is constrained.	Must Start on Date Specified	Overrides All Other Scheduling	
	Must Finish on Date Specified	Parameters	

before activity *B* can finish. The four types of dependencies, as they would appear on a Gantt or PERT chart, are shown in Figure 5-11.

Dependency relationships can also be controlled through the use of lead and lag times. Lag time can be specified to introduce a delay between the completion of one activity and the start of its successor. Lead time is the opposite, creating an overlap of activities so that the successor actually begins before its predecessor finishes.

Several types of constraints may also be applied to tasks in the network to control actual start or finish dates. These are summarized in Table 5-3. Constraints having limited degrees of flexibility work with the network dependencies to make activities occur as early or late as the combination will allow. Inflexible constraints lock in the beginning or end of a task to the date specified, overriding all other scheduling parameters.

Durations

A what-if analysis of task durations can be performed by using PERT analysis only if the tasks are linked. Microsoft Project makes this type of analysis very simple by allowing you to specify optimistic, likely, and pessimistic durations for the tasks in your network. These activity durations may be used individually to calculate earliest, most likely, and latest completion dates for the project. You can also have the program calculate a single, weighted average using parameters set up to suit your particular situation.

Resource Allocation

For the type of project with which we are concerned, resources are primarily people, the members of the project team and those on the client side who are needed for information and approval. In general, resources may be defined as specific individuals, equipment, facilities, utilities, or other contractors. Resources of whatever type have costs associated with them. As we saw earlier, project staff of different experience levels most often have different hourly pay rates (Figure 5-9), which are considered unit costs. Certain kinds of equipment resources have fixed or variable costs depending on whether or not the cost of the item is dependent on the extent of its use. Microsoft Project allows for two kinds of resource costs:

1. *Work resources:*   People and equipment that expend time to complete project activities, which have a periodic rate.

2. *Material resources:*   Supplies or other consumable items used in the completion of project tasks. Costs are usually measured in terms of unit quantities: boxes, square yards, pounds, and so on. Depending upon whether material usage is based on time or not, costs may be fixed or variable.

In Project, an inventory of resources is created using a resource sheet, after which the resources can be associated with the appropriate activities.

An important property of resources is their availability. If a given member of the team is only available half of the time, which may have an effect on the schedule's time line, you can adjust the maximum units that can be allocated to that resource on a task-by-task basis. Thus, you can schedule an individual full time for certain tasks and part time for others, either by percentage or using the appropriate time units.

Monitoring the Project

It should be apparent that using the network approach to managing a project offers many more ways of understanding and controlling progress than simply drawing a Gantt Chart and shoving the bars around. Project management, like programming and facility planning, is first and foremost a thinking process. Resolving issues and conflicts and identifying opportunities are essential project-management activities, and the thought process benefits from organization and structure in direct proportion to the size of the job.

There is a substantial difference between project administration and project management. Looking at the task list and the time line and marking off incremental progress are administrative functions to be sure. Setting up the schedule structure and maintaining the network in a way that will use all of your resources most effectively are management. The resiliency of your project plan will be tested the first time a crisis occurs or a major issue comes to the fore. How the project manager and the team deal with the first few crises will set the tone throughout the endeavor. The manner in which issues are resolved will often point up opportunities, chances to change some aspect of the project structure in order to achieve a benefit. The project plan represents the potential issues and opportunities in the project; and questions with regard to timing, resources, or operational procedures should be approached in an analytical frame of mind. When an issue or an opportunity presents itself, it is helpful to remember the following guidelines:

- Make sure you understand the issue or opportunity clearly and are not confusing it with consequences or symptoms. Resist the temptation to be reactive, especially if it is a crisis.
- Evaluate the consequences of possible actions or of taking no action. How will the overall project be affected by a particular course of action or inaction? If an action is necessary, should it be taken now or later?
- Once you decide on a course of action, determine whether you can resolve other issues at the same time. If the action involves a change to the schedule, you want to minimize how frequently such disruptions occur.

QUALITY AND FACILITY MANAGEMENT

As we mentioned earlier, traditional quality control in manufacturing consisted of the inspection of finished goods in order to identify defective products. Defective items were discarded or reworked in order to meet the standards in effect. Then it was determined from experience that products that did not go straight through the process tended to be more damage prone once they reached the marketplace.

This kind of limited quality control is no longer acceptable in manufacturing and by its very nature could never be applied effectively to the facility planning and design process. It is just not sufficient to find flaws and defects after the fact in order to correct them. By then, it is too late. The client has not been well served and credibility has been lost.

Total quality management is a commitment to maximizing the percentage of product that goes straight through the process. In this case, the product consists of rational planning and design solutions, together with the necessary documentation to get the project built. The principles of TQM help to open up lines of communication and foster the kind of analytical thinking that can recognize false data, allowing a potential failure to be recognized as such before it turns into a disaster. Total quality management requires that we

Figure 5-12   Total Quality Management Diagram and Entry Point.

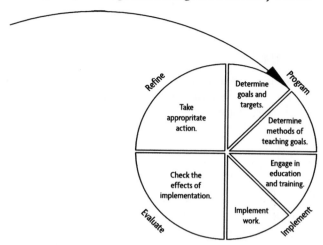

understand the potential sources of defects and flaws, so that they may be removed from the process. In that way, the process is improved on a continuing basis, possible causes of failure are foreseen, and costs can be kept under effective control by minimizing the necessity of rework.

The chart in Figure 5-12 is based on the Deming quality management cycle pictured at the beginning of this chapter. In this chart, we have attempted to summarize the TQM approach as it pertains directly to planning and design. The arrow at the top of Figure 5-12 shows one possible entry point into the cycle. One of the important reasons for having an ongoing facility management program within an organization is so that current planning information is available when it is needed. It naturally follows that facility programming is a logical point at which to enter the cycle and establish such a program.

Determine Goals and Targets

The checklist in Appendix A lists a number of issues that may need to be addressed in developing the project plan for a programming and planning analysis. The particulars will differ of course, depending upon specific conditions and the requirements of each project. Considerations should include the following:

- Examine the factors that comprise the interior and exterior environment, including the relationships among individual items; formulate a program statement that will aid in making capital resource commitments.
- Develop the program statement in adequate detail to thoroughly document the planning and design requirements of the project so that appropriate construction and engineering drawings can be prepared to accomplish the defined objectives.
- Synthesize the information into an organized format that the client, the architect, and other consultants or contractors may use as the basis for completing the work.
- Maintain a feedback loop so that the experience and information gained in the course of the project is not lost but is incorporated into an ongoing analytical process.

Determine Methods of Reaching the Goals

Once again, process is a collection of causal factors. The network model suggests that there is an endless variety of methods. Standardized methods are those that have been proven through repeated use in different situations, are useful to the organization as a whole, and are free of known deficiencies.

Uniform effects do not always occur, however, even as a result of standardized procedures. The same person may make the same measurements and use the same materials and equipment to produce something, but the effects will vary because the number of causal factors is limitless. The planning process must be controlled through its most important causal factors. Significant obstacles should be anticipated before they are met, which constitutes vanguard control.

Engage in Education and Training

Quality management depends upon education. In order to meet defined goals and targets, all participants must understand what is required at each stage. This requires that we establish effective two-way communications at the outset and maintain them throughout the project. Only in this way can a thorough understanding of unique requirements and problem areas be gained.

Implement Work

As data are developed, the advantages and disadvantages of alternative planning and design approaches must be studied and evaluated in relation to the client's functional and aesthetic goals. Potential solutions are presented and discussed and the client's comments carefully noted. Conclusions and recommendations are synthesized from these discussions and feedback.

Check the Effects of Implementation

This is the point at which it is important to make a special point of evaluating any exceptions to the usual process that may have occurred so that appropriate changes can be made. Another axiom is, "If standards are not revised in six months, it is proof that no one is seriously using them."°

There are two ways to determine whether or not work is being implemented smoothly:

1. *Check the causes.* Each process may be checked to determine if causal factors are clearly understood in accordance with established standards. The five categories of causal factors are: methods, measurements, materials, machines, and people.
2. *Check through the effects.* If the effects, or products (i.e., layouts, design concepts, reports), are not satisfactory, it is an indication that something unusual is happening in some of the processes.

Take Appropriate Action

The closing segment of the TQM cycle deals with the use of performance data to refine the process, removing or modifying the causal factors that have been responsible for any exceptions. In this way, quality assurance is maintained on two levels:

- Resolution of immediate issues on a short-term basis, as they may occur
- Ongoing refinement of the process through the removal of the causes for exceptions

In this way, the highest level of quality assurance can be maintained through constant feedback and fine-tuning of the planning, design, and production processes.

CLOSING WORDS

Understanding management issues and concerns is an essential part of effective programming. Organizational theory, quality control, and the knowledge of product lifecycles all need to be part of the analyst's vocabulary. The facility planner should have some perspective on how business practices evolved over the past century as a basis for understanding how a particular enterprise may be changing today. Organizational culture, management's plans for future structural changes, and new business systems being put in place are all forces that will shape the built environment. The support of all kinds of networks is a critical element, both in terms of person-to-person interaction and at a machine level.

Systematic management of the project at hand requires the development of a project plan and monitoring of the implementation process stage by stage. The principles of quality management should be applied to the process, and programming usually offers a natural point of entry into the quality cycle.

In the next chapter, we concern ourselves with the mechanics of putting together a facility program. We look at the kinds of information that must be collected and methods for carrying out the collection process. Different kinds of space need to be classified for allocation and accounting purposes. Certain types of areas, such as circulation, are not programmable except through the use of factors that must be added to the net space inventory.

Structuring the facility program requires that management's views on the organization hierarchy be thoroughly understood. Locations of numerous individual components of the organization must be charted in ways that will allow different layout options to be evaluated. The open network, a tree structure, is among the useful organizational mechanisms for modeling the facility-management database. We look at other related aspects of the process such as the development of space-assignment guidelines, use of a planning module, and the detailed analysis of adjacency relationships.

TIME CAPSULE   The Pantheon

The Romans did not invent the arch, but they are credited with its development. Earlier civilizations preferred to use the column and beam, but the arch and by extension the vault, together with concrete and the buttress, were all building technologies refined to a high art by Roman engineers. Revolving the arch through half a circle produces a dome, of which the Pantheon's was the largest for nearly four hundred years.° With an exterior diameter of more than 180 feet, it encloses over twenty-six thousand square feet, to which the portico adds another ten thousand.

The existing portico stands on the site of the original Pantheon, which was begun in 27 BCE by Marcus Vipsanius Agrippa, statesman and lifelong friend and son-in-law of Caesar Augustus. The building stands in an area known as the Campus Martius, a level district where the Tiber bends in what was once the center of Imperial Rome.

A round structure of considerable magnitude was built to the south of this first temple eight years later in 19 BCE. It has been theorized° that this rotunda, also developed by Agrippa, belonged to a complex of thermal baths. If this was the case, it explains why the exterior of the present rotunda is rather roughly finished, not in keeping with its exquisitely refined interior space. It would have been encapsulated within the existing *thermae* and therefore concealed.

One hundred years later, the baths and the original Pantheon were destroyed by fire. This happened during the time of the emperor Domitian, who restored them both. Once again the building was destroyed after being struck by lightning in the year 110 AD, during the reign of Trajan. From 120 to 124, the present rotunda was constructed by the emperor Hadrian along with its portico, which was rebuilt on the site of the original structure. The first temple faced south and had ten columns along its north elevation. Hadrian's north-facing portico is fronted by a row of eight gray Egyptian granite columns. Behind the first, third, sixth, and eighth columns of this octastyle colonnade are two columns of red granite, defining three aisles. Colored marbles were often used for the column shafts of Roman porticos, *unfluted* so as to emphasize their texture and veining. The Corinthian capitals are made of white Pentelic marble, the same as that used in the Parthenon. The widest central aisle leads to the entrance of the present rotunda.

As its name has come to signify, the Pantheon was dedicated to all the ancestral gods. The interior of the rotunda is lined with marble and porphyry, and around the perimeter are seven large recesses, four rectangular and three circular, which would once likely have contained statues of the planetary deities: the sun, the moon, Mercury, Venus, Mars, Jupiter, and Saturn. A statue of Julius Caesar also stood within, with statues of Augustus and Agrippa in each of the large recesses flanking the rotunda's entrance. The place of Zeus, sky god and father, is marked by the oculus at the top of the dome and the shaft of light that passes across its interior during the course of the day. The ancients believed that the sun was the eye of Zeus.

The base of the rotunda is a brick cylinder constructed in two tiers, buttressed by a third tier around the perimeter. These walls, more than twenty feet thick, are not solid, as evidenced by the recesses visible in the plan. All the forms of buttressing later used in medieval building are employed here, tied together with semi-domes at the top of each tier, with the singular difference being their great mass and the fact that they are largely hidden from view.° Further resistance against the outward thrust of the assembly is provided by a series of concrete rings that ascend two-thirds the height of the dome.

The construction of the dome itself shows a mastery of the use of concrete in accordance with its function in that the dome's thickness, as well as its density, decreases from bottom to top. From the twenty-foot thickness where it rises from the rotunda, the dome decreases to a thickness of four feet where it surrounds the unglazed twenty-seven-foot oculus, or eye, at its crown. Five rings of coffers encircle the interior, embellishing the dome with ornament while diminishing its weight. The circular steps that reinforce it can be clearly seen when looking at the exterior of the roof.

The Pantheon's universality is symbolically expressed in the austere simplicity of its geometry. Its rotunda is the archetype of a *centralized* building in which the major controlling axis is *vertical*. The Ramesseum, by contrast, epitomizes the *horizontal* plan in which the main entrance and the sanctuary are located at opposite ends of its longitudinal axis, about which the building is completely balanced. The Parthenon is likewise symmetrical about its longitudinal axis, with entrances at both ends into two separate chambers. As we have seen, the Parthenon has very strong symmetries about its transverse axis as well. Both ends of the building outside the square region defined by the overlap of its golden section rectangles (Figure TC2-2a) are identical except for the columns within the interior chambers.

Originally, one entered the Pantheon after mounting several steps and then traversing the eighty-five-foot-deep colonnaded portico. Over the centuries, the ground level surrounding the building has risen so the grade change is now minimal. Passing through the forty-three-foot-high opening with its ancient bronze doors, one enters the twelve-story-high rotunda, finished in green, white, yellowish, and reddish brown marbles. The principal source of illumination is sunlight from the oculus, which produces an atmosphere of great solemnity.

(continued)

The height of the dome's interior from the floor of the rotunda is precisely equal to its diameter; thus, it is symmetrical in three dimensions, containing a perfect sphere, as shown in Figure TC4-1. The curvature of the lower, cylindrical portion is relieved by the seven recesses into the exterior wall, which are interspersed with eight *pedimented* temple fronts. The entrance from the portico marks the position of the eighth recess. If one imagines a cube encapsulating the sphere, having one side centered on the entrance, the recesses are symmetrically placed on the axes and diagonals of a square base. Rotating the base of the cube $22\frac{1}{2}$ degrees ($\frac{1}{16}$) gives all the alignments of the projecting *tabernacles*. Hence, the rotunda is articulated into sixteenths on a basis of the square. Each of the recesses and projections has two Corinthian columns, which would total thirty-two but for the two missing at the entrance. The two columns opposite the entrance, set in front of the niche they flank, are larger than all the rest, giving emphasis to the north-south horizontal axis.

Between each pair of columns at the front of the four square recesses lie the corners of a square placed on the floor of the rotunda. A mirror copy of this square coincides with the outside of the portico (Figure TC4-1). The dimensions of the squares are such that a cube of similar height encloses the portico and is two-thirds the height of the rotunda. The base of the portico's pediment is aligned at half the height of this same cubic module. Within the rotunda, the pattern of the floor is composed of squares and circles within squares. The floor grid is laid out in checkerboard fashion, aligning with the square axes. A circle marks the exact center of the rotunda; and a grid of circles, viewed along the diagonal axes, aligns with the four recesses at their ends.

Vitruvius wrote, of course, about the relationships between the square, the circle, and the human form during Agrippa's time, as rendered in Leonardo da Vinci's drawing (Figure 1-4). All these geometric relationships provided the very foundation of Roman architecture—the radii of circles, the sides and diagonals of squares, and therefore the square root and powers of 2: *Ad Quadratum*, as we have seen.

There are numerous examples of round vaulted buildings throughout all Roman architecture, through the Renaissance in the works of Palladio and Serlio to the nineteenth century in Jefferson's Monticello and his library at the University of Virginia. The form continues to appear in ecclesiastical, academic, and civic buildings, wherever a recognizable symbol of authority and hopeful stability is required by church or state.

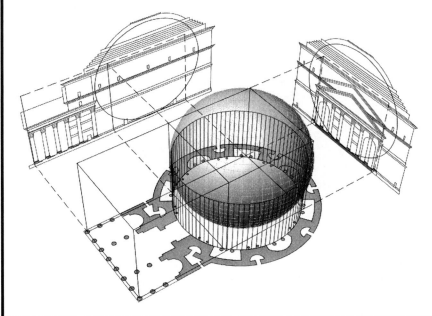

Figure TC4-1 Plan, Elevations, and Spherical Geometry of the Pantheon.

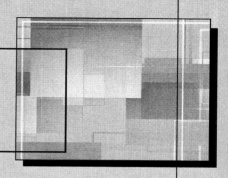

CHAPTER **6**

Workplace Programming

*Capitalism has brought about the emancipation of collective humanity
with respect to nature. But this collective humanity has itself taken on
with respect to the individual the oppressive function formerly exercised by nature. . . .
Question: can this emancipation, won by society, be transferred to the individual?*
—*Simone Weil*°

One hundred years ago, Max Weber° wrote of the causal factors that led to the rise of Gothic architecture in Western civilization. Weber observed that the pointed arch had been used elsewhere as decoration but that it was only in Europe during the Middle Ages that the structural principle of the great cathedrals was applied. The technical basis of Gothic architecture came from the Orient, a fact to which Weber gives due credit, but the convergence of the pointed arch with the cross-arched vault as a rational method of distributing compressive forces and enclosing great soaring spaces did not occur elsewhere. Weber credits the "rational, systematic, and specialized pursuit of science, with trained and specialized personnel," only as it has existed in the West, with the development of a culture in which all the "political, technical, and economic conditions of life" are absolutely dependent on organizations with specially trained "officials."

Literature "designed *only* for print and only possible through it" powered the Renaissance and the development of Western culture, including business and commerce, in the centuries that followed. As we discussed in Chapter 1, quoting Victor Hugo, the printing press was the singular development that led to the decline of architecture as a principal expression of "man in his different stages of development." Weber goes on to say that . . .

> . . . the State itself, in the sense of a political association with a rational, written constitution, rationally ordained law, and an administration bound to rational rules or laws, administered by trained officials, is known, in this combination of characteristics, only in the Occident, despite all other approaches to it. . . . And the same is true of the most fateful force in our modern life, capitalism.°

Weber specifically did *not* mean capitalism as "unlimited greed for gain," in the sense that led to Dos Passos' scathing characterization of F. W. Taylor. Rather,

> . . . capitalism is identical with the pursuit of profit, and forever *renewed* [emphasis Weber's] profit, by means of continuous, rational, capitalistic enterprise. For it must be so: in a wholly capitalistic order of society, an individual capitalistic enterprise which did not take advantage of its opportunities for profit-making would be doomed to extinction.°

BEYOND THEORY X AND THEORY Y

Up until the seventeenth and eighteenth centuries, the authority of organizations was largely vested in individuals, either by virtue of inherited status determined by precedent and custom or by charismatic qualities that set them apart. It was during this period that large bureaucratic organizations began to develop, first in government and then with the onset of the Industrial Revolution, in business and commerce.

Weber defined the bureaucracy-based organization as a rational-legal framework in which

- Official functions are bound by written rules setting out defined areas of specific competence—*specialization.*
- Functions are structured into offices having technical rules and standards for which training is required—*hierarchy.*
- Ownership and management are separated from administration and production, and written records of the administration's actions and methods are maintained—*depersonalization.*°

To many people's way of thinking, all this has a decidedly negative flavor to it; but in comparison with what had gone before, bureaucratization replaced the mercurial, irrational, and unquantifiable with a systematic method characterized by *professionalism.* This became the model for the early twentieth-century corporation, of which the greatest example was General Motors, and remains the way large segments of government function to this day.

The second half of the twentieth century brought increased awareness and sensitivity to the fact that people are not cogs in a machine. We now are able to recognize that bureaucracy's assets were also its greatest liabilities.

- Officials are circumscribed by written definitions of their authority, which limits their ability to be responsive to changing conditions—*depersonalization.*
- Their methods of processing information and filtering it on its way to the top makes them unwieldy and slow to respond—*hierarchy.*
- The emphasis on defined procedures makes handling special circumstances difficult because individuals are geared to treat all people and situations the same—*specialization.*

Since the 1970s, there have been many forces driving the changes that are taking place in the modern enterprise, some of which are discussed in Chapter 5. Accelerating technological growth not only leads the list but is the primary causal factor behind many of the others:°

- Higher customer expectations and demand for higher quality and better service at a lower price.
- Loss of traditional loyalties as a result of people becoming simultaneously the most valuable and most expendable assets, especially in the service sectors.
- An influx of newcomers to the rapidly growing information technology industry with a corresponding decrease in the value of experience, as accelerated change lessens its importance for the future.
- Deregulation of industries such as energy, telecommunications, financial services, and transportation. Enterprises classed as "utilities" tended to become monopolistic before technology provided cost-effective means for open competition as well as greater transparency.
- Changes in the nature of command and control structures as workforces become more empowered, mobile, and geographically dispersed, while organizations become less obviously structured.

The process of facility programming and planning is much more complex as we enter the twenty-first century than it was even thirty years ago. It used to be possible to walk into a corporate headquarters and ask for an organization chart that identified each level of management, clearly portrayed as an inverted tree starting with the president and/or chairman at the top and branching out into divisions, departments, and so on. Now we use descriptions such as chief executive officer; chief operating officer; chief financial officer; and, perhaps most significant, chief *information* officer.° These are functional titles that may overlap and in some cases replace traditional departments.

No two organizations are exactly alike, and it is difficult to generalize about common factors shared even by those in similar fields of endeavor. In the last quarter of the twentieth century, managers began to speak of organizations in which reporting, communication, and team relationships much more resembled our picture of the open network (Figure 5-5) than the traditional hierarchy. New organizational paradigms borrow terminology from scientific disciplines such as biology and mathematics, and we speak of organizations that are *amoebic* and *fractal*. One of the earliest pieces of software for prioritizing physical-adjacency relationships among departments was based on a mathematical, vector-analysis technique developed by a biologist to quantify the degrees of separation among animal species.°

It may well be that we are on the verge of a cultural inflection point as far-reaching as the one promulgated by the invention of printing. Six decades passed between Gutenberg printing the Bible from movable type and Luther nailing his theses to the door; but the rest, as they say, is history. The convergence of mechanical and electrical relays with binary arithmetic that resulted in the first digital computer occurred in 1937, and *hypertext* has since become commonplace.

The Facility Program and the Balance Sheet

In the next section, we describe the characteristics of several types of organization for which large-scale facility programs are often prepared and which are likely to institute on-going facility-management systems. First, though, let us underscore one of the themes of the last chapter, *globalization*, by listing the categories a major business magazine used in ranking the two thousand largest public companies worldwide. It is not without significance that 2004 was the first time that the magazine's ranking was global in scope, having only included American companies since its inception thirty-five years previously. Fully half of the ten types of organization we discuss presently, for which office space is a major consideration, fit within the categories listed in Table 6-1.

A key document in the modern corporation's financial statement is a *balance sheet*, a two-column tabulation of the organization's assets and liabilities. It is so named because the balance sheet has a standardized format in which the totals at the bottom of each column are always equal. The equalizing factor is called *stockholders' equity*, which represents ownership's current stake in the company.

The relative proportions of the components of both columns, including the bottom line, change incrementally. This reflects Max Weber's image of the enterprise over time in terms of "continually renewed profit." The balance sheet is a slice in time, a snapshot of the condition of the company at a given moment and, as such, is representative of its success or failure.

Table 6-1   Classifications of the Two Thousand Largest Public Corporations Worldwide°

Aerospace & Defense	Household & Personal Products
Banking	Insurance
Business Services & Supplies	Materials
Capital Goods	Media
Chemicals	Oil & Gas Operations
Conglomerates	Retailing
Construction	Semiconductors
Consumer Durables	Software & Services
Diversified Financials	Technology Hardware & Equipment
Food, Drink, & Tobacco	Telecommunications Services
Food Markets	Trading Companies
Health-Care Equipment & Services	Transportation
Hotels, Restaurants, & Leisure	Utilities

Similarly, the facility program is a snapshot of an organization's physical space requirements at a specific point in time. The act of making a physical layout requires some concrete expression of organizational structure, representing whatever form may actually exist. If the organization is a traditional one with a clearly defined hierarchy, the programming process is simpler. Anticipated personnel and space requirements can be organized and tabulated according to the formalized division/department structure. Then the necessary physical-adjacency relationships among the various elements—offices and support spaces—can be analyzed.

Performing a space-requirements analysis becomes more complex in direct proportion to the level of chaos (in the fractal sense°) in a particular organizational structure. An appropriate physical expression for one of today's business organizations can be an elusive goal when the company is in a constant state of change. An extreme example is the fashion industry, in which the term *anarchical network* has been used to describe the characteristic structure or, rather, the lack thereof. In "creative" businesses, a chaotic nonstructure can allow for unfettered experimentation at the risk of being vulnerable to more disciplined competition. How well the facility suits its defined purpose may have a significant effect on the success of the enterprise.

This realization, along with the considerable costs associated with physical change, is largely the reason that facility management has become an integral part of the enterprise resource-planning and asset-management process. We are charged with applying a rational planning process to species of organizations that Weber would scarcely have recognized.

Types of Organizations

Let us attempt to generally describe some of the physical (and cultural) features that characterize space in the facilities of ten different types of organizations. Although the last three organization types included are not offices, per se, these kinds of facilities—cultural institutions; schools, colleges, and universities; and health-care facilities—may well be the subject of long-range, facility-planning analysis or programming studies; and facility planning is an important part of the design process. Another level of complexity for these organizations is that they most often need to be planned to accommodate the public, which makes them subject to special scrutiny by code authorities and accrediting organizations. These types of institutions are often users of CAFM technology, both to manage their facilities and to provide necessary documentation to governing bodies.

Industrial Corporations

Offices comprise a relatively small part of the facility, which may largely consist of factories and production plants; research laboratories; and shipping, receiving, and maintenance areas. The organizational structure may tend toward the more traditional type, with clearly defined areas for marketing and sales; human resources and training; finance; information technology (IT) and, of course, executive offices and a boardroom if the location is the company's headquarters.

Corporate headquarters are often located in major cities that are also financial centers, but many are found in campus settings where they may be combined with one of the company's primary manufacturing or research and development (R&D) centers. Depending on the size of the operation and the company's sensitivity to economic factors, security, and other issues, headquarters may be in suburban locations or office parks.

New technology is often a matter of vital concern; and periodic mergers, acquisitions, divestitures, and restructurings prompt reconfigurations involving offices and plants as well. The headquarters will want to have a character reflecting its self-image and, if it is an established concern, appropriate ways of displaying its history as well as its products.

High-Technology Companies

Most often found in extraurban locations, the work atmosphere is often fast-paced, if not frenetic, but informal. Strong physical distinctions between general office and executive areas are the exception. Work areas tend to be configured to support constantly changing project teams.

The technology that is its primary focus is a prominent feature of the company's infrastructure, and customer showcase as well as training areas may occupy a significant portion of the facility. Personnel are highly mobile, particularly those involved with sales, marketing, training, and technical support. They may divide their time between the office, customer sites, home offices, and work in transit. Virtual or part-time offices may be provided using *hot-desk* or *hoteling* schemes, in which the same physical facilities are used by different people at different times. Informal teamwork and recreation areas are often significant amenities to a fast-moving but usually dedicated workforce.

Financial Institutions

Large retail and commercial banks, investment brokers, insurance companies, and the like are often located in urban financial centers, although the companies may maintain numerous satellite offices linked through secure computer networks. Financial services firms in today's business environment are completely immersed in computer technology, in which lies their ability to maintain the velocity of money.

Banks catering to both the retail and commercial markets maintain customer-contact *platform* areas and meeting rooms where the design treatment will usually convey a sense of financial solidity and history. These areas are supported by various kinds of back-shop facilities for transaction processing (electronic and check); wire-transfer departments; and electronic security-trading operations.

Insurance companies once were veritable paper factories, and, in some ways, they still are; but their use of electronic data-processing systems has increased enormously. Incoming claims are verified and processed electronically along with payments to policyholders.

Consulting and Accounting Firms

Like the high-technology companies, the largest of the international accounting firms develop, use, and promote electronic processing and analysis of financial and other data. They are often headquartered in downtown corporate and financial centers but often maintain multiple facilities in suburban locations as well.

Unlike their high-tech counterparts, however, offices tend to be more highly structured in terms of expressing hierarchy. In that sense, they more closely resemble the traditional corporation in presenting visual cues as to where a member of the firm stands in her career path, advancing from trainee to auditor or consultant to partner. As the companies' stock-in-trade often involves creating new methods of doing business, consulting firms are usually precedent breakers in their extensive use of systems furniture and sharing of workstations among junior personnel who are in the field a great deal (hot-desking and hoteling).

A recent trend has been the separation of accounting and consultancy operations into independent companies. This is often ascribed to conflicts of interest arising between the requirements of the traditional accounting and auditing functions on the one side, the strategic-planning focus on the other, and the development of new systems and ways of doing business. There were some spectacular failures as we crossed into this new millennium, but the TQM cycle appears to be operating as creative minds continue to look for effective ways to respond to change.

Advertising and Public Relations Firms

Again, these firms are comprised of creative minds but with a focus on sales, marketing, and image making. Offices are often located outside the central business district, where space is cheaper, but near the urban centers, where their corporate clients are. The interior environment is usually light and dynamic with a good deal of freedom given for individual expression.

The image of the firm is often strongly projected from the front entrance throughout reception areas, conference rooms, and presentation rooms. In these areas, high-tech graphics are integral to the operation. The firm's creative output is usually produced in flexible studio areas designed to support changing teams and fluid working structures.

There are also more traditional support areas dedicated to the "business" operations of the company—financial and IT; but, in general, there is little interest in an expression of hierarchy, with the possible exception of a few senior members of the firm.

Law Firms

Legal offices have traditionally been—and remain—paper factories! Private offices remain the order of the day, especially for partners and associates. As in accounting firms, office size and quality of furnishings are usually a reflection of a person's status in the firm. Tastefully appointed meeting rooms for clients are an important part of the facility program, along with a legal library and file rooms. File cabinets and book shelving are often built in to corridor walls and shared support spaces.

Group offices for paralegals and clerical support staff are often found in close association with file rooms and research areas. Open plan spaces with systems furniture, together with team workrooms for special projects, are becoming more frequent in the legal departments of large corporations, but expanses of open office areas are rarely found.

Large law firms are most commonly located in urban corporate and financial centers and, of course, where there are substantial enclaves of governmental offices and judicial facilities.

Governmental Organizations

If there is an office archetype that reflects the last bastion of bureaucracy as described by Weber, it is to be found in our centers of government. Federal offices are managed in the United States by the General Services Administration (GSA), which maintains uniform standards for space allocation and assignment. Offices of varying sizes and types are assigned according to General Service (GS) grade except where special needs can be amply demonstrated. The sizes and types of furnishings that may be used are also regulated by the GSA, which is responsible for procurement and relocation management.

Federal offices are found within centralized areas in large cities, and agencies that directly serve the public have satellite offices in smaller cities and towns. These include many branches of the U.S. Treasury (Internal Revenue Service—IRS; Alcohol, Tobacco & Firearms—ATF), the FBI, Housing & Urban Development—HUD, and so on. State and local government offices are located in the areas that they service.

Many older government buildings were designed along classical lines, but among the newer ones are many models of modern architecture and functional design. Public accountability usually dictates that furniture and furnishings are functional in nature, without ostentation. Many manufacturers' lines of system furniture are found on federal purchasing schedules.

A great deal of space remains devoted to paper records storage in government agencies of all types and at all levels. Electronic data-retrieval systems continue to gain widespread acceptance, even in the IRS where tax forms are now routinely processed electronically. The GSA is heavily committed to computer-aided facilities management (CAFM) in all its regions.

Cultural Institutions

This is a broad classification of facilities, most of which fall into what building codes categorize as "assembly spaces." Performing arts facilities such as theaters constitute one subtype, while galleries, libraries, and museums are classed under the heading of *amusement, recreation, and worship* (which also includes churches and courtrooms). Most all cultural institutions must be publicly accessible. Their staffs generally fall into three categories: those who interface with the public, those who manage the institution, and those who are responsible for the maintenance of the facility itself and its contents.

Offices as such are generally provided for the management personnel and their immediate staffs, which may handle the accounting, public relations, and other such functions. Whether the management and administrative office are in a contiguous area or scattered often depends on the age of the building and how the institution has grown. Offices may be found in some rather *inventive* locations.

The facility and maintenance staff often operate out of conservatorial or shop space adjacent to areas for storage, shipping, and receiving. The people who have the most contact with the public are usually found in the performing, exhibit, or study areas, which are the principal focus of the organization's activities. Storage of paper media (and the lack of space for same) may be a major issue. In such facilities, and where artwork is stored, environmental control is essential.

Schools, Colleges, and Universities

Classrooms, laboratories, and their associated support areas are, of course, the principal space types here, along with assembly, exhibition, recreation spaces, and other specialized ancillary facilities. Office spaces, including general administrative areas, are usually organized on a departmental basis, and private offices are the norm. Hierarchical and territorial identification are important for senior academic personnel. Junior lecturers and teaching and research assistants may utilize shared offices or open plan areas. Space-allocation standards are often maintained following the *Postsecondary Education Facilities Inventory Classification (PEFIC) Manual guidelines.*° These guidelines are discussed in Chapter 9.

Book shelving and filing are ubiquitous as are computer networks and communications infrastructure. The universities were, after all, the second heaviest users of the Internet (after the defense agencies). It has been said that the use of technology in the university environment is generally five-plus years ahead of industry.

Health-Care Facilities

Building codes treat health-care facilities under the "institutional" classification, which describes them as being planned for occupants with physical limitations due to age or health. These include hospitals, nursing homes, and mental health centers.

Offices as such are a secondary consideration except for areas concerned with administration and the maintenance of the facility. Departments are extensively compartmentalized for hygienic and security reasons as well as for fire and smoke protection. Areas having specialized or hazardous uses are subject to other special separation requirements. Maintenance of the technical infrastructure, which includes process gas piping and specialized mechanical and electrical systems, is a crucial part of the ongoing facility-management process.

Performance standards for hospitals and other health-care-delivery institutions are administered by the Joint Commission on Accreditation of Healthcare Organizations (JCAHO). This organization conducts full surveys of each facility every three years as well as periodic performance reviews that are often unannounced. The survey emphasis is heavily upon what are termed *tracer* activities in which the field surveyors look at how care is actually being delivered. One important aspect is a building evaluation in terms of life-safety-code compliance, for which the documentation provided by an in-place facility-management system can be invaluable.

THE PROGRAMMING PROCESS

In the last chapter, we discussed the importance of understanding the client's management objectives. Understanding the particular nature of the organization and its structure is equally important, as both give meaningful direction and form to the facility-requirements analysis. Most if not all data gathering and refinement is done through face-to-face discussions or interviews at the client's workplaces, supplemented by the use of survey worksheets to gather preliminary data.

How is this process begun? The first step is to establish—with senior management and those who are to be directly involved in the project on a day-to-day basis—overall goals and objectives. Whatever ongoing processes that are in place related to administering the existing facilities need to be understood at the outset. A working project plan and schedule need to be developed, and the plan should clearly outline project control and communication procedures.

Whether or not there are manual or computerized facility-management procedures already established, it is necessary to construct as clear a picture as possible of the current organizational structures. As we mentioned initially, this may not be as simple as asking for an organization chart. These may only exist in fragmented form, if at all. There may be individual, departmental charts available, and there may be a diagram of the top-level hierarchy; but they may not always agree, and there will be inevitable gaps. Often, this is due to their having been prepared at different times.

There are other potential sources of organizational data that can be extremely helpful. Many companies are organized for accounting purposes by *profit* or *cost centers*. These reporting units often (but not always) relate to functional units within the organization that have a physical existence and location. A computerized ERP system will have an established alphanumeric coding system to represent these accounting entities.

Even telephone directories can be a useful source of organizational information. Such listings often contain information related to physical location and job titles along with individual names and telephone numbers. Human resources, in an organization of any size, usually can provide up-to-date listings of personnel with useful data such as job or pay grade classifications. The latter are occasionally tied in to established standards for area allocation, as, for example, GSA grades in the case of the federal government.

All other available documentation that will help to represent the organization as it currently exists should be sought out at this time, certainly including

- Plans of the existing facilities at some scale, which should be clearly marked up to show existing departmental locations and space utilization. These may be cross-referenced with profit center codes if they are found to be at least partially representative of the functional operating structure.
- Historical background information relating to past growth and consulting studies in process or recently completed, as they may affect organizational development or the use of facilities, present and future. If there has recently been an organizational restructuring or if one is in the offing, the responsible personnel or consultants should be an integral part of the initial data-gathering effort.

Generating Facility-Requirements Data

One of the important objectives of the initial familiarization and data-gathering activities is to identify key individuals who will be expected to provide detailed requirements information pertaining to the functional areas for which they are responsible. Most of these will be found in that fuzzy region that goes by the name of *middle management*.

Such individuals are seldom found at the same level in the organizational hierarchy. The selection process should seek out managers and supervisors who are close enough to the day-to-day operations of their organizational units that they can supply detailed information in several categories including the following:

- Numbers of people, starting from those on board at present and projected over some defined time frame, according to descriptive job function.
- What kinds of auxiliary facilities, furnishings, equipment, and machines are needed in order for the people in this group to perform their primary functions.
- How people and support spaces should ideally be grouped, again according to function.
- Finally, what physical-adjacency relationships are important, first among the elements of their own groups and then between their areas and functions elsewhere in the organization.

Figure 6-1 diagrammatically shows a structure that can be used to develop ten specific kinds of data to work with. This is the format of the space-requirements worksheet detailed in Appendix B.

The upper part of the worksheet in Figure 6-1, in which column 2 is labeled "Work Space," inclusive of columns 1, 2, and 4 through 8, is directed to office-type spaces that are assigned to specific individuals, most often on a full-time basis. Section 3, at the bottom and to the right as far as column 8, is directed to "support" or *auxiliary facilities*, that are used in common by any number of people within the group.

Classifying Programmable Area

Both office-type areas and support areas fall within the classification that ASTM/IFMA° (one of several area measurement standards that are discussed in Chapter 9) calls *facility assignable area*. This standard defines *facility usable area* as "the floor area of a facility that can be assigned to occupant groups."° For our present purposes, we could use the term *programmable net area*, which includes either usable space that is assigned to and occupied

Figure 6-1   Programming Information-Gathering Structure.

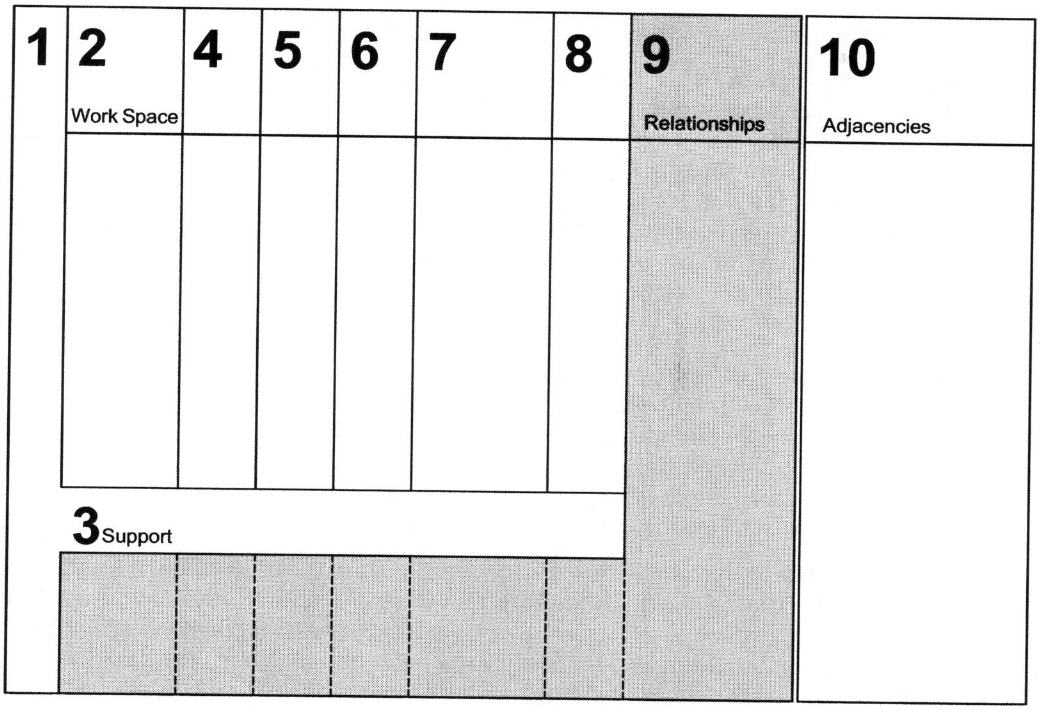

by specific individuals (*work space*) or space allocated to specific *activities* but used by a number of people or departments (*support*).

It is necessary to be precise when talking about space classifications, for reasons that become abundantly clear in Chapter 9. The very last section of the ASTM/IFMA document identifies a category of space it calls *common support areas*, which it defines as follows:

Facility assignable area includes the area devoted to common support services. *Common support area* is the portion of the facility usable area not attributed to any one occupant but provides support for several or all occupant groups. Examples of common support areas are cafeterias, vending areas, auditoriums, fitness facilities, building mailrooms and first aid rooms. These may be separately identified as a sub-category of facility assignable area if required.°

Common support area is sometimes referred to as *nonassignable*, because it cannot be attributed to a single department or "occupant group." If we use the word *occupiable* to mean that a space is assigned to a person on an exclusive basis, we then can classify programmable net area into three categories, the last two of which are usually within the scope of the worksheet structure in Figure 6-1 (Appendix Figure B-1 and Figure B-3):

• Common support area, or *nonassignable area*, which must be apportioned or prorated to the floor, building, or site for cost-allocation or *charge-back* purposes.
• Assignable but nonoccupiable area—*support space* belonging to a particular organizational unit or tenant, which is not apportioned or prorated.
• Assignable and occupiable area—*all office spaces* including private offices with full-height walls and doors, semienclosed areas including cubicles, and open workstations assigned to specific individuals on a full-time basis.

To recap: *Work space* is assignable and occupiable and is associated with the descriptive job functions or position descriptions listed in column 2 of the worksheet's upper part. *Support* is assignable but nonoccupiable shared space that belongs to that organizational unit. These spaces are also listed in column 2, but in the lower part of the worksheet labeled section 3. *Common support spaces* are accounted for separately according to where they are classified by area of responsibility. Essentially, they are treated as independent departments that can be located wherever necessary, as defined by physical-adjacency priorities, to maintain accessibility by the organizational units that need them.

Circulation: Nonprogrammable Area

There are two formal categories of circulation space called *primary* and *secondary*. All the space-measurements standards define a hierarchy of classifications used, beginning with gross area and working inward (or downward) toward secondary circulation. *Secondary circulation* is essentially the space that remains after all other spaces, enclosed or otherwise, have been classified. It therefore falls at the very bottom of the space hierarchy. In developing block layouts, secondary circulation is included within each block of space representing a department or other defined grouping of programmed spaces.

The ASTM standard defines primary circulation as follows:

> *Primary circulation* area is the portion of the building that is a public corridor or is required for access by all occupants on a floor to stairs, elevators, toilet rooms, building entrances, and tenant entry points on multi-tenant floors. . . . [It] does not necessarily include all circulation required for life safety access and egress. However, if dedicated circulation required for egress can serve no normal secondary circulation function, it [is] considered primary circulation.°

It should be evident that the way in which circulation may be classified is somewhat dependent on its location. (Issues related to life-safety access and egress are addressed in Chapter 13 and Chapter 14.) For programming purposes, however, because an actual plan with physical walls and dividing elements is yet to be developed, a tare factor needs to be added to the *programmable net area* to account for all circulation. This factor normally includes all secondary circulation but may not account for all passageways in the primary class if public corridors are defined. Most often this is the case where a floor is configured for multitenant use where permanent corridors are part of floor common area. The *circulation factor* is thus an estimate of the amount of all corridors or aisles that will be required as part of the total usable area. It is applied to the net area as a conversion factor and usually expressed as a percentage. By convention, the circulation factor is defined with respect to gross so the calculation is made as follows:

$$\text{Net area} / (1 - \text{circulation factor}) = \text{gross area}$$

Analyzing the Data Worksheet

As the people (by job description) and auxiliary spaces are listed in column 2 of the worksheet, they should be grouped in a way that will indicate which positions ought to be located together. These may be formal subunits that relate to the organization structure or informal task groupings for which physical-adjacency relationships are important as a function of communication and work flow. These clusters of related work spaces and support areas should be bracketed and identified in column 1.

Figure 6-2 shows an illustration of two such subgroups and part of a third, which have been marked up to the left of column 2. In this example, which is taken from a GSA project, column 2—"Positions"—has been modified to include GS grade designations and whether each position listed is supervisory or not. These are highly relevant qualifications for making area assignments. Job grade classifications, where they exist, are helpful in developing space-allocation guidelines. If this information is not available from the managers and supervisors supplying the facility-requirements data, it may be available from the human resources or payroll departments.

In addition to the bracketing and identification of work space and support area clusters, Figure 6-2 also shows three sets of alphanumeric codes that uniquely identify each cluster and line entry. These "Level" and "Sequence" codes are used together to structure the requirements data when they are tabulated. Another code can be seen in the shaded column titled "SAR Key" (space-allocation reference), which can be used to calculate the actual area assignments based upon furniture and equipment requirements. The shaded

Figure 6-2   Programming Data-Analysis Worksheet.

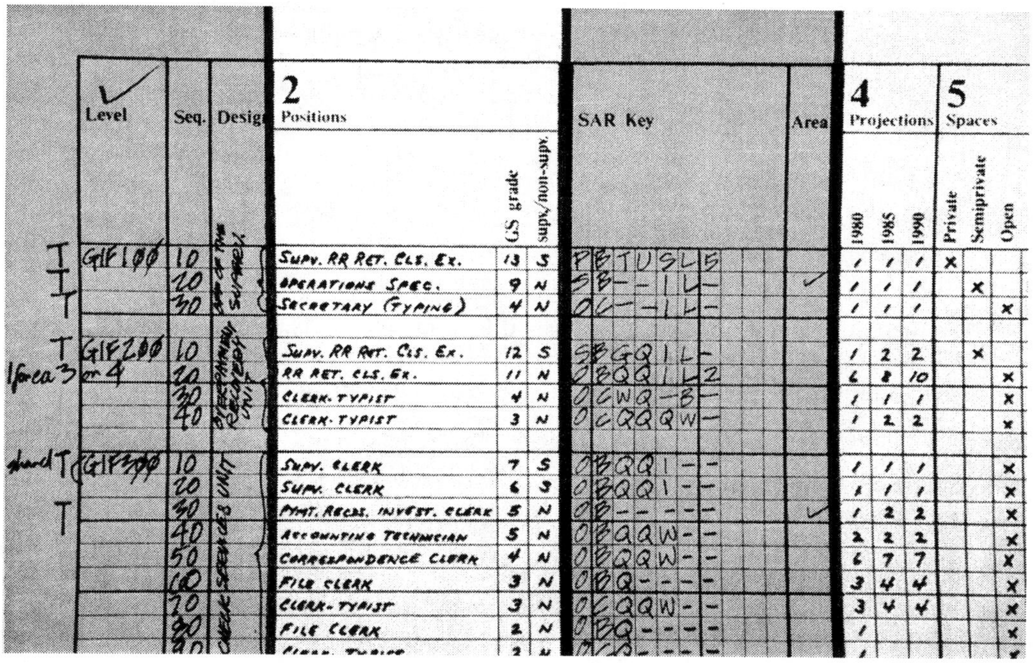

columns as shown are overlays to the actual worksheet that are used during the analysis process. We speak more of the SAR key shortly.

Referring again to the worksheet in Figure 6-1 (Appendix Figure B-2), column 4 through Column 8 provide input that will be the basis of assigning area to every workstation and support space: "Projections," "Spaces," "Furnishings," "Storage," and "Machines." Column 4 and column 5 can be seen at the right of the worksheet overlay in Figure 6-2. Each column addresses a specific type of information or physical component:

- *Column 4: Projections.* In the case of *work space,* the numbers of people currently in each position are shown, which are then projected forward over time. Projections are made beginning with the present condition, defined as positions either filled or budgeted as of a defined base year. Then manpower requirements are estimated for future time periods. If, for example, short- and long-range projections are made, the short-term target years may define a move-in condition two or three years hence. The long-term figures will then be targeted three to five years beyond that on the assumption that space will be planned to accommodate some growth after initial occupancy. *Support* space requirements are projected in a similar manner, except that the quantities are not directly related to headcount. Longer-term projections may also be developed, bearing in mind that this becomes more difficult (and less accurate) over time, especially at the level of detail required for space-layout purposes.
- *Column 5: Spaces.* At the initial stage of data gathering, *spaces* indicate a relative level of enclosure or privacy, from fully enclosed (a private office with a door) to fully open. When the stated requirements are analyzed and evaluated, privacy requirements provide part of the basis for developing assignment guidelines.
- *Column 6: Furnishings.* Part of the strategy of generating facility-requirements data in this manner is to avoid directly asking anyone how much space he or she needs. This focuses attention on the necessary components of the workstation or support area and helps to separate needs from wants. Column 6 deals with individual items of office furniture, plus seating that is needed for each position listed. In the case of support areas such as conference or training areas, the amount and type of seating are the primary considerations. A menu of standard furniture components is usually supplied from which a selection can be made (see Appendix Figure B-2).

- *Column 7: Storage.* Required numbers of drawers of filing, lineal feet of shelving, and other special storage requirements are estimated. Once again, the focus is upon quantities and type of material being stored as a determinant of the requirement for space.
- *Column 8: Machines.* In most cases, this section is concerned with machines that require floor area or special electrical, mechanical, or telecommunications services. In facilities in which there are very complex support areas, which can encompass anything from a suite of computing equipment to a technical library to an operating room, the specialized requirements are obviously beyond the scope of any worksheet. In such cases, it is sufficient to identify the fact that the areas exist and to identify to which department they belong. The detailed planning requirements then need to be further defined through visits to existing facilities and work sessions with staff.

The two final sections of the worksheet address proximity relationships. Column 9, titled "Relationships" (once again referring to Figure 6-1), is concerned with establishing the locations of support elements in relation to the office areas represented on each worksheet. Unless there are specific requirements to the contrary, it is usually assumed that all the components of a group represented on a single worksheet would be colocated, so prioritizing these adjacencies is not a major consideration. Column 9 provides a checklist that indicates which support spaces are most often utilized by specific individuals occupying work space within the group.

Column 10, "Adjacencies," is designed to generate input as to proximity relationships that should be established between *this* group and other groups within the organization. The manager completing the worksheet is asked to list other departments or physical facilities with which his or her group needs to have direct contact, then indicate the importance of maintaining direct physical adjacency. Such proximity relationships are prioritized using a numeric rating scale of *1* to *5*. A priority rating of *1* means that close proximity is essential, whereas a rating at the other end of the scale indicates that closeness is *not* desirable. Inverse or negative priorities often indicate past or current issues such as quiet or high-security functions being too close to high-traffic or noisy public areas. This kind of input at the worksheet level provides a checklist of important questions that need to be followed up later on, when such details are discussed in planning work sessions.

Structuring the Data

There are many ways in which facility data can be structured; and, at the programming stage, a computerized database is particularly useful because the information can be examined from different points of view simultaneously. All large-scale CAFM systems provide two principal organizational hierarchies: *management* and *location*. Names of the typical levels for both are listed in Table 6-2.

Not all projects will entail a global location hierarchy as shown in the left column of the table. A corporate or university campus may have several complexes of individual

Table 6-2 Facility Organizational Hierarchies

Location	Management
Country	Group
Region	Division
State	Department
County	Section
City	Unit
Site	Subunit
Complex	
Building	
Floor	

buildings, though, that need to be accounted for individually and as groups for accounting chargeback purposes. The costs of many spaces used in common, those in the *support* category, often need to be allocated among many different users. This may need to be done at different levels (building, complex, etc.) depending on their purpose and use. A building that functions primarily as a conference and training center, for instance, may have its costs allocated to departments across an entire site.

A third potentially informative data structure is *space type*. Being able to sort the information by privacy requirement (private, semienclosed, or open, for example) and to look at similarities and differences in furniture and equipment is an essential part of developing standards. This is one of the functions of the space-allocation reference (SAR key) mentioned earlier.

Numerous patterns are implicit in the raw data at an early stage of analysis. Support spaces usually require a disproportionate amount of attention in direct relation to their complexity. Planning support functions such as conventional filing or storage rooms and libraries is largely a function of accommodating a determinate quantity of material. Much more complex, however, are specialty areas such as classrooms, laboratories, IT machine areas, and certainly medical facilities.

The most that a worksheet line entry can do with regard to support spaces is identify that specific areas exist so that their needs can be analyzed in greater detail. The area of a conference and training center is determined by the size, frequency, and duration of events, as well as by the level of usage by visitors together with regular personnel. Areas involving specialized equipment and support infrastructure (mechanical, plumbing, electrical, telecommunications, etc.) need analysis commensurate with their intricacy. Checking through the program requirements by *space type* ensures that none of these components is ignored and that each item in the space inventory is given the required level of attention.

ORGANIZATIONAL MECHANISMS

When we spoke of project management in the last chapter, we used the picture of two trees to represent the idea of a closed network (Figure 5-6). A single tree—an open network—is one of the most fundamental data structures. A graph of a tree, characteristic of the traditional organization chart, is illustrated in Figure 6-3. We can represent the same arrangement mathematically using parentheses, as follows:

$$(A(B\ (E(I\ J\ K))\ (D(I)\ ))\ (C\ (F(Y))\ (G(V\ W\ X))\ (H(Z))\ )))°$$

Figure 6-3   Representation of a Tree Structure. °

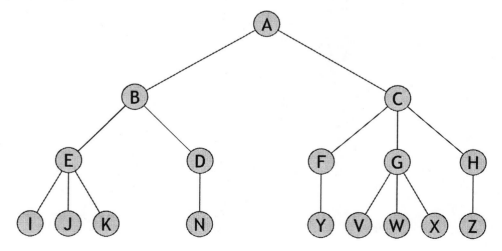

or in an indented outline format,

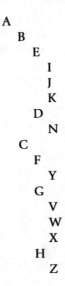

which is the format we see in the *folders* window of a file-management tool such as Windows Explorer. The Explorer model uses plus and minus signs adjacent to the folder name at each level to allow us to open and close each folder, thereby *drilling down* to the file we are interested in.

Another way of looking at the tree is as nested sets of objects, or containers. This is an appropriate image to use with respect to space. Figure 6-4 shows an organizational hierarchy of numbered containers. The position of a number represents an organizational unit's place within the overall scheme, whether it is a department within a group or a floor within a building. Common denominators among space types can be resolved by using a similar structure to represent assortments of furniture and equipment allocated to each workstation or support area.

Figure 6-4   Organizational Hierarchy.

Codes

Codes of one sort or another can appear *cryptic* from a human point of view but are necessary in representing a structure that does not lend itself to the essentially arbitrary sequence of the alphabet. Listing names alphabetically is useful in compiling a telephone directory, but even in that case reversing the common name order works best. Telephone numbers, from the exchange switch's point of view, are simply codes that the communication system can interpret. The switching mechanism has no use for or knowledge of user names, but we do. The directory just serves as a convenient cross-reference.

Transmission Control Protocol/Internet Protocol—*TCP/IP*—uses a numeric addressing system to uniquely identify nodes in the Internet. TCP/IP addresses are of the form *63.240.93.147*. The code numbers are cryptic to be sure, but the domain-name system provides an automatic lookup function through the uniform resource locator—*URL*—that supports human-language names such as *phptr.com*. In this sense, the Internet provides its own built-in cross-reference, which is also its means of organizing pages of data that exist as files on millions of servers.

Alphabetically listing department names gives us very little information about how functions are related, so we must resort to code designations to organize facility data in a meaningful way. Because one of our principal interests is looking at the data from different perspectives, we can take advantage of such data structures to help us understand and analyze relationships as well as the quantitative information.

Enterprise resource-planning applications use alphanumeric coding systems to represent cost or profit centers and other significant entities for accounting and forecasting purposes. Often, it is found that these classification codes also relate to space utilization; and, in such cases, it may be desirable to adopt these codes for space-planning purposes. Such coding systems, as useful as they may be in controlling the sequencing of information, require some initial familiarization on the part of the user. Adapting elements of a structure already in place can make the system of organization much more immediately apparent to the user.

Management Structure

The container paradigm in Figure 6-4 is appropriate for the *level* and *sequence* codes discussed earlier, examples of which appear in Figure 6-2. The *level* code G1F in Figure 6-2 represents *Department F* within *Division 1*, which belongs to *Group G*. Using the nomenclature from Table 6-2, G1F1, G1F2, and G1F3 each represent *sections* within *Department G1F*.

A worksheet set up this way with *Microsoft Excel*, for example, can automatically generate totals for *headcount* and *area* using the *Subtotals* selection from the *Data* pull-down menu, as shown in Figure 6-5. At each change in the *code*, Excel's *Sum* function is used to generate totals for any fields selected in the *Add subtotal* list box. In a database application, each character in the *level* code can be linked to a separate table in order to produce a fully

Figure 6-5   Excel Dialogue for Data Subtotals.

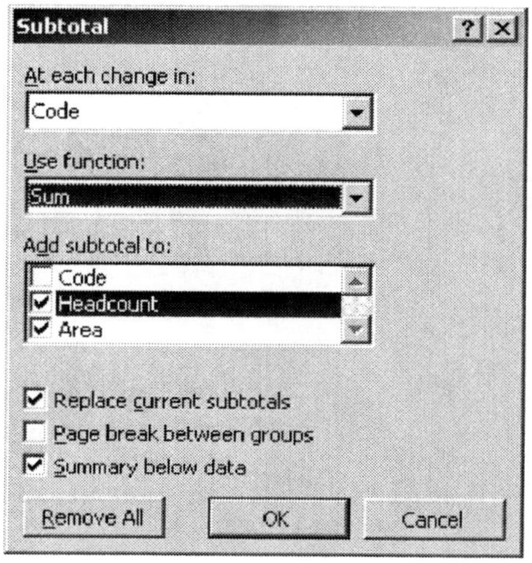

detailed report with subtotals at every level of the organizational network. A facility-programming report application of this kind, using *Microsoft Access*, is introduced in the next chapter.

Sequence codes, again referring to the example in Figure 6-2, simply control the order of line entries within each organizational unit. Work-space line entries can be given sequence codes beginning with 10 and incremented by 5 or 10 so that additional lines can be inserted in the proper order. Support areas are listed after work spaces if they are similarly coded using a higher number sequence, starting at 100 for instance.

The combination of *level* and *sequence* codes gives each line entry a unique address that identifies the item as well as its position within the management structure. *Level* codes can usually be assigned on a basis of whose responsibility area each organization unit belongs to. Common names for the different levels in a particular organization may vary from those listed in Table 6-2, so whatever designations are most meaningful to client and planner alike should be used on all reports.

Depending on the formality of the actual organization structure, the management hierarchy represented by the facility program may or may not reflect the official one at all levels. This is particularly the case with a loosely structured enterprise or one with a matrix organization. Typically, the level codes will more closely represent formal hierarchical lines of authority toward the top (group, division, department) and distinctions that are less formal—work clusters, project or task groups—toward the bottom (section, unit, subunit). The important things are that significant physical groupings are represented for space-planning purposes and that the reporting structure is reflected well enough to allow the data to be reviewed effectively by management.

Space Allocation

The space-allocation reference key (SAR key) mentioned earlier offers another type of arrangement that can be used to create estimates of area requirements based upon furniture and equipment. This method uses alphabetic codes assigned to individual items of furniture and support components (individually or in groups) to determine an initial area allocated to each space. Usually formulated in terms of conventional furniture (desks, tables, file cabinets, etc., as opposed to office system components), the SAR key can first be used to calculate required area and then as a basis for standardizing assignments and evaluating the variations in workstation types.

The *cruciform* model pictured in Figure 6-6 shows how the calculation is made. Each workstation typically has one item of furniture that is the primary focus of its occupant. In the figure, the space allocated to this component, including an allowance for occupant seating, is labeled as position *1*. The component in position *1* is facing toward us in the illustration. Position *2* represents guest seating that is situated in front of and usually facing the desk, and position *3* is reserved for a second furniture component placed behind the

Figure 6-6 Cruciform Model of Work-Space Components.

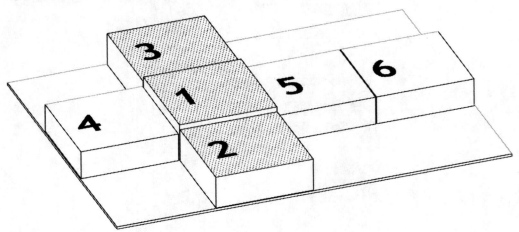

Figure 6-7   Furniture Component Selection Matrix.

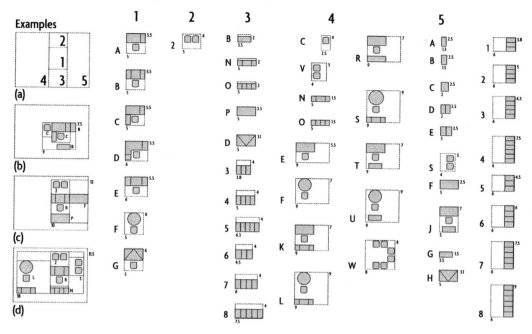

desk. (These three positions are shaded in the illustration.) Positions *4*, *5*, and *6* can be used to allocate additional space for furniture components to the left and right of the primary position as required.

A furniture component selection matrix is shown in Figure 6-7. Columns numbered 1 through 5 correspond to each position in the cruciform model, and each has front-to-back and side-to-side dimensions associated with it. For example, item *B* in column *1* (the primary component) is a thirty-by-sixty-inch double-pedestal desk. Three feet are allowed behind the desk for a chair so the area is five feet side to side and five feet six inches front to back, or a total of 27.5 square feet.

The examples at the left of Figure 6-7 show the application of this method. Figure 6-7*a* illustrates the cruciform model (only five positions are shown in this case, one to the left and one to the right of the principal component). Plan *b*, plan *c*, and plan *d* show the front-to-back and side-to-side dimensions generated by three different combinations of furniture. Plan 6-7*c*, for instance, shows a work space having the double-pedestal desk as the primary component, two guest chairs in front, and two work tables, one behind and one to the right of the desk. The overall dimensions for this combination are nine feet side to side and seven feet six inches front to back, or 67.5 square feet minimum.

If a choice of components is expressed in terms of the letter codes from the selection matrix, the configuration in Figure 6-7*c* would be represented as *B2P-F* (the hyphen is a placeholder for position 4, which is not being used). This is the SAR key representation of the work space. A more complex, executive-type work area is represented in plan 6-7*d* as *B2NLS*, which requires a minimum area allocation of 207 square feet.

When work-space and support area components are represented in this way, the area each work space occupies can be computed automatically, even using a relatively simple Excel macro. First, the possible front-to-back and side-to-side component dimensions are summed and then the maximum values multiplied together to determine the minimum area required.

Space-Allocation Guidelines

Calculation of individual areas using the cruciform model is usually done very early in the analysis process and based on worksheet data exclusively. We said earlier that the detailed personnel requirements, equipment needs, intradepartmental groupings, and adjacency priorities that are gathered in the early stage of facility-requirements data generation represent the perspectives of middle management. The assumption is that the managers and supervisors at this level are best able to respond in detail as to the needs of their staffs.

As valuable as this information is, it needs to be reviewed by management at a higher level. One reason is that in any large organization there is likely to be a great deal of variation from department to department in the work-space definitions for similar positions. Also, upper management is likely to have its own views as to how costly space should be allocated. The analysis process usually requires several iterations up the management ladder and down again before everyone is satisfied that individual needs are satisfied within a defined set of permitted variations. This process of reviewing the raw worksheet data with a higher level of management also helps to point the programming analysis in the right strategic direction. There may well be initiatives underway that will affect how departments are structured, how they operate, and how they may relate to other organizational units in the future.

Carrying the SAR key representation of the work-space components along into the detailed tabulations of programming data can be very helpful in the process of developing space-allocation standards or *guidelines*. We prefer the implied flexibility of the term *guidelines* in this context, remembering the TQM axiom about standards not being revised in six months being proof that no one is following them.

The space-allocation reference key allows us to sort the data by *space type*, which gives us the ability to compare similarities and differences in allocations among departments according to job function.

Job Evaluations and Descriptions

Earlier we referred to the *General Services* standards that are used for space allocation by the federal government. Job evaluations are one of the principal human resources functions in any sizable organization. This process is most simply defined as a systematic method of establishing pay differentials among different categories of jobs. The GS grade system is an example of one of four major categories of job-evaluation methods known as *classification*. Using this method, a series of standard classes is established that covers the range of jobs. Grades are assigned by comparing each job to the defined standard. For example, GS-9 is an intermediate grade, the definition of which goes something like: *to perform, under general supervision, very difficult and responsible work along special technical or administrative lines*. The analysis worksheet in Figure 6-2 shows the job description "Operations Specialist" as a GS-9, and the position is noted as nonsupervisory. Immediately above it, a "Supervisor" position is classified as GS-13. Standard federal job grades begin with GS-1 and range up to GS-15.

Two other job-evaluation methods are *ranking* and *factor comparison*. *Ranking* is generally used only in small organizations and simply orders written job descriptions from highest to lowest based on a basis of the perceived value of the position. *Factor comparison* assesses jobs relative to each in terms of how many of certain desired factors each possesses.

The fourth method, widely used in industry, is called *point-factor job evaluation*. Virtually all companies utilize some form of job evaluation. According to the Hay Group, one of the leaders in human resources consulting, nearly one-third of all companies use the point-factor method. The job family or classification method is second at about 25 percent. Nearly half of the largest one thousand companies in the United States and half of the top five hundred industrials in Canada use the *Hay point system*, a variation of the point method.°

Point systems share three basic characteristics: (1) Job elements that are of value to the organization—that help it pursue its strategy and objectives—are identified. (2) These elements are weighted in terms of their relative importance, and (3) factors are applied that are numerically scaled. A typical point scale may range from one hundred to three or four thousand points.

The *Hay point system* uses three primary factors in evaluating a job: know-how, accountability, and problem solving.° Problem solving addresses the degree of creative or proactive thinking required by the job for reasoning, analyzing, evaluating, and arriving at and making conclusions. To the extent that issues are referred to others for decisions or are circumscribed by standards, the problem-solving aspect is diminished and greater emphasis is placed upon know-how.

Implementation

The detailed description of workstations and support facilities by *space type* using the SAR key offers a clear opportunity for comparing and normalizing space allocations and job functions. Many organizations have established guidelines related to job grades or points that can be used to determine who qualifies for a private office versus a cubicle, as well as the amount of area allocated. Space allocation on the basis of job status tends to be made mostly at higher-management levels; hence, executive offices in a large organization usually represent a relatively small percentage of the total facility. How much area is allocated for the greater numbers of staff is more often determined by function and whatever physical and budgetary constraints exist. If a company maintains any kind of job-evaluation system, chances are that relationships can be established between job grade and tasks performed that can be helpful in equitably and functionally allocating space.

Figure 6-8 shows a set of space-allocation guidelines that may be derived from the individual component selection matrix pictured in Figure 6-7. Basic workstation types increase in size and complexity from left to right, and typical variations within each type are shown from top to bottom. Configuration *AA* at upper left is simply a thirty-by-sixty-inch desk with a chair, most likely in an open work area. At the opposite end, *MA* would be expected to be a fully enclosed executive office.

Several SAR keys are illustrated in Figure 6-9 that are based on the component selection matrix in Figure 6-7, plus some additional information in the first two character positions. The letter *X* in the first column (0) is used to differentiate support (or auxiliary) areas from work spaces. This column is left blank if the area is a workstation assigned to a specific individual. The code thus distinguishes between personnel headcount data and numbers of support spaces when projection figures are tabulated.

Figure 6-8   Space-Allocation Guidelines.

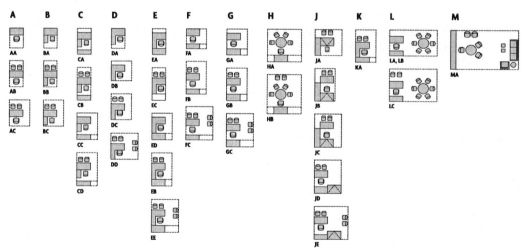

Figure 6-9   Space-Allocation Reference Keys.

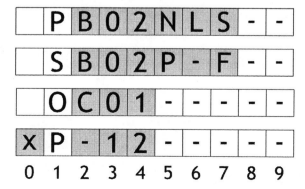

The second column (1) designates the type of space required, in this case:

- **P**—Private
- **S**—Semiprivate (partial-height walls, no door, or panel-system cubicles)
- **O**—Open

Column 2 designates the primary workstation type, columns 3 and 4 the quantity of guest seating (pull-up chairs) required, and the remaining columns additional furniture corresponding to the components of the cruciform model. The last row in Figure 6-9 (*xP-12*) represents a private conference room seating twelve.

Just as the level, sequence, and location designations are used to structure the evolving facility database in terms of management and location hierarchies, the SAR key allows proposed work-space components to be compared across departmental lines and in terms of job grades. This helps to identify atypical situations that may require more space than a rigid standard would allow due to special equipment or the need for some type of additional support space.

If there is a point-based job-evaluation system in place, point ranges can also be evaluated against the SAR keys in the process of normalizing space-allocation guidelines. This has proven to be an effective way of evaluating the correlation between job levels on a point basis and equipment requirements by function in terms of the effectiveness of overall space utilization. For example, Table 6-3 shows several ranges of normalized space types. In this way, the most frequently used configurations can be identified and variations in workstation arrangements reduced to a manageable number. The use of different transition points between job factor ranges can also be evaluated to assess the effect upon the total area requirements.

Systems furniture components using modular panels are employed most efficiently in situations in which a certain amount of repetition occurs in the workstation configurations. Individual areas divided by panels can be grouped into clusters having compatible dimensions. This results in cost savings when powered panels are used by reducing the number of required power and telephone connections into the building utility systems. Developing a vocabulary of workstation and support area clusters having common geometric elements also provides the groundwork for an orderly, *eurythmic* floor layout.

Figure 6-10 shows several typical workstation clusters based on modular furniture. Once space-allocation guidelines have been established on a basis of conventional furniture and typical arrangements identified that will satisfy the greatest variety of functional requirements, equivalent modular clusters can be developed for those job categories in which their deployment is planned.

Building Module

Most office layouts include some combination of private, semiprivate, and open spaces. Wall-location possibilities may be limited by ceiling conditions including lighting and building mechanical systems. The presence of a strong modular grid must be taken into account in developing space-allocation guidelines for all fully enclosed spaces where partitions intersect the ceiling. This is particularly the case with exterior offices where walls must also intersect window mullions.

A five-foot square module, which is arguably the most common in modern office buildings, limits available room sizes in specific increments if the module plan is followed. The possibilities include 10-foot square, 10 feet by 15 feet, 15-foot square, and 15 feet by 20 feet for rooms ranging from 100 to 300 square feet. This progression of modular sizes can be seen in Figure 6-11 up to 375 square feet. The limitations of the module can provide a useful discipline in establishing symmetry and order in the arrangement of the layout elements.

Table 6-3   Space-Type Allocations by Job Factor (Hay Point) Ranges

Space Type	A	B	C–D	E–F	G–H, K	J	L	M
Hay Points	50–100	100–200	200–300	300–500	500–800	300–700	800–1,500	1,500+

Figure 6-10   Workstation Clusters Using Systems Furniture.

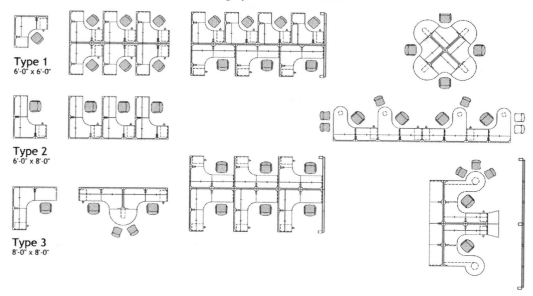

Type 1
6'-0" x 6'-0"

Type 2
6'-0" x 8'-0"

Type 3
8'-0" x 8'-0"

Figure 6-11   Incremental Room Sizes Using a Five-Foot Square Module.

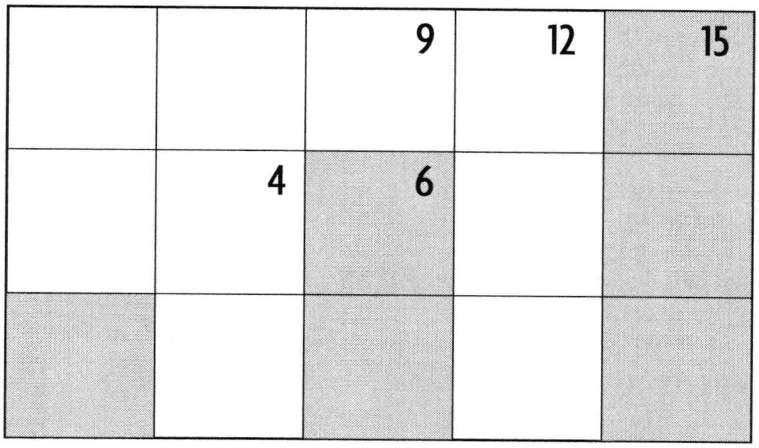

SYNTHESIS

So far, this chapter has focused primarily on the analysis of detailed requirements data generated through the use of a worksheet completed by supervisory personnel. Once tabulated, these preliminary data and the evident planning strategies they suggest need to be reviewed with upper management for strategic direction. Areas needing follow-up and more detailed study must be identified from the data assembled up to this point and a focused plan developed for interviews and work sessions. Discussion content will include questions identified during the analysis of the survey worksheets, clarification of incomplete or ambiguous items, further dialogue regarding adjacency priorities and work flow among the organization's various functional units, and any other issues that are considered of special importance.

It is during the interviews and work sessions that the synthesis process actually begins. Patterns begin to emerge as soon as the preliminary data have been assembled, and initial assumptions need to be verified. Face-to-face meetings with the midlevel managers and administrative people, either one-on-one or in small groups, are essential to test these assumptions, flesh out additional detail, and identify possible dead ends. Users even farther down in the organizational hierarchy may become involved in the process from this point forward, especially when there are special technical or equipment requirements to be considered. Detailed requirements for all the spaces classified as support—all the functional elements that are not work spaces or offices—must be thoroughly understood by the time the work sessions, site visits, and other data-gathering activities have been completed.

The personal involvement of decision makers and other key people can go a long way toward marshaling support of far-reaching changes that may be occurring within the organization. Frequently, a major replanning or relocation effort takes place within the context of a reorganization, downsizing, or period of major structural and cultural change. With proper input from upper management, information can also be disseminated during this process that can alleviate stress and help build internal support for changes taking place or pending.

Close communication with management becomes increasingly important as the facility-planning data are assembled and promising solutions begin to take shape. Changes in organizational structure, departmental functions, work flow, and adjacency relationships are all intimately related. A workable planning strategy must support all these causal factors and more, and management's special knowledge and perspective are essential. The planner can give shape to the program, but it is ultimately the users who must live with the operational result.

Adjacency Relationships

The final column of the data worksheet we have been describing deals with relationships between defined work groups and other physical areas within the organization. During initial data gathering, individuals are queried as to their distinct points of view, prioritizing relationships between their areas and other functional elements within the organization. These are often other departments but can also include physical support facilities such as conference rooms; training facilities; or required points of access for customers or service, such as a public entrance. Adjacency requirements and work flow can thus be analyzed on several levels.

In Chapter 2, we used a simple adjacency matrix to chart the functional relationships among the rooms in a home. Raw adjacency information can be consolidated using matrices such as the ones pictured in Figure 6-12. Adjacency charts using symbols or numeric ratings give a basic representation (Figure 6-12*a* and Figure 6-12*b*) and are similar to the one shown in Figure 6-12.

Figure 6-12   Adjacency-Relationship Matrices.

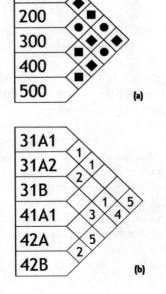

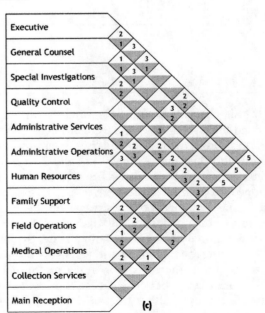

If a sizable number of adjacency responses have been developed from many different groups, a more advanced use of the matrix consists of matching opposing responses to determine whether they reinforce or contradict each other. Department *A* saying it is essential that it be adjacent to department *B* when *B* says it should be remote from *A* definitely identifies a matter for further discussion! Divergent responses such as these provide valuable substance for interviews at the middle- and upper-management levels. A more complex variation of an adjacency-matrix format appears in Figure 6-12c. Here, one set of responses is represented in the upper half of each coordinate, reading the function list from top to bottom, while complementary or opposing ratings are shown in the bottom (shaded) half, reading the list from bottom to top.

It should be recognized that the proximity matrix, while providing an effective tool for organizing, consolidating, and analyzing adjacency data from many different sources, is not a particularly effective presentation format. Interpretation is needed so that significant relationships and planning strategies can be easily visualized. The matrix data need to be synthesized into more concrete, albeit abstract, representations such as the bubble diagram and block layout. Both of these techniques were introduced along with the matrix concept in Chapter 4 and are useful tools even for very large-scale projects. Bubble diagrams, in particular, can be used to represent alternative planning strategies within different organizational models, which are discussed further in the next chapter. Similarly, the stacking diagram, a three-dimensional extension of the block layout, is an essential tool for working out multifloor and multibuilding planning strategies.

The Sandpaper and Army Blanket Technique

In strategy sessions with upper levels of management, some distinctly *low-tech* information-processing methods can be used effectively. The so-called sandpaper and army blanket technique derives from a battlefield tactic of using whatever materials are at hand to develop and communicate an action plan. It is essentially the same approach as a football coach's use of a locker-room blackboard.

Assembling a detailed adjacency matrix from numerous individual points of view represents a bottom-up approach. *Sandpaper*, in the form of scaled components representing departments or other functional elements, provides a top-down method that has proven successful in work sessions with higher levels of management. The components can be as simple as construction paper circles, ellipses, or polygons, proportionate in area to the organizational units they represent. The geometric shapes provide a sense of scale in differentiating one area from another, and color coding can aid in keeping the major organizational structures (finance, marketing, production, for example) clearly in mind.

Documentation is almost an automatic product of this method if the "bubbles" are taped to a large backdrop hung on a convenient wall. Lines and other symbols can be marked up to indicate special relationship priorities and work-flow constraints. Alternatively, different arrangements can be photographed as possible schemes are worked out. In either case, a great deal of management thinking can be pictorialized in a short period of time by this method. Many other related aspects of the facility-requirements analysis process can be discussed in the context of this kind of exercise. Results can later be compared with other sources of adjacency and work-flow data and synthesized into more concrete representations of the planning options that are considered viable. Working at this level of detail, the conceptual planning of several hundred thousand square feet of space can be accomplished in a few hours.

PROGRAM CASE STUDY

The two illustrations that follow, concluding this chapter, show a set of clearly prioritized functional relationships in the form of a bubble diagram and a set of block plans for a three-story building. This program was prepared for an association of banks to prepare for the development of a new office building for credit-card processing in a suburban location.

Using the methodology we have described, a facility-requirements analysis for an initial five-year time frame established a gross area requirement slightly in excess of fifty thousand square feet. The organizational structure consisted of eight principal groups, as follows:

1. Executive
2. Administration
3. Marketing
4. Operations
5. Member services
6. Data processing
7. Systems development
8. Special projects

The bubble diagram in Figure 6-13 shows adjacency relationships prioritized as *primary*, *secondary*, and *tertiary*, each priority level being indicated by a different line type connecting the circles. The proportionate areas of individual departments are, of course, indicated by the relative sizes of the bubbles. In a few cases in which a proximity relationship is implicitly understood due to an organizational relationship but not specifically prioritized, the organizational units are placed where they logically belong without showing

Figure 6-13   Bubble Diagram of Adjacency Priorities (see Plate Four).

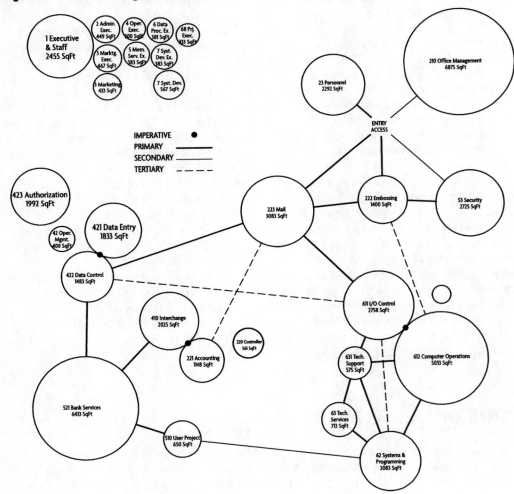

an explicit connection. Also, there are three cases in which juxtaposition is considered mandatory. These are indicated using a heavy black dot, essentially a *superprimary* relationship. So, in total, there are actually five priority levels graphically depicted. Departments needing entry access for one reason or another are also identified and those relationships are prioritized (three primary and two secondary).

Because the processing of both incoming and outgoing mail is a critical function, the mail-handling center (Department 223) serves as an operational hub. In addition to the executive and upper-management functions, which appear as a cluster at upper left, there are three principal groupings to be seen (reading clockwise): *Administration* (Group 2), *Data Processing* (Group 6), and *Operations* (Group 4). *Bank Services* (Department 521) has a strong affinity with several of the operations units. *Security* (53) has a direct relationship with the functions that have visitor access, although both it and Bank Services are part of the same group.

A number of studies were done to evaluate planning schemes with different numbers and sizes of floors before arriving at the representation shown in Figure 6-14. The block diagrams show three floors and suggests an optimal floor size of approximately eighteen thousand usable square feet. In this scheme, the *public* functions occupy the first floor, all having direct access to *Mail* (223).

The second and third floors are essentially secure areas with very little public access, and the second will be the most densely populated of the three floors. The principal operations and financial functions occupy the second floor. *Bank Services*, as the main

Figure 6-14   Three-Floor Functional Block Diagram (see Plate Five).

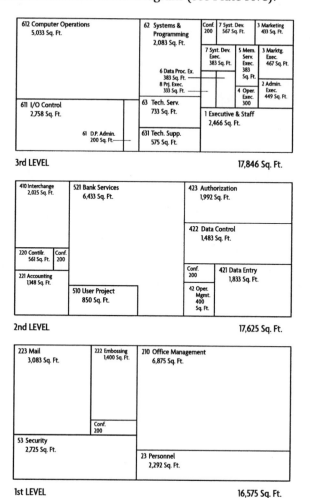

department interfacing with bank customers who are members of the association, is physically the hub of the card-authorization and transaction-processing functions located on that floor.

The top floor is shared between the information technology (IT) functions and an executive area for upper management and related staff functions. A conference facility is provided on each of the three floors.

We hasten to point out that the areas used in both of these diagrams represent *program gross*, which is simply the sum of work space and *dedicated* support areas (program net), plus an allowance for primary and secondary circulation, defined earlier as the *circulation factor*. In this case, neither *common* support nor building core and service areas are included. Food-service facilities, for example, which might be available as a public amenity if these three floors were located in a high-rise, have not been considered. These may need to be added in a stand-alone facility.

On the other hand, there is no building mail room, which would be *common* support if one existed. Since the central mail room is an integral part of this organization's operation, it is included as *departmental* support, what ASTM/IFMA terms *assignable but nonoccupiable* area.

CLOSING WORDS

In this chapter, we have looked at some of the mechanics of developing a facility program, beginning with a look at some of the types of organization one is likely to encounter in planning for business ventures. We discussed the kinds of information needed, suggested methods for going about gathering it, and identified ten categories that can be used to describe virtually all work spaces and support areas.

A significant part of the development of facility-management databases requires precise classification for allocation and accounting purposes. Giving the facility program a structure requires that the management hierarchy be represented. A tree structure such as the open network is among the most useful mechanisms for organizing the facility-management database.

We concluded this chapter with a short case study illustrating the use of some basic graphic tools to visualize organizational and adjacency relationships within a medium-size company. More case studies follow. Three strategic master occupancy plans are compared following the next chapter to show how the tools of facility-requirements analysis can be applied in different situations.

Chapter 7 concentrates on several basic tools for documenting facility-requirements data, focusing on the use of a general-purpose *Microsoft Access* database application, a spreadsheet, and *AutoCAD*. Emphasizing ways in which these applications can be used together, we will see how the elements of the *program→implement→evaluate→refine* cycle can be used to develop the facility program and, if desired, lay the foundation for a computer-aided facilities management (*CAFM*) database.

TIME CAPSULE   Cathedrals and the Medieval Program

Although the Roman Empire remained officially polytheistic until almost a hundred years after Constantine, his conversion to Christianity was a pivotal event in the emperor's rise. The Council of Nicea in 325 and the *Peace of Constantine* that followed marked the beginning of the church as an institution. Over the next several centuries, the bishops of Rome gradually established their power to define official doctrine and exercise discipline within the church. The ultimate decline of the Roman Empire created the power vacuum in Western Europe that was filled by the Carolingian monarchs, of whom the first and best known was Charlemagne, king of the Franks. Crowned as Roman emperor on Christmas Day in the year 800, Charles the Great was regarded by the Greeks as a usurper. His coronation by the pope was seen as a rebellious act that precipitated the lasting break between Rome and the Eastern church, while establishing the foundation of medieval Christian unity.

The Carolingian dynasty persisted until 987 and was both secular and sacerdotal in its conception of kingship. On the secular side, the monarch provided for his subjects by maintaining the peace and administering justice as best he was able. One of the greatest sources of wealth during this era resulted from periodic raids on surrounding tribes and principalities, in which all forms of plunder: coins, jewels, weapons, animals—and people—were carried off. Those who were kidnapped were either sold as slaves or held for ransom.

In his priestly capacity, the monarch acted as intermediary between the common people and God, and Constantine created an educated clergy that served as instruments of secular administration. The mechanics of medieval kingship were not fundamentally different from those of the god king in ancient Egypt, with the notable exception that the ruler of heaven became the singular Christian God. It was considered fitting that there be one who reigns here on Earth, under him, but the institutional church was subordinated to the king. The king's role as the embodiment of God was manifested in the form of conspicuous sacrifice, now largely in the form of great works of art. To this end, the most accomplished craftsmen were employed under the patronage of the sovereign. The common people expected the king's acts of sacrifice to be opulent, conveying to God the high regard in which the king and his people held him, as the sacrificial offerings were considered a determinant of the ongoing happiness and well-being of the realm.°

The first great Gothic church was the Abbey of St. Denis, built in France under the direction of Abbot Suger beginning in 1137. In less than two hundred years, hundreds of structures in the Gothic style were constructed across Europe and England. They were the result of highly complex building programs driven by the bishops of the church and the monarchs of state. Towering cathedrals and abbeys became political as well as theological symbols, made possible by a thousand years of social, economic, and technological development.

Ancient rituals of sacrifice had usually involved animals being offered to the gods in fire before an altar. Later offerings of objects denoting wealth typically required their entombment or destruction, effectively removing them from circulation. As Christianity spread over the course of the next half millennium, the Carolingian empire gradually gave way to countless smaller political units under rival kings. By the time work on the Abbey of St. Denis began, the power exercised by the bishops and other ecclesiastical leaders had greatly increased in response to the fragmentation of secular power. The church had long been working to limit the destabilizing plundering campaigns directed against itself and the poor, and its efforts culminated in the *Peace of God* at the opening of the second millennium. This decree established a tripartite division of society: those who fought (the secular order); those who prayed; and, of course, the peasants who were, as usual, expected to work to support the other two groups.

Under the *Peace of God*, and the *Truce of God* that followed it in 1040, alliances were formed between the monastic orders, notably the Benedictines, and the feudal princes, knights, and other nobles. The priestly orders came to be recognized as the new mediators of the God-man relationship, and as a consequence controlled a substantial portion of the royal wealth formerly earmarked for sacrifice, so that it might be appropriately administered "for the glory of God." It also became common practice for the secular heads of state to ensure their entry into heaven by contributing to the church; and the religious orders, for their part, used the wealth to construct the monastic establishments needed for their liturgical practices. Although those who prayed produced little in the way of tangible goods, this new method of transferring wealth kept it in circulation, which was one of Christianity's important contributions to economic growth throughout the Middle Ages.°

From this time through the thirteenth century, a change took place by degrees in the balance of power between the feudal lords, the monastic orders, and the common people. New orders such as the Dominicans and Franciscans were formed in the early 1200s, whose outlook was radically different from that of the earlier orders. Rather than following a monastic way of life, emphasizing salvation through isolation from the secular world, the new orders focused on the idea of God manifested in the human form, their members endeavoring to follow Christ's teachings as apostles. Good works in this life became the first qualification for salvation in the next, and the members of the new orders focused their efforts in the newly developing urban centers, living and laboring alongside the poor to whom they ministered.

(continued)

(*continued*)

In addition to working to alleviate the suffering of the poor and for their salvation, the church actively promoted worldly activities such as the Crusades, the first of which was ordered by Pope Urban II in 1095. Nearly every year from 1185 to 1285 saw a Crusade being planned or actually carried out. The Crusades were not restricted to forays into the Holy Land, as evidenced by the suppression of the Albigensian "heresy" in the second quarter of this century, within a triangle roughly bounded by Lyon, Marseille, and Toulouse in what is now southern France.° Thirty years into the thirteenth century, Pope Gregory IX established an institution with explicit responsibility for discovering and investigating cases of heresy. The Inquisition thus came into being, its members largely drawn from the Dominican order. For the next 450 years, the Inquisition would function as the church's ecclesiastical court.

Numerous orders of knights were founded to carry on the Crusades, such as the Order of the Knights of Solomon's Temple, known as the Knights Templar, which was formed in 1119, ostensibly to protect pilgrims visiting the Holy Land. The line between history and myth becomes easily blurred when considering the Templars and the medieval masons, but it is well documented that they had a substantial role in the construction of numerous castles and churches throughout Europe. During the period of their ascendancy, the Templars established a vast organization and banking system extending from Jerusalem to the west coast of Spain and from Ireland to the Baltic Sea.

The historian Malcolm Barber describes the Templar network as a "complex back-up organization for the frontline,"° which provided for a constant two-way logistical connection between East and West. Barber delineates in some detail the importance of the Templars within the body politic of Latin Christendom during the twelfth and thirteenth centuries, the building programs that were an integral part of life in the order, even extending to its extensive collection of holy relics "used to strengthen links with potential patrons in the West and to maintain interest in the affairs of the Holy Land." This has led many to speculate that these educated knights, as a result of their constant travels between Christian and the Muslim worlds, were the means by which the mathematical, scientific, and engineering knowledge was brought west, making the construction of the Gothic cathedrals possible.

The Knights Templar ultimately outlived their usefulness to the church, and in 1312 Pope Clement V officially abolished the order by "an inviolable and perpetual decree."° Its leaders, many of whom had already been arrested by the Inquisition beginning in 1307, were accused of heresy and other offenses; and the last of them were burned to death in 1314.

Within this setting, development of the large urban centers in which the great cathedrals came to be located was an immensely important economic factor. Centuries of uncompromising and often brutal exploitation of the peasants ultimately resulted in civil unrest and loss of productivity, which gradually led in turn to the feudal lords adopting more enlightened methods of managing their lands. Agricultural changes and improved land management and farming practices in due course led to an improved standard of living for the peasants and accelerated population growth. Once the living conditions of the general population were allowed to rise above subsistence level, fairs and markets developed in which a new class of traders and merchants sold a portion of the goods that they produced. Towns and cities formed around these market centers, most often in close proximity to the secular and religious courts that were their greatest patrons. The growing use of money led to increased literacy as a result of people learning to count in order to determine the value of their goods and services. The emergence of a new merchant class along with the craft guilds in the context of an economy based on capital, together with the organizational skills and increased wealth of landowners—kings, counts, and ecclesiastic leaders—created the conditions that made possible the great cathedrals and palaces.

ORDER IN THE UNIVERSE

The seven planetary deities once represented in the recesses of the Pantheon became the basis of the Ptolemaic model of the universe. Each of the seven planets was contained in its own sphere, beyond Saturn was the *Stellatum*, the sphere containing all the stars. The outermost unseen sphere was the *Primum Mobile*, the prime mover, the final frontier, full of pure light. The medieval vision of the whole universe was of its being illuminated in its entirety by the sun, night being only a local phenomenon caused by Earth's shadow. As Christianity became the prevailing religious belief system in what had been the Roman Empire, the outermost sphere *became* heaven, the ineffable world beyond space or time.

Our modern vision of the heavens, the dark cosmic grid to which we referred earlier, "may arouse terror, bewilderment, or vague reverie." The spheres of the medieval model presented us with an object in which the mind could rest, overwhelming in its greatness but satisfying in its harmony.° So said C. S. Lewis in describing the now-discarded image held by medieval man. To put yourself in this frame of mind, Lewis suggests taking a walk at night with this medieval astronomy in mind and "conceive yourself looking up at a world lighted, warmed, and resonant with music."°

Medieval science envisioned all matter as having certain inherent sympathies or antipathies. All things were thought to naturally incline toward their rightful place. Lewis quotes Chaucer in naming the so-called *four contraries*, the "ultimately sympathetic and antipathetic properties in matter"°: hot, cold, moist, and dry. These combine in pairs to form the four elements: earth = cold + dry, air = hot + moist, fire = hot + dry, and water = cold + moist. The human body was likewise considered to be constituted out of the same contraries forming the four temperaments. Melancholic = cold + dry, in the Middle Ages meaning thoughtful and meditative. Blood = hot + moist: the person of sanguine character, easily roused to short-lived anger but generally hopeful and cheerful. Choleric = hot + dry: quarrelsome and vindictive. And finally, phlegmatic = cold + moist: dull, sluggish, and pale.°

Lewis also tells us, as these classifications show, that medieval man was neither a dreamer nor a wanderer and that there was nothing he liked or did better than sorting out and tidying up. Lewis describes him as "an organizer, a codifier, a builder of systems" and speculates that "of all our modern inventions . . . [he] would most have admired the card index."° One must wonder what he might have thought of the relational database!

For the great majority of people, life was uncertain, difficult, and in many cases terrifying. The threats to their daily lives came from every direction, including violence and brutality, malnutrition and disease. In the early fourteenth century, a seven-year famine decimated a population of thirty million people across all of Europe and the British Isles. At the middle of the same century, a periodic outbreak of the Black Death killed as many as one-half of the people living in urban population centers. The average peasant had a life expectancy of thirty years.

Always a source of concern, as today, was fire. Most structures in the Middle Ages were built of wood with thatched roofs and floors covered with straw. The only sources of heat and light were the open hearth and an occasional candle. A single spark was sufficient to set off a conflagration, and there was very little that could be done except to let it run its course.

EARTHLY ORDER

Our modern sense of the heavens as a dark, infinite void inspires a certain amount of apprehension, but postmodern civilization does not often lack for light. When we do, it is a major disruption revealing the delicacy and vulnerability of our technological networks. Throughout most of history, though, and even today in nonindustrialized regions, people have lived in enveloping darkness. To the medieval mind, though, the necessities of basic survival here on Earth, as well as salvation in the afterlife, were completely dependent on the will of God. When the church emerged as the uncontested mediator of the relationship between man and God, a function it retained until the time of Martin Luther, it exercised sole responsibility for making sure that God maintained a benevolent attitude toward mankind. The role of the secular authorities was to provide a common defense and enforce adherence to God-given rules of order.

The church established itself in society through its teachings, the establishment of doctrine, and the works performed in the community by its various orders. The performance of liturgy according to annual, weekly, and daily cycles provided the visible manifestation of the priests' role as masters of the sacred in their communities. We use the word *sacerdotal* to describe the liturgical rituals because they were designed to foster an atmosphere of awe and mystery. The rituals were performed in Latin within the most sacred inner spaces of the church, from which the sounds emanated to be shared with the common people standing in the outer nave. It was the function of the Gothic church to serve as a backdrop for the ceremonial activities that took place within it.

Virtually all art was held sacred, and this included the places of worship themselves, as well as the paintings, sculpture, and decorations within. The purpose of all was sacrificial, to secure God's grace and pacify his anger. Nothing was spared in the material configuration of these unprecedented structures to maximize their symbolic effect, which made them "powerful instruments of propaganda in aid of the church's great struggle against heresy." In *The Gothic Enterprise*, Robert Scott goes on to say that each new cathedral was a "concrete display and representation of orthodox Catholic theology. This theology portrayed God as light and the universe as a luminous sphere that radiated outward from God, infusing the body of Christ, who as both God and man linked ordinary humans and the divine. Cathedrals were the vectors through which this process worked, and bishops, whose seats were housed in these great buildings, were the spiritual masters of the orthodoxy."°

Crossing the threshold of a cathedral, one left the secular world and entered into a space where every detail, most particularly the light, conveyed medieval man's sense of the cosmic order.

SALISBURY

The subtle organization of Geometry in the gothic cathedral attains an unprecedented fusion of horizontal and vertical elements. English Gothic cathedrals can generally be characterized as long and narrow compared to their loftier French counterparts; however, the spire at Salisbury is the highest in England at just over four hundred feet.

In the case of Salisbury, one enters the main entrance at the west end and passes through the nave, the central crossing of the main transept, around the choir, presbytery, and main altar as one approaches progressively more sacred spaces. From the central crossing eastward, there are chapels on either side as one traverses a second transept before reaching the Lady Chapel known as Trinity Chapel at the extreme eastern end. Construction of Salisbury began at this point in 1220 and progressed westward until its substantial completion one hundred years later.

Viewed from the west end in Figure TC5-1, the pointed arches of Salisbury clearly define its central axis extending nearly 470 feet from the great main entrance to the Lady Chapel. The vaulted bays are in two heights from the floor to the apex of each arch, either forty-two or eighty-two feet. The heart of the cathedral's geometry is centralized around its nearly thirty-nine-foot square central crossing bay; accordingly, it fills a volume defined by two stacked cubes. Some bays are square, others rectangular, but the cubic module is repeated throughout the space.

Ten tall bays define the central aisle of the nave between the entrance and the central crossing and are flanked by side aisles comprised of ten of the shorter bays on either side. East of the central crossing, there are seven tall vaults under which are the choir and chancel. The sanctuary containing the altar and presbytery were typically enclosed by a lattice or railing known as *cancelli*, hence the term *chancel*.

(continued)

(continued)

At Salisbury there are two transepts, one separating the nave from the choir and the other between the choir and chancel. Each arm of the main western transept contains four tall bays, and the eastern arms consist of three. On the eastern side of both arms is an equal number of the shorter bays that serve as chapels. The chapel bays are in the same proportion as those flanking the nave; both are one-half the dimension of the main central crossing module.

The side aisles join beyond the east end of the presbytery forming a processional path. All of the vaults at the east end are of the forty-two-foot height. Three small central bays at the east end form Trinity Chapel, again flanked by smaller rectangular bays of one-half the area. Each of the side aisles also terminates in a chapel, a single vault beyond the path that joins them.

Figure TC5-1 and Figure TC5-2 illustrate the low and high vaults of Salisbury, showing how the primary circulation and all the chapels under the smaller bays encapsulate the primary and transverse axes of this monumental and flawlessly organized structure. The stacked cubes of the central crossing of course locate the centralized *vertical* axis that continues upward to the pinnacle of the cathedral's four-hundred-foot spire.

The awe and mystery of the inner recesses of Salisbury cathedral provide constant points of visual and spiritual connection with the circulation areas that are conceived and rationalized explicitly for processions. Even today, one can gain a sense of what a person of that time must have felt in such a space by participating in one of the three great processional services marking Advent, Christmas, and the Epiphany. The modern Advent service begins in total darkness and silence as a single candle is lit at the west end of the nave. Two great processions move around the entire cathedral accompanied by music and readings, until at the conclusion of the service the space is illuminated by over 1,200 candles.

In earlier versions of the liturgy, the initial black silence was suddenly broken by a dissonant blast from the organ accompanied by shrieks and clanging of percussion instruments, all calculated to evoke the terrifying chaos of hell. As the service progressed, the noise and chaos gradually gave way to light and divine consonance. The metrical tonalities of sacred music reinforced in the medieval mind the harmony of the celestial spheres.

Scott offers an interpretation of the role of processions in periodically renewing the sanctification of the heart of the building and the objects contained there.

Figure TC5-1 Gothic Vaults of Salisbury Cathedral.

Figure TC5-2 Low and High Vaults of Salisbury Cathedral.

Figure TC5-3   Palm Sunday Procession Route according to the Sarum Use Liturgy.

Figure TC5-3 shows the route of the annual community procession on Palm Sunday, conducted according to the *Sarum Use* rules set down by Osmund, a local bishop who had been chancellor in the time of William the Conqueror (1066–72), which were followed until the time of the Reformation. The procession began in the choir area and was led by the bishop through the main transept, around the cloister, out through a side entrance where it circumnavigated the building entirely. Entering the cloister once again through an entrance adjacent to the octagonal chapter house,° the procession proceeded once again toward the main west entrance, finally returning up the nave's center aisle to the altar where it began. This portrayal of Christ's entry into Jerusalem symbolically renewed the consecration of the sacred space within while marking the walls of the cathedral as the boundary between it and the profane outer world.°

Ad Quadratum

Medieval man, as Lewis portrays him, was not a dreamer, yet he was capable of raising the principle of the arch to its highest expression simply by pointing it heavenward.

Banister Fletcher describes the adoption of this form from Muslim art, employed first over wall openings before 1100, after which it began to appear in vaults.° This innovation allowed the Gothic masons to develop complex networks of stone ribs supporting thin stone panels that satisfied the need, both functional and spiritual, for natural interior light. The role of the master mason in accomplishing this is described as follows:

. . . The architects—the master masons—worked within the framework of an organic tradition, that of the mason's craft. . . . Medieval designers seldom indulged in radical innovations nor did they seek artificial renaissances of the architectural standards of bygone ages. There were changes in design but they were continuous and evolutionary rather than radical or reactionary. Furthermore, since medieval architects of stone buildings were themselves masons they knew their building material at first hand. This helps to account for the great skill of these designers for both decorative and structural purposes. Their ability in combining form with function . . . should occasion no surprise

(continued)

(continued)

Figure TC5-4   The Ad Quadratum Module Grid of Salisbury.

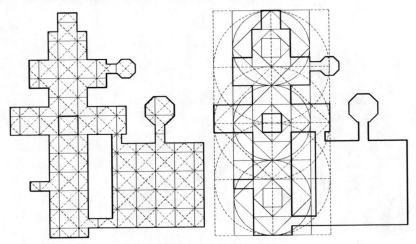

that the great medieval tradition of building in stone was developed in a period when the architects themselves were thoroughly imbued with an empirical knowledge of the materials and forms available to them . . . The training and background of the medieval designer meant that there was not that separation between the architectural idea and the execution of the idea that is characteristic of modern building . . . °

Simple geometric ratios such as the square root of 2, documented by Vitruvius, govern the layout at Salisbury. In the introduction to his ninth book, Vitruvius, referring specifically to Plato, outlines a method for determining an area exactly double that of a given square (and in the next breath tells of Pythagoras' discovery that a triangle having sides in the proportion of 3, 4, and 5 always yields a right angle). The central crossing bay at Salisbury is the starting point of a modular grid that determines the major features of the plan. Diagrammed in the *Meno* of Plato and known since Roman times as *Ad Quadratum*, the relationship between the dimension of a square and its diagonal had special significance to the medieval mind, and practical applications as well (Figure TC5-4).

With very little deviation, *Ad Quadratum* determines the arrangement, in plan, of the masonry at Salisbury, including its cloister and chapter house. Beginning at the central crossing, squares set on the diagonal of the one before determine key alignments in the plan, determining the length of the cathedral, the width of its main aisles, and both transepts. The square at the central crossing divides the longitudinal axis in half and, mirrored toward

the east and west ends, by three as well, carrying the symbolism of the Trinity while determining the placement of the high altar within the chancel area.

It is tempting to believe, as some medieval builders did, that regular geometry alone is capable of solving all planning and structural concerns. This is decidedly not the case. Earlier, we gave the example of Beauvais, the tallest cathedral in France with a nave height almost twice that of Salisbury's. Beauvais took nearly three centuries to complete and suffered a disastrous collapse of its choir vaults fifty years after construction began in 1230.

Medieval builders occasionally made the mistake of thinking that because their structures aspired to reflect divine order that structural stability would be ensured simply by their geometric regularity. When the cathedral at Milan showed signs of increasing instability in the 1390s, it was argued that the problem could be solved by *increasing* the height so that its section would be exactly square, conforming to *Ad Quadratum*. Fortunately, the designers chose the alternative of increased buttressing.°

Modern structural engineering has shown us that the geometry is not always so simple. But it is when the functional program has been completely satisfied and one also has a sense of order and balance that the experience of place rises above the mundane. In a historical context completely different from our own, Salisbury addresses basic spatial issues such as shape, qualities of enclosing surfaces, points of visual connection, and, of course, light, while illustrating flawless integration of function and circulation in a modular plan.

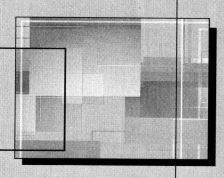

CHAPTER 7

Space Allocation

Every architect who's any good, no matter what they say,
is trying to make some kind of personal mud pie.
—Frank O. Gehry

In the last two chapters, we described several different kinds of workplaces to which advanced programming techniques may be applied, along with some important influences on their organization and development. We also discussed the network as being a fundamental structure for the division of labor, the management of work (projects, programs, and companies), and the architecture of information.

Most of the preceding chapter was an introduction to the programming process as applied to the workplace, the basic kinds of organizational and facility data that must be assembled, and how such information can be generated and classified. We looked at several ways in which alphanumeric codes can be used as a shorthand to represent management structure and facilitate the allocation of space. *Level* and *sequence* codes were developed to represent management's organizational hierarchy, and the space-allocation reference key (*SAR key*) was employed to arrive at initial area assignments based upon individual workstation components.

We ended the chapter with a discussion of several ways in which adjacency relationships can be visualized, including matrix charts, the bubble diagram, and block diagrams.

FROM PROGRAMMING TO FACILITY MANAGEMENT

In Figure 5-12, we presented an adaptation of the Deming quality management cycle as applied to facility management and proposed that the development of a facility program provides a logical entry point into that cycle. Figure 7-1 shows the process flowing from the development of the facility program through several stages of space planning to the importation (pipelining and linking) of finalized space layouts and associated data into a facility-management database.

If a *computer-aided facilities management* (CAFM) system is already in place, of course, it may be used to help generate the facility program for a particular project. In that case, the requirements data are extracted from the CAFM database; the necessary planning, design, reconfiguration, or relocation is implemented; and the database is updated with the new facility information, completing another planning cycle. Whether or not CAFM is used, though, space allocation and planning follow a similar path and the larger the project, the more important the decision making at each stage becomes. There is a great deal of information to be assembled in programming; some backtracking is inevitable as alternatives are evaluated (and some discarded); and, from a project-management standpoint, it is imperative that the amount of detail be limited at each successive phase of work. Otherwise, a great deal of costly effort can be wasted.

Figure 7-1 Flow of Planning Data Among Applications.

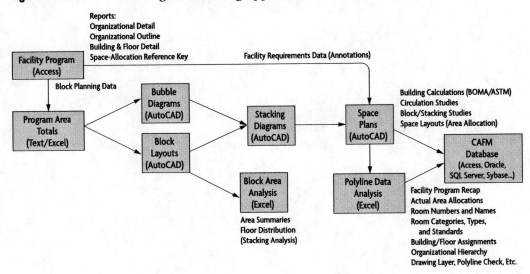

Some of the specific capabilities of CAFM are discussed in Chapter 11. Suffice it to say at this point, though, that while a CAFM database excels at tracking and helping to manage the many small changes that take place in a facility on a day-by-day basis, it often does not provide effective tools for working from a strategic perspective. Planning a new facility, evaluating different candidate locations, and determining the best fit of a specific set of requirements into a particular building envelope—these are all conditions that require the planner to step back from the day-to-day detail and take a higher-level view. When a sizable facility-planning effort is at hand, it is necessary to move from facility management to a wider and usually longer-range view in order to identify and evaluate alternatives that may not be immediately evident. Outside consultants (design, management, real estate, construction), as well as top management, may be heavily involved as we have seen; so the effort usually is clearly identified as a major project.

In Chapter 6, we discussed the gathering and organization of program-requirements data. This chapter focuses on a number of tools that can be used to help manage these data and to visualize and help analyze different planning approaches. As we look at Figure 7-1, we can observe that there are three fundamental ways of representing planning data, corresponding to an equal number of common computer applications:

1. The *database* is a highly structured means of keeping track of and updating various kinds of related information, producing numeric tabulations and pre-defined reports.

2. The *spreadsheet* is a flexible tool for manipulating and analyzing lists of data, sorting them, summarizing them, and creating ad hoc reports.

3. The *computer-aided-design* or *CAD* program allows the spatial manipulation of graphic information, which can be annotated with text data derived from the program at the beginning of the cycle and that flows into (or back into) a CAFM database at its conclusion.

Computer-aided facilities management systems almost universally use CAD drawings linked directly to a database to store, manipulate, and validate facility data. When a major planning effort is underway, some of the constraints of the linked and constantly validated relational database can seriously inhibit the thinking (or *what-if?*) process of planning and design. This occasionally results in the abandonment of computer technology to a significant degree (if not entirely) during this and other phases of the project, leading to lost time and duplicated effort.

What is needed is a method by which all the planning data can be conveniently moved from one application to another as the project develops, taking advantage of the capabilities of each *when* it is required. We use the term *pipelining* to describe this process.

Data Pipelining

Bearing in mind that the planning and design process remains essentially the same if all the data tabulations and area calculations are done with a scratch pad and calculator, the reports typewritten, and the plans drawn with a *2H* pencil or *rapidograph* on drafting film, the examples that are shown in the next several chapters were all produced using personal computer applications of the kind we have listed. It should be evident by now that it is our conviction that technology is with us for the indefinite duration and that, used effectively, it can only make our job easier and less prone to error.

Looking more closely at Figure 7-1, we see that reference is made to three specific applications that are market leaders among today's PC-based software:

1. *Microsoft Access*—Small-scale relational database
2. *Microsoft Excel*—Computerized spreadsheet
3. *AutoCAD*—A general-purpose, computer-aided-design program

Other than the fact that they are, for the moment, ubiquitous, there is no compelling reason to recommend these particular applications. The marketplace being what it is, they may be replaced by other more capable software in the future. What they do offer, however, is an accessible, underlying structure that allows for the free exchange of data among them. Each of the applications has access to a common *component object model*° that allows it to communicate directly with the others using relatively simple procedures called *macros*. No programming knowledge is required to *use* such macros, which facilitate pipelining the data, changing their form from one application to the next, and effectively tailoring each application to our space-planning needs.

A facility program can be created and maintained and a variety of reports produced using *Access*. Personnel and area totals can be extracted by organizational unit and transformed into graphic form so that bubble diagrams of the kind we have seen can be produced using *AutoCAD*. These preliminary adjacency studies can then be further refined as stacking diagrams (if required) and space plans are developed using the CAD program. At any point along the way, text and numeric data can be extracted to *Excel* for further analysis and reporting in whatever format may be required. This applies to making area takeoffs and verifying all rentable and usable area calculations. At the end of each project, the final CAD drawings are available for use as facility-management documents in a CAFM system. At the same time, the significant alphanumeric text information in the drawings can be used to create an *SQL*° script that automatically pipelines the information directly into a CAFM database of any scale, from Access to Oracle.°

How this is accomplished is discussed stepwise throughout the balance of this chapter and the next. The process is described using a model database designed for the purpose using Access. Along the way, some additional case studies are also presented. We look at how the bubble diagram can be used to clearly represent interdepartmental relationships and work flow from different points of view and how a building should be analyzed in terms of its circulation patterns and the potential impact of its fixed elements before detailed planning begins.

PROGRAMMED AREA REQUIREMENTS

We use the word *schema* to refer to the layout of a database. True databases are *relational* in nature, which means that they are built using multiple tables that have clearly defined relationships to each other. Figure 7-2 illustrates the schema of the Access database used to produce the facility-planning reports pictured on the next several pages.

Figure 7-2   Facility-Program Database Schema.

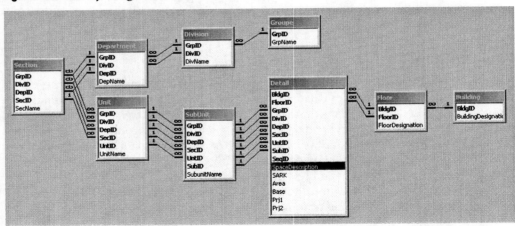

Near the center of the relationship diagram is a table called *Detail* to which all the other tables are linked, either directly or indirectly. The ancillary, or validating, tables are connected following the two hierarchies discussed in Chapter 6: *Management* and *Location*. The management or organizational structure, as indicated by the table names, is the same as that listed in Table 6-2. Only part of the *Location* hierarchy, *Floor* and *Building*, is included in this schema. The links indicate each table's position in each hierarchy relative to the tables immediately above and below it.

One of the limitations (and advantages) of the relational structure is that data must be entered into the database starting with the uppermost table in each hierarchy and working downward. For example, the *Building* table must contain valid data before the *Floor* table can be populated, and no building or floor may be specified in the *Detail* table if it does not have a corresponding entry in both tables above. The same applies to all six levels in the *Management* sequence.

Populating the Database

Populating the database means to fill it with data; recent versions of Access make this very easy to accomplish, as Figure 7-3 illustrates. For example, the *Level* code for the Vendor Hearings section of the General Counsel department, which is part of the Executive group, is *10C2*. After opening the database in Access and then the *Group*° table (select *Tables* under database *Objects*), we can enter the first character into the table and drill down as far as necessary to complete the other level designations. The zero code at the division level indicates that it is not being used and no personnel or area totals will be generated for that level in any reports.

Figure 7-4 highlights the individual line items for this section in the *Detail* table. Column 3 through column 6 as a group contain the *Level* code for section *10C2* (actually *10C200*, but the unit and subunit levels are also inactive here). The *Sequence* codes ensure the uniqueness of each of the six line entries in this section and determine the order in which they are listed. If support areas are numbered beginning with *100*, they will always be listed following the workstations in each organizational unit. Column 1 and column 2 contain the *Location* codes, indicating that the Vendor Hearings section is to be placed on the sixth floor of building A (*A-06*) in this example.

As the program-requirements data are developed from the survey worksheets, interviews, and discussions at the project site, they are entered into these database tables beginning with the *Group* table and working downward to the *Detail* table in the order shown in the schema (Figure 7-2). Similarly, *Building* and *Floor* codes may not be entered into the *Detail* table until there are entries in those tables that will validate them.°

Figure 7-3   Entering Data into Access Validating Tables.

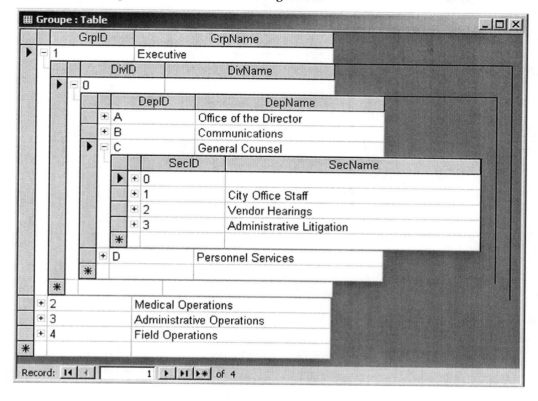

Figure 7-4   Entering Program Requirements into the Detail Table.

BldgID	FloorID	GrpID	DivID	DepID	SecID	UntID	SubID	SeqID	SpaceDescription	SARK	Area	Base	Prj1	Prj2
A	06	1	0	C	1	0	0	065	Office Clerks	O	80	5	7	10
A	06	1	0	C	1	0	0	100	File Area	X	500	1	1	1
A	06	1	0	C	1	0	0	110	Conference Room (per current layout)	X	600	1	1	1
A	06	1	0	C	2	0	0	010	TA IV	E	120	1	2	2
A	06	1	0	C	2	0	0	020	Hearings Supervisor	S	100	1	2	2
A	06	1	0	C	2	0	0	030	Office Coordinator	S	100	1	2	2
A	06	1	0	C	2	0	0	040	Hearing Referee	S	100	7	9	10
A	06	1	0	C	2	0	0	100	Hearing Rooms	X	300	4	5	5
A	06	1	0	C	2	0	0	110	File Area	X	800	1	1	1
A	06	3	1	A	7	3	6	100	Conference Rooms	X	300	2	2	2
A	06	3	1	A	7	3	6	110	Smoking Rooms	X	300	1	1	1
A	06	3	1	A	7	3	6	120	Vending Machine Area	X	300	1	1	1

The six remaining fields to the right of the *Sequence* code in the *Detail* table are:

- *Space description*—A text description of the work-space occupant or support space (up to about thirty characters);
- *Space-allocation reference key* (SAR key)—Described in the [a]"Space-Allocation Guidelines" section of Chapter 6—please refer to Figure 6-9 (up to ten characters);
- *Area*—The area assigned to the work space or support space, usually in square feet (net area); and
- *Base, Prj1, Prj2*—Number of people or spaces in the present or base year and projected for the future, short and long term (see Chapter 6, "Analyzing the Data Worksheet").

Management and Planning Reports

Microsoft Access provides highly flexible and easy-to-use mechanisms for tabulating and reporting the contents of the database. The process of finding and analyzing patterns in the data is made easier if they can be looked at from different points of view without having to spend the tedious and time-consuming effort of retabulating them. Once a report

Figure 7-5 Facility-Planning Reports (Selection Form).

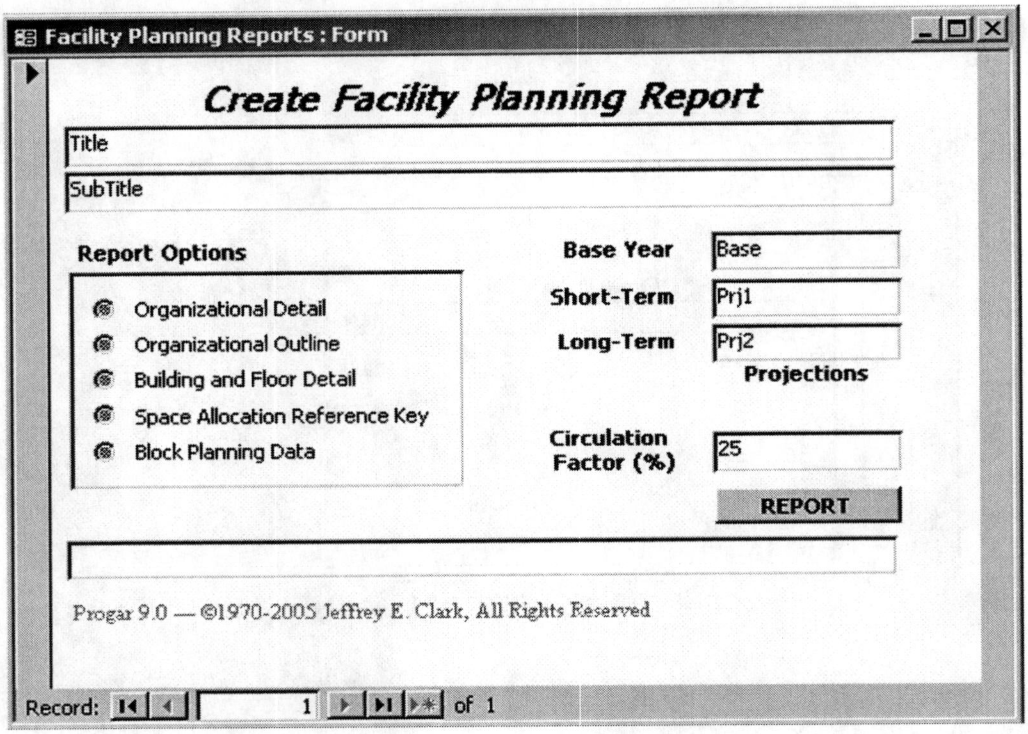

has been designed, it can be produced using a form such as the one pictured in Figure 7-5. (After opening the database, select *Forms* under database *Objects*.)

Forms can also be designed to facilitate data entry; but with facility programs up to a few hundred thousand square feet, it is generally as easy to work directly with the actual tables. Much of the preliminary worksheet analysis and preparation of input from effort expended in the field can often be prepared using spreadsheets. This includes interviews with client managers and support personnel, surveys of present conditions, and so forth. If the space-allocation data are properly formatted following a column layout similar to that shown in Figure 7-4, they can easily be imported into the database *Detail* table simply by cutting and pasting. °

When you open an Access database, you will see that in addition to *Tables, Forms*, and *Reports*, there are several other database objects such as *Queries*. Queries are virtual tables that present different views of the data in the actual tables according to specific parameters and selection criteria. In most cases, queries select certain data from the tables and present them to the user, most often in the form of reports. How this process actually works is the subject of Chapter 10, because a deeper understanding of database theory is helpful in working with large-scale CAFM systems.

The form in Figure 7-5 allows the user to select from any of four different reports and also to prepare the accumulated results for export to a CAD program such as AutoCAD. Examples of the four reports are illustrated in Figure 7-6 through Figure 7-9.

Besides the radio buttons for selecting report options, the form consists chiefly of text boxes for entering titles and column headings that are to appear on each report. Two lines are provided for titles and subtitles.

The "Facility-Planning Detail Report" illustrated in Figure 7-6 shows the years 2005, 2008, and 2014. These dates head the columns containing personnel requirements for the current or base year and the short- and long-term projections, in this case looking ahead three plus an additional six years. There is a text box for specifying the circulation factor to be used (please refer to the section on programmable versus nonprogrammable area in Chapter 6) and one at the bottom of the form for any additional identification text desired (company name, etc.).

Figure 7-6   Facility-Planning Detail Report.

Title
SubTitle

Levels	Description	Allocation	Sq.Ft	2005	2008	2014		
1 **Executive**								
10A Office of the Director								
10A000-010	Office of the Director	E	576	1	1	1		
10A000-015	Office of the Deputy Director	E	324	1	1	1		
10A000-020	Executive Offices	E	120	6	10	11		
10A000-030	Staff Semiprivate	S	100	2	3	3		
10A000-040	Staff Open	O	80	3	5	5		
10A000-100	Expansion Offices	X	120				5	5
10A000-110	Large Conference Room	X	864				1	1
10A000-120	Small Conference Room	X	288				1	1
10A000-130	Video Room	X	200				1	1

Department Total	10A	Office of the Director		15	20	21
		Staff:				
		Percentage Increase			33%	40%
		Net Usable Square Feet			4,752	4,872
	(Circulation is 25% of Gross)	Gross Usable Square Feet			6,336	6,406
		Percentage Increase				3%
		Gross Square Feet per Person			317	309

Figure 7-7   Facility-Planning Report Index.

Title
Subtitle

Levels	Description
1	**Executive**
10A	Office of the Director
10B	Communications
10C	General Counsel
10C1	City Office Staff
10C2	Vendor Hearings
10C3	Administrative Litigation
10D	Personnel Services
10D1	Office of Human Resources
10D11	City Office Staff
10D12	Office of Labor Relations
10D13	Equal Employment Opportunity
10D2	Selection & Recruitment
10D3	Employment & Training
10D4	Metro Support
10D5	City Staff Development

The basic management report shown in Figure 7-6 is subtotaled according to the six-level organizational hierarchy we have described. Each room or space is identified by name or description, and its SAR key is shown immediately to the right in the column titled *Allocation*. Each line item shows the net area assigned and, if it is a work space, the number of personnel requiring offices of this type. The quantities of support spaces are indicated at the far right and tabulated separately from personnel so as not to distort staffing and occupancy projections.

A full display of statistics is shown at the group, division, and department levels. In the illustration, staff totals appear along with the percentage increases anticipated with respect to the base year, both short and long term. The net usable area is shown together with gross, which is calculated by applying the factor for circulation. Finally, the projected rate

Figure 7-8   Facility-Planning Building/Floor Report.

Title
Subtitle

Levels	Description	Allocation	Sq. Ft	2005	2008	2014	
			Floor A07 Seventh Floor				
1 Executive							
10A Office of the Director.							
10A000-010	Office of the Director	E	575	1	1	1	
10A000-015	Office of the Deputy Director	E	324	1	1	1	
10A000-020	Executive Offices	E	120	6	10	11	
10A000-030	Staff semiprivate	S	100	2	3	3	
10A000-040	Staff Open	O	60	3	5	5	
10A000-100	Expansion Offices	X	120			5	5
10A000-110	Large Conference Room	X	864			1	1
10A000-120	Small Conference Room	X	288			1	1
10A000-130	Video Room	X	200			1	1
Department Total 10A	Office of the Director	Staff:		15	20	21	
		Gross Usable Square Feet:			6,336	6,496	
10B Communications							
10B000-010	Executive Offices	E	160	1	1	1	
10B000-020	Executive Office	E	120	8	10	11	
10B000-030	Open Work Area	O	60	3	3	4	
Department Total 10B	Communications	Staff:		12	14	16	
		Gross Usable Square Feet:			2,144	2,411	
10C General Counsel							
10C000-010	General Course	E	324	1	1	1	
10C300-010	Technical Advisor II	S	100	2	2	2	
10C300-020	Technical Advisor III	E	120	3	3	3	
10C300-030	Technical Advisor IV	E	120	1	2	2	
10C300-040	Technical Advisor V	E	120	1	2	2	
Section Total 10C3	Administrative Litigation	Staff:		7	9	9	
		Gross Usable Square Feet:			1,387	1,387	
Department Total 10C	General Counsel	Staff:		8	10	10	
		Gross Usable Square Feet:			1,818	1,819	
Division Total 33	Support Services	Staff:		25	23	37	
		Gross Usable Square Feet:			5,007	6,167	
Group Total 3	Administrative Operations	Staff:		25	33	37	
		Gross Usable Square Feet:			7,207	7,767	
4 Field Operations							
40A Office of the Deputy Director							
40A000-010	Office of the Deputy Director	E	324	1	1	1	
40A000-020	Executive Office	E	168	1	3	3	
40A000-030	Executive Office	E	120	1	1	1	
40A000-040	Open Work Area	O	60	1	2	2	
Department Total 40A	Office of the Deputy Director	Staff:		4	7	7	
		Gross Usable Square Feet:			1,477	1,477	
Group Total 4	Field Operations	Staff:		4	7	7	
		Gross Usable Square Feet:			1,477	1,477	
Floor Total 07	*Seventh Floor*	Staff:		66	87	94	
		Percentage Increase:			32%	42%	
		Net Usable Square Feet:			14,721	15,481	
	(Circulation is 25% of Gross)	Gross Usable Square Feet:			19,628	20,615	
		Percentage Increased:				5%	
		Gross Square Feet per Person:			226	219	

of increase in space between the short- and long-term condition is given along with the area per person, or utilization rate, for that level. An abbreviated display of personnel and gross area totals is shown for the three lower organizational levels—section, unit, and subunit—in the report examples in Figure 7-8 and Figure 7-9.

The example database illustrated in Figure 7-6 and Figure 7-8 uses a single letter to distinguish between enclosed (private) offices, semiprivate offices, and open workstations. The letter *X* is used to distinguish all support spaces. The use of the letter *X* somewhere in the SAR key is the only formatting requirement in the use of this field, as it determines whether the quantity coefficient represents people or spaces. In the example in Figure 7-9, the key is fully utilized to describe the elements of each space as discussed in the preceding chapter (see Figure 6-9). In this case, the first letter of the key is either *X* or *blank*, so that the second character can be used to indicate the privacy requirement for both offices and support spaces.

The listing in Figure 7-7 is simply an outline of the six-level organizational hierarchy showing the *Level* code for each department or functional grouping. The example includes *10A—Office of the Director*, detailed in Figure 7-6, and *10C2—Vendor Hearings*, used in the Figure 7-3 example. The index provides an overview of the database structure and serves as a reference to the other reports.

Figure 7-9 Space-Allocation Detail Report.

Title
Subtitle

Levels	Description	Allocation	Sq. Ft.	2005	2008	2014		
		Floor bf Front Office						
2	Finance							
202	Telecommunications							
202000-120	latex	XP—	50				1	1
Department Total 202	Telecommunications	Staff:		0	0	0		
		Gross Usable Square Feet		67	67			
203	Budgets & Forecasts							
203000-050	Budget Clerk	OF2C-1k-	75	1	1	1		
203000-030	Budget Accountant	OF2d-1K-	75	1	1	2		
203000-040	Budget Analyst	OF2d-1K-	75	1	1	2		
203000-010	Budget Manager	PC1bo3ek-	150	1	1	1		
203000-020	Asst. Budget Mgr.	SE2dg2k-	100	1	1	1		
203000-060	Sr. Budget Acct.	SE2dn1k-	100		1	1		
203000-100	Copier	XO—	0				1	1
Department Total: 203	Budgets & Forecasts	Staff:		5	6	8		
		Gross Usable Square Feet:			767	967		
20a	Data Processing							
20a300-040	System Operator	O-b-k-	0	1	1	1		
20a200-040	Clerk Typist	OG2c-k-	60		1	1		
20a100-020	Secretary	OH2c—	38		1	1		
20a100-010	Director of M.I.S.	PB1bo4egk	225	1	1	1		
20a200-010	Data Processing Mgr.	PC1bo4fgk	150	1	1	1		
20a300-010	Asst. D.P. Mgr	PC1bo4fgk	150	1	1	1		
20a300-020	Documorenist	PC10dgtis-	150	1	1	1		
20a300-060	2nd Shift Supr.	S-b-k-	0			1		
20a200-020	Programmed Analyst	SE2bp1km-	100	4	5	5		
20a200-030	System Analyst	SE2bp1km-	100	2	2	2		
20a200-030	Data Control Clerk	SF1d-k-	75	1	1	1		
20a300-030	Data Entry Optr.	SG1a-km-	50	2	7	2		
20a400-050	Art Area	XO—	38				1	1
20a501-140	Distribution Center	XO—	50				1	1
20a300-200	Documentation Storage	XO—	100				1	1
20a400-110	Tape Storage	XP—	100				1	1
20a400-120	Forms Storage	xP—	150				1	1
20a400-100	Computer Room	xP—net	600				1	1
20a400-130	Conference Room	xS—10—	225				1	1
Department Total 20a	Data Processing	Staff:		14	17	18		
		Gross Usable Square Feet:			3,868	3,868		
20b	Controller							
20b350-030	Swbd./Reception	O—	0	1	1	1		
20b360-040	Mail Clerk	O—	0	1	1	2		
20b110-030	M Cost Accountant	0-dp 1kr.	150	2	3	4		
20b330-040	Alpha Clerk	O-dp-kr.	100	5	6	6		
20b110-060	Cost Clerk	OF2cq-r-	75	1	1	1		
20b320-040	Clerk (Crod. & Col.)	OF2d-l-	75	1	2	2		
20b320-030	Sr. Clerk	OG2dP—	75	2	3	3		
20b340-020	Billing Clerk	OG2c-	50	3	3	2		
20b330-070	File Clerk	OG2dp—	50	1	1	1		
20b110-050	Sr. Cost Clerk	OG2dp—	50	1	1	1		
20b350-020	Serial Number Clerk	OG2d-r-	50	2	2	1		
20b330-060	Clerk Typist	OH20—	38	1	1	2		
20b230-020	Clerk	OH2d—	38	2	2	2		
20b330-050	Vouching Clerk	OH2d—	38	2	2	2		
20b010-010	Controller	PB2do6ev.	225	1	1	1		
20b110-010	Cost Acctg. Mgr.	PC1ba20k.	150	1	1	1		

Planning Reports

The "Facility-Planning Building/Floor Report" uses both facility hierarchies to organize the database. For a project utilizing multiple floors and/or multiple buildings, this report shows the program detail organizationally within each spatial unit. It therefore can be used to provide statistical backup to block layouts, proposed stacking configurations, and space layouts. It is a useful report against which to balance area takeoffs from space plans to ensure that all elements of the program have been correctly accounted for and allocated.

The example in Figure 7-8 is a composite of several pages from the building/floor report showing several departments from a single floor and the personnel and area totals at several levels.

Finally, Figure 7-9 illustrates the "Space-Allocation Detail Report." This tabulation also is first grouped by building and floor but only uses the top three levels of the management hierarchy: group, division, and department. The individual line items within each department are sorted by the space-allocation reference. This arrangement facilitates comparing space allocations on a basis of furniture and equipment requirements and job description. If job grades or points have been included in the description or allocation key, those criteria can be used in normalizing allocations with job functions as well.

Figure 7-10 Block-Planning Data (Unedited).

	A	B	C	D	E	F
1	1	Executive	214	52,568	235	55,332
2	10A	**Office of the Director**	20	6,439	21	6,602
3	10B	**Communications**	14	2,179	16	2,450
4	10C	General Counsel	88	16,856	100	18,320
5	10C1	**City Office Staff**	63	9,805	74	11,133
6	10C2	**Vendor Hearings**	15	5,203	16	5,339
7	10C3	**Administrative Litigation**	9	1,409	9	1,409
8	10D	Personnel Services	92	27,093	98	27,961
9	10D1	Office of Human Resources	26	5,501	28	5,827
10	10D11	**City Office Staff**	18	4,119	20	4,444
11	10D12	**Office of Labor Relations**	4	921	4	921
12	10D13	**Equal Employment Opportunity**	4	461	4	461
13	10D2	**Selection & Recruitment**	18	3,015	18	3,015
14	10D3	**Employment & Training**	5	759	5	759
15	10D4	**Metro. Support**	9	1,355	9	1,355
16	10D5	**City Staff Development**	34	16,463	38	17,005
17	2	Medical Operations	57	9,187	63	10,027
18	20A	**Office of the Deputy Director**	3	656	3	656
19	20B	**Program Analysis**	2	325	2	325
20	20C	**County Medical Programs**	8	1,187	10	1,485
21	20D	**Hospital Services**	1	163	1	163
22	20F	**Long-Term Care**	2	336	2	336
23	20G	Comprehensive Health Services	36	5,501	40	6,043
24	20G1	Outreach	16	2,260	20	2,802
25	20G2	Unit II	4	656	4	656
26	20G3	HMO	5	705	5	705
27	20G4	**Hospital Services**	11	1,881	11	1,881
28	20H	**Disability Services**	5	1,019	5	1,019
29	3	Administrative Operations	263	55,698	280	57,893
30	31	General Services	91	26,827	1??	?? ???
31	31A	Administrative Services	44			
32	31A1	**Facilities Management**				
33	31A2	Work C?				
34	31A3					

Exporting Program Area Totals

Data for block planning can be generated using the fifth and final report option. Figure 7-10 shows a partial tabular listing of the organizational units for all levels. Column C and column D show personnel and gross area totals for the short-term target date. Column E and column F show the long-term projections.

Using Access and Excel, this must be done in two steps.° The tabular data are first exported using *File→Export* to the *Text Files* (*.txt*) format in Access, and then imported to a workbook using Excel's *File→Open→Files of type* (*.txt*) menu option. Excel provides a text-import wizard that takes care of formatting the data, which has three steps:

1. *Original data type*—*Delimited* or *Fixed Width* (the wizard should choose *Fixed Width*, which is correct);

2. *Set field widths*   (this should require no adjustment);

3. *Set data format*   of each column of data (shown in Figure 7-11): the first column, containing the *Level* codes, should be set to *Text* format.

The necessity of step 3 is due to the fact that if the letter *E* is used in a *Level* code (which is not unlikely), Excel formats the code as a number using exponential notation. This makes something of a mess of your carefully formatted data!

Once the program area totals have been successfully imported into Excel, they may be sorted and edited as required. Generally, two kinds of editing are necessary. First, personnel and area totals for all organizational levels are initially present in the worksheet. This is redundant for space-planning purposes and will lead to duplicated areas. The edited list

Figure 7-11   Excel Text-Import Wizard (Final Step).

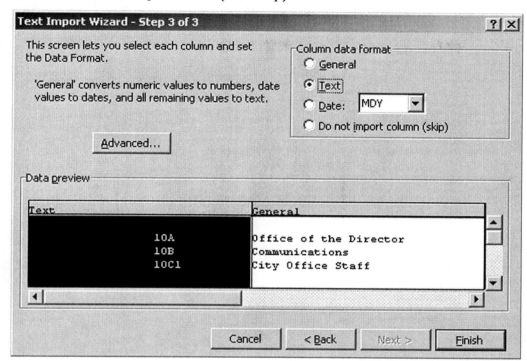

of block-planning data is shown in Figure 7-12. The disused rows are highlighted in gray in Figure 7-10 (please compare), and these may simply be deleted from the worksheet. Generally, the lowest-level totals should be kept as these provide the greatest flexibility in the adjacency-diagramming or block-planning process.

The second choice that must be made is whether to use the short- or long-term projections to generate bubbles or blocks for planning. Either or both, but only one at a time, can be used. The short-term totals have been deleted from the edited worksheet in Figure 7-12.

AREA-ALLOCATION TOOLS

The area-allocation toolbar provides an interface to a set of *VBA*-enabled AutoCAD utilities that carry out many of the pipelining functions shown in Figure 7-1. Visual Basic for Applications (VBA) is the language in which macros can be written for all three applications. The toolbar itself (Figure 7-13) is activated using an AutoCAD macro that is loaded using the *Tools→Macro→Load Project* . . . menu selection in AutoCAD. (Detailed instructions for obtaining, setting up, and loading the macro procedures and their related files are found in Appendix C.)

Each of the ten buttons on the toolbar executes a macro procedure or AutoCAD command that performs a specific function in preparing bubble diagrams, block layouts, and stacking configurations; annotating space layouts; or exporting text and numeric data from CAD drawings for further analysis and reporting. Their use is referred to throughout the remainder of this and the next chapter, and their respective functions are as follows, reading the toolbar from left to right. (*Help strings* are informative messages that appear in the status line at the bottom of the AutoCAD's drawing window whenever a menu item is highlighted.)

Figure 7-12   Block-Planning Data (Edited).

	A	B	C	D
1	10A	Office of the Director	21	6,602
2	10B	Communications	16	2,450
3	10C1	City Office Staff	74	11,133
4	10C2	Vendor Hearings	16	5,339
5	10C3	Administrative Litigation	9	1,409
6	10D11	City Office Staff	20	4,444
7	10D12	Office of Labor Relations	4	921
8	10D13	Equal Employment Opportunity	4	461
9	10D2	Selection & Recruitment	18	3,015
10	10D3	Employment & Training	5	759
11	10D4	Metro. Support	9	1,355
12	10D5	City Staff Development	38	17,005
13	20A	Office of the Deputy Director	3	656
14	20B	Program Analysis	2	325
15	20C	County Medical Programs	10	1,485
16	20D	Hospital Services	1	163
17	20F	Long-Term Care	2	336
18	20G1	Outreach	20	2,802
19	20G2	Unit II	4	656
20	20G3	HMO	5	705
21	20G4	Hospital Services	11	1,881
22	20H	Disability Services	5	1,019
23	31A1	Facilities Management	22	2,992
24	31A2	Work Simplification	2	325
25	31A3	Cashiers Service	5	1,450
26	31A4	Mail Processing	9	1,477
27	31A5	Network Control	3	1,328
28	31A61	City Office Staff	9	1,138
29	31A62	Central Photo I.D.	3	325
30	31A71	Main Reception	0	1,220
31	31A72	Employee Food Service	0	1,626
32	31A731	First Floor Support	0	1,626
33	31A732	Second Floor Support	0	1,626
34	31A733	Third Floor Support	0	1,626
35	31A734	Fourth Floor Support	0	1,626
36	31A735	Fifth Floor Support	0	1,626
37	31A736	Sixth Floor Support	0	1,626
38	31B1	City Office Personnel	27	3,496
39	31B2	Technical Recovery	13	1,653
40	31C	Research & Analysis	7	976
41	31D	Information Systems	2	474
42	32A	City Office Staff	2	271
43	32B	Medical Quality Assurance	84	12,770
44	32C1	Special Investigations	30	6,279
45	32C2	Fraud & Traffic	17	2,770
46	32D	Internal Affairs	8	1,301
47	33A	Child Care & Development	15	2,484
48	33B1	Refugee Program	9	1,940
49	33B2	Immigrant Service	3	434
50	33B3	Advocacy	3	407
51	33B4	Prevention & Services	7	1,003
52	40A	Office of the Deputy Director	7	1,501
53	41A1	Quality Assurance	9	1,220
54	41A2	Local Office Performance	4	542
55	41B1	City Office Staff	2	401
56	41B2	Contract Unit	12	1,762
57	41B3	AFDC/JOBS - County Office Staff	11	1,951
58	41B4	F.S.E.& T. - Monitoring & Program	18	2,721
59	41B5	F.S.E. &T. - Earnfare	20	2,710
60	41C1	City Office Staff	12	2,198
61	41C2	Metro. Prog. Serv.	9	1,644
62	42A	Office of the Ombudsperson	32	4,347
63	42B	Planning	5	932

Import Block Data

Help String: *Input block area data from an Excel worksheet*

This procedure imports the data from an Excel worksheet in the format shown in Figure 7-12. A dialog box appears (Figure 7-14) with which you can specify the form of the area data as you want them to be represented in your drawing: *Blocks*—golden section rectangles, *Bubbles*—circles, or *Ellipses*.

Next, a standard *File Open* dialog box appears with which you select the workbook containing your data. The objects chosen are created in your drawing in the same order that they are listed on the worksheet and are precisely scaled in area. The three formats are shown in Figure 7-15.

Figure 7-13   Area-Allocation Toolbar.

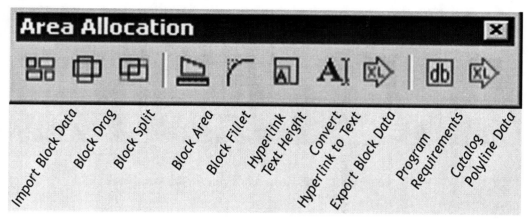

Figure 7-14   Block-Style Selection Dialog Box.

Figure 7-15   Scaled Departmental Area Objects in AutoCAD.

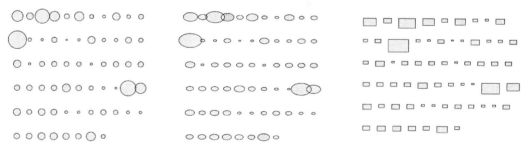

Each object has a *hyperlink* attached containing the text from the first three columns of the worksheet: level code, description, and number of personnel. This text becomes visible whenever you pass your cursor over the object (Figure 7-16) so each bubble or block can easily be identified as the layout is developed.

Block Drag

Help String: *Adjust shape of planning blocks while maintaining constant area*

This procedure functions only with rectangular blocks but allows you to change the shape of a block while maintaining the area of the object. You can select one corner of the block using the "Snap to Intersection" *Object Snap*° and drag it to a new position, changing the length or width of the rectangle. The other dimension automatically changes in the opposite direction to compensate, keeping the area constant.

Figure 7-16 Departmental Area Object Hyperlink.

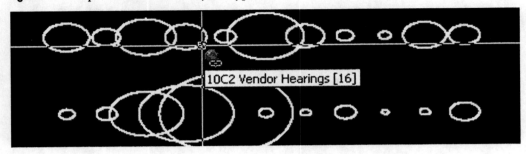

Figure 7-17 Proportionate Splitting of Area Blocks.

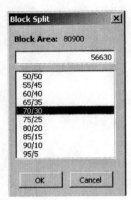

Figure 7-18 Use of the Block-Split Procedure.

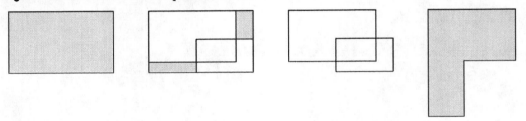

Block Split

Help String: *Split a planning block into two blocks of equivalent area*

Block split uses a dialog box to specify the proportion in which an area rectangle is to be split into two (see Figure 7-17).

Figure 7-18 shows an area object before and after the Block-split procedure has been executed and after the original rectangle has been deleted (reading from left to right). At far right, the smaller of the two rectangles (30 percent) has been rotated ninety degrees and joined to the larger one using AutoCAD's PEDIT° command. This is useful when it is necessary to create an irregularly shaped but accurately scaled area object. The original text hyperlink is associated with each of the two resultant blocks and remains after they have been spliced back together, so long as one of them is not EXPLODED.°

Block Area

Help String: *Calculate the area and perimeter of objects*

This button invokes AutoCAD's AREA command with its *Object* option selected. When this option is used, the AREA command calculates the area and perimeter of the selected object—in this case, circles, ellipses, and *polylines*° (block-area objects).

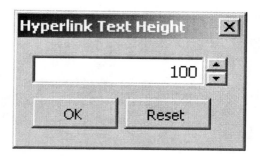

Figure 7-19   Setting Hyperlink Text Size.

Block Fillet

    Help String: *Fillet all corners of a planning block*

This button uses the AutoCAD FILLET command to apply a specified radius to the corners of rectangular block objects.

Hyperlink Text Height

    Help String: *Set block-label text height*

The hyperlink attached to each block-area object can be converted to visible, single-line text. This dialog box sets the desired text size (Figure 7-19).

Convert Hyperlink to Text

    Help String: *Convert hyperlink block labels to single-line text*

This procedure converts a hyperlink object into single-line text to annotate your drawing. The currently active text style is used and the hyperlink data remain intact so that the block may be identified if the text is separated or deleted.

Export Block Data

    Help String: *Export area data to an Excel worksheet*

This procedure exports block-object data from your drawing to an Excel worksheet. The worksheet output format is very similar to the one shown in Figure 7-12, with the addition of the AutoCAD layer name° on which the object is placed; the entity type of the block object; and whether or not polylines, if used, are closed. Open polylines can give incorrect area results. °

As bubble diagrams, block layouts, and stacking configurations are developed, the block-area objects are usually moved onto different layers. Extracted to the worksheet, the layer names can be used as sort criteria by floor or whatever organizational scheme is being used. This can speed up and simplify the numeric analysis of personnel and area data after they have been spatially arranged in the drawing. (One method of accomplishing this is discussed in Chapter 6—see Figure 6-5.)

Program Requirements

    Help String: *Annotate polylines using program-requirements Access database*

The program-requirements dialog box (Figure 7-20) allows you to annotate your block diagram or space layout with the detailed facility-requirements information from your Access database. When you select the desired group using the drop-down menu, or *ComboBox*, in the upper left-hand corner of the dialog box, a list of the departments within that group appears in the list box below. When you select a department, all the personnel and support spaces assigned to it are displayed in the *Program Detail* list box at the bottom of the form. As soon as you select a line item in that list, its position within the management hierarchy is indicated by the codes for each level and the names of the

Figure 7-20 Program-Requirements Text-Annotation Dialog Box.

Facility Requirements Data

Group	Administrative Operations		

Allocate | Suspend | Exit

| Division | General Services | 3 1 |

Department			
Administrative Services	3 1 A
Collection Services	3 1 B
Research & Analysis	3 1 C
Information Systems	3 1 D

Section | General-Use Support Areas | 7
Unit | Main Reception | 1
Subunit | | 0

Program Detail

3	1	A	6	2	0	010	Office Assistant Option I	O	80	1	1	1
3	1	A	6	2	0	020	Office Clerk Option I	O	80	2	2	2
3	1	A	7	1	0	100	Public Waiting Area	X	600	1	1	1
3	1	A	7	1	0	110	Main Receptionist Area	X	300	1	1	1
3	1	A	7	2	0	100	Employee Cafeteria	X	1200	1	1	1
3	1	A	7	3	1	100	Conference Rooms	X	300	2	2	2

container units in which it belongs (please see Figure 6-4 and the accompanying text in Chapter 6).

When you click the *Allocate* button in the dialog box, AutoCAD prompts for a text-insertion point. Four lines of text are inserted into your drawing at this point: the level and sequence codes of the item you selected, the item's description, the space-allocation reference key, and programmed square footage.

The *Suspend* button allows you to temporarily dismiss the dialog box, return to your drawing, and continue where you left off the next time you invoke the dialog box. The *Exit* button terminates the dialog box, resetting its drill-down position in the database.

The use of the program-requirements text-annotation dialog box is discussed in more detail in the next chapter. An example of a space layout annotated with facility-requirements data is shown in Figure 8-11*a*.

Catalog Polyline Data

Help String: *Catalog polyline text to an Excel worksheet*

This routine exports several different kinds of information from a space layout to an Excel worksheet. Its use is also discussed in Chapter 8 after some important additional concepts relating to area measurement using *polylines*, as well as the use of *layers*, have been presented.

CLOSING WORDS

This chapter began with a close look at the process of transforming the facility program into progressively more detailed space-allocation diagrams, taking into account adjacencies and other functional requirements. A number of tools were introduced to assist the planner in analyzing quantitative data and creating summary reports suited to management review and evaluation. Refining space layouts in progressive stages requires that the program information be easily converted between graphic and tabular formats. Optimal spatial relationships can thus be determined as different layout approaches are evaluated, while constantly checking to ensure that important elements are not overlooked.

A database tool such as Access can be used to prepare detailed program reports and summaries, from which selected data can be extracted for different purposes and converted directly into multidimensional models, such as stacking and blocking diagrams and detailed CAD layouts. Polylines on the plans are then used to make accurate area calculations and can later be directly linked or pipelined into a CAFM database, from which accurate allocation and utilization reports will be produced. If a new building is being planned, the program will influence the organization of the circulation system and possibly even the shape of the building envelope. There may be many alternative configurations, which must be evaluated so that a meaningful connection is maintained between the facility and its occupants.

We often use the word *wayfinding* to refer to the perception of an orderly flow when traveling from one place to another. It is the implementation of the program that determines the arrangement of physical space, which becomes the network through which, at any given moment, a sense of place is experienced.

Three case studies follow this chapter, each illustrating a different way in which blocking diagrams can be used to represent different interpretations of an organization's business plan as reflected in the facility program. The last of the studies carries the process through a detailed stacking analysis that is followed into space planning in the next chapter. In addition to examining the development of wayfinding, Chapter 8 further details the use of polylines to delineate space and categorize its utilization.

CASE STUDY   Three Facility Programs

The three case studies that follow illustrate the use of the programming methods discussed in this chapter, including block planning and bubble diagrams. Most of the chapter was focused on the tools themselves. Now we turn our attention to ways in which these tools can be used to help visualize functional relationships among organizational elements. These types of diagrams are the first step toward developing space layouts that will satisfy specific management strategies and planning goals.

BLUE CROSS AND BLUE SHIELD

The Blue Cross system began as a prepaid hospitalization insurance plan in the late 1920s. By 1940, the system had grown to national scale under the leadership of the American Hospital Association (AHA). The Blue Cross symbol was adopted by the AHA in 1930 to identify plans meeting specific performance standards. In 1960, with continued growth, the sponsorship and coordination of Blue Cross plans nationwide were taken over by the Blue Cross Association (BCA). Shortly thereafter, BCA built a headquarters immediately adjacent to the AHA building in Chicago, Illinois.

The roots of Blue Shield extend back even farther, to the beginning of the twentieth century, beginning with prepaid medical service bureaus formed by groups of doctors. In 1948, the Blue Shield symbol was adopted by nine such groups, which ultimately became known as the National Association of Blue Shield Plans (NABSP). Finally, in 1982, NABSP merged with BCA to form the Blue Cross and Blue Shield Association.

It is important to understand that throughout most of their existence, the plans in the Blue Cross and Blue Shield system have functioned as independent, non-profit companies, each serving a defined geographic area, generally by state. In recent years, a number of state plans have merged to achieve economies of scale. Blue Shield handles claims related to physician services, and Blue Cross processes claims for hospitalization costs, each on a regional basis. The BC/BS Association continues to fulfill a standards-setting role and also handles interplan transfer or payment of claims from one state to another. This ensures that if you have Blue Cross/Blue Shield coverage in New York, you can be hospitalized in San Francisco and still be covered.

When the federal Medicare and Medicaid programs were created in 1965, another functional area was added to the organizational structure of BC/BS. Medicare is a publicly funded health-insurance program for the elderly and disabled. Medicaid provides health insurance for the poor. Both programs are administered by the U.S. Department of Health and Human Services, but many of the plans in the Blue Cross/Blue Shield system act as *fiscal intermediaries* under contracts with the government, processing the hospital and medical claims of eligible residents.

Phase I

At the beginning of the project we are describing (more accurately, a long-term consulting relationship, as the program took more than ten years to fully implement), the president of a midwestern Blue Cross plan perceived the need for a new headquarters building. The plan's corporate offices at that time were in a well-maintained but aging four-story building on the outskirts of a medium-size city. To start off with, a detailed facility-requirements analysis was performed to develop area projections and other requirements data on which the acquisition of a suitable site could be based. Numerous department heads and staff were interviewed and personnel growth projections developed, projected over a ten-year time frame.

This process began at a time when Medicare and Medicaid had been in place for only a few years; consequently, growth as a result of these new programs was in a rather volatile state. There was a considerable margin for error in the area requirements that were projected initially. In addition, the Blue Shield component was at that time a separate organization entirely, administered by the county medical society.

From the outset, it was a stated goal of this chief executive to build a new headquarters tower downtown. There were several motivations for his wanting to do this. There had been a great deal of recent development on the east side of the river dividing the central business district, including a first-class bank tower designed by a leading architect. He believed that a large-scale project west of the river would spur additional development in an area that had largely fallen into decay. Of course, like an architect quoted at the beginning of the preceding chapter, the prospect of marking his tenure as chief executive by being the driving force behind such a project was not an unattractive proposition.

Phase II

It took several years to assemble the necessary parcels of land to create a usable site; once that had been accomplished, the facility program was extensively updated to reflect changes that had taken place in the intervening period. It was during this second round of requirements analysis that the diagram shown in Figure CS2-1 (one of many) was developed. At this time, an architect of

Figure CS2-1 Functional Adjacency and Work-Flow Diagram (see Plate Six).

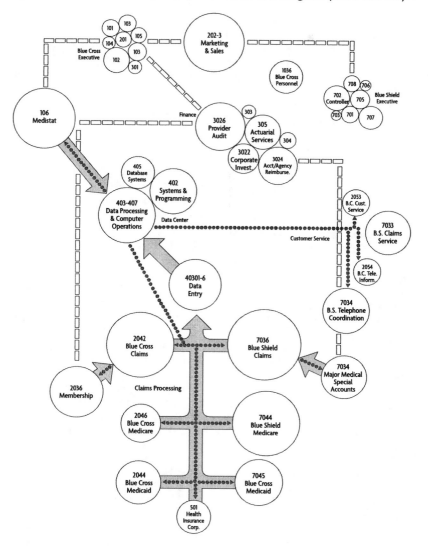

international stature was engaged to design a thirty-two-story office tower totaling a half-million square feet, of which 250,000 square feet would be occupied initially by Blue Cross and Blue Shield. The first six floors of the building were to have been considerably larger than those in the tower so that the great number of claims-processing and data-entry personnel, as well as direct customer-service areas, could be kept near the entrances and food-service facilities.

In spite of economy measures such as these, the projected construction cost of such a building proved to be prohibitive. Unlike commercial insurance companies that regularly invest substantial portions of their assets in real estate, the public perception of Blue Cross's status as a nonprofit organization made this kind of project something of a community-relations issue. In addition, national

health insurance, which is still a volatile political issue that surfaces from time to time, seemed to be a likely prospect in the near future. This meant that Blue Cross needed to maintain a competitive position with respect to commercial health insurance providers in order to win the government contracts that would be offered to administer these new programs, much like Medicare and Medicaid.

Phase III

Let us examine the organizational structure of this BC/BS plan in greater detail, as it is portrayed in Figure CS2-1.

Most organization charts show a single executive group, but this one was somewhat unique in that there were two. Although from a facility-planning standpoint the

(continued)

(*continued*)

offices and infrastructure of both organizations were to be colocated, their management structures had not been fully integrated at this point. They shared several public-interface functions such as marketing and sales and personnel. The financial function—including actuarial services, which statistically determine risk factors and set rates—constituted a Blue Cross division that supported both organizations. The other main centralized function was data processing and computer operations, which, like finance, was originally a Blue Cross division that had fully integrated its services to both organizations.

Virtually all of the departments shown in the lower half of the bubble diagram, with the exception of membership and major medical special accounts, perform some type of processing function relating to approving and paying claims. The wide, vertical arrow up the center of the diagram shows the work flow of claims as they are processed and sent to data entry for input into the computer system. The smaller, dotted, black arrows represent access to the computer database by the claims and other departments. Membership and major medical are responsible for making new accounts operational in the system.

Hospitalization (Blue Cross) claims are shown on the left and major medical (Blue Shield) on the right. In both cases, the claims functions are segmented vertically into regular private and group accounts, Medicare, and Medicaid accounts. Customer-service functions are shown at the right of the diagram opposite the data center. These functions require a great deal of customer contact with both walk-in and telephone traffic.

For the reasons we have outlined, plus the fact that the very structure of the two organizations was still in a state of flux, the Blue Cross president made the difficult decision to abandon his monumental tower. A different set of strategic options became apparent that are pictured in this diagram.

There were no fewer than nine related but essentially separate functional areas, all heavily clerical, all needing communication links with the data center, but none with any compelling need to be adjacent to one another: the six claims departments; the membership departments on both sides; and HIC, a small subsidiary shown at the very bottom of the diagram. Another special program, Medistat, was determined to be a potentially separable element, making ten.

The decision was made to build a more modest headquarters on the downtown site that would house the executive offices of Blue Cross and Blue Shield, the data center, and all of the customer-service and staff functions. The remaining organizational elements, the ten densely populated processing areas, were relocated to several different remote locations. Facilities were leased to provide for expansion and growth on a very cost-effective basis. In one case, more than two hundred claims personnel were set up in a remodeled, one-story facility that had formerly been a supermarket.

Phase IV

It took several more years to finally effect a formal consolidation of the two organizations, during which time this BC/BS plan was among the first to automate its claims processing using remote telecommunications technology. The company also gained enough experience with its government programs to be able to predict ongoing growth with greater precision, and national health care continues to be debated. Finally, after amortizing a substantial part of the cost of its headquarters, the company constructed an operations center on the remainder of its site, into which it consolidated all of its claims-processing and membership functions.

AN INTERNATIONAL PHARMACEUTICAL COMPANY

The headquarters of this publicly traded company is located on a suburban campus of several hundred acres in the midwestern United States. A master facility-planning effort was undertaken in response to a perceived immediate need for expansion space. The company was very much in a growth mode, and a matrix organizational structure had recently been put in place by the company's new chief executive.

At the outset, there were six principal buildings totaling approximately 340,000 square feet, among them a large, single-story manufacturing complex of nearly 150,000 square feet and a monumental, low-rise corporate headquarters and administration building. There were a number of different business groups with overlapping interests and objectives, among them pharmaceuticals and chemicals, both domestic and international; an agricultural division; a laboratory and research division; and pharmaceutical manufacturing, plus the usual executive and staff functions including marketing, planning, finance, and administrative services.

The matrix organizational structure was instituted to promote better communication and collaboration across the traditional business group lines. A stated objective of the facility study was to produce an in-depth analysis of ways in which the various functions, regardless of organizational location or reporting relationship, might be grouped spatially to promote a higher level of information interchange and avoid relocations that would inhibit such interaction.

The basic information-gathering and space-documentation activities initially took place over a sixty-day period during which time a series of questionnaire worksheets relating to personnel projections, equipment, and adjacency requirements were completed by a cross section of company personnel. There were many interviews and follow-up meetings to refine the initial input data and develop a detailed picture of special area requirements. Space-allocation guidelines for workstations were developed along the lines pictured in Figure 6-8.

Analysis of Existing Conditions

Any program that must deal with the reuse of existing space needs to distinguish between areas that are potentially reassignable and those that are not. For reasons of construction, function, or cost, certain areas in this complex of buildings needed to remain fixed. These included mechanical equipment rooms and shafts, vertical transfer systems such as stairs and elevators, toilet rooms, and food service and other facilities involving special equipment and mechanical and plumbing systems. In addition, there were certain operations that were defined as being "outside the scope of the study" and therefore not subject to change, such as the pharmaceutical manufacturing production areas and the executive wing of the main headquarters building.

Another type of space that is often classed as unavailable under such circumstances is *arterial* circulation, generally, walled corridors providing for major egress and main building lobbies. These areas are classified as building common for facility-management purposes and are not the same as corridors and aisles within the various office areas, accounted for in the circulation factor, the differential between *programmed* net and gross areas. (Once again, we refer to the discussion of programmable versus nonprogrammable area in Chapter 6.)

Spatial Planning Models

In an initial presentation to the company's officers and directors, three hypothetical scenarios were offered as ways in which the organization might be viewed:

- Spatial grouping A—The company is a multinational organization; therefore, the focal point of potential interdependency relationships should be the established international structure and the manner in which it relates to the rest of the company.
- Spatial grouping B—The company is primarily a manufacturing organization that produces and sells, on a worldwide basis, a broad range of pharmaceutical products for human, animal, and plant use. Interrelationships should be developed within this framework.
- Spatial grouping C—Function—The company's present distinction between domestic and international sales will continue to be viable, and interrelationships should be sought by juxtaposing the similar functions of the domestic and international divisions.

Grouping C was generally agreed to be an acceptable model, but A and B were judged to be less so, at least as a determinant of future spatial relationships. As a result of these discussions, two additional models were proposed:

- Spatial grouping D—Market—The company is a marketing organization selling pharmaceutical products, chemicals, and other such items around the world; therefore, interrelationships should be built around the marketing functions as a central hub.
- Spatial grouping E—Product—The company is a chemical manufacturing company producing mostly, but not solely, health-care products. Therefore, juxtapositions should be sought that reinforce the basic nature of the business, allowing it to grow different product lines in different markets without the corporate organization imposing artificial constraints.

Figure CS2-2 shows spatial grouping *E*, the *product* matrix, in the form of a spreadsheet. The executive administrative and manufacturing functions bracket all of the line and staff disciplines. All are shown with their projected area requirements individually and totaled by group. Figure CS2-3 and Figure CS2-4 illustrate the *function* and *market* matrices, spatial groupings *C* and *D*. These are shown using the bubble-diagram format, and it is remarkable that they present such different management views of the same company.

Implementation and Phasing

From a long-term perspective, a master occupancy plan must provide the kind of internal flexibility that allows the company to reassess its goals and make slight relationship changes periodically. It should be able to do this without upsetting the overall pattern of functional adjacencies. If, in this case, the mix of certain products changes in several years or completely new ones are added, there will inevitably be a shift in working relationships.

In the final analysis, the *product* grouping strategy was judged to be the most advantageous on a continuing basis. A grouping by *function* would be strongly affected by change because of the strong separation of the two major markets—domestic and international. With internal functions divided according to the markets they serve, adjacencies would have to be altered throughout the matrix in order to add elements related to a new product. A *market* grouping would be similarly affected because any significant shift in the marketplace would dictate a change in departmental relationships.

Similar functions can be clustered by discipline using a *product* grouping. The disciplines, in turn, can be allocated space according to the line and staff responsibilities they have for a specific product or group of products. Changing external events may precipitate a shift in emphasis that may radically affect the business of and the adjacencies within any of the disciplines, but adjustments can be made as needed with minimal effects on other disciplines. Although all three adjacency groupings promote interchange among the various business groups, the product orientation allows the desired kind of information interchange to occur on many more levels, with less potential disruption, than either of the other alternatives.

(continued)

(continued)

Figure CS2-2   Adjacency Relationships—Product Matrix.

Admin	Chemicals Administration			14,707
	A1	Intl Agriculture-Veterinary	1,402	
	A2	Fine Chemicals	1,261	
	A3	Agricultural Division	9,526	
	A4	Pharmaceutical Admin	2,518	
Line Staff Functions	Corporate Services			96,671
	C1	Intl Employee Org Devel	4,027	
	C2	Personnel	15,758	
	C3	Personnel Admin	2,176	
	C4	Engineering & Maintenance	33,221	
	C5	Treasurer	2,231	
	C6	Corporate Purchasing	4,080	
	C7	Office Services	35,178	
	Research			19,278
	D1	Intl Pharm R & D	5,877	
	D2	Pharmaceutical Research	7,551	
	D3	Laboratory Division	1,815	
	D4	Clinical Bioavailability	2,682	
	D5	Clinical Epidemiology	1,353	
	Planning			96,694
	E1	Intl Legal & Business Devel	3,496	
	E2	Intl Administration	38,168	
	E3	Fine Chem Matls Planning	2,520	
	E4	Legal Administration	3,338	
	E5	Domestic Pharm Med Affairs	29,524	
	E6	Public Relations	19,648	
	Finance			80,166
	F1	Intl Pharm Finance Admin	5,833	
	F2	Corporate Auditing	3,486	
	F3	Corporate Controller	23,033	
	F4	Corporate Finance	3,248	
	F5	Information Syst & Computer Se	44,566	
	Marketing			52,926
	M1	Intl Pharm Marketing	9,474	
	M2	Intl Chem Marketing	3,005	
	M3	Fine Chemicals Marketing	2,890	
	M4	Pharmacy Marketing Admin	1,454	
	M5	Lab Operations-Mktg Serv	1,415	
	M6	Domestic Pharm Sales	2,090	
	M7	Hosp & Govt Sales	7,397	
	M8	Domestic Pharm Marketing	4,663	
	M9	Marketing, Planning & Promo	20,538	
Manufacturing	Production Control			104,558
	P1	Intl Engnr, Distrib, & Mfg	5,673	
	P2	Pharmaceutical Manufacturing	27,585	
	P3	Pharmaceutical Mfg Admin	3,228	
	P4	Pharm Plan, Dist, & Inv Control	4,087	
	P5	Utilities, Plant Serv, & Security	1,761	
	P6	Plant Maint & Area Eng	4,488	
	P7	Control (General)	6,225	
	P8	Control (Records)	5,859	
	P9	Chemical Manufacturing	45,652	
	Total			465,000

Figure CS2-3   Adjacency Relationships—Function Matrix.

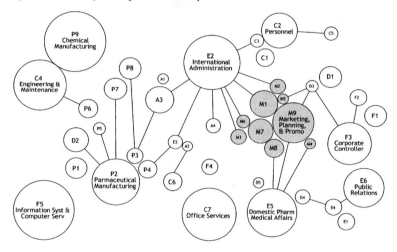

Figure CS2-4   Adjacency Relationships—Market Matrix.

In the end, it was determined that by implementing the new guidelines for space allocation and making more effective use of existing space, the company had no immediate need to construct a new building. In fact, projections ten years forward indicated that there would likely be an increase in office personnel from 10 to 15 percent annually. This was judged to be a moderate enough rate of expansion to allow areas to be systematically prepared so that major groups could be moved intact and then only once.

Initially, the decision was made to move the information-systems and computer-services functions out of the main manufacturing building to an available block of space in one of the other buildings on the campus. This allowed engineering and maintenance to vacate its undesirable location on a mezzanine above the manufacturing floor and occupy the space that formerly housed information systems and computer services. This, in turn, permitted many functions directly related to pharmaceutical manufacturing and production to occupy the mezzanine, such as manufacturing administration, industrial engineering, pharmaceutical planning, distribution, and inventory control. These functions were more directly related to manufacturing operations.

Three years would pass before the company needed to engage in any major new construction on its headquarters campus. When this finally occurred, a fifty-five-thousand square-foot modular wing of a new building was built, not so much to directly accommodate increased population but to provide the necessary staging area for other moves needed to implement the long-range configuration. This was the first of several modular expansions that were made as ongoing business trends were evaluated and appropriate changes implemented.

(continued)

(continued)

Causal Factors

We shortly turn our attention back to some of the tools and techniques of programming and planning. Before doing so, however, let us make a few observations regarding the underlying thought process we have been trying to expose. As we said in an earlier chapter, no two organizations are alike, even those that may be in a similar area of business. Indeed, the differences between two organizations in the same basic business may well determine why one becomes a leader and the other fails. Obviously, the success of an enterprise depends upon a great many factors other than the space it occupies, but the role of facility management as a key *causal factor* should not be underestimated. A Blue Cross/Blue Shield plan might have found itself committed to a monumental and costly headquarters project had it proceeded according to its CEO's original concept. He was, however, presented with other palatable options. The capital commitment that was *not* made, in this case, kept costs under control while a phased consolidation provided needed flexibility over several years while some significant organizational and management issues were worked out.

Identifying not only projected space requirements but also alternative configuration options and relationship priorities can help management identify strategic options and better plan for the future. A well-thought-out adjacency diagram should not just picture the physical relationships but should explore alternative visualizations of the way the company functions or might optimally function in the future. The facility planner, the designer, and the architect all bring a unique perspective to the table. The planner does not make policy, but her basic understanding of organizational structure and how the enterprise functions, combined with a sympathetic comprehension of her client's management concerns, should enable her to assemble, present, and put into perspective some important pieces of a larger puzzle.

In the case of the pharmaceutical company, the space analysis itself answered an important immediate concern that no immediate building program needed to be undertaken. Several well-chosen departmental moves freed up enough space to resolve some short-term problems. More important, having a clear picture of the company's several businesses from a physical-relationship point of view provided a strategic model for incremental changes needed to support long-term growth.

A State Government Agency

The third case study is derived from a project involving the relocation of a state public welfare agency. The motivation for the move was the need to consolidate various parts of the agency that, over many years, had become scattered among multiple buildings. This was not an organization experiencing extraordinary growth but one whose facilities had suffered from a lack of any coherent planning effort over an extended period of time.

Figure CS2-5 is an ellipse diagram of the agency's major groups showing many clusters of smaller functional elements within them. This diagram is based on the program examples presented at the beginning of this chapter, and the ellipse objects were generated from the block-planning data shown in Figure 7-12. In addition to the relationships expressed by the proximity of the various departmental bubbles, the need for accessibility to outside traffic is also indicated by the arrows at right. We follow this project through the next chapter as we examine how it progresses to the next level of detail, the space layout, along with related calculations and other documentation.

Projected personnel and area requirements were developed using questionnaire worksheets, as described earlier, followed by interviews and planning sessions with the managers of virtually all the groups. Space-allocation guidelines were developed on the assumption that a substantial amount of furniture and equipment would be reused. For that reason, a simple set of assignment guidelines established a range of four private office sizes, two sizes of semiprivate cubicles, and a similar number of open work areas. Because the agency intended to evaluate a variety of candidate buildings, the standards were not based on a specific building module but were worked out to cover the most commonly occurring workstation configurations. These are illustrated in Figure CS2-6.

Due to budgetary considerations, it was proposed that semiprivate workstations be constructed of partial-height drywall partitions (Figure CS2-6, items *E* and *F*). This allowed existing furniture to be utilized in these office areas as well, saving the cost of installing and maintaining furniture panels and components. A principal objective was to make the layout as uniform as possible to facilitate moving people rather than physically reconfiguring the work areas. Each standard configuration was allocated an appropriate number of electrical and telecommunications outlets for estimating purposes. Open workstations were to be serviced by outlets placed in adjacent partitions.

The Building Envelope

After evaluating several candidate buildings as well as real estate market conditions, the decision was made to construct a new building on the periphery of the central business district. Accessibility to the agency's client base was an important consideration as it serves the entire city and surrounding county. The location chosen was within two to three blocks of several public-transportation hubs but out of the higher-cost downtown area.

Figure CS2-5   Ellipse Diagram of General Agency Relationships.

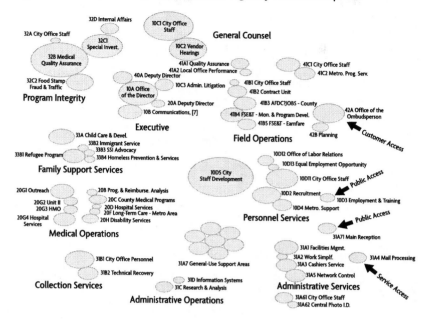

Figure CS2-6   Basic Space-Allocation Guidelines.

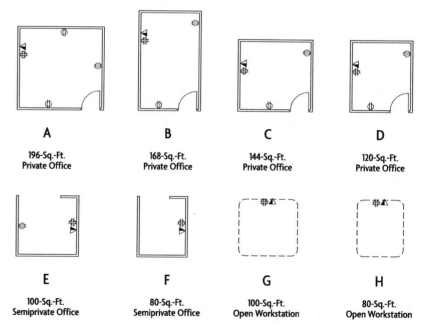

A

196-Sq.-Ft.
Private Office

B

168-Sq.-Ft.
Private Office

C

144-Sq.-Ft.
Private Office

D

120-Sq.-Ft.
Private Office

E

100-Sq.-Ft.
Semiprivate Office

F

80-Sq.-Ft.
Semiprivate Office

G

100-Sq.-Ft.
Open Workstation

H

80-Sq.-Ft.
Open Workstation

Figure CS2-7 shows the key elements of a typical floor plan of the agency's building. The area outlined includes all the floor's *facility usable area* plus *primary circulation* (according to the ASTM/IFMA definitions). Strictly speaking, there is an elevator lobby within the building core that is also primary circulation (see Figure 8-6), but the core is not shown here.

The building module was nominally five feet square within a thirty-foot structural bay (the columns and their centerlines are shown in the figure). There were some variations in the module in the building's end bays, however, due to the fact that the windows in both were four feet six inches wide on the front and rear facades (the longitudinal sides—the service core is located at the

(continued)

(*continued*)

Figure CS2-7   Typical Floor Plan—Occupiable Area.

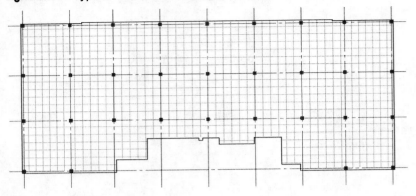

Figure CS2-8   Development of Block Plans.

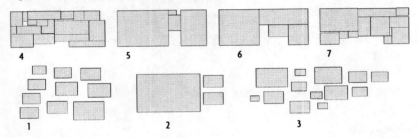

rear). The second bay from each end is narrower than the others, resulting in a split module, and the designers of the building also added half modules along the front and rear. In spite of these anomalies, the module grid was followed in laying out virtually all partitions, for both the full-height and the partial-height cubicles.

We make a point of the peculiarities of this specific building envelope only to emphasize that all such conditions need to be taken into account when developing a rational floor-planning framework. We further discuss such constraints and the relationship of building module to primary-circulation pathways at the beginning of the next chapter.

Blocking and Stacking

With the building envelope established, a clear picture of the departmental adjacency requirements in hand, and an AutoCAD drawing containing the blocks generated from the program, block planning and stacking studies can be prepared. The occupiable area of a typical floor, as shown in Figure CS2-7, is approximately 21,500 square feet. Based on the program requirements, the building was designed with seven floors containing a total of 150,000 usable square feet. The gross building area totaled 164,000 square feet, so about 91.5 percent of the gross area was allocable to programmed functions and circulation.

Many different blocking configurations were evaluated before and after the building was designed. Figure CS2-8 illustrates the development of one such study in a state of partial completion. As shown, the first-, second-, and third-floor elements are simply grouped according to their adjacencies in an attempt to arrive at a proper *fit* within the available space. If the blocks representing departments on each floor are placed on different layers, their areas can easily be exported to a spreadsheet and tabulated using the area-allocation tools described earlier.

The blocks on the fourth through seventh floors have been grouped and reshaped to fit the floor perimeter as closely as possible. In Figure CS2-9, we see an isometric view of the stacking arrangement derived from the block layouts in two stages of refinement. The stack on the left shows the departmental blocks exactly as arranged in the completed block studies, with the building utility core added (shaded blocks). The individual plans on the right have been further polished by adjusting the departmental boundaries and labeling the various functions for presentation and discussion purposes.

The mechanics of creating a 3-D view of the block layouts that we refer to as a *stacking diagram* could not be simpler using an application such as AutoCAD. The only requirement is that the information pertaining to each floor be separated on different layers. From that point, each layer can be elevated to a different height,

Figure CS2-9   Development of Stacking Diagrams (see also Plate Seven).

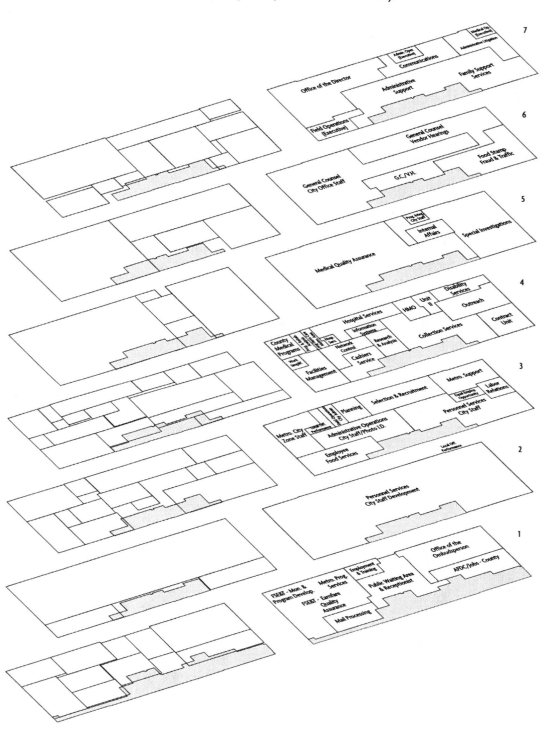

thereby separating the floors vertically. Then, any of several commands are used to rotate the model so it can be viewed from an oblique angle.° In Figure CS2-9, the floors are shown one hundred feet above one another—monumentally high ceilings, to be sure, but desirable for the sake of readability.

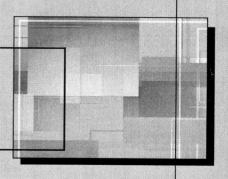

CHAPTER 8

Space Planning

*What splendid buildings our architects would be able to execute
if only they could finally be less obedient to gravity!
Just imagine . . . a dodecahedral church
dedicated to the twelve apostles, for example. . . .*
—Maurits Cornelis Escher°

Once the program requirements are thoroughly understood and documented, adjacency priorities defined, and a satisfactory blocking and stacking approach (or options analysis) agreed upon, planning can begin on a more detailed level. The program itself will likely still be evolving and refinements being made at this point, and the *options analysis* may mean that there are several buildings under serious consideration. In such cases, the initial space plans may focus on areas or floors in the candidate buildings using a portion of the program that is considered typical. The object at this point is to determine which of the buildings will best satisfy the planning requirements functionally, economically, and sustainably.

WAYFINDING

If preliminary block layouts have been taken to the level of detail shown in Figure CS2-9, the controlling parameters relating to the size and shape of the floors, placement of fixed structural elements, and building service cores should be well-defined. Just as the program determines the functional requirements that need to be fitted into the space, the development of major circulation pathways determines the way in which the individual elements relate to one another and the clarity of the resulting space layouts.

Primary Circulation

The term *wayfinding* is commonly used to refer to the manner in which people orient themselves within a space. Formally, wayfinding bears directly upon means of *egress*, the means by which people can safely remove themselves from an area should an emergency arise. Egress and accessibility are the subjects of later chapters in a building-code context, in which elements such as occupancy load, egress path width, signage, and illumination are discussed. These are the life-safety considerations of wayfinding, but we should not lose sight of the sense of place that is gained through the perception of an orderly flow as we move through the space. The regulating lines, planes, and masses that we spoke of earlier, defined by boundaries and edges, largely determine the circulation system throughout the environment.

It is tempting to compare the circulation through planned space to the circulation system within the human body, but there is an essential difference. We speak of *arterial* circulation in reference to major thoroughfares, particularly main streets and highways. The body, though, is a closed system, with arterial circulation leading from a central organ (the heart) into smaller and smaller tributaries. After finally reaching the smallest capillaries, circulating blood enters the *venous* part of the system through which it returns by way of the lungs to the heart, where the cycle begins once again. If we look at the picture of the analysis/syntheses network in Figure 5-6 and imagine one end looping back around to the other, we have a clear picture of a closed network. The flow is always in one direction in this kind of loop, in which no question of wayfinding exists.

The map of the Internet in Figure 5-5 presents quite a different picture. We have a network that is both open and closed. It has both multiple paths and dead ends with which we must contend, making the problem of wayfinding much more complex. We can compare the multiple backbones of the Internet to the interstate highway system at the arterial level. At the other end of the scale, the corridors and passageways in a building are similar to the coaxial cabling that connects our computers: We leave by going back essentially the same way we came in. We may use an alternate route, but most interior circulation systems are two-way, and all building codes specifically limit the length of dead-end conditions for safety reasons.

We can provide exit signage, special illumination, and all sorts of alarm systems to cope with emergency exiting necessities, but these should reinforce an intuitive wayfinding system that is evident in the logic of the plan. As we pass corners, intersections, surfaces, and objects, we build our experience of a place. Shape, color, changes in the lighting, surface textures, ceiling height, floor elevation, and materials all contribute to our sense of place as we become familiar with our surroundings. A piece of sculpture or a fountain or a group of large plants may provide a focal point, define a formal entryway, or in some other way distinguish levels of formality and control as we pass from one space to another. All these points of reference determine not only our learned knowledge of the space for finding our way but also our sense of the quality of the environment.

Corridors and Aisles

Depending on the layout flexibility offered by a subject building, there may be a few or a great many possibilities for establishing primary circulation patterns. Figure 8-1, Figure 8-2, and Figure 8-3 all show variations of main aisle patterns that may be applied across an office floor. These plans are of a Mies van der Rohe building designed around 1970. The columns are spaced thirty feet apart in both directions. This spacing establishes a five-foot square planning grid, as shown on the drawings.

A centrally located core provides for vertical penetrations containing stairs, elevators, and utility shafts, as well as common areas shared among the floor occupants: toilets and utility closets (electrical, mechanical, telecommunications, janitorial, etc.). There are four stairways in this building for emergency egress located in the extreme corners of the core. Two stairs are thus accessible at both ends of the core on either side. The dark-shaded area of the plan in Figure 8-1c shows the minimum corridor necessary to provide access to all four stairways around and within the core. The crossbar of the *H*-shaped backbone corridor is the floor's elevator lobby. Plans in Figure 8-1a and Figure 8-1b show this main corridor extended to encircle the core guaranteeing maximum accessibility from any direction.

- 8-1a—The lighter-shaded circulation aisles are five feet wide and follow the modular grid to divide up the floor in a conventional *orthogonal* pattern. This could be a single-tenant or a multitenant floor with open aisles or enclosed corridors. The interior spaces can be subdivided into enclosed offices using the modular incremental sizes shown in Figure 6-11 (for example, four module—ten by ten feet; six module—ten by fifteen feet; nine module—fifteen by fifteen feet; and so on).
- 8-1b—Defined circulation areas are ten feet wide, allowing for traffic along one side or the other or down the center. The halfwidth not used as aisle can be assigned as open work area, or allocated as thirty-inch-deep (half-module) strips for banks of files, book shelving, or similar support space. The presence of walled corridors, of course, implies that the unshaded portions of the plan are either allocated as private offices as or larger partitioned spaces used as open office area or subdivided by furniture panels.
- 8-1c—The corridors on either side of the core are extended all the way to the exterior curtain wall at both ends of the building. These main public corridors are nearly seven feet wide, but the office area they define is nearly column-free.

Figure 8-1   Major Circulation in a Late Twentieth-Century Office Building.

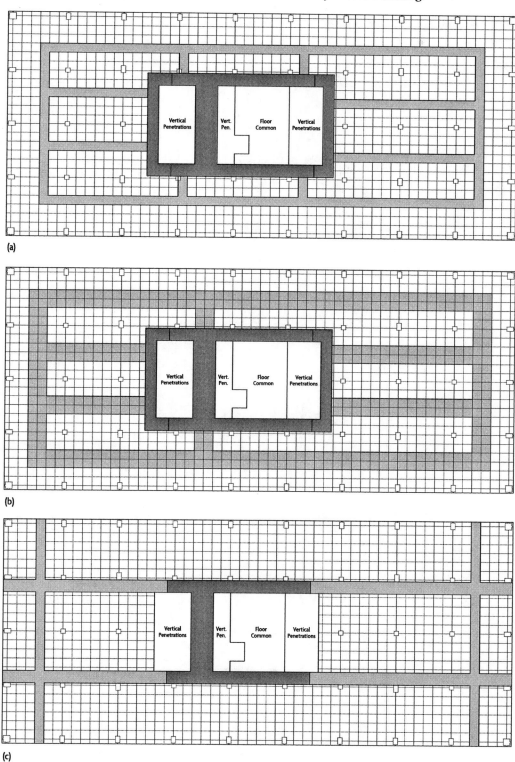

(a)

(b)

(c)

- 8-2a—This layout is a variation on 8-1c with a defined aisle following the module grid around the entire perimeter of the building. Such a configuration provides substantially column-free office areas that may be planned in many different ways. This type of configuration offers advantages when sustainability is a consideration, because daylight and views, particularly for occupants who are seated, have potential

Figure 8-2   Major Circulation in a Late Twentieth-Century Office Building.

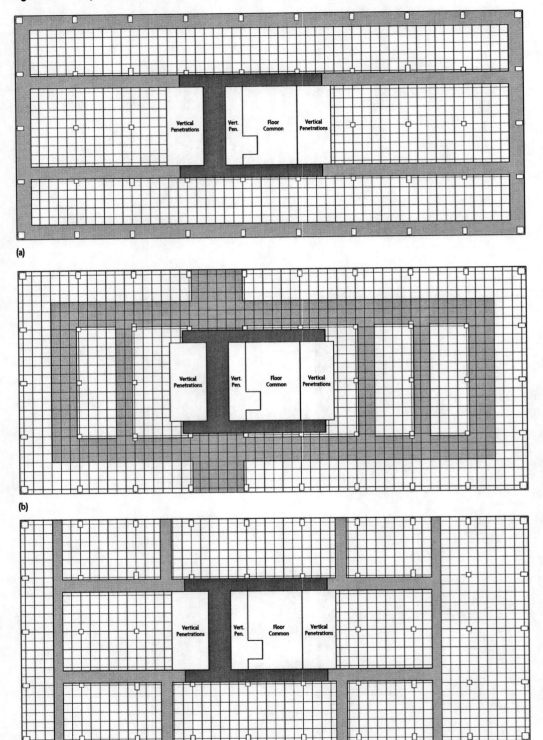

value toward LEED (Leadership in Energy and Environmental Design) certification (see Chapter 12).

- 8-2*b*—The light-shaded area in this case defines approximately fifteen-foot-wide open areas that can contain an *amorphic* (irregular or serpentine) circulation aisle,

Figure 8-3   Circulation Patterns "Off the Grid."

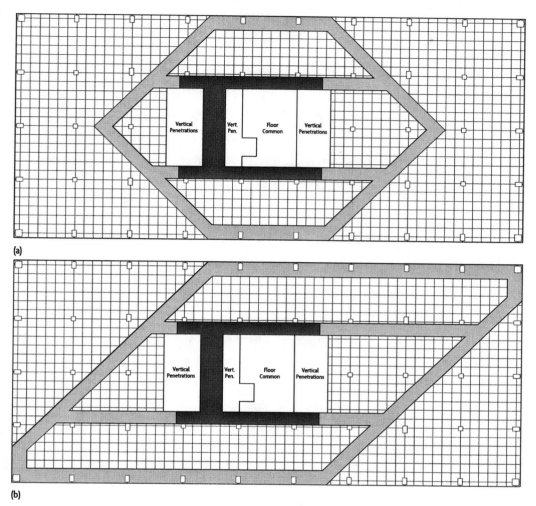

(a)

(b)

the remaining space being assigned to open workstations, small cubicles, and support spaces. Cross aisles bounded by interior columns on one side and modular gridlines on the other may be used in the same fashion. Such a configuration lends itself to full-floor tenant occupancy, in which the expanses of space adjacent to the elevator lobby can serve as formal reception or conference space highly accessible to visitors. The unshaded office areas around the perimeter and those encapsulated by the defined circulation pattern can be a combination of enclosed spaces with full-height walls, partial-height partitions, systems furniture, or open office areas.

• 8-2c—This is another variation on the corridor schemes with multiple aisles extending to the perimeter glass curtain wall. Extending circulation to the exterior in this manner brings daylight deep into the floor and helps occupants to orient themselves by being able to focus on exterior views while traveling to their destination.

The plans illustrated in Figure 8-3 show two possible ways of subdividing the floor using diagonal aisles. A planning approach such as this is likely to be more efficient in terms of utilization rate (area per person) when there is a relatively smaller number of exterior private offices needed. Private offices with windows can be planned on the ends of the building, with cubicles or systems furniture making a transition toward more open spaces adjacent to the interior aisles or corridors.

Figure 8-4   Comparative Plans of 2005 and 1927 Office Buildings.

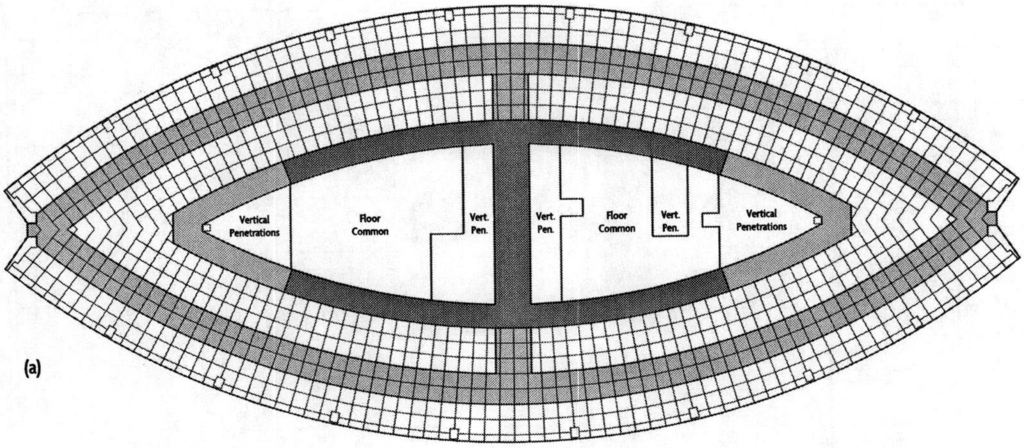

Building Character

Each building envelope has unique possibilities and limitations that dictate circulation patterns. Office floor plans of two other structures are pictured in Figure 8-4. These buildings were designed approximately sixty-five years apart, and the one in Figure 8-4a has just been completed as this is being written.°

The building in Figure 8-4b was completed in 1927 and extensively updated in 1986. Part of the renovation involved an addition, which now accounts for nearly one-third of the typical floor. (The portion of the building with the bay windows at left is the addition.) The full-height addition also contains new elevators and mechanical systems necessary to upgrade the building to contemporary standards.

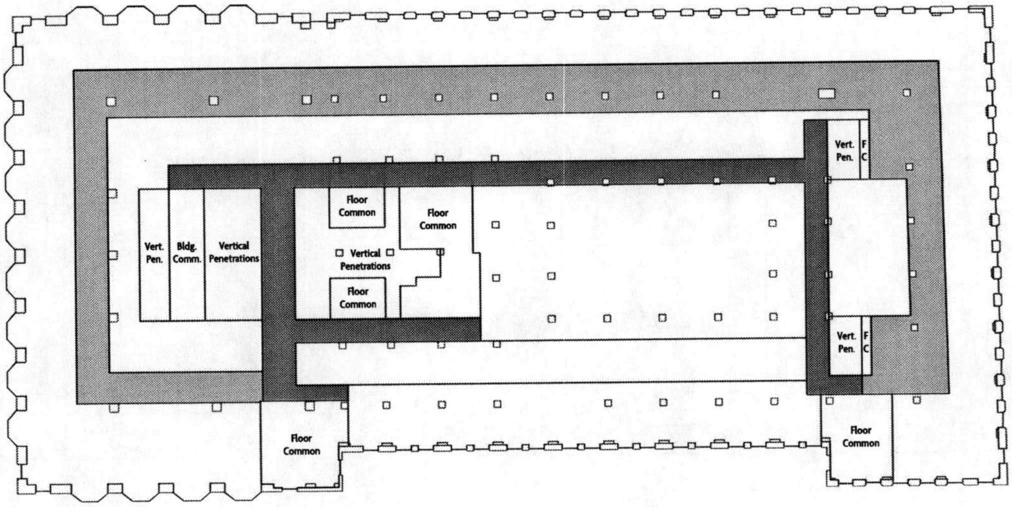

Fire stairs are located in the central area labeled *Vertical Penetrations* and in the two similarly designated core areas at the right end of the plan. There is no planning module, and an irregular column spacing can be seen that is typical of many early to mid-twentieth-century office buildings. It is necessary to maintain a rather extensive primary corridor system to connect the various parts of the floor to the exit stairs. Part of the rationale for the longitudinal corridor immediately adjacent to the central core is that it conveniently fits between the existing columns, thereby maximizing unobstructed space. The secondary circulation throughout the central part of the floor must logically connect into this main axis in a manner that meets exiting requirements according to the building code.

The area designated with light shading is a ten-to-fifteen-foot-wide zone that can contain orthogonal or irregular aisles interspersed with workstations and support spaces. The unshaded areas around the perimeter of the building can be used for private offices, as required, opening into these aisles. If enclosed offices are built in the interior space, circulation must be provided to connect with the main corridor axis.

By contrast, the building pictured in Figure 8-4a was completed in 2005, and its interior is entirely column-free. The distance between the perimeter columns is thirty-eight feet four inches, defining a module slightly less than five feet at the perimeter that tapers to about four feet as it nears the core. The radial distance from the exterior curtain wall to the central service core is forty-five feet.

As the dark shading indicates, access to all fire stairs can be easily maintained with a rather minimal primary corridor that can be extended around the ends of the core if desired. A multitenant floor can be divided as required by locating demising partitions along the radial module lines and providing entrances where necessary.

On a single-tenant floor, an internal circulation pattern can be defined (here shown ten feet wide by the lighter shading), equally divided between corridor and workstations/support along one side. A wider internal aisle can provide for workspace along both sides with the remaining (unshaded) space allocated for use as private offices and enclosed collaborative spaces if necessary. This type of configuration is suggested by the plan in Figure 8-5.

The perimeter of this office building is formed by two overlapping circles (compare with Figure TC2-1). One method of establishing a linear circulation pattern is to impose a polygon on the circular plan. Figure 8-5 shows that three sides of a fourteen-sided polygon can be exactly overlaid onto every third perimeter column, creating a trapezoidal grid that provides for straight partitions of substantial length.

Figure 8-5   Superimposing a Linear Grid on a Curved Floor Plan.

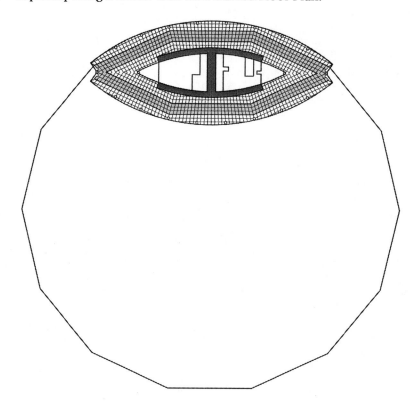

THE SPACE LAYOUT—PARTITION PLAN

Figure 8-6 illustrates the location of primary circulation in relation to the building core for the state government agency project that we began discussing at the end of the preceding chapter. We follow the development of this plan from blocking and stacking through the next several stages, including comparative area calculations (actual layout to program) and the classification of rooms by category, type, and space standard.

As in earlier plans, the dark shading indicates the elevator lobby within the core, a longitudinal corridor connecting the two fire stairs at either end of the core with the lobby, and a central corridor across the floor. Lighter shading shows a typical circulation pattern for the remainder of the floor. The precise location of the secondary corridors may vary by a module or so depending on the requirements for a specific floor, but the presence of a common pattern helps to make wayfinding intuitive and consistent from one floor to another.

Each of the plans in Figure 8-7 shows the stages in developing the space layout itself as well as the necessary documentation.

- 8-7*a*—Building outline with interior columns and service core, gross building area (*exterior*) and gross measured area (*interior*) polylines.
- 8-7*b*—Seventh-floor block layout (typical—compare with the topmost floor in the right-hand column of Figure CS1-4) with service core. Vertical penetrations in the core (stairways, elevators, and shafts) are shaded.
- 8-7*c*—Seventh-floor space layout (typical) showing polylines defining all assignable spaces and secondary circulation (light shading)—what ASTM/IFMA calls facility usable area (see the section titled Classifying Programmable Area in Chapter 6). In addition, primary circulation (dark shading) and the building core are shown, which all add up to *facility interior gross area* as defined by ASTM/IFMA.°

Note that the secondary circulation on the seventh floor generally follows the pattern defined in Figure 8-6, although in 8-7*c* we also see the space necessary to enter each fully enclosed office. A secondary longitudinal aisle has been added to service the relatively large number of private offices and cubicles on this floor, which is an executive area. Notice also that the primary transverse aisle has been shifted one module out of alignment with the elevator lobby so that it falls along a departmental boundary. In general, the corridor and aisle locations on all floors follow the typical circulation pattern, but some variations may be necessary to accommodate specific program requirements and organizational alignments.

Figure 8-6 Floor Plan Showing Typical Circulation Pattern.

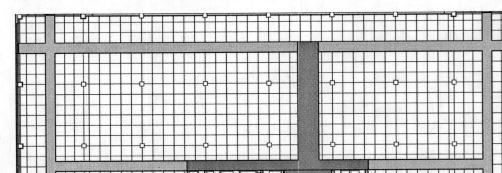

Figure 8-7   Plans Showing Development of a Typical Floor Layout.

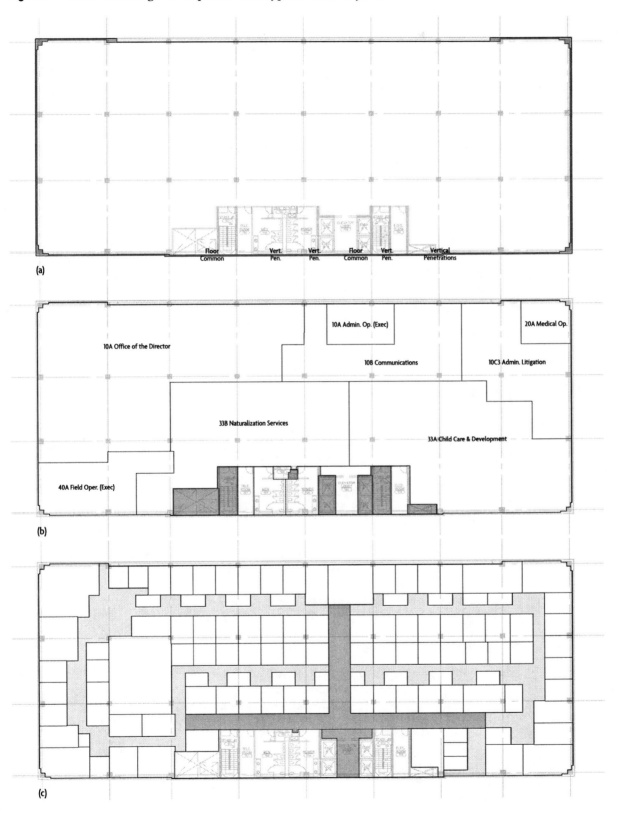

(a)

(b)

(c)

Figure 8-8   Floor Plan Showing Only Full-Height Partitions.

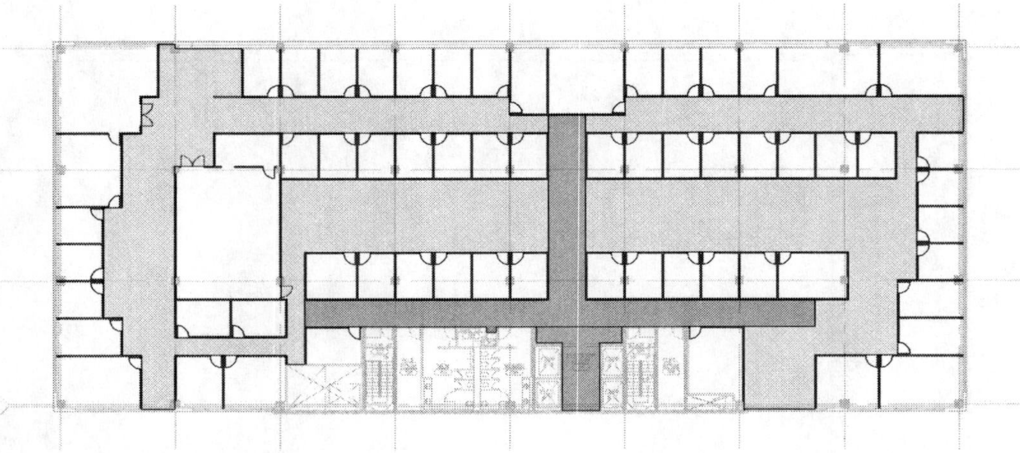

Figure 8-8 shows the same plan with the light shading indicating all secondary aisles plus the areas occupied by low-height cubicles and open workstations (not shown here). The full-height partitions form the principal visual boundaries of the interior space. The secondary aisles actually continue to the exterior windows at four points for visual orientation and connection with the exterior, as well as to let daylight into the interior space.

AREA-ALLOCATION ANALYSIS

Once the space plans have been completed, we have the opportunity to pipeline the layout data out of the drawing for analysis and validation using an Excel worksheet. Looking back at the flowchart in Figure 7-1, we are now at the stage of the process labeled *Polyline Data Analysis*.

Earlier, we saw how polylines can be used to automatically convert numeric area-requirements data into AutoCAD objects (blocks or bubbles) with which to begin planning layouts. Because polylines enclose space and possess the area property, we can reverse the process, using them to make accurate square footage measurements. The fact that they are also containers allows us to place text annotations on the drawing describing the ways in which spaces are being utilized. AutoCAD layers provide a convenient mechanism by which we can use these annotations to classify our space allocations for further analysis. Ultimately, the data extracted from our drawings can be used to populate tables belonging to a CAFM database that is being built or extensively updated.

Drawing Area Polylines

Area polylines provide the primary link between the floor plan and the facility database. Polyline area calculations allow accurate reporting of space utilization by organizational unit, cost center, and whatever other criteria are specified. Such a reporting system can calculate and prorate support space as well as common areas at the floor, building, site (campus), and even regional levels.

Polylines must be drawn with care or the spaces they represent will not be meaningful, no matter how accurate the CAD drawings. It is essential that the polylines register precisely with the underlying floor elements. For example, AutoCAD provides object snaps that can snap directly to the centerline of a wall intersection. AutoCAD also provides a command (BPOLY) that will automatically place a polyline around an enclosed space from a single reference point chosen within the space.

Area polylines on a floor plan should be drawn working from the *outside in*, beginning at the floor perimeter, allocating boundary-wall thicknesses to the spaces in each successive category. It is important to proceed in the proper sequence as specified by the numbers of the following sub-headings and in Figure 8-10. The polylines created at each stage provide clear boundaries for those that are added in the next step.

Floor Polyline (1)

Begin by drawing a polyline around the entire floor, following the inside face of the exterior wall. The inside face is either the solid portion or the glass, whichever constitutes more than half of the floor-to-ceiling height at any given point. This *dominant portion* may change going around the floor, so the polyline may move in and out according to the configuration at each point. Ignore all columns and other projections; the polylines should pass through them. The shading in Figure 8-9*a* indicates gross measured floor area.

Vertical Penetrations (2)

Vertical penetrations include *stairs, elevator shafts, flues, pipe shafts, vertical ducts, atria,* and *light wells.* Columns and small floor openings for cables or pipes are excluded. Polylines around vertical penetrations should include their surrounding walls (Figure 8-9*b*).

Figure 8-9   Floor Area, Vertical Penetrations, and Floor Common.

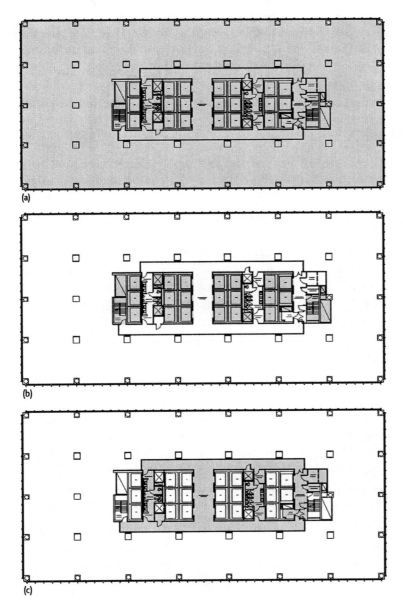

If a vertical penetration is adjacent to an exterior wall, draw the polyline coincident with the floor polyline at the inside face of the exterior wall. Several penetrations may be grouped in a single polyline if they are contiguous, unless it is necessary for accounting purposes to identify individual spaces within this classification.

Floor Common or Service Areas (3)
Typical floor common areas are shown in Figure 8-9c. Service areas used in common on a floor include *toilets; mechanical, electrical, communications,* and *maintenance closets; main lobbies,* together with *public or exit corridors* (primary circulation); and any voids (any inaccessible spaces that are not shafts).

If a *floor common* space adjoins a vertical penetration or an exterior wall, draw the new polyline to the already polylined boundary of the higher-level category; otherwise, service-area polylines should include their surrounding walls. Pipe chases within or between toilet rooms should be included with floor common.

Several floor common areas may be grouped within a single polyline if they are contiguous, except when special requirements dictate accounting for the spaces individually.

Primary Circulation (4)
Primary circulation is a subcategory of floor common, which should always be polylined *last* of the common areas and separately identified. Once again, surrounding walls are included where they have not already been allocated to vertical penetrations or other common (service) areas. Thus, polylines around primary corridors follow the inside face of the wall on the tenant side, the side *away from* the corridor.

Usable Area (5, 6, 7)
Usable area is the remaining space between the floor polyline and the *inside finished surface* of the public corridor or other permanent walls in any of the preceding categories. Usable is the lowest classification level specifically defined by the Building Owners and Managers Association (BOMA).°

The matrix in Figure 8-10 illustrates the classification levels with which we are usually concerned in constructing a facility database. The numbering indicates the order in which they should be polylined; numbers *1* through *4* have already been discussed. The arrows indicate that the space in the higher category always *gets the wall,* which is why it is important to proceed from outside in, in the sequence shown. When spaces with the same classification adjoin at a wall, the polyline should be located at the centerline of the wall.

The usable area on a floor may be subclassified according to occupancy, that is, whether the floor contains multiple tenants or several organizational units (divisions, departments, cost centers, etc.) within an enterprise. Polylines dividing one occupancy from another should be drawn at the centerlines of any demising partitions, equally allocating the wall between them. Usable area within a particular occupancy may be further broken down as follows:

* *Unassignable (5)*—Support space shared by several tenants or organizational units, which may be apportioned or prorated to the floor, building, or site for cost-allocation (*charge-back*) purposes.
* *Assignable/unoccupiable (6)*—Support space belonging to a particular organizational unit or tenant, which is not apportioned or prorated.
* *Assignable/occupiable (7)*—Private offices with full-height walls and doors, semi enclosed areas including *cubicles,* or open workstations that are assigned to specific individuals on a full-time basis.

Secondary Circulation (8)
Secondary circulation is the space remaining after all deductions have been made. It consists exclusively of the aisles leading to public corridors, exits, or main lobbies and, by definition, should include no boundary walls.

Area-Allocation Layers

To facilitate the classification of space, we have defined a set of names for layers on which polylines are to be drawn, as well as layers for the text objects to be placed within

Figure 8-10   Area Classifications for Polylining.

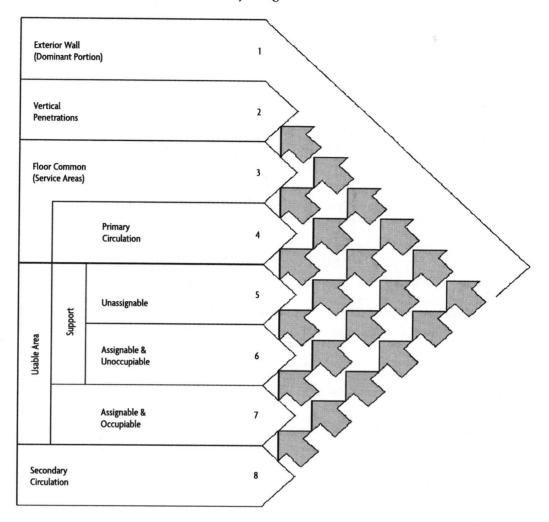

the polyline boundaries. All of the dozen and a half layers follow the *AIA*° layer-naming convention, beginning with *A-AREA*, plus a minor node denoting the use for which the layer is intended.

The polyline layers follow the area definitions used by the Building Owners and Managers Association (BOMA) and the International Facility Managers Association (IFMA). Both organizations have established national standards that complement each other, although each has a distinct focus. BOMA's primary interest is in the calculation of rentable area to be written into a lease, while IFMA provides a detailed method for classifying spaces *within* owned or leased space, detailing how that space is used. These standards and their application are discussed in more detail in Chapter 9, "Area Measurement Standards." In situations in which a space category belongs principally to a BOMA or IFMA classification, it is so noted. (Please note that most of these minor node names are not part of the "official" AIA convention but have been created in order to provide the level of detail necessary for area allocations according to the BOMA and IFMA standards.)

Polylines

- A-AREA-ASGN   Assignable (Blocks-Tenants)
- A-AREA-CIRC   Circulation (Primary)
- A-AREA-COMM   Common (BOMA)
- A-AREA-CSEC   Circulation (Secondary)

- A-AREA-GRSB   Gross Building (BOMA-Exterior)
- A-AREA-GRSM   Gross Measured (BOMA-Interior)
- A-AREA-PKNG   Parking (IFMA)
- A-AREA-ROOM   Rooms
- A-AREA-SERV   Service (IFMA)
- A-AREA-VERT   Vertical Penetrations
- A-AREA-VOID   Voids (IFMA)

Text Annotations
- A-AREA-IDEN   Identification (Blocks-Tenants)
- A-AREA-PROG   Program Requirements
- A-AREA-RCAT   Room Category
- A-AREA-RMID   Room ID (Number)
- A-AREA-RNAM   Room Name
- A-AREA-RSTD   Room Standard
- A-AREA-RTYP   Room Type

Table 8-1 lists the *A-AREA* layers alphabetically, and the dots in each column indicate their primary uses. The text-annotation layers are shown in the table in italics.

Making rentable and usable area calculations under the BOMA and IFMA standards (column 2 of Table 8-1) is covered in Chapter 9. Block diagrams (column 3) were discussed

Table 8-1   Layer-Use Convention for Space-Allocation Drawings

Layer Name	BOMA/ASTM	Block Calcs.	Space Diagrams	Layouts
A-AREA-ASGN	•	•		Assignable (Blocks-Tenants)
A-AREA-CIRC	•	•	•	Circulation (Primary)
A-AREA-COMM	•			Common (BOMA)
A-AREA-CSEC	•		•	Circulation (Secondary)
A-AREA-GRSB	•			Gross Building (BOMA-Exterior)
A-AREA-GRSM	•			Gross Measured (BOMA-Interior)
A-AREA-IDEN		•	•	*Identification (Blocks-Tenants)*
A-AREA-PKNG	•			Parking (IFMA)
A-AREA-PROG			•	*Program Requirements*
A-AREA-RCAT			•	*Room Category*
A-AREA-RMID			•	*Room ID (Number)*
A-AREA-RNAM			•	*Room Name*
A-AREA-ROOM			•	Rooms
A-AREA-RSTD			•	*Room Standard*
A-AREA-RTYP			•	*Room Type*
A-AREA-SERV	•			Service (IFMA)
A-AREA-VERT	•			Vertical Penetrations
A-AREA-VOID	•			Voids (IFMA)

in the preceding chapter. The layers designated by the dots in column 4 are used to annotate space layouts as described in the next section.

Allocation Process

1. Each usable space or room is assigned a room number, which is placed on the *A-AREA-RMID* text layer, as shown in Figure 8-11.

2. Each space is annotated with its program-requirements data using the dialog box pictured in Figure 7-20. The annotations consist of the level-sequence codes, space description, SAR key, and Area (Program) fields taken from the Access database, as shown in Figure 8-11*a*. They provide the means by which the layout data may be checked against the program to ensure that all spaces have been properly accounted for and are placed on the *A-AREA-PROG* text layer.

3. Figure 8-11*b* shows annotations used to classify the spaces for facility-management purposes. Each item is placed on its own layer using (in AutoCAD) standard single-line text objects, as follows:

Room Number	*A-AREA-RMID*
Room Category	*A-AREA-RCAT*
Room Type	*A-AREA-RTYP*
Room Name	*A-AREA-RNAM*
Room Standard	*A-AREA-RSTD*

All rooms belonging to the *usable area* classification have their polylines drawn on the *A-AREA-ROOM* layer. Exterior walls, vertical penetrations, and common and circulation areas each have their own layers to facilitate accurate area calculations (please refer to Table 8-1). The classification by layer provides the means by which the data-export procedure separates the annotations. The text may be of any size or style, the only qualification being that it must be placed entirely within the polyline boundaries.

Figure 8-12 shows a portion of the floor with the building core (vertical penetrations are highlighted—this is the same typical floor layout depicted in Figures 8-6, 8-7, 8-8, and 8-11).

The format of the program requirements, area allocations, and other relevant data, as they are exported from AutoCAD to the polyline data-analysis worksheet, are illustrated in Table 8-2. The letters in the first column of the table correspond to the columns of the Excel worksheet.

- A–D—Original Program-Requirements Data
- E—Actual Area Allocations (calculated from the polylines on the plan)
- F–J—Text Annotations indicating Space Utilization (if polylines on the *A-AREA-ASGN* layer are present, text on the *A-AREA-IDEN* layer is exported to the worksheet so that block-plan designations or tenant names are included)
- K–L—Building/Floor (Space Hierarchy)
- M–O—Group/Division/Department (Organization Hierarchy)
- Q—Layer Name (From this column to the right, the exported data are included for verification purposes)
- R–X—Room Numbers and the number of text items within the polyline
- Y–Z—Polyline type and closure status (AutoCAD properties)
- AA–AB—Drawing name and polyline entity handle (AutoCAD properties)

Figure 8-13*a* and Figure 8-13*b* illustrate an Excel worksheet created using the *Cataloged-Polyline-Data* procedure. The data are taken from the layouts shown in Figure 8-11 and Figure 8-12.

Figure 8-11   Space-Allocation Plan with Text Annotations.

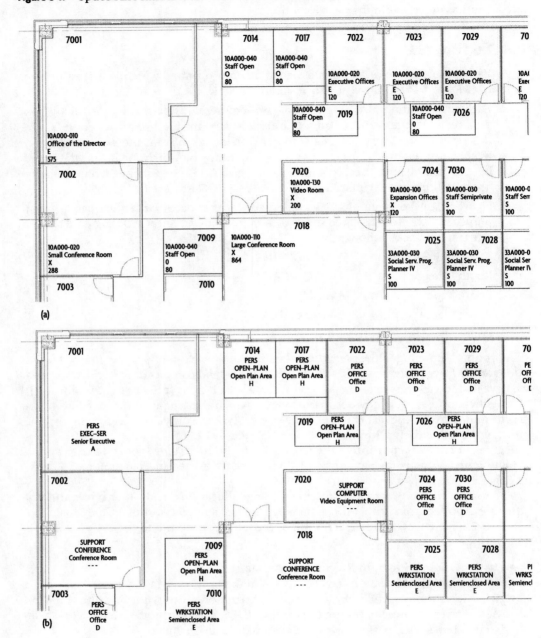

FACILITY-MANAGEMENT DATA

Once the data have been converted from drawing annotations to the tabular format of a worksheet, they can be visually and programmatically checked and edited before being input (or uploaded) to a CAFM system. Here, as elsewhere, sorting the data by selected columns provides an effective means of performing consistency checks. This can save a great deal of time if it is done prior to pipelining the data into a relational database. Once in the relational environment, invalid items or format inconsistencies will cause numerous entry errors.

For example, codes used to assign spaces to a particular building and floor (and by organizational unit) may differ in format from those used during the development of the

Figure 8-12 Space-Allocation Plan Showing Building Core.

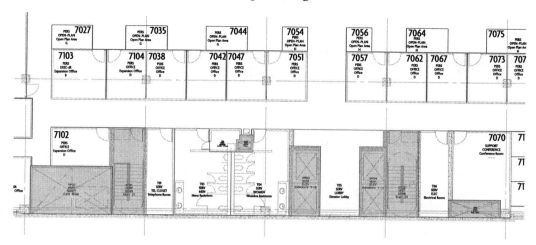

Table 8-2 Spreadsheet Format of Data Exported from Drawings

	Column Heading	Polyline Layer	Text Layer
A	Level/Sequence	ROOM	
B	Description	ROOM (GRSB,M)	
C	SAR Key	ROOM	PROG
D	Area (Program)	ROOM	
E	Area (Allocated)	All Layers	Calculated from Polylines
F	Room Number		RMID
G	Room Name		RNAM (IDEN)
H	Room Category	ASGN, CIRC, COMM, CSEC, PKNG, ROOM, SERV, VERT, VOID	RTYP
I	Room Type		RCAT
J	Room Standard		RSTD
K	Building	Space Hierarchy Codes (for Facility Management)	
L	Floor		
M	Group	Organization Hierarchy Codes (for Facility Management)	
N	Division		
O	Department		
Q	Layer Name	Polyline Layer for Current Row of Data	
R			
S	sRMID		
T	aIDEN		
	aPROG		
U	aRMID	Room Numbers and Numbers of Text Items within Current Polyline (Internal Flags)	
V	aRNAM		
	aRCAT		
W	aRTYP		
X			
Y		Polyline Type (LWPOLYLINE/POLYLINE)	
Z		Closed Polyline (TRUE/FALSE)	
AA		Drawing Name (.DWG)	
AB		Entity Handle	

Figure 8-13a   Cataloged Polyline Data Exported from Drawings.

Level/Sequence Description	SAR Key	Area (Program)	Area (Allocated)	Room Number	Room Name	Room Category	Room Type	Room Standard
10A000-010 Office of the Director	E	576	599	7001	Senior Executive	PERS	EXEC-SR	A
10A000-120 Small Conference Room	X	288	348	7002	Conference Room	SUPPORT	CONFERENCE	---
10A000-020 Executive Offices	E	120	124	7003	Office	PERS	OFFICE	D
10A000-040 Staff Open	O	80	79	7009	Open Plan Area	PERS	OPEN-PLAN	H
10A000-030 Staff Semi-Private	S	100	103	7010	Semi-Enclosed Area	PERS	WRKSTATION	E
10A000-040 Staff Open	O	80	88	7014	Open Plan Area	PERS	OPEN-PLAN	H
10A000-040 Staff Open	O	80	88	7017	Open Plan Area	PERS	OPEN-PLAN	H
10A000-110 Large Conference Room	X	864	1,000	7018	Conference Room	SUPPORT	CONFERENCE	---
10A000-040 Staff Open	O	80	61	7019	Open Plan Area	PERS	OPEN-PLAN	H
10A000-130 Video Room	X	200	163	7020	Video Equipment Room	SUPPORT	COMPUTER	---
10A000-020 Executive Offices	E	120	132	7022	Office	PERS	OFFICE	D
10A000-020 Executive Offices	E	120	132	7023	Office	PERS	OFFICE	D
10A000-100 Expansion Offices	X	120	127	7024	Office	PERS	OFFICE	D
33A000-030 Social Serv. Prog. Planner IV	S	100	98	7025	Semi-Enclosed Area	PERS	WRKSTATION	E
10A000-040 Staff Open	O	80	61	7026	Open Plan Area	PERS	OPEN-PLAN	H
33A000-040 Social Serv. Prog. Planner III	S	100	63	7027	Open Plan Area	PERS	OPEN-PLAN	G
33A000-030 Social Serv. Prog. Planner IV	S	100	100	7028	Semi-Enclosed Area	PERS	WRKSTATION	E
10A000-020 Executive Offices	E	120	132	7029	Office	PERS	OFFICE	D
10A000-030 Staff Semi-Private	S	100	127	7030	Office	PERS	OFFICE	D
10A000-030 Staff Semi-Private	S	100	127	7032	Office	PERS	OFFICE	D
33A000-030 Social Serv. Prog. Planner IV	S	100	100	7033	Semi-Enclosed Area	PERS	WRKSTATION	E
33A000-040 Social Serv. Prog. Planner III	S	100	64	7035	Open Plan Area	PERS	OPEN-PLAN	G
33B400-010 Exec. II	E	120	128	7038	Office	PERS	OFFICE	D
33B300-020 MPA I	E	100	128	7042	Office	PERS	OFFICE	D
33A000-040 Social Serv. Prog. Planner III	S	100	64	7044	Open Plan Area	PERS	OPEN-PLAN	G
33B300-010 Exec. III	E	120	128	7047	Office	PERS	OFFICE	D
33B200-010 Exec. II	E	120	130	7051	Office	PERS	OFFICE	D
33A000-040 Social Serv. Prog. Planner III	S	100	39	7054	Open Plan Area	PERS	OPEN-PLAN	H
33B300-030 Office Associate	O	80	39	7056	Open Plan Area	PERS	OPEN-PLAN	H
33B100-050 Exec II	E	120	130	7057	Office	PERS	OFFICE	D
33A000-050 Accountant III	E	120	128	7062	Office	PERS	OFFICE	D
33B100-070 Office Coordinator	O	80	64	7064	Open Plan Area	PERS	OPEN-PLAN	H
33A000-050 Accountant III	E	120	128	7067	Office	PERS	OFFICE	D
33B100-100 Conference Room	X	300	270	7070	Conference Room	SUPPORT	CONFERENCE	---
33A000-020 Supervisor - Exec. III	E	120	128	7073	Office	PERS	OFFICE	D
33A000-060 Secy. - Office Assoc. Option 2	O	80	64	7075	Open Plan Area	PERS	OPEN-PLAN	H
33A000-020 Supervisor - Exec. III	E	120	128	7077	Office	PERS	OFFICE	D
			236	7101	Expansion Office	PERS	EXEC-SR	A
			145	7102	Expansion Office	PERS	OFFICE	D
			175	7103	Expansion Office	PERS	EXEC-JR	B
			128	7104	Expansion Office	PERS	OFFICE	D
			67	7107	Expansion	PERS	OPEN-PLAN	H
			67	7108	Expansion	PERS	OPEN-PLAN	H
			67	7109	Expansion	PERS	OPEN-PLAN	H
			1,443	C01	Primary Circulation	SERV	PRIMCIRC	
			2,289	C02	Secondary Circulation	SUPPORT	SECNCIRC	
			1,594	C03	Secondary Circulation	SUPPORT	SECNCIRC	
			205	C04	Secondary Circulation	SUPPORT	SECNCIRC	
			191	ST01	Stair 01	VERT	STAIR	
			187	ST02	Stair 02	VERT	STAIR	
			161	T01	Telephone Room	SERV	TEL CLOSET	
			42	T02	Janitor Closet	SERV	JANITOR	
			284	T03	Mens Restroom	SERV	MEN	
			296	T04	Womens Restroom	SERV	WOMEN	
			271	T05	Elevator Lobby	SERV	LOBBY	
			184	T06	Electrical Room	SERV	ELEC	
			254	VP01	Duct Riser	VERT	SHAFT	
			17	VP02	Flue	VERT	SHAFT	
			141	VP03	Elevators 3-4	VERT	ELEV	
			118	VP04	Elevators 1-2	VERT	ELEV	
			58	VP05	Duct Riser	VERT	SHAFT	
			561		10A Admin. Op. (Exec)			
			6,153		10A Office of the Director			
			2,108		10B Communications			
			2,015		10C3 Admin. Litigation			
			433		20A Medical Op.			
			4,760		33A Child Care & Development			
			3,657		33B Naturalization Services			
			1,283		40A Field Oper.(Exec)			
GROSS BUILDING AREA			23,540					
GROSS MEASURED AREA			23,173					

initial space program. New group, division, and department codes can be assigned to each room entry using the worksheet and automatically checked against validation tables. Columns *K* through *L* and *M* through *O*, as shown in Figure 8-13*b*, provide space for the creation and replication of these initial entries in whatever format may be needed.

The space-allocation information: *Room Number, Name, Category, Type,* and *Standard* (Figure 8-13*a*, columns *F* through *J*) also need to be validated. This can be accomplished using Excel macros to check the line items for the proper data and formatting. Figure 8-14 shows some of the error messages that are used to flag inconsistencies during an automatic scan of the worksheet. Cells containing incorrect data can also be marked for later correction simply by having the macro change the font or background color. Room numbers, for example, are likely to be key fields and must therefore be unique. If numbering

Figure 8-13b   Cataloged Polyline Data Exported from Drawings.

Building	Floor	Group	Division	Department	...	Layer Name	sRMID	aIDEN	aPROG	aRMID	aRNAM	aRCAT	aRTYP	Polyline Type	Closed Polyline	Drawing Name	Entity Handle
						A-AREA-ROOM	7001	0	4	1	1	1	1	AcDbPolyline	TRUE	AllocationPlan.dwg	3BC6
						A-AREA-ROOM	7002	0	4	1	1	1	1	AcDbPolyline	TRUE	AllocationPlan.dwg	3BC5
						A-AREA-ROOM	7003	0	4	1	1	1	1	AcDbPolyline	TRUE	AllocationPlan.dwg	3BC4
						A-AREA-ROOM	7009	0	4	1	1	1	1	AcDbPolyline	TRUE	AllocationPlan.dwg	3E0D
						A-AREA-ROOM	7010	0	4	1	1	1	1	AcDbPolyline	TRUE	AllocationPlan.dwg	3D40
						A-AREA-ROOM	7014	0	4	1	1	1	1	AcDbPolyline	TRUE	AllocationPlan.dwg	3E12
						A-AREA-ROOM	7017	0	4	1	1	1	1	AcDbPolyline	TRUE	AllocationPlan.dwg	3E11
						A-AREA-ROOM	7018	0	4	1	1	1	1	AcDbPolyline	TRUE	AllocationPlan.dwg	3BEA
						A-AREA-ROOM	7019	0	4	1	1	1	1	AcDbPolyline	TRUE	AllocationPlan.dwg	3E10
						A-AREA-ROOM	7020	0	4	1	1	1	1	AcDbPolyline	TRUE	AllocationPlan.dwg	3BEB
						A-AREA-ROOM	7022	0	4	1	1	1	1	AcDbPolyline	TRUE	AllocationPlan.dwg	39B2
						A-AREA-ROOM	7023	0	4	1	1	1	1	AcDbPolyline	TRUE	AllocationPlan.dwg	39B1
						A-AREA-ROOM	7024	0	4	1	1	1	1	AcDbPolyline	TRUE	AllocationPlan.dwg	3CD4
						A-AREA-ROOM	7025	0	4	1	1	1	1	AcDbPolyline	TRUE	AllocationPlan.dwg	3CF1
						A-AREA-ROOM	7026	0	4	1	1	1	1	AcDbPolyline	TRUE	AllocationPlan.dwg	3E0F
						A-AREA-ROOM	7027	0	4	1	1	1	1	AcDbPolyline	TRUE	AllocationPlan.dwg	3E1D
						A-AREA-ROOM	7028	0	4	1	1	1	1	AcDbPolyline	TRUE	AllocationPlan.dwg	3CF2
						A-AREA-ROOM	7029	0	4	1	1	1	1	AcDbPolyline	TRUE	AllocationPlan.dwg	39B0
						A-AREA-ROOM	7030	0	4	1	1	1	1	AcDbPolyline	TRUE	AllocationPlan.dwg	3CD3
						A-AREA-ROOM	7032	0	4	1	1	1	1	AcDbPolyline	TRUE	AllocationPlan.dwg	3CD2
						A-AREA-ROOM	7033	0	4	1	1	1	1	AcDbPolyline	TRUE	AllocationPlan.dwg	3CF3
						A-AREA-ROOM	7035	0	4	1	1	1	1	AcDbPolyline	TRUE	AllocationPlan.dwg	3E1C
						A-AREA-ROOM	7038	0	4	1	1	1	1	AcDbPolyline	TRUE	AllocationPlan.dwg	3D13
						A-AREA-ROOM	7042	0	4	1	1	1	1	AcDbPolyline	TRUE	AllocationPlan.dwg	3D12
						A-AREA-ROOM	7044	0	4	1	1	1	1	AcDbPolyline	TRUE	AllocationPlan.dwg	3E48
						A-AREA-ROOM	7047	0	4	1	1	1	1	AcDbPolyline	TRUE	AllocationPlan.dwg	3D11
						A-AREA-ROOM	7051	0	4	1	1	1	1	AcDbPolyline	TRUE	AllocationPlan.dwg	3D10
						A-AREA-ROOM	7054	0	4	1	1	1	1	AcDbPolyline	TRUE	AllocationPlan.dwg	3E47
						A-AREA-ROOM	7056	0	4	1	1	1	1	AcDbPolyline	TRUE	AllocationPlan.dwg	3E46
						A-AREA-ROOM	7057	0	4	1	1	1	1	AcDbPolyline	TRUE	AllocationPlan.dwg	3B34
						A-AREA-ROOM	7062	0	4	1	1	1	1	AcDbPolyline	TRUE	AllocationPlan.dwg	3B33
						A-AREA-ROOM	7064	0	4	1	1	1	1	AcDbPolyline	TRUE	AllocationPlan.dwg	3E45
						A-AREA-ROOM	7067	0	4	1	1	1	1	AcDbPolyline	TRUE	AllocationPlan.dwg	3B32
						A-AREA-ROOM	7070	0	4	1	1	1	1	AcDbPolyline	TRUE	AllocationPlan.dwg	3D2A
						A-AREA-ROOM	7073	0	4	1	1	1	1	AcDbPolyline	TRUE	AllocationPlan.dwg	3B31
						A-AREA-ROOM	7075	0	4	1	1	1	1	AcDbPolyline	TRUE	AllocationPlan.dwg	3E44
						A-AREA-ROOM	7077	0	4	1	1	1	1	AcDbPolyline	TRUE	AllocationPlan.dwg	3B30
						A-AREA-ROOM	7101	0	0	1	1	1	1	AcDbPolyline	TRUE	AllocationPlan.dwg	3C9D
						A-AREA-ROOM	7102	0	0	1	1	1	1	AcDbPolyline	TRUE	AllocationPlan.dwg	3D2B
						A-AREA-ROOM	7103	0	0	1	1	1	1	AcDbPolyline	TRUE	AllocationPlan.dwg	3D15
						A-AREA-ROOM	7104	0	0	1	1	1	1	AcDbPolyline	TRUE	AllocationPlan.dwg	3D14
						A-AREA-ROOM	7107	0	0	1	1	1	1	AcDbPolyline	TRUE	AllocationPlan.dwg	3E77
						A-AREA-ROOM	7108	0	0	1	1	1	1	AcDbPolyline	TRUE	AllocationPlan.dwg	3E76
						A-AREA-ROOM	7109	0	0	1	1	1	1	AcDbPolyline	TRUE	AllocationPlan.dwg	3E75
						A-AREA-CIRC	C01	0	0	1	1	1	1	AcDbPolyline	TRUE	AllocationPlan.dwg	4E14
						A-AREA-CSEC	C02	0	0	1	1	1	1	AcDbPolyline	TRUE	AllocationPlan.dwg	4E7C
						A-AREA-CSEC	C03	0	0	1	1	1	1	AcDbPolyline	TRUE	AllocationPlan.dwg	4EC7
						A-AREA-CSEC	C04	0	0	1	1	1	1	AcDbPolyline	TRUE	AllocationPlan.dwg	4F61
						A-AREA-VERT	ST01	0	0	1	1	1	1	AcDbPolyline	TRUE	AllocationPlan.dwg	4398
						A-AREA-VERT	ST02	0	0	1	1	1	1	AcDbPolyline	TRUE	AllocationPlan.dwg	439D
						A-AREA-SERV	T01	0	0	1	1	1	1	AcDbPolyline	TRUE	AllocationPlan.dwg	43A2
						A-AREA-SERV	T02	0	0	1	1	1	1	AcDbPolyline	TRUE	AllocationPlan.dwg	446E
						A-AREA-SERV	T03	0	0	1	1	1	1	AcDbPolyline	TRUE	AllocationPlan.dwg	43AE
						A-AREA-SERV	T04	0	0	1	1	1	1	AcDbPolyline	TRUE	AllocationPlan.dwg	43B0
						A-AREA-COMM	T05	0	0	1	1	1	1	AcDbPolyline	TRUE	AllocationPlan.dwg	43B2
						A-AREA-SERV	T06	0	0	1	1	1	1	AcDbPolyline	TRUE	AllocationPlan.dwg	4D41
						A-AREA-VERT	VP01	0	0	1	1	1	1	AcDbPolyline	TRUE	AllocationPlan.dwg	4397
						A-AREA-VERT	VP02	0	0	1	1	1	1	AcDbPolyline	TRUE	AllocationPlan.dwg	43A8
						A-AREA-VERT	VP03	0	0	1	1	1	1	AcDbPolyline	TRUE	AllocationPlan.dwg	439B
						A-AREA-VERT	VP04	0	0	1	1	1	1	AcDbPolyline	TRUE	AllocationPlan.dwg	439C
						A-AREA-VERT	VP05	0	0	1	1	1	1	AcDbPolyline	TRUE	AllocationPlan.dwg	439E
						A-AREA-ASGN		1	0	2	0	0	0	AcDbPolyline	TRUE	AllocationPlan.dwg	53EB
						A-AREA-ASGN		1	68	29	17	17	17	AcDbPolyline	TRUE	AllocationPlan.dwg	5063
						A-AREA-ASGN		1	0	15	3	3	3	AcDbPolyline	TRUE	AllocationPlan.dwg	507F
						A-AREA-ASGN		1	0	12	0	0	0	AcDbPolyline	TRUE	AllocationPlan.dwg	5451
						A-AREA-ASGN		1	0	2	0	0	0	AcDbPolyline	TRUE	AllocationPlan.dwg	509A
						A-AREA-ASGN		1	36	27	13	13	13	AcDbPolyline	TRUE	AllocationPlan.dwg	5443
						A-AREA-ASGN		1	44	19	15	15	15	AcDbPolyline	TRUE	AllocationPlan.dwg	5444
						A-AREA-ASGN		1	0	7	0	0	0	AcDbPolyline	TRUE	AllocationPlan.dwg	5072
						A-AREA-GRSB		1	1	1	1	1	1	AcDbPolyline	TRUE	AllocationPlan.dwg	551A
						A-AREA-GRSM		8	148	4	48	48	48	AcDbPolyline	TRUE	AllocationPlan.dwg	450A

assignments have been duplicated on the plan, the invalid fields can easily be identified in this fashion.

Room Categories and Types

Categories are broad classifications that differentiate rooms according to the functions they serve. Typical room category codes may include the following:

ADMIN	Administrative
LAB	Laboratory

Figure 8-14   Error-Message Dialog Boxes in Polyline Catalog Procedure.

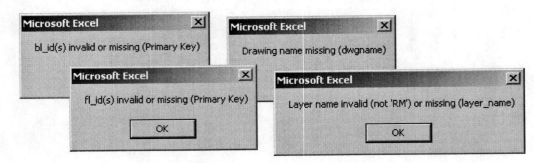

PERS	Personnel Area
PROD	Production Area
SERV	Service Area Rooms
STORAGE	
SUPPORT	
VERT	Vertical Penetrations

In a CAFM system, room categories determine how spaces are to be treated when calculating charge-backs or making accounting classifications. Room types further break down these room categories for space-inventory-accounting purposes. Subgroupings of service and support areas, for example, allow the use of shared spaces to be prorated by floor, building, or other allocation method. Even room names are sometimes standardized and need to be validated, as in the case of federally owned or leased space managed by the GSA (see Chapter 9, Table 9-4).

Table 8-3 shows a basic set of room categories and types, many of which are associated with this chapter's space-plan examples and exported worksheet data.

Table 8-3   Basic Room Categories and Types

Room Category Category Description	Room Type	Type Description
ADMIN Administrative	FILE	File Room
	MAIL	Mail Room
	SECY	Secretary Station
LAB Laboratory	ANIMAL	Animal Quarters
	BENCH	Bench Lab
	CLASS	Classroom Laboratory
	DRY	Dry Lab
	RESEARCH	Research Lab
	WET	Wet Lab
PERS Personnel Area	EXEC	Executive
	EXEC-JR	Junior Executive
	EXEC-SEC	Executive Secretary
	EXEC-SR	Senior Executive
	OFFICE	Office
	OPEN-PLAN	Open Plan Area
	WRKSTATION	Workstation

Table 8-3   *(continued)*

Room Category Category Description	Room Type	Type Description
PROD Production Area	ASSEMBLY	Assembly Area
	CLEAN-ROOM	Clean Room
	FABRIC	Fabrication
	SERV-CENT	Service Center
	SHIP-REC	Shipping/Receiving
	STAGING	Staging
SERV Service Area Rooms	CORRIDOR	Corridor
	ENTR FACIL	Entrance Facility
	EQPM ROOM	Equipment Room
	HALLWAY	Hallway
	JANITOR	Janitor/Custodial Closet
	LOBBY	Lobby
	MECH	Mechanical Closet/Room
	MEN	Men's Restroom
	PRIMCIRC	Primary Circulation
	SERVICE	Service Area
	TEL CLOSET	Telecom Closet
	WOMEN	Women's Restroom
STORAGE Storage	CHEM	Chemical Storage
	STORAGE	General Storage
	WAREHOUSE	Warehouse
SUPPORT Storage	AUDITORIUM	Auditorium
	CAFETERIA	Cafeteria
	COAT	Coat Room
	COMPUTER	Computer Room
	CONFERENCE	Conference
	COPY	Copy Room
	KITCHEN	Kitchen Area
	LIBRARY	Library/Reading Room
	LOUNGE	Lounge Area
	MECH	Mechanical Room
	SEC-CIRC	Secondary Circulation
	SECURITY	Security Station
	TRAINING	Training/Classroom
VERT Vertical Penetrations	ELEV	Elevator
	PIPE	Pipes
	SHAFT	Shaft, Duct
	STAIR	Stairs
	VERT	Vertical Penetration

Data Pipelining

Macros can also be used to write scripts to pipeline the finalized worksheet data into a CAFM database. The exact format of the script depends on the relational-database system being used, but it is usually written in a dialect of SQL. Relational databases, including Structured Query Language or SQL, are the subject of Chapter 10. A fragment of one such script is shown in Example 8-1.

```
INSERT INTO rm (bl_id, fl_id, rm_id, area,
       dp_id, dv_id, dwgname, ehandle, layer_name,
       name, rm_cat, rm_type, rm_std)
   VALUES ('A', ' 7', ' 7001', 598.8407,
       (Null), (Null), 'DwgNam', ' 10001', 'RM',
       'Senior Executive', 'PERS', 'EXEC-SR', 'A');

INSERT INTO rm (bl_id, fl_id, rm_id, area,
       dp_id, dv_id, dwgname, ehandle, layer_name,
       name, rm_cat, rm_type, rm_std)
   VALUES ('A', ' 7', ' 7002', 348.3333,
       (Null), (Null), 'DwgNam', ' 10002', 'RM',
       'Conference Room', 'SUPPORT', 'CONFERENCE', '. . .');

INSERT INTO rm (bl_id, fl_id, rm_id, area,
       dp_id, dv_id, dwgname, ehandle, layer_name,
       name, rm_cat, rm_type, rm_std)
   VALUES ('A', ' 7', ' 7003', 123.75,
       (Null), (Null), 'DwgNam', ' 10003', 'RM',
       'Office', 'PERS', 'OFFICE', 'D');
```

Example 8-1 Partial SQL Script for Populating a CAFM Database Table

Structured Query Language is an extremely verbose language, very much suited to scripts such as this. In the script shown in Example 8-1, each INSERT INTO statement begins with the name of the table, is followed by a list of the field names for each item to be inserted, and concludes with a list of the VALUES to be inserted into each field. Each such statement creates a record in the table and then populates its fields according to the order of the items in the lists.

CLOSING WORDS

We began this chapter with a discussion of circulation patterns and wayfinding in relation to several different building configurations. The character of each building is a primary determinant of the ways in which space can be configured within its boundaries. Apart from the functional program, major circulation options should be examined before detailed space planning is undertaken.

Once an overall pattern has been established, the location of each required space must be determined. A particular building layout may offer many options, and it is at this point that different approaches can be most effectively and economically studied. The efficiency with which occupiable space can be utilized and the logical organization of the circulation system need to be worked out, each in relation to the other. If a new building is being planned, both of these key elements must be considered in arriving at proposed architectural options.

As space layouts are developed and annotated with specific allocation requirements, each category of space can be represented by a polyline. In Chapter 8, we established a set of layer names for polylines consistent with the AIA guidelines, together with a methodology for drawing them according to recognized standards. Finally, we saw how spaces can be classified according to category and type and how the data can be pipelined into a CAFM database.

In Chapter 9, we examine several standards for the classification of space and the measurement of area. There is a large and confusing array of nomenclature associated with building-area measurement. ANSI/BOMA is the leading standard for calculating space in commercial office buildings and focuses on determining *rentable* area, which is the figure most often written into leases. ASTM/IFMA takes this process to the next level of detail, providing terminology for subcategorizing rentable space according to functional use. In the next chapter, we make a detailed examination of the BOMA method for calculating and tabulating area measurements using AutoCAD polylines.

Chapter 9 also considers the way in which the federal government, through its General Services Administration, applies the BOMA methodology in classifying the space it manages. We also look at an area-classification system widely used by postsecondary educational institutions.

TIME CAPSULE   Cathedrals of Today

The nineteenth century ended and the twentieth began with the development of the *high-rise,* a relative term given that the record-setting height increased by an order of magnitude over the course of the latter century. Actually, by current code definition, the medieval cathedral would be considered a high-rise building.

Chicago's Reliance Building, designed by Daniel H. Burnham & Co., marked the end of the nineteenth century at 202 feet and fourteen stories. Shortly thereafter, in 1903, the same firm's Flatiron Building in New York was completed at twenty-one stories and 285 feet. The Empire State Building, built to a height of 1,250 feet in 1930–31, was the world's tallest building for forty years, until the construction of the dual towers of the World Trade Center. All made use of progressively more advanced engineering techniques, high-grade steel, and reinforced concrete. These were significant advances in building technology—as significant as the Gothic arch had been from the Greek column and beam.

A detailed typological analysis of Frank Lloyd Wright's buildings classifies a substantial number of his works into three archetypes, the first two being oriented around the hearth or the atrium and the third being the tower.° Wright's characteristic approach to towers eschewed the rectangle in favor of the trapezoidal and circular forms. Earlier, we discussed his circular museum in the form of a spiral ramp. To this example, we can add, among others, his proposed mile-high skyscraper for Chicago (late 1950s), trapezoidal in plan but strongly triangular in its expression (Figure TC6-1).

The Johnson Research Tower (1944), with its alternating circular and square floors, and the lesser-known Price Tower in Bartlesville, Oklahoma (1952), each have distinctive geometries.

Mies van der Rohe brought the rectangular, glass-enclosed office building to its apotheosis with his bronze-clad and finely detailed Seagram Building in New York City (1954–58), along with numerous other examples, such as One Illinois Center in Chicago (1970). Until the advent of postmodernism in the 1980s, the glass-curtain wall in the International Style appeared in numberless variations, most highly efficient from a space-utilization standpoint but some unwelcoming and lacking in human scale.

Two of our principal themes in Chapter 5 were the convergence of technologies and globalization. We tabulated some of the most significant technological advances that have found commercial applications since 1870, and it is no coincidence that the rise of our contemporary cathedrals coincided with these events. The "rational, systematic, and specialized pursuit of science" that Max Weber spoke of began in Europe in the eighteenth century; migrated the Atlantic to the New

Figure TC6-1   Ideographic Plan of Wright's Mile-High Tower.

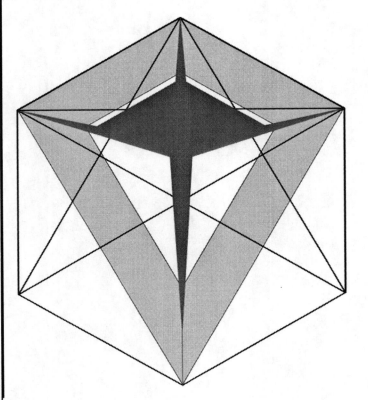

World during the centuries following; and, particularly since the mid-twentieth century, has continued to encompass the globe. There is little doubt that capitalism is indeed one of the most fateful forces in our lives, and it is in the geometry of these monuments that these forces are expressed. Tragically, we need only remember the events of September 11, 2001, to be reminded of the symbolic significance of such buildings in our recent history.

The twin towers of the World Trade Center in New York's financial center were designed by Minoru Yamasaki and completed in 1973. The two virtually identical towers were 1,365 feet in height; and, inclusive of several smaller buildings on its plaza, the complex contained nearly ten million square feet of space. Completed during the first major oil crisis, it took almost ten years for the World Trade Center to be profitably rented. The period of the mid-1980s marked the beginning of nearly twenty years of worldwide economic growth.

Although the twin towers were not universally loved due to their severity and scale, their profile became an icon in both the local and international landscapes. When the Trade Center was first bombed by terrorists in 1993,

it became clear that the towers had become an international symbol as well. Eight and one-half years later, nearly three thousand lives were lost in the collapse of both towers after each was struck by a fully-fueled, heavy commercial aircraft.

Yet the drive to assemble structures of this magnitude—symbols of economic development and national pride—continues on an unprecedented international scale. Commercial high-rises first appeared in dense urban areas as a means of attaining higher rates of space utilization, but this is now often only part of the reason for their development. When Frank Lloyd Wright proposed his mile-high skyscraper for Chicago, he said such buildings should be placed far enough apart that they may cast their shadows on the ground. Professor Hilberseimer asserted a similar objective in his rules for sun penetration. Two of the four archetypical towers that we cite here unquestionably dominate their surroundings with dramatic effect (see Figure TC6-2). Even the Sears Tower, standing as it does among many high-rises that would dominate any other landscape, remains a clearly distinguishable element of the Chicago skyline.

Figure TC6-2 Four late Twentieth-Century Towers.

(continued)

(continued)

Figure TC6-3 Sears Tower, Chicago, and Bank of China, Hong Kong.

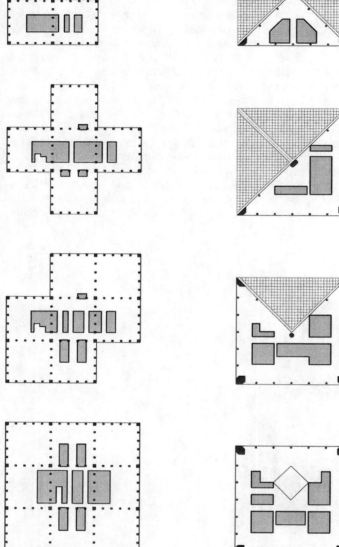

SEARS TOWER

Not much more than a year after its completion, in 1974, New York's World Trade Center lost its position as the world's tallest skyscraper to the Sears Tower at 1,450 feet. Each of the Sears Tower's first fifty stories contains over fifty thousand square feet, comprised of nine column-free squares seventy-five feet on a side. The substantial base of the building, illustrated at lower left in Figure TC6-3, is 225 feet square and occupies most of a city block. The architect was the firm of Skidmore, Owings, & Merrill, and the structural model was that of its engineer Fazlur Kahn,

who also conceived the x-braced John Hancock Center several years earlier.

Two of the 5,625-square-foot tubes terminate at the fifty-first floor, two more at the sixty-seventh, and three at the ninetieth floor, so that the floors from 91 up to 108 each contain slightly more than eleven thousand square feet. Because the unarticulated, black façade with its gray-tinted windows continues unbroken throughout the rise of half the building, one must stand a considerable distance away to clearly see the setbacks. Only in the west elevation along Wacker Drive can one face of the building be observed rising the full 108 stories. To reach the upper floors, it is necessary to change elevators at a transfer floor.

Figure TC6-4   Torres Petronas, Kuala Lumpur, and Torre Espacio, Madrid.

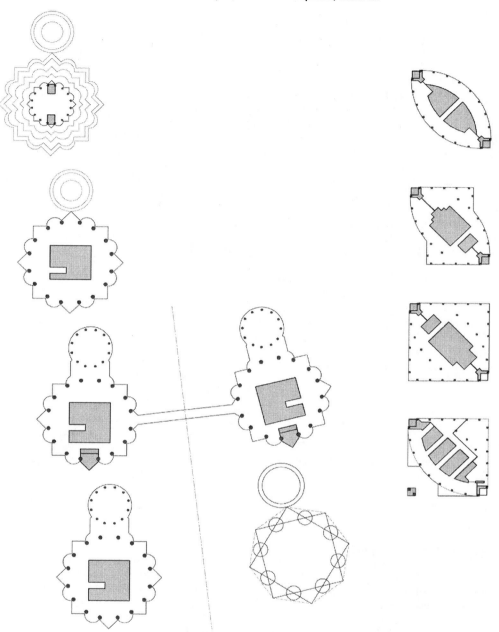

Although the Sears Tower's hard, rectilinear form gives the building a weighty and static presence from the exterior, it is comparatively efficient on the interior from a planning standpoint. The service cores containing common areas, vertical utility shafts, exit stairs, and elevators are centrally located and reduce in size or terminate as one rises within the building. The column-free interior of each tube allows for many options in the placement of corridors and office partitions. The challenge when implementing a facility program requiring a high percentage of enclosed spaces is to avoid creating a confusing maze of interior corridors.

BANK OF CHINA

The Bank of China, designed by I. M. Pei & Partners and completed in 1999, did not break any height records at 1,205 feet and seventy-two floors. Its structure was topped out on August 8, 1988, a date considered the most auspicious of the twentieth century by the Chinese people. The building's unconventional, geometrically clean form based on the triangle gives it a compelling sense of lightness and grace.

Remarkably, this Hong Kong monument uses a structural concept similar to that of the Sears Tower, that of the

(*continued*)

(continued)

bundled tube; except, in this case, the tubes are also trusses derived from the triangular form in all three dimensions. The base of the tower is a square divided by its diagonals into four right triangles, an inversion of the *Ad Quadratum* (Figure TC6-3, right). There are massive supports at the four corners, which, together with a central column, are joined by diagonal supporting elements. Setbacks occur at the twenty-fifth, thirty-eighth, and fifty-first floors, as shown in the right half of Figure TC6-3, but, unlike the Sears Tower, which steps back at ninety-degree angles, a sloping, triangular glass surface gradually closes each tube against those tubes remaining. Only one of the triangular towers rises the full height of the building to form a glass prism at the top.

The Bank of China is an exceptionally graceful structure that effectively projects through its height and location the economic vitality of the region. Interior planning offers some challenges, however, due to necessary placement of major structural elements in the center of the building and at its four corners. Corner office views are only feasible on the upper floors due to the large corner columns—and then only when rooms are large. Building core locations change from low-rise to high-rise floors, making vertical transportation less direct than with a central elevator core. There are also several areas where acute angles created by the service cores form cul-de-sacs that are difficult to plan contiguously with adjacent spaces.

TORRES PETRONAS

The Petronas Towers in Kuala Lumpur, Malaysia, were designed by Cesar Pelli & Associates. Completed in 1998 at a height of 1,483 feet, they broke the Sears Tower's height record by thirty-three feet while having twenty fewer floors.

Developed to serve a variety of public and commercial uses by Malaysia's national petroleum company, this city within a city contains offices and residential space combined with hotels, retail space, and cultural facilities with direct access to rail transportation and parking for several thousand vehicles. Conceived by their architect as a symbolic gateway, the Torres Petronas were the first buildings outside of North America to hold the record as the world's tallest. The symmetrical twin towers are linked by a two-story bridge connecting sky-lobbies at the fourty-first and fourty-second floors. The structure of the bridge is isolated from that of the two towers, which move independently of each other.

The elegantly proportioned floor plan is based on two squares offset at a fourty-five-degree angle (Figure TC6-4, left and center). Where the squares intersect, eight circles are superimposed, forming a sixteen-lobed façade of alternating right-angle and circular segments. The main towers are buttressed up to the sky-bridge levels by smaller cylindrical *outrigger* towers that increase the typical floor areas by some 25 percent. From that point the main towers rise alone to their full eighty-eight-story height. Their principal means of support are the 150-foot diameter concrete perimeter tubes formed by the outer rings of sixteen columns that are tied into the inner concrete core. The curtain walls of polished stainless steel and glass are independent of the main support structures.

The polished surfaces of the towers not only reflect the light from their surrounding environment but also symbolize the development of a new global economic order. At the same time, the plan and many smaller details were designed in accordance with traditional Islamic geometry. Similar modular patterns that inspired Frank Lloyd Wright, which Owen Jones described as *Moresque,* have been employed to make these towers fit into the Malaysian landscape (see Plate One).

TORRE ESPACIO

One of the newest international towers is the Torre Espacio in Madrid, designed by Pei Cobb Freed & Partners, the successor firm to the designers of the Bank of China. At its completion in 2007, Torre Espacio will be the tallest building in Spain at fifty-six floors and 718 feet. Constructed of high-resistance concrete in the Madrid Arena complex, the building contains primarily office space and includes a gymnasium, restaurant facilities, and underground parking for over a thousand cars.

This tower is quite unique in its geometry in that its plan rises from a square base and changes continuously throughout its height into a shape resembling a convex lens at the top. The building thus appears to have a pronounced twist, with virtually every floor plan being different from the one below (Figure TC6-4, right).

SUMMING UP

The Renaissance tradition followed classical forms and in its aftermath became preoccupied with the orders and ornament in the Beaux-Arts tradition. It was not until the late nineteenth century that architects such as Louis Sullivan began to adopt the principle that "form follows function." However, the principles of classical geometry as a means of establishing order and maintaining proportion remain valid. We can see in these four towers some of the infinite variations that are possible when these principles are applied imaginatively.

It is worth reminding ourselves that the medieval mind revered the Ptolemaic, Earth-centered image of a universe constructed of nested spheres within the Stellatum, itself inside an outer sphere of pure illumination. The astronomer Johannes Kepler in 1595 built of the five regular Platonic solids an inverted universe based on the Copernican model with the sun at its center (Figure TC6-5). Kepler refined this picture when he discovered that the planets actually move in elliptical orbits that have the sun at one focus. Isaac Newton later explained Kepler's laws through gravitation—that the planets move faster as they near the sun, sweeping out equal areas in every unit of time.

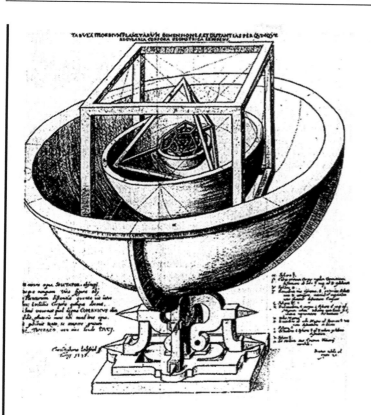

Figure TC6-5 Kepler's Heliocentric Model Universe.

At the end of the nineteenth century, Sullivan returned to nature to find the basis for his system of ornament: the *cotyledon* that is at the heart of all plant life.

Owen Jones expressed concern that his *Grammar of Ornament* would invite still more transplantation of the forms peculiar to bygone eras, yet Wright saw in it a contemporary geometry that would strongly influence much of his work. Le Corbusier directly incorporated the proportionality of the Fibonacci series and the *mean and extreme ratio* into his Modulor. Mies van der Rohe and Wright both shared with the medieval masons the idea that the true understanding of construction materials lay "in the fingers," and Mies particularly embraced the principle of parsimony.

There is a certain programmatic continuity between the plan of the Ramesseum and the vertical envelope of a modern skyscraper. One enters the temple through a narrow, easily controlled opening. There is a monumental entrance court incorporating sculpture, from which one passes (horizontally) through a second court and several progressively more secure halls before reaching the pharaoh's sacred space.

The sense of place of a corporate CEO or any worker in the Torres Petronas, for example, is largely determined by the path she follows from the time she approaches the building to the point at which she reaches her office or workstation. In the ground-floor lobby there is likely a security checkpoint. Then an elevator ride takes her to the appropriate floor where she may pass through a main reception area. At least one—possibly several—primary and secondary corridors lead to her organizational unit where she may pass by several work areas occupied by support people before reaching her own personal space.

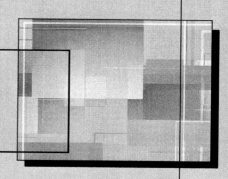

CHAPTER 9

Area-Measurement Standards

The Naming of Cats is a difficult matter . . .
—*T. S. Eliot*°

This chapter examines several widely used standards for area measurement, two of which are highly compatible. The BOMA standard, published by the Building Owners and Managers Association International, dates from 1915 and has become an American National Standard for determining rentable area in office buildings. It is therefore the basis for many lease documents, as well as being followed by owners and managers of owner-occupied buildings. The federal government uses the BOMA standard to account for and allocate space managed by the General Services Administration (GSA).

A second standard, widely promoted by the International Facility Managers Association and published by ASTM, is also an American National Standard. The ASTM standard in effect picks up where BOMA leaves off, defining classifications of assignable and common support areas that facilitate the breakdown of rentable area into elements that can be programmed and developed into detailed space layouts. We ultimately see how these two standards can be used to arrive at largely complementary results.

In his introduction to *Old Possum's Book*, Mr. Eliot avers that the Naming of Cats "isn't just one of your holiday games." This declaration aptly describes the virtual *Babel* of measurement standards, which present a daunting lexicon of area terminology. I once had a client, an office manager, who habitually referred to usable area as "net net," confounding the discussion of area with a term borrowed from the vocabulary of gross, net, and triple-net leases! A biblical commentator once observed that, prior to the attempted construction of the noted Babylonian high-rise, "both the learned and the unlearned spoke 'the same words,' [and] there was no philosophic or technical 'jargon' to separate people from each other."° Understanding area-measurement jargon is only slightly less daunting.

At some risk of adding to an already debased vocabulary of area terminology, we also look at *PEFIC*, a standard used widely in the academic community. There are certain common denominators to be found here, along with a useful set of room-use classifications. The area-measurement system to which one is led using the PEFIC definitions, however, is basically incompatible with the other two standards. This illustrates, if nothing else, that one's purpose(s) in setting up such a system must be thoroughly understood and carefully defined before proceeding. Otherwise, the area calculations and summary data resulting therefrom may not be what one intended.

ANSI/BOMA

The focus of the ANSI/BOMA Standard for Measuring Floor Area in Office Buildings is the determination of *rentable area*. Rentable area, as defined in lease documents, includes more space than can actually be *used* by the tenant or occupant of the building. This is because rentable area includes a fair share of what are termed *floor common area* and *building common area*. These relationships can be expressed by several formulas, which have in common the ratio of rentable to usable area: the *R/U ratio*.

R/U ratio = rentable area / usable area   (a)

R/U ratio = floor R/U ratio × building R/U ratio   (b)

Looking at these two formulas, we immediately see that there are several different kinds of R/U ratios. These derive from BOMA's expansion of the standard from measuring space strictly on a floor-by-floor basis to allocating "building-wide amenities" proportionately into rentable area.

The current standard, occasionally referred to as "new BOMA," was published in 1996.° This was largely in response to the incorporation into new office buildings of such features as multistory atria, day-care centers and health clubs, as well as conference and training facilities, none of which were specifically addressed by the previous 1981 edition.

The concept of defining building common area in addition to floor common area also provides a consistent basis for prorating common areas among multiple buildings if necessary, although BOMA does not explicitly define a *site common* category. Extending the methodology in this way is often desirable for facility-management purposes, as we see later in the chapter. BOMA does, however, define R/U ratios for both floor and building, each of which can be calculated in a couple of different ways:

Floor R/U ratio = floor rentable area / floor usable area   (c)

Floor R/U ratio = basic rentable area / usable area   (d)

Building R/U ratio = rentable area / basic rentable area   (e)

Building R/U ratio = building rentable area / (building rentable area − basic rentable area of building common area)   (f)

Now that we have defined several different levels of R/U ratios, just what do we mean when we say "rentable area?" What, indeed, do we mean by "*basic* rentable area?" In order to have a clear understanding of what these area classifications are and how they interrelate, we need to define the BOMA area categories with some precision, beginning with *usable area*.

BOMA Usable Area

It is important to remember that ANSI/BOMA is only interested in allocating a pro rata share of all the common areas in a building to the usable areas belonging to (or leased by) the building's occupants. Other than the distinction made between *office area* and *store area*, which are each components of *usable area*, BOMA offers no subcategories of space within the usable classification. For a further breakdown of usable area, we ultimately need to look beyond the definitions that BOMA offers.

- *Office area*—*Usable area* occupied by a tenant, generally used to house personnel, furniture, and other office equipment, *which can be directly measured*.
- *Store area*—*Usable area* within an office building that is suitable for occupancy by a retail establishment. Store area is also measurable in a manner similar to office area, except where it has street frontage, where the measurement is made to the building line (the finished outside surface of the exterior building wall).
- *Common area*—Spaces within an office building that are directly measurable and provide specific services to tenants of the building, which are not included in the office area or store area classifications. Common areas are further categorized as *floor common* (available for the use of the occupants of a particular floor) or *building common areas* (available for *general* use by the building's tenants). Examples of common areas are listed in Table 9-1.

BOMA Rentable Area

Of course, the story does not end here! As you can see in the preceding formulas, there are no fewer than four different kinds of rentable area with which we must contend.

Table 9-1   Floor Common and Building Common Areas

Floor Common Area	Building Common Area
Public Corridors	Main and Secondary Lobbies
Elevator Lobbies	Atria (at the Finished Floor Level)
Toilet Rooms	Conference Facilities
Electrical & Telephone Rooms	Food Service, Vending, & Lounge Facilities
Janitors' Closets	Health/Fitness Centers, Locker, & Shower Facilities
Mechanical Rooms	Day-Care Centers
	Central Mail Rooms
	Building Core/Service Areas (Mechanical, Equipment, Fire-Control Rooms)
	Courtyards (Fully Enclosed but Outside the Exterior Walls)

- *Rentable area*—The sum of all office areas and/or store areas together with a proportionate share of floor common and building common areas.
- *Basic rentable area*—The usable area of a particular office area, store area, or building common area together with its proportionate share of the floor common area *on that floor*. The basic rentable area can be calculated using the floor R/U ratio using the following formula (a transposition of formula *d*).

Basic rentable area = usable area × floor R/U ratio   (g)

Adding together all the *basic rentable areas* on a floor gives the *floor rentable area* of that floor. If a tenant occupies multiple floors of a building, its total basic rentable area is the sum of the basic rentable areas on each of the floors it occupies.

- *Floor rentable area*—The *gross measured area* of a floor less the *major vertical penetrations* through that floor yields the floor rentable area. It is seldom affected by changes in corridor configuration and is therefore *fixed for the life of the building*.

Floor rentable area = gross measured area – major vertical penetrations   (h)

- *Building rentable area*—The summation of all floor rentable areas.

Area Measurement

Using the terminology we have defined thus far, we can now complete our definitions of *usable area* and go on to discuss *how* these areas are measured.

- *Floor usable area*—The sum of all office areas, store areas, and building common areas of a floor. As floor layouts are reconfigured and as corridors change, the available floor usable area is *variable over the life of the building*.

It is important to note that the ANSI/BOMA standard *specifically excludes* floor common area from this definition, as well as major vertical penetrations, parking space, and portions of loading docks that fall outside the building line.

Floor usable area is measured from the *dominant portion* of the inside *finished surface* of the exterior wall to the *finished surface* of the *office-area* side of public corridors or other permanent walls.°

- *Major vertical penetrations*—Elevator shafts, stairs, pipe shafts, vertical air ducts and returns, flues, and so on, *together with the walls that enclose them*.

Relatively smaller openings for electrical or telecommunications cabling and plumbing, as well as supporting columns, are not included when measuring vertical penetrations. Also not included are penetrations constructed for the exclusive use of a tenant occupying office area on multiple floors, such as internal staircases.

Gross Area

- *Gross building area*—This is the total constructed area of a building, as defined by BOMA, not used for leasing purposes except when the entire structure is leased to a single tenant. Gross building area is measured "to the *outside finished surface* of permanent outer building walls,"° with no deductions.
- *Gross measured area*—The total building area, measured to the *inside finished surface* of the exterior building wall. It is also not customarily used by itself for leasing purposes and excludes loading docks (or portions thereof) outside the building line as well as parking areas.

Dominant Portion

The ANSI/BOMA standard is very specific as to how gross area is to be measured, introducing the concept of *dominant portion* when referring to the inside finished surface of the exterior wall. First we need to define *finished surface*:

- *Finished surface*—A wall, floor, or ceiling surface of any permanent construction material including glass. For the purposes of area measurement, finished surfaces *exclude* the thickness of special surfacing materials and interior construction such as furring strips, wall panels, carpeting, and suspended ceilings.

The concept of *dominant portion* is specific to ANSI/BOMA; however, it has become the de facto standard for virtually all systems for measuring of interior space. As defined by BOMA, "Dominant Portion shall mean the portion of the inside Finished Surface of the permanent outer building wall which is 50% or more of the vertical floor-to-ceiling dimension, at the given point being measured as one moves horizontally along the wall."°

The number of measurements required around the inside perimeter of a building is determined by the materials used to construct its outer wall. For example, a building has an exterior wall constructed of masonry with glass above it. The glass is recessed with an interior sill and the ceiling height is eleven feet six inches. If the height of the sill above the floor is five feet nine inches, the dominant portion is determined to be the inside face of the masonry. Were the sill one inch lower, the dominant portion would be measured to the inside face of the glass.

To the extent that the construction changes as measurements are made going around the building, the line denoting dominant portion moves in and out following those changes to the extent required. As with vertical penetrations, structural columns are ignored when determining dominant portion. The building line determines dominant portion for store area with street frontage.

Figure 9-1   Exterior Wall Sections Illustrating *Dominant* Portion.

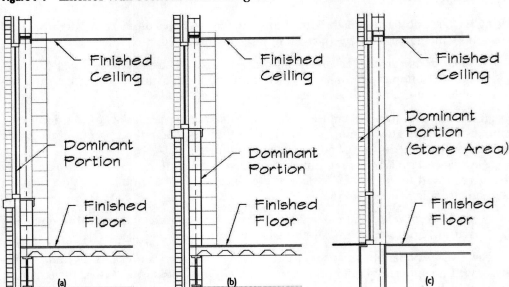

Finally, if the exterior building wall is not vertical, or if no single finished surface constitutes at least 50 percent of its vertical dimension, measurements are taken at floor level wherever that condition exists.

In Figure 9-1a, the *Dominant Portion* of the exterior building wall is the window system, because it constitutes more than 50 percent of the vertical floor-to-ceiling distance. The *Dominant Portion* is the nonglass material in Figure 9-1b. In Figure 9-1c, where the exterior wall bounds a *Store Area* with street frontage, the *Dominant Portion* is the building line.

Step by Step
Rentable area as defined by BOMA is calculated on a floor-by-floor basis using the following steps:

1. Determine each floor's gross measured area using the preceding definitions of finished surface and dominant portion.
2. Calculate floor rentable area by subtracting the major vertical penetrations through each floor from the floor's gross measured area.
3. Determine each floor's floor usable area by measuring the usable area of every office area, store area, and building common area on that floor.
4. Calculate each floor's floor common area by subtracting its floor usable area from the floor rentable area.

Figure 9-2 Base Building Area Calculations.

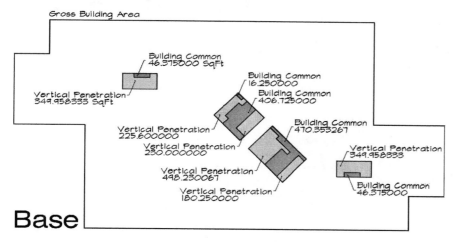

Figure 9-3 Typical Architectural Floor Plan.

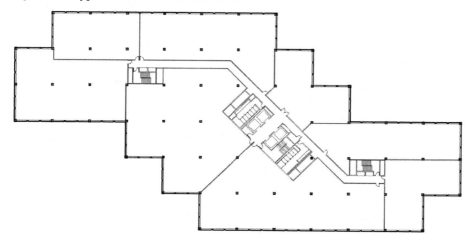

Figure 9-4   Floor 1, 2, and 3 for BOMA Building Calculations.

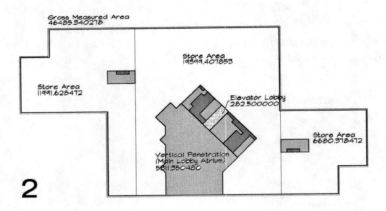

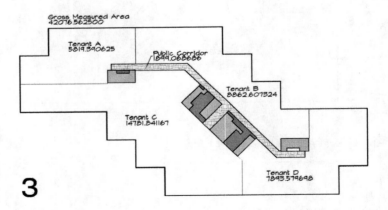

5. Calculate the basic rentable area of each floor, allocating a pro rata share of its floor common area by multiplying each usable area by the floor R/U ratio (see formula *c* and formula *d*).

6. Calculate the building R/U ratio using the following two steps. First, add together the total basic rentable areas of all floors to obtain the building rentable area. Second, subtract the sum of the basic rentable areas of building common areas from the building rentable area, dividing the result back into the building rentable area.

7. Calculate the rentable area of each floor, allocating a pro rata share of the building common area by multiplying the basic rentable area by the building R/U ratio (see formula *e* and formula *f*).

Figure 9-5 Floors 4, 5 through 7, and 8 for BOMA Building Calculations.

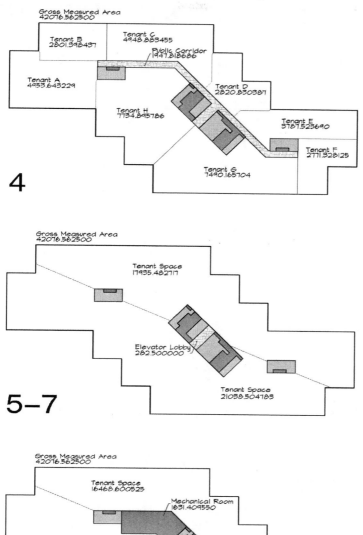

Figure 9-2 shows a typical floor of an office building that we will use to develop a set of calculations using the BOMA method we have outlined. We begin by measuring the elements of the base building: the core areas containing vertical penetrations and building common. In this case, we have a central core containing the elevators, stairs, toilets, duct shafts, and some mechanical spaces. There are two auxiliary cores containing fire stairs, duct shafts, and electrical and telecommunications closets.

A simplified architectural floor plan showing the main building elements is illustrated in Figure 9-3. Demising walls that divide up the floor among eight tenants are also shown. The *International Building Code* requires that demising walls be *fire partitions*

Figure 9-6 Building Calculations Using the BOMA Method.

1	2	3	4	5	6	7	8	9	10
				3 – 4					7 + 8 + 9
						Usable Areas			
Floor	Gross Building Area	Gross Measured Area	Major Vertical Penetration	Floor Rentable Area	Space I.D.	Office Area	Store Area	Building Common Area	Floor Usable Area
01	53099.41	52626.96	1834.00	50792.97			44151.98	6358.49	50510.47
02	47884.00	46485.34	7145.35	39339.99			38071.41	986.08	39057.49
					a	5819.39			
					b	8862.61			
					c	14781.84			
					d	7893.58			
03	42484.00	42076.56	1834.00	40242.57		37357.42		986.08	38343.50
					a	4953.64			
					b	2801.40			
					c	4948.88			
					d	2820.83			
					e	3787.52			
					f	2771.33			
					g	7490.17			
					h	7734.90			
04	42484.00	42076.56	1834.00	40242.57		37308.67		986.08	38294.75
05	42484.00	42076.56	1834.00	40242.57		38973.99		986.08	39960.07
06	42484.00	42076.56	1834.00	40242.57		38973.99		986.08	39960.07
07	42484.00	42076.56	1834.00	40242.57		38973.99		986.08	39960.07
08	42484.00	42076.56	1834.00	40242.57		37342.58		2617.49	39960.07
	355887.41	351571.68	19983.32	331588.35		228930.63	82223.39	14892.45	326046.47

with a one-hour rating, but some jurisdictions (such as Chicago) require that they be of two-hour construction.

Figure 9-4 and Figure 9-5 show the area classifications outside of the building core for all the floors of the building. Beginning with floors *1* and *2*, we see the main building lobby, which is a two-story atrium containing an escalator (not shown). Both of these floors are calculated as *Store Area*, with measurements along the front portion of the first floor being taken to the outside finished surface of the exterior wall (see Figure 9-1*c*) because this side has street frontage. Measurements in the rear of floor *1* and all of floor *2* are made as shown in Figure 9-1*b*, as the *Dominant Portion* of the exterior wall is the masonry below the windows. Note also that the atrium space on the second floor is treated as a *Vertical Penetration*.

The remaining floors (*3* through *8*), where the windows are the *Dominant Portion*, are measured as shown in Figure 9-1*a*. Floor *3* and floor *4* are multitenant floors with a central corridor connecting all three building cores containing exit stairs. The third floor has four tenants, and the fourth floor (the same floor pictured in Figure 9-3) has eight. Floor *5* through floor *8* are single-tenant floors. Note that there is more *Usable Area* on these floors due to the elimination of the central corridor, but the amount of *Rentable Area* is the same from floor *3* on up. There is a mechanical room on floor *8*, which is allocated to *Building Common*.

Setting Up the Calculations

Figure 9-6 illustrates the necessary building calculations using the BOMA method. (The Excel workbook pictured is included on the accompanying CD, or may be downloaded from prenhall.com.) The columns are numbered from 1 to 20, and the relationship formulas are shown in the row beneath the column numbers. Working from left to right, both types of *Gross Area* are recorded, floor by floor, along with *Major Vertical Penetrations*. Then the *Floor Rentable* area is calculated (column 2 through column 5, steps 1 and 2 outlined earlier). Almost in passing, BOMA recommends that gross building area be calculated "for record keeping."°

Figure 9-6   *(continued)*

11	12	13	14	15	16	17	18	19	20
5 – 10	5 / 10	7 * 12	8 * 12	9 * 12	Σ14 + Σ15	16 / (16 – Σ15)	13 * 17	14 * 17	18 + 19
		Basic Rentable Areas					**Rentable Area**		
Floor Common Area	Floor R/U Ratio	Office Area	Store Area	Building Common Area	Building Rentable Area	Building R/U Ratio	Office Area	Store Area	Total Rentable Area
282.50	1.0056	0.00	44398.92	6394.05			0.00	46497.06	46497.06
282.50	1.0072	0.00	38346.78	993.21			0.00	40158.93	40158.93
1899.07	1.0495	39207.65	0.00	1034.92			41060.48	0.00	41060.48
1947.82	1.0509	39206.33	0.00	1036.23			41059.10	0.00	41059.10
282.50	1.0071	39249.52	0.00	993.05			41104.32	0.00	41104.32
282.50	1.0071	39249.52	0.00	993.05			41104.32	0.00	41104.32
282.50	1.0071	39249.52	0.00	993.05			41104.32	0.00	41104.32
282.50	0.9774	38422.25	0.00	2558.45			40237.96	0.00	40237.96
5541.89		234584.78	82745.70	14996.01	332326.49	1.0473	245670.49	86655.99	332326.49

Measurements of *Office Area*, *Store Area*, and *Building Common Area* on each floor are recorded in column 7 through column 9 (step 3). From these measurements, individual *Floor Usable Area*, *Floor Common Area*, and *Floor R/U Ratio* can be calculated (column 10 through column 12, step 4).

The *Basic Rentable Areas* of each floor are then calculated (column 13 through column 15) by multiplying the *Office, Store,* and *Building Common Areas,* respectively, by the *Floor R/U Ratio*. This prorates the *Floor Common Areas* among the *Basic Rentable Areas* (step 5).

All the *Basic Rentable Areas* are added together, yielding the *Building Rentable Area* (column 16). This figure is divided by *Building Rentable* less the sum of all *building Common Areas* (the total of column 15). The result is the *Building R/U Ratio* (column 17, step 6).

The *Rentable Office* and *Store Areas* on each floor are calculated (column 18 and column 19) by multiplying the *Basic Rentable Office* and *Store Areas* (floor by floor) by the *Building R/U Ratio* (step 7). This prorates the *Building Common Area* among the *Rentable Areas.* Finally, columns 18 and 19 are summed in column 20, giving the *Total Rentable Area* by floor and for the entire building.

Analysis

At this point, we need to make several observations. First, reiterating a statement with which we began this section: The focus of the ANSI/BOMA Standard for Measuring Floor Area in Office Buildings is the determination of *rentable area.* We observed that BOMA offers no guidance in categorizing facilities *within* the *usable area* classification, such as dedicated offices versus support areas belonging to the building occupant, or traffic aisles within tenant-occupied areas. At the risk of stating the obvious, BOMA's primary interest is in building management. The building owner and manager have a limited interest in how usable area actually is utilized. However, such subcategorization of space is of *critical* importance to the facility manager.

Finally, the BOMA methodology specifies what *floor common area* includes, but does not call for the *measurement* of these spaces directly (see Table 9-1 and step 4). *Floor common* is taken as the difference between *floor rentable* and *floor usable area*. It is often desirable, from a facility-management perspective, to have accurate area figures available for these spaces individually. This is particularly so in the case of "public" corridors, which can change relatively often and which the facility manager may wish to *charge-back* to her departments. (We discuss *charge-backs* in Chapter 11.)

ASTM-E1836 (IFMA)

We said earlier that we ultimately need to look beyond BOMA for a further breakdown of *usable area*. It was to the ASTM-E1836° (American Society of Testing Materials) standard to which we were referring. This standard was developed with the cooperation of IFMA (the International Facility Managers Association). Within this document, "Annex A1, Classification of Building Floor Area Measurements in Offices, Research, Laboratory, and Manufacturing Buildings and Building-Related Facilities" furnishes us with a comprehensive set of guidelines for measuring space for facility-management purposes. The ASTM/IFMA standard is similar to the ANSI/BOMA method; however, its terminology is different in several respects, not the least of which is that it breaks down usable area in greater detail.

Facility Area Relationships

The ASTM/IFMA standard specifies the following set of relationships among floor area measurements:°

Interior gross area = exterior gross area − exterior walls	(i)
Rentable area = interior gross area − major vertical penetrations, interior parking, and void areas	(j)
Usable area = rentable area − (building core and service area and primary circulation)	(k)
Assignable area = usable area − secondary circulation	(l)

Let us look at each of these categories, comparing them with BOMA.

The definition of *exterior gross area* is consistent with BOMA's definition of *gross building area*, except it is more specific about what is *excluded*.

- *Exterior gross area*—On a floor-by-floor basis, all the floor areas that are completely enclosed within the building (including basements), measured to the outside face of the exterior walls. Building exterior gross area is the sum of the gross areas on all floors. Elements extending beyond the wall face, such as balconies, buttresses, canopies, cornices, pilasters, and enclosed but unroofed courtyards, are excluded.

There are also some guidelines for potential conflict situations identified by E1836, such as common walls between adjoining buildings relative to property-line locations. If a property line is located *within* such a wall, the exterior gross area is measured to the property line. Otherwise, the measurement is taken to the centerline of the structural portion of the common wall. Fully enclosed structures connecting multiple buildings are included in exterior gross measurements, including bridges and tunnels.°

- *Interior gross area*—Exterior gross area minus the thickness of the exterior walls yields interior gross area (see formula *i*).°

This definition is rather troublesome because it would be very difficult to directly measure the thickness of the exterior wall. In fact, the "Annex" offers another definition:

The measurement of area "at the intersection of the plane of the finished floor and the finished interior surface of the walls."°

This definition effectively solves one problem while creating another. It uses the term *finished surface*, which E1836 defines identically to BOMA.° However, by specifying that all measurements are to "be made along the plane of the floor to the points where floors and walls intersect,"° the ASTM standard contradicts BOMA.

There is no concept of *dominant portion* here. Where the exterior wall construction material is uniform and vertical, measurements will be the same under both standards. With mixed construction, however, the figures are likely to differ. For example, if the same masonry wall described earlier has a sill at a height of two feet six inches with nine feet of glass above, the glass is the BOMA boundary while the measurement under ASTM would be made to the masonry at the floor plane.

Facility Rentable Area

ASTM declares that *facility rentable area* "can be calculated for any building, whether leased or owner-occupied," and is therefore "useful as a consistent basis of comparison with other buildings." At the same time, however, this standard states that rentable area "as defined in this classification is not necessarily the basis for lease agreements."°

Formula *j* encapsulates ASTM rentable as interior gross less vertical penetrations, void areas, and interior parking. BOMA does not specifically mention void areas; however, both definitions *exclude* interior parking as well as vertical penetrations from the rentable classification. E1836 includes columns and building projections in facility rentable,° and its definition of *vertical penetrations* is virtually identical to that of BOMA.°

But for the difference in the way the two standards specify how the interior building perimeter is to be measured, their definitions of *rentable area* (specifically, in the case of BOMA, *floor rentable area*) would be essentially identical. It is, in fact, common practice for facility managers and their companies to use the BOMA method in determining the gross measured area of their facilities. Doing this ensures that the area figures generated are consistent with their leases where applicable and can also be accurately compared with other buildings.

Building Core and Service Area
Building core and service area are the same kinds of spaces that BOMA defines as *floor common area* (see Table 9-1). If such spaces are dedicated to the use of a particular tenant, however (telecommunications, electrical, and mechanical rooms and toilets), they are then included in the occupant's *usable area*. Loading docks are included in this classification if they serve the whole building; but, if they are dedicated to a single user, they are considered part of the tenant's usable area.

Primary Circulation
ANSI/BOMA does not specifically define *primary circulation* but includes "elevator lobbies and public corridors"° in *floor common* (Table 9-1). Main building lobbies and atria are classified as *building common areas*. The ASTM-E1836 standard includes elements of both building common and floor common areas in its definition of *primary circulation*. ASTM defines *secondary circulation* as well.

In general, ASTM defines *primary circulation* as including all lobbies and public corridors necessary to provide unobstructed access to entrances and exits, elevators, stairs, toilet facilities, and entrances to individual tenant spaces on multitenant floors. The amount of area dedicated to primary circulation is determined by the establishment of reasonably direct routes connecting these building elements, maintaining adequate clearances for ingress and egress as required by applicable building safety codes.

Atria, bridges, and tunnels used for public access are considered to be primary circulation even though they may also serve other functions. Space required for access to building telecommunications, electrical, and mechanical equipment rooms is primary circulation unless such rooms serve a single tenant exclusively.

ASTM specifies that where several possible routes are available connecting occupant spaces with stairs and exits, one principal route should be dedicated as primary circulation. Auxiliary routes required for life-safety access should be classified as secondary circulation unless the layout dictates that they cannot serve any normal secondary circulation function, in which case they must be considered primary.

Measurement Methods

Here, as with the measurement of interior gross area, ASTM muddies the waters a bit by defining two ways of measuring building core and service area elements.° These have to do with the handling of the wall thickness when measuring adjacent areas of different classifications. The preferred method is as follows:

- Include the wall surrounding a major vertical penetration in its measurement where it adjoins a building core or service area.
- Include the wall surrounding a building core or service area in its measurement where it adjoins facility usable area.
- Include the interior (demising) walls enclosing primary circulation in its measurement where it adjoins facility usable area.

These spaces are otherwise measured to the wall centerlines.

The alternative method that E1836 suggests is to measure *all* walls in this category to their centerlines, making note of this fact when reporting the results. This method is not recommended for the same reason that it is common practice to use the BOMA method (dominant portion) to delineate *gross measured area*. Building area calculations made using ASTM's *preferred* method can be kept compatible with BOMA, at least down to the *rentable-area* level, as we shall see.

Facility Usable Area

ASTM-E1836 defines *facility usable area* as "the floor area of a facility that can be assigned to occupant groups."° Usable area excludes all those types of areas, such as vertical penetrations, that are excluded from *facility rentable area*. Also excluded from usable area are the building core and service areas and primary circulation, just discussed. This relationship is set forth in formula *k*.

Facility usable area can be subdivided into two main categories: *assignable area* and *secondary circulation*. Here, it may be appropriate to begin referring to E-1836 as the ASTM/IFMA standard, because it is the allocation of assignable space with which IFMA is primarily concerned.

ASTM (or IFMA) describes *facility usable area* as a basis for programming and planning space. This is a level of detail that BOMA does not address. It is space that can be assigned to personnel and to accommodate furniture and equipment, plus interior circulation. Adjoining spaces of a similar classification are measured to the centerlines of the boundaries between them.

Common Support Areas

Common support areas are a third source of potential confusion, a classification that ASTM defines somewhat ambiguously compared to BOMA.

If we say that *office* area is space assigned to specific individuals, then we can define *common support areas* as spaces that provide support to multiple individuals or business units but are not assigned to any one person. This category may include conference and meeting rooms, auditoriums, food-service and vending areas, copy centers, mailrooms, and the like.

The ambiguity arises depending upon whether a particular support space serves the entire building or a particular floor, in which case BOMA treats it as *building common* or *floor common area* (see Table 9-1). Only if a support area serves a specific tenant (or business unit) does BOMA consider it part of *usable area*. In the latter case, such spaces are considered part of BOMA's *basic rentable area* and not calculated individually at all.

From the ASTM/IFMA point of view, this is not so much a distinction affecting how these spaces are *measured* as how they may be *classified*. E-1836 simply says that they "may be separately identified as a subcategory of facility assignable if required."° Support areas belonging to a particular business unit are part of the assignable-area category along with office area. This is consistent with the BOMA definition of *office area*, which includes offices within *store area*.

Assignable Area

So, *facility assignable area* is space that can be allocated to specific users, their furniture, and equipment; it is space that can be subjected to detailed programming, utilization planning, and layout.

As defined by ASTM/IFMA, assignable area includes the structural columns, projections, and interior partitions that occur within its boundaries. It *excludes* all the categories of space that occur above it in the hierarchy defined by both the BOMA and IFMA standards and secondary circulation.

Assignable area is measured to the outside face of the walls that enclose it (if any). This includes boundaries established by system furniture panels. The exception, as usual, is that where spaces of the same category adjoin, the measurement is made to the centerline of the panel or partition.

Secondary Circulation

About *secondary circulation*, what can we say? It is what remains.

Usable area is the sum of assignable and secondary circulation (formula *l*). Secondary circulation may or may not be delineated by walls or panels, but these boundaries are never included in the measurement of the area itself. We calculate secondary circulation by measuring all assignable spaces and subtracting them from facility usable area.

GSA/PBS

The General Services Administration, through its Public Buildings Service, manages all facilities owned or leased by the U.S. government. The GSA maintains a computerized database linked to what it calls *assignment drawings*, which document the occupancy information of each building occupied by federal offices. Assignment drawings are updated continuously and include plans of all floors, the site, and the roof showing specialized equipment such as antennas, overlaying architectural plans as a base.

The links to the database contain occupancy information such as space classification and usage. The plans are often accompanied by tables showing the space assignments of each tenant, generated by a two-way database link.

Architectural plans used as base drawings show the building's shell and core information, including columns and other structural elements. A labeled grid indicates column centerlines, elevator cores, stairs, all entrances and exits, and other door locations. Building-core and shell elements are labeled to clearly establish the use of each area.

GSA Area Measurement

A detailed methodology for area measurement and documentation is spelled out in a document titled *PBS Assignment Drawing Guidance*.° This document is used in conjunction with PBS CAD standards° and BOMA Z65-1. Since the GSA guidelines follow ANSI/BOMA nomenclature very closely, we will not reiterate the basic definitions. However, GSA does offer a number of procedural specifics that have potential application elsewhere.

Area Hierarchy

The GSA document adds a number of specifics as to how and in what sequence area measurements should be made:

1. The gross building area (outside building gross line) and gross measured area (inside construction) should be determined *first*.

2. All major vertical penetrations should be located. These measurements should include the thickness of their enclosing walls but not the exterior walls of the building.

Table 9-2   General Services Administration Space Categories

	Space Category	
01	Assigned New	Tenant Agency Assignment to New Space
02	Building Common	See ANSI/BOMA.
03	Building Joint Use	Usable Space Available to All Federal Agencies in a Building; Used to Assign Area for Billing Purposes°
04	Committed	Reserved Space for an Agency
05	Committed under Alteration	Reserved Space under Renovation
06	Facility Common	Building Common Serving Several Buildings in a Facility (Complex)
07	Facility Joint Use	Building Joint Use Serving Several Buildings in a Facility (Common); Available to All Agencies in the Multibuilding Complex
08	Lease Common	Building Common in Leased Space
09	Structured Parking	Inside or Covered Parking
10	Unmarketable	Construction Area and Vertical Penetrations
11	Vacant	Unassigned Space
12	Under Construction	Space under Construction and Not Yet Assigned
13	Backfill	Space Assigned to an Agency Other Than the Original Occupant
14	Zero Square Feet	Used to Track Nonspatial Assets°
15	Lease Joint Use	Facility Joint Use in a Leased Space

3. Building common and floor common areas should be delineated, again including their enclosing walls. Outside construction and vertical penetrations are excluded.

4. Finally, office space is measured to the boundaries established by the three higher-level area categories.

Vertical Penetrations

Spaces normally included in *vertical penetrations* are elevator shafts, public and/or multi-tenant stairs, HVAC supply and return chases larger than nine square feet, chimney flues, and atrium spaces that are not amenities belonging to a single tenant. Major penetrations must be of sufficient size to allow a person to fit through comfortably (approximately nine square feet). Electrical, telecommunications, plumbing chases, and sleeved floor slabs are not considered vertical penetrations.

Slab penetrations or voids designed and constructed for a specific tenant are considered part of usable area and measured as if the slab were intact. Such voids, in PBS usage, include private stairs and elevators outside the building core, two-story courtrooms, auditoria, stages with fly galleries, dumbwaiters, and special air shafts for laboratory hoods. These are classified by PBS as *tenant floor cut* (space-type TFC), *assigned new*, or *backfill* (space category 01 or 13).

Building Common

Building common includes telecommunications, electrical, and mechanical rooms servicing multiple floors and not belonging to a single agency. Toilet rooms available for common use but not required by code are considered building common. All plumbing chases associated with public toilets are part of the area allocated to the toilet rooms; they are not vertical penetrations. Main public corridors used by all building tenants are building common.

Floor Common

Floor common includes telecommunications, electrical, and mechanical rooms servicing a single floor and not belonging to a particular agency. These kinds of spaces are treated as office space if they exclusively support a specific agency.

Public toilets required by code are allocated to the floor on which they are located. Principal corridors servicing multiple agencies on a particular floor are considered floor common.

Table 9-3   General Services Administration Space Types

		Space Type
Office Area	ADP	Automated Data Processing
	AUD	Auditorium
	CAF	Cafeteria
	CFT	Conference/Training
	CLD	Child Care
	CRJ	Courtrooms/Judiciary
	FDS	Food Service
	FIT	Fitness Center
	GNS	General Storage
	HUT	Health Unit
	INS	Light Industrial
	JCC	Judges' Chambers, U.S. Courts
	JHR	Judicial Hearing Room
	LAB	Laboratory
	QRR	Quarters & Residence
	SNK	Snack Bar
	STC	Structurally Changed
	TFC	Tenant Floor Cut
	TTO	Total Office
	WRH	Warehouse
Common Area	CRH	Circulation Horizontal
	CST	Custodial
	MCH	Mechanical
	TLT	Toilet
	CRV	Circulation Vertical
	ANT	Antennas
	BDK	Boat Dock
	BRG	Bridge
	DHS	Double-Height Space
	LND	Land
	OTH	Other
	RRC	Railroad Crossing
	WYD	Wareyard

Occasionally, space classified as floor common may be assigned to a tenant as usable area if the space is controlled and used exclusively by that tenant. Such conditions may include an elevator lobby used as a tenant-reception area or a tenant securing a portion of a corridor leading to or through his offices (assuming it is not required by code for emergency egress). These spaces are classified by PBS as *total office* (space-type TTO).

Table 9-4 GSA-Approved Room Names

ATM	Entry Lobby	Lobby	Supply
Attic Space	Entry Vest.	Lockers	Tele.
Auditorium	Equip. Room	LOG	Teller
Box Lobby	Evidence	Mail Room	Toilet
Break Area	Exam	Mech.	Total Gross
Break Room	Exercise Room	Men	Training
Cafeteria	File/Storage	Office	Vault
Canopy	File/Supply	Open Office	Vending
Classroom	Files	Open to Below	Vert. Pen.
Closet	Freight Elev.	Parking	Vest.
Computer	Frt. Elev. Vest.	Print Room	Waiting
Conference	Full Serv. Cntr.	Ramp	Warehouse
Control Booth	Furring	Reception	Weight Room
Copy	Garage	Residence	Wet Area
Copy/File	Hearing	Robing Area	Women
Copy/Storage	Hldg. Cell	Robing Room	Work Room
Corr.	Interview	Roof 1, 2, 3, etc.	
Courtroom	Judge's Chamber	Roof A, B, C, etc.	
Cust.	Jury Assembly	Sallyport	
Elec.	Jury Room	Shop	
Elev. Lobby	Kitchen	Snack Bar	
Elev. Vest.	Laboratory	Stair 1, 2, 3, etc.	
Elevator	Law Clerk	Stair A, B, C, etc.	
Elevator Pit	Library	Storage	
Elevators	Loading Dock	Storage/Supply	

Office Space

The office space classification includes primary office space and related storage areas occupied and used by a particular tenant, together with telecommunications, uninterrupted power supply (UPS), mechanical rooms, and private toilets. Space provided for the use of all government agencies in a building or complex are classed as office area, such as conference and training centers, central libraries, cafeterias, and fitness/wellness centers.

Space Categories and Types

In several of the preceding sections, we made reference to space *categories* and space *types*. GSA, like many organizations with substantial amounts of area to manage, requires a more detailed classification system than either ANSI/BOMA or ASTM offers. The *PBS Assignment Drawing Guidance* manual details a system of categories, types, and standardized room names that suggest a method (if not the precise designations) that can be used by many companies and other space users.

GSA's space categories generally classify areas by occupancy status or their availability for assignment at a given point in time. These standard categories are listed in Table 9-2.

The standard space types listed in Table 9-3 are more specific in establishing how each area is used and are grouped here according to whether they fall into BOMA's office area or common area classifications. (Some fall into neither.)

Finally, the GSA-approved room names (Table 9-4) standardize the designations given to individual spaces (rooms or open areas) in the inventory. This list, of course, reflects the space needs of governmental organizations, including courtrooms, holding cells, and the like. Other designations are more general, including file/storage, reception, and training areas.

One of the attractive features of a detailed room-by-room area inventory is that it allows reporting to whatever level of detail is required, based upon a completely flexible method of classifying occupancy data. Standard designations such as these can be set up to suit any kind of organization, be it a corporation, health-care facility, or educational institution, with two main goals in mind:

1. Tracking the condition and utilization of occupiable and nonoccupiable areas (as opposed to just distinguishing between rentable and usable space
2. Establishing a verifiable method for charging departments for the use of both occupiable and nonoccupiable facilities

PEFIC

Lest we be tempted to assume that all institutions follow the space-measurement and classification standards set forth by ANSI/BOMA or ASTM-E1836, let us now turn to one other widely used set of guidelines, the *Postsecondary Education Facilities Inventory and Classification Manual* (PEFIC manual).°

It is interesting to note that these guidelines are based upon definitions and standards established for the Federal Construction Council in the 1960s and published by the National Academy of Sciences. They were originally intended to be used by federal agencies but have since been widely adopted by colleges and universities. At the same time, many federal agencies—at least those whose facilities are managed by the GSA—gravitated toward ANSI/BOMA.

Once again, let us begin with the basic relationships among floor-area measurements:

Gross area = net usable area + structural space (GSF) (m)

Assignable area = sum of area designated by the ten assignable major room use categories (ASF) (n)

Nonassignable area = sum of the area designated by three nonassignable room use categories (o)

Net usable area = assignable area + nonassignable area (p)

Structural area = gross area − net usable area (NUSF) (q)

The five main categories refer to the total area on all floors of a building, defined as follows:

- *Gross area*—Total area measured to the outside faces of the exterior walls, including all floor penetrations, "however insignificant."° Buttresses, cornices, pilasters, and so on, extending beyond the wall faces are ignored, as are areas having a ceiling height less than six feet. Also excluded are portions of floors eliminated by any spaces rising above a single-floor ceiling height, light wells, and other open areas such as parking lots and playing fields.

  Included are excavated basements, mezzanines, attics and penthouses, enclosed porches, balconies, and corridors whether walled or not as long as they are within the perimeter of the building as measured to the roof drip line. Also included are what PEFIC calls *building infrastructure*: the footprints of stairways, elevator shafts, and ducts.

- *Assignable area*—Areas that are or can be allocated to a specific occupant for a particular use. Measurements are made to the inside faces of the surfaces defining the boundaries of the area. Building columns and projections are ignored. Structural, mechanical, building service, and circulation areas are excluded, together with areas having a ceiling height under six feet six inches.

   A subdivision of one of the ten major room-use categories should be used to describe each assignable space in the inventory: classrooms, labs, offices, study facilities, special use, general use, support, health care, residential, and unclassified.

- *Nonassignable area*—Areas not available for allocation to a specific occupant or use but essential for the operation of the building. Measurements are made in the same way as assignable area, with similar exclusions. There are three nonassignable room-use categories: building service, mechanical, and circulation.

- *Net usable area*—The sum of all assignable and nonassignable areas. All spaces in the three nonassignable and the ten assignable major room-use categories are included. Building columns, projections, and structural areas are excluded.

- *Structural area*—Also referred to as *construction area*,° floor space that cannot be utilized due to structural conditions. Areas in this classification include exterior walls, firewalls and other permanent partitions, areas in basements or attics not usable because of inadequate ceiling heights or other such restrictions, and unexcavated areas.

   Structural area is not measured; it is calculated by subtracting the measured net usable area from gross area (see formula *q*).

Nonassignable Categories

All nonassignable spaces are measured from the *inside faces of surfaces that form the boundaries of the designated areas*. Areas having a ceiling height less than six feet six inches are generally excluded.° The three nonassignable room use categories are defined by PEFIC as follows:

- *Building service area*—The total of all floor areas used for public toilets, janitorial closets and related custodial areas, and trash rooms (nonhazardous waste) used by the building occupants in common. Special-purpose storage areas and maintenance rooms (such as linen closets and housekeeping rooms in residence halls) and central shop areas are excluded.

- *Mechanical area*—All areas dedicated to *shafts* (i.e., vertical penetrations), mechanical equipment, and utility services. PEFIC specifically defines mechanical area as including "central utility plants, boiler rooms, mechanical and electrical equipment rooms, fuel rooms, meter and communications closets, and each floor's footprint of air ducts, pipe shafts, mechanical service shafts, service chutes, and stacks."°

- *Circulation area*—The total of areas necessary for public physical access to all subdivided areas of the building whether defined by partitions or not. Private corridors or aisles used only within an organizational unit's working area (secondary circulation) are excluded.

   Here again, PEFIC is quite specific about what this category includes: "public corridors, fire towers, elevator lobbies, tunnels, bridges, and each floor's footprint of elevator shafts, escalators and stairways, [and] receiving areas, such as loading docks."° Uncovered portions of loading docks are excluded from both circulation and gross building areas. Loading docks also used for storage are classed as assignable area.

Room-Use Codes

In the breakdown of assignable area, PEFIC's "Room-Use Category Structure"(RUCS) is also quite different from GSA's space types and categories, although it serves a similar purpose. The room-use codes provide a guideline for classifying space within campus facilities on a basis of use. While PEFIC asserts that its recommended structure "should provide a dimension of standardization and compatibility for comparisons across institutions and states,"° it states, almost in the same breath, that a coding scheme such as this should be *flexible*. The coding systems of individual institutions may be augmented or expanded using subcategories, leaving a great deal of latitude for enhancement.

It is important, especially if a computer-aided facilities management (CAFM) system is employed, that the coding scheme provide for the aggregation of related areas in reporting. For instance, PEFIC uses codes ending in 5 to denote *support space* belonging to a specific functional area with a code ending in 0.

Expanding the codes to encapsulate additional information may be accomplished by adding suffixes. For instance, the basic PEFIC structure does not make a distinction between offices for instructors and offices for research personnel or, for that matter, between private offices with walls and a door and semienclosed spaces such as cubicles. This may be accomplished by creating such designations as *310P* for private offices and *310W* for workstations or cubicles.

100	**Classroom Facilities**
110	Classroom
115	Classroom Service
200	**Laboratory Facilities**
210	Class Laboratory
215	Class Laboratory Service
220	Open Laboratory
225	Open Laboratory Service
250	Research/Nonclass Laboratory
255	Research/Nonclass Laboratory Service
300	**Office Facilities**
310	Office
315	Office Service
350	Conference Room
355	Conference Room Service
400	**Study Facilities**
410	Study Room
420	Stack
430	Open-Stack Study Room
440	Processing Room
455	Study Service
500	**Special-Use Facilities**
510	Armory
515	Armory Service
520	Athletic or Physical Education
523	Athletic Facilities Spectator Seating
525	Athletic or Physical Education Service
530	Media Production
535	Media Production Service
540	Clinic
545	Clinic Service
550	Demonstration

555	Demonstration Service
560	Field Building
570	Animal Quarters
575	Animal Quarters Service
580	Greenhouse
585	Greenhouse Service
590	Other (All Purpose)
600	**General-Use Facilities**
610	Assembly
615	Assembly Service
620	Exhibition
625	Exhibition Service
630	Food Facility
635	Food Facility Service
640	Day Care
645	Day-Care Service
650	Lounge
655	Lounge Service
660	Merchandising
665	Merchandising Service
670	Recreation
675	Recreation Service
680	Meeting Room
685	Meeting Room Service
700	**Support Facilities**
710	Central Computer or Telecommunications
715	Central Computer or Telecommunications Service
720	Shop
725	Shop Service
730	Central Storage
735	Central Storage Service
740	Vehicle Storage
745	Vehicle Storage Service
750	Central Service
755	Central Service Support
760	Hazardous Materials
765	Hazardous Materials Service
800	**Health-Care Facilities**
810	Patient Bedroom
815	Patient Bedroom Service
820	Patient Bath
830	Nurse Station
835	Nurse Station Service

840	Surgery
845	Surgery Service
850	Treatment/Examination
855	Treatment/Examination Service
860	Diagnostic Service Laboratory Support
865	Diagnostic Service Laboratory Service
870	Central Supplies
880	Public Waiting
890	Staff On-Call Facility
895	Staff On-Call Facility Service
900	**Residential Facilities**
910	Sleep/Study without Toilet or Bath
919	Toilet or Bath
920	Sleep/Study with Toilet or Bath
935	Sleep/Study Service
950	Apartment
955	Apartment Service
970	House
000	**Unclassified Facilities**
050	Inactive Area
060	Alteration or Conversion Area
070	Unfinished Area
	Nonassignable Area
WWW	Circulation Area
XXX	Building Service Area
YYY	Mechanical Area
	Structural Area
ZZZ	Structural Area

Other PEFIC Classifications

The preceding categories are intended to be assigned on the basis of primary use. Where a room is used for multiple purposes and the system will accommodate it, the uses should be prorated on a basis of time and multiple codes assigned. There are several classifications spelled out in PEFIC, such as the following:

1. Definitions of functional categories (taxonomy of functions)
2. Coding for academic disciplines
3. Architectural features related to room use (physical characteristics)
4. Room suitability (for designated use)°

For specific classification of rooms other than by use, additional coding structures are often developed. These may reflect the management hierarchy of departments or other organizational units, function, or academic discipline.

COMPARISON

ANSI/BOMA's current Standard, designated Z65.1-1996, "is a building-wide method of measurement, allowing spaces that benefit all the building occupants to be measured and allocated on a pro-rata basis."° Oriented primarily to the needs of the building owner, manager, and real estate professional, BOMA defines a multitude of specific classifications and relationships that rely on the ratios of rentable to usable area in order to fairly allocate shared facilities. BOMA's paramount interest in the accurate determination of revenue-generating area is demonstrated by the fact that there are no fewer than four different kinds of rentable area and three levels of R/U ratios in the specification.

BOMA very specifically spells out how areas are to be measured, introducing the concept of *dominant portion*, its unique method of differentially measuring the interior perimeter of a building's exterior wall. Within the building's perimeter, common areas are shared either on a floor-by-floor basis or throughout the entire structure.

At the same time, BOMA offers no vocabulary for classifying areas with greater precision than simply as *usable*. In fact, BOMA offers no definition of *circulation*, referring only to public corridors, the areas of which can change over time.

ASTM/IFMA

IFMA is the largest professional organization representing facility managers, whose interests and expertise cut across the disciplines of architecture, engineering, interior design, communications, finance, personnel management, information technology, and building systems. In cooperation with IFMA, ASTM developed the specification known as E1836, which expands upon and is largely compatible with BOMA. The ASTM/IFMA standard provides necessary definitions of both *primary* and *secondary circulation* and also introduces the concept of *assignable* area.

Categorizing usable space as assignable office, assignable support, or secondary circulation derives from the notion of programmable area, space that can be allocated, planned, and accounted for on a basis of organizational and functional requirements. If we view the IFMA nomenclature alongside BOMA's, however, we now have no fewer than three classes of shared service area: *building common*, *floor common*, and *assignable support* within the envelope of usable space.

GSA's methodology specifically references the ANSI/BOMA method but also adds many of the same classifications used by ASTM/IFMA through a system of space categories and types. Horizontal circulation is defined within the *common* area type; and, at the macro end of the scale, a facility joint-use category makes possible the proration of usable common among multiple buildings in an office complex.

PEFIC

By contrast to the two preeminent standards concerned with classifying commercial office space, PEFIC has no concept of *rentable* area. Its categories are derived from space utilization, and it simply defines *net usable* as the sum of *assignable* and *nonassignable* area. The difference between gross area and net usable is simply called *structural area*. Vertical penetrations, a separate category identically defined under BOMA and IFMA, does not exist under PEFIC either. Stairways, elevators, and so on are referred to as part of *building infrastructure*, classified as nonassignable: circulation area.

Yet another significant difference from the other two specifications is that PEFIC measures both assignable and nonassignable areas from the *inside faces* of their defining walls. By default, this relegates the floor space occupied by all elements of interior construction to the structural-area classification.

In summary, area measurements made using the PEFIC standard are fundamentally incompatible with ANSI/BOMA and ASTM/IFMA. If, on the other hand, the concept of dominant portion is followed in determining the inside building perimeter—gross measured area (BOMA) and interior gross area (IFMA), the remaining area classifications are largely complementary with IFMA continuing to greater levels of detail where BOMA leaves off.

CLOSING WORDS

As we have seen, there is great variation in the calculation methods and nomenclature associated with area measurement in buildings. ANSI/BOMA is the leading standard for calculating space in office and other commercial buildings, focusing as it does on the determination of *rentable* area. ASTM/IFMA takes this process to the next level, providing terminology for subclassifying rentable space according to function. These are complementary, national standards. This chapter also considered the method used by the federal government's General Services Administration to classify the space it manages by category, type, and room designation. We concluded with a brief look at the room-use classifications generally used by postsecondary educational institutions.

Chapter 10 returns to the subject of application technology with a discussion of relational-database concepts. Relational databases are among the most advanced and elegant approaches to information management in use today and are used by virtually all CAFM systems. It is therefore quite important for the space planner and facility manager to have a working understanding of the basic theory.

The idea of the relational database stems from algebraic number theory and relies on the simple concept of linked tables, or *relations*, using clearly defined sets of data. These data structures provide a *nonredundant* means of storing information so that it may be efficiently input, retrieved, and updated. The relational model was first proposed in 1970, and there are currently numerous large-scale, commercial database systems available from companies such as Microsoft, Sybase, and Oracle.

CASE STUDY Global-Technology Company

This case study presents the implementation of a CAFM solution for a global provider of technology-based products and services to consumers, small-and medium-size business, and large enterprises. The company recently expanded significantly due to the acquisition of a major computer manufacturer, so its offerings now include personal computing devices, enterprise servers and storage, imaging products, and consulting and integration services.

The company, which has been in business for nearly sixty years, owns or leases over sixty five million square feet of space worldwide, of which nearly one-quarter is office space devoted to sales and support activities. The remaining space encompasses administrative facilities, research and development, manufacturing, and warehouse space. More than 1,300 facilities are distributed throughout 170 countries, and altogether the company has an estimated 140,000 employees.

Over the years, a number of different systems were implemented using a variety of approaches to tracking space utilization and real estate information. Dozens of independent databases came to be used, each reporting on different, sometimes redundant, parts of the business. This made the standardization of business processes and accurate reporting extremely difficult. System maintenance and training of personnel on numerous legacy systems had long since been proven to be costly and inefficient.

Scope of the Project

The acquisition referred to earlier triggered an initiative to consolidate the company's real estate portfolio, which in turn motivated the consolidation of space-related data into a single comprehensive CAFM database system. All the preexisting facility data needed to be migrated from the existing systems, converted to a uniform format, and thoroughly checked for accuracy. This was accomplished using a phased process that extended over a two-year period. Many Latin American, European, and Asian facilities previously had no space-management system at all; for these locations, the necessary drawings and space-utilization data needed to be assembled from scratch.

The firm retained a consulting and systems-integration firm specializing in facility-management applications° to manage and execute the conversion project. The target system chosen by the company early in the process was *Archibus/FM* (see Chapter 11).

Project Management and Consulting

Both the systems consultant and the client firm assigned project managers who would serve as primary contacts

for the duration of the engagement; they followed a six-phase methodology:

1. ***Planning (consulting)***—Establish guidelines for the project; project task schedule, communications and transition plans, and a project collaboration Web site where e-mail communications, key schedule postings, drawings, and other relevant data could be stored.
2. ***System specification and design (consulting)***— Develop data and reporting standards, data dictionary and conversion mapping for each existing source application, business rules, and system-configuration details.
3. ***Prototyping and system testing (conversion)***— Using a significant subset of the company's data, a prototype version of the new database was set up. Conversion processes for drawings and data from each legacy platform were tested and documented. The conversion process was carried all the way through using sample data from each platform so that results could be evaluated before committing to the full-scale conversion.
4. ***System build (conversion)***—In the full-scale build phase, the facility data were migrated from each platform into the Archibus/FM database. Simultaneously, floor plan drawings were cleaned up in preparation for linking to the new database. After drawings were linked, the system was subjected to extensive quality-control checks and exception reporting.
5. ***Transition (training and support)***—System users were trained; and the new, full-scale system was rolled out. The transition to operational status proceeded in orderly, sequenced subphases rather than all at once, to ensure that all quality-control issues were properly resolved.
6. ***Operation (support)***—Continuing support, follow-up training, and ongoing improvements in the application were provided after the full system went live.

Assembling Drawings and Data

The division of responsibilities between owner and consultant was clearly spelled out at the beginning of the project. In this case, the client firm shouldered the responsibility for data and drawing maintenance at each site. Transition plans for each platform and site were established jointly among the team members so that the owner's staff could be properly trained in the operation of the new system. Training for each site was typically done immediately preceding the delivery of its completed site drawings and data.

(continued)

Because of the consultant's extensive experience in managing projects of similar scope, this conversion project was looked upon as being a *logistically* challenging rather than a *technically* challenging endeavor. This was due to the thousands of drawings and a daunting amount of data that needed to be converted from at least five legacy CAFM platforms, as well as the sheer number of different sites involved.

DRAWING-CONVERSION MANAGEMENT

The drawing-conversion-management process is diagrammed in Figure CS3-1. Once the occupancy data and drawings were converted to the new standards, they were exported from the native CAFM systems and uploaded via the project Web site for final cleanup and linking to the new Archibus/FM database. The step-by-step process by which this was accomplished follows:

1. *Submit work*—Space drawings were exported from native CAFM systems with room-identification information. Architectural base plans were attached to these drawings as external references. Drawings and data were then uploaded to the project Web site.
2. *Check-in*—Space drawings were checked in to the *FAST* drawing-management system (see discussion that follows) so that information was available to all members of the project team.
3. *Audit and verify content and information*—Drawing information was reviewed for completeness and accuracy. Basic information such as column lines and bubbles, walls, doors, windows, stairs, elevators, plumbing fixtures, and room names were field verified as required. Room IDs were created for unidentified spaces where necessary.

4. *Drawing conversion and polyline*—Drawings were cleaned up and reformatted to the layer standard required by Archibus. Correspondences of room IDs and polylines were checked to prepare for automatic linking. Additional polylines were added to fill gaps in converted drawings or to create new space drawings from AutoCAD backgrounds or paper scans.
5. *Link polylines and update data*—As rooms were linked, room records and areas would immediately appear on the FAST Web site.
6. *Quality audit*—After linking, new areas were compared against old and exception reports produced for reconciliation. The issue that most typically needed manual intervention was cross-linked polylines.
7. *Publish to Web site*—Drawings and database information were continuously updated to the project Web site for review by all project team members.
8. *Notification and acceptance*—The user was responsible for formally accepting or rejecting completed work according to agreed-upon standards within a specified time frame.

FAST Web Interface

The *FAST* (Facility Asset and Space Tracking) Web interface provides a flexible, graphic, drill-down capability to various CAFM reporting systems.° A collage of FAST screens is shown in Figure CS3-2, illustrating the graphic and explorerlike drill-down capability of the interface. From a countrywide map, a region can be selected, progressing through city and building, until an individual room is detailed. Several reports can be produced showing personnel, equipment, and space-utilization

Figure CS3-1   Drawing-Conversion-Management Process (see Plate Nine).

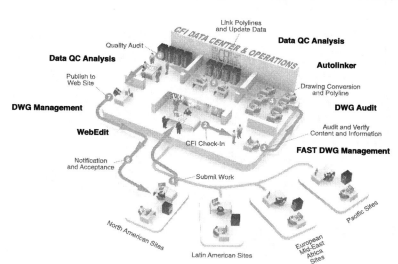

(continued)

(continued)

statistics at each level of detail. FAST space report calculations are made in accordance with BOMA and IFMA standards.

FAST is also capable of producing stacking diagrams showing overall space allocations and adjacencies by floor, as illustrated in Figure CS3-3.

Figure CS3-2   Facility Asset and Space Tracking (FAST).

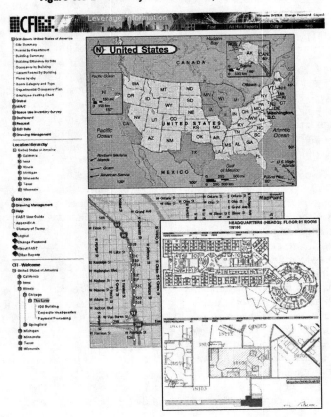

Figure CS3-3   Departmental Stacking Plan (FAST).

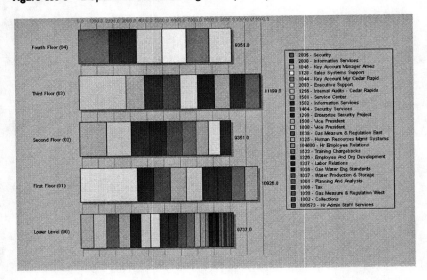

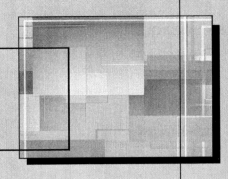

CHAPTER **10**

Relational Databases

> *There was nothing which medieval people liked better, or did better, than sorting out and tidying up. Of all our modern inventions I suspect that they would most have admired the card index.*
> —C. S. Lewis°

The relational-database model was proposed in a paper titled "A Relational Model of Data for Large Shared Data Banks," by E. F. Codd in 1970.° Codd was then working at the IBM Research Laboratory in San Jose, California. A mathematician, Codd believed that a set of operations could be defined that would enable the manipulation of stored data the same way that algebraic operators work on numbers. Codd's mathematics-based model provided the foundation for essentially all relational-database systems developed since.

Codd's principles of database design stipulated that data must be accessible through logical queries based on the stored data themselves and not be tied to any visible file structures or physical locations. A self-sustaining data-management system must also provide a language that facilitates communication with the database, supporting its administration as well as data definition and manipulation.

The development of such a language, which came to be known as *SQL* or *Structured Query Language*, began almost simultaneously at another IBM facility in Yorktown Heights, New York, and was completed at San Jose. Raymond Boyce and Donald Chamberlin were the original designers of the language, which was largely implemented by a team led by C. J. Date. The first commercial product was marketed under the name *SEQUEL*, for *Structured English Query Language*. There still remains in the industry a mild disagreement over whether the proper pronunciation of *SQL* is [*ess-que-ell*] or [*sequel*].

It did not take long for other companies to aggressively pursue the relational-database market.

ORIGINS OF THE RELATIONAL MODEL

In the 1950s and 1960s, data were most often stored in a flat file format that could be sequentially or randomly accessed, depending largely on the physical nature of the storage medium. In those early days of computing, before security was much of a consideration, computer rooms were often fitted out with large windows through which one could view the operator's console flanked by row upon row of floor-standing tape drives. Data were actually processed directly from large reels of half-inch-wide magnetic tape. Donald Knuth's classic *Art of Computer Programming* devotes no small amount of space to data-processing methods designed around the inherently sequential limitations of tape.°

Like accessing a musical selection on an audiocassette, one had to choose between playing the cassette from beginning to end or fast-forwarding the tape, searching for a desired piece of information. Sequential files consisted of blocks of tape containing data in some defined record format that was subdivided into fields. An individual field might contain a name, an address, or some particular numeric quantity. Once the records were put in the desired order, reports could be produced by the central processor, making calculations as each magnetically stored record passed across the read head of the tape drive. Sorting and searching, as Knuth describes them, were mechanically intricate operations involving multiple passes of multiple tapes resulting in the resequenced information being copied to yet another output tape.° Several copies of the same data were often maintained, sorted in different sequences.

An even earlier concept was that of the *unit record*, a data structure usually designed within the framework of the eighty-column tab-card format that was standard at the time. Individual fields within the record were represented by a defined number of holes punched in the card using a coding system known as *binary-coded decimal (BCD)*. Formats such as BCD were not dissimilar to methods for encoding data still in use such as the ASCII character set.°

The idea of the unit record is a useful model in understanding the database concept of *atomic values*: individual pieces of data that are specifically and unambiguously defined with respect to one another. Relationships are set up such that the fields in a given record appear to the user in a consistent physical layout and all the records in a group contain data of a similar kind. Grouping data in this manner naturally leads to the idea of a table in which the rows represent individual records and the columns—or *domains*, as they are called in database parlance—define the nature of each field.

If a flat file could be represented as a table and provided with a key field (in some cases, simply a record number) by which any record could be identified and accessed, auxiliary tables could then be used as indexes. Multiple indexes pointing to a master table would obviate the necessity of sorting the data and maintaining multiple copies. This was attractive, once random-access, disk-storage devices became available, both from a data-management standpoint and because random online storage was extremely expensive.

A simple table of names in random sequence accompanied by an alphabetical index illustrates the concept in Figure 10-1. After the master table has been appended, deletions made, or records modified, the index can be regenerated by making one sequential pass over the data. Only the relatively smaller index is then sorted so that it can be efficiently searched, facilitating the retrieval of matching records using the designated key field.

Hierarchies

The word *hierarchy* appears in one form or another on just about every other page of Chapter 6 and on some pages more than once. It was used in Max Weber's context to refer to functions being structured into offices with rules and standards that led to unwieldy bureaucracies "filtering" information on its way up the organizational tree. This *Theory X* model gave way to *Theory Y* organizations, often featuring matrix structures more closely resembling networks than hierarchies. A similar progression can be seen in the development of databases.

In the mid-1960s, IBM's IMS was one of the first commercial database systems to use a hierarchical model. This model suited the representation of one-to-many relationships as portrayed in Figure 6-3 (*A* relates to *B* and *C*, *B* relates to *D* and *E*, and so forth). But hierarchies have a problem representing many-to-many relationships, such as *D and E*

Figure 10-1 Alphabetical Index and Master Data Files.

Name	Key		Key	Name	Other Data
Bill	05		01	Rich	Other Data . . .
Billie	08		02	Teresa	Other Data . . .
Carl	03		03	Carl	Other Data . . .
Don	04		04	Don	Other Data . . .
Gina	10		05	Bill	Other Data . . .
Mary	06		06	Mary	Other Data . . .
Rich	01		07	Sue	Other Data . . .
Scott	09		08	Billie	Other Data . . .
Sue	07		09	Scott	Other Data . . .
Teresa	02		10	Gina	Other Data . . .

reporting to *A and B,* for many of the same reasons that such chains of command can be problematic: redundancy is introduced into the structure.

Networks

Networks such as those pictured in Chapter 5 became the preferred model toward the end of the 1960s. Networks allow many-to-many relationships and can perform extremely efficiently because direct links between related physical-data locations are stored along with the data. This eliminates the need for sorting and indexing. The analysis/synthesis network in Figure 5-6 abstractly pictures one-to-many relationships expanding on the left side and many-to-one relationships on the right, narrowing to a conclusion. The PERT diagram in Figure 5-7 illustrates a project network with both kinds of connections representing a process from beginning to end.

On a large scale, however, the kind of optimization required can result in inflexibility. The PERT network depends upon hardwired connections among the tasks in calculating the critical path. Similarly, large data networks require hard pointers: cross-references from one part of the database to another. Accessing the data can require substantial programming, and the data structures are largely inflexible. Reorganizing a network database is generally a major operation.

Relational Databases

Codd continued his work in the development of relational theory throughout the 1970s and 1980s and published numerous papers that are generally accorded the authority of gospel. It was a firm named Relational Software, Inc., founded in 1977, that developed the first commercial SQL-based product. Oracle Corporation, as the firm was renamed, delivered the first commercial SQL relational-database-management system (RDBMS) in 1979 and remains the market leader with over forty thousand employees.°

IBM brought its first commercial product to market three years later. Several other software developers entered the market including Sybase, Inc., in 1986. Microsoft Corporation, in partnership with Sybase, produced SQL Server in 1992. Shortly thereafter, the partnership ended with Sybase largely pursuing the UNIX market.

All of these products used SQL, which became the de facto standard language of relational databases and, subsequently, the official standard as well. In 1982, ANSI convened a database committee, which adopted IBM's SQL vernacular (with minor modifications) as a national standard in 1986. This version, SQL/86, became an international standard as well when it was accepted by the ISO in 1987.

CODD'S TWELVE RULES

E. F. Codd set down twelve rules defining the characteristics of an *ideal* relational database in two articles published in 1985.° These rules encapsulate most of the principles that enable relational databases to function properly. Although there is no database that maintains perfect adherence to these principles, they can be viewed as guides to progress. Understanding them can help us to gain a fairly complete picture of the framework in which an RDBMS works.

Rule 1: Information

A relational database is more than just data stored in tables. Mathematical principles of relational algebra provide the logical basis for defining the relationships among them. Codd's first rule states that *all* information must be presented to the user in one and only one way: as *tables.*

Three sets of nomenclature are shown in Table 10-1. The format of a spreadsheet consists of rows and columns, which, in the case of Microsoft Excel, is called a worksheet.

Table 10-1 Names of Table Components

Excel Data	Flat-File Data	Formal Database
Worksheet	Table	Relation
Column (Cell)	Field (Datum)	Domain (Attribute)
Row	Record	Tuple

Excel files are called *workbooks*, which are collections of worksheets. Elements of data (including formulas) are stored in cells formed by the intersections of each column and each row. An *Excel* worksheet is limited to slightly less than 1.7×10^7 cells in a table containing 256 columns and 65,536 rows. Cell references can be defined from one worksheet to another, providing for links of sorts; yet spreadsheets are not relational. Any kind of data can be entered into a cell whether it has a logical relationship to data in other cells or not.

As we have seen in earlier chapters, Excel worksheets can be used as flat files of data, making them particularly useful when analyzing those data or converting them into other forms. In this case, some kinds of relationships are usually defined among the cells in a row. In the case of the Access facility program example we used in Chapter 7, we could just as well have prepared the detail table pictured in Figure 7-4 using an Excel worksheet. Each of the table's fifteen columns would have had the same significance as fields, each row containing related information qualifying it as a record. As a matter of fact, when we prepare a space-allocation program, we often use such a worksheet to consolidate the information from questionnaires and interviews with the results of our requirements analysis. Such a table can then be pipelined into Access or some other database.

But why do that?

Before we answer this question, we need to look at column 3 of Table 10-1. Using formal database terminology, we refer to a table as a *relation*, because the data it contains is set up so that every value (or *datum*—the word *data* is plural) is logically related to the other values in the same table. This rigorous process of structuring the data is known as *normalization*.

Normalization is a formal process for designing a database so that repeating elements are eliminated and the atomic value of each is stored in one and only one location. There is an accurate parallel between data elements being comprised of atomic values and chemical elements having atomic form. We now know that matter at the atomic level is made up of ever smaller particles (protons, neutrons, electrons, quarks, and such). An element of data may consist of arbitrary numbers of alphanumeric characters, or it may be an object such as a picture; but at that level, the individual characters are simply building blocks common to all elements.

Rule 2: Logical Accessibility

Within a relation, each *column* or *field* is known as a *domain*. Each *row* or *record* is designated with perhaps the most unfamiliar-sounding piece of vernacular yet: a *tuple*.° Each individual *cell* contains a single *datum*, the atomic value that we call an *attribute*. A database constructed of multiple, normalized tables allows complex data structures to be represented by the links among the tables, using key fields.

Codd's second rule says that every attribute must be guaranteed unambiguous access using a combination of the table name, column reference, and a key consisting of one or more fields within each record. In short, every *attribute* within a *relation* must be retrievable using only the name of its *domain* and its *key*. Relationships among tables are purely logical; there are no physical or hierarchical relationships among them, nor are the *tuples* maintained in any fixed order.

Rule 3: Null Values

The third rule is that fields must be able to be empty, which requires the support of a null value. *Null* means that the value is unknown, which is not the same thing as zero or blank.

Plate One Morsque No. 1 (Plate XXXIX) Owen Jones, *The Grammar of Ornament* (see Figure 2-12 and notes to pages 25-27).

Plate Two   Vitruvian Principles and Attributes (Figure 3-8).

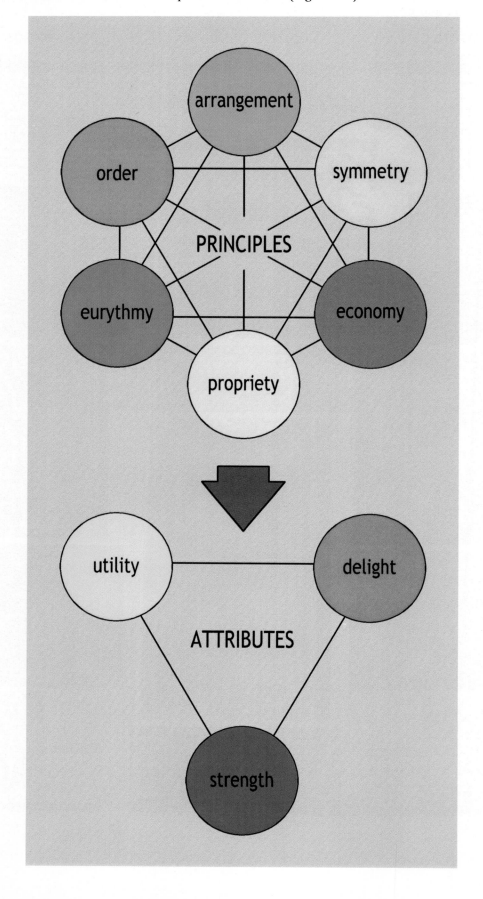

Plate Three   Interior Elevation and Perspective of Finished Kitchen (Figure CS1-2).

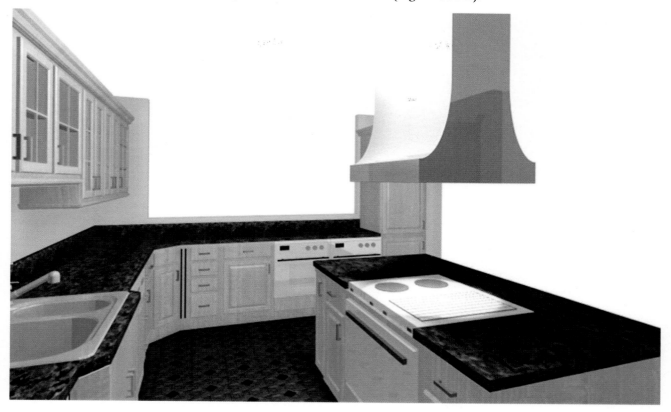

Plate Four   Bubble Diagram of Adjacency Priorities (Figure 6-13).

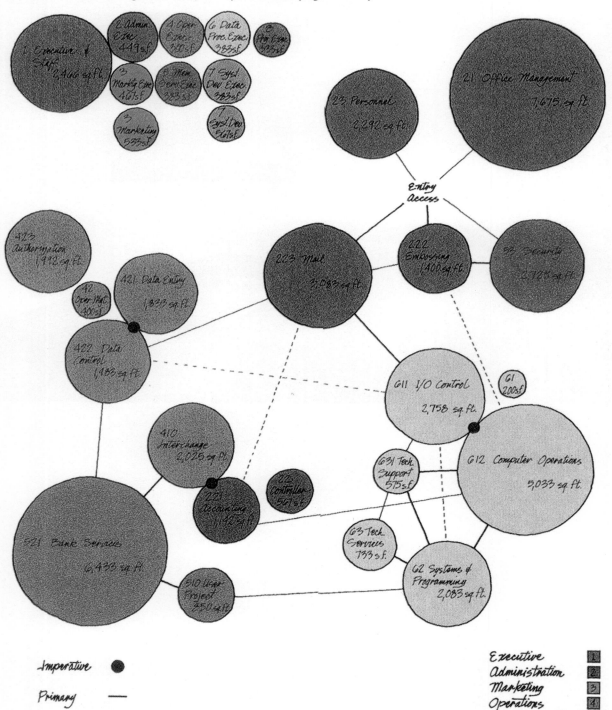

Plate Five   Three-Floor Functional Block Diagram (Figure 6-14).

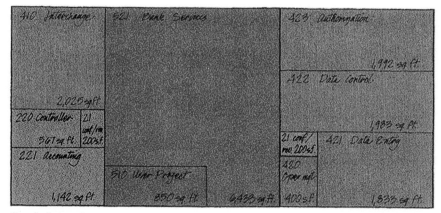

3rd Level   17,846 sq ft

612 Computer Operations — 5,033 sq ft.
611 I/O Control — 2,758 sq ft.
62 Systems & Programming — 2,083 sq ft.
63 Technical Services — 1,308 sq ft.
21 conf.rm 200 s.f.
7 Systems Development 567 s.f.
3 Marketing 533 s.f.
7 Syst. Devel. Exec. 383 s.f.
8 Proj. Exec. 383 s.f.
6 Data Proc. Exec. 383 s.f.
8 Member Devel. Exec. 383 s.f.
3 Marketing Exec. 467 s.f.
4 Oper. Exec. 300 s.f.
2 Admin. Exec. 449 s.f.
1 Executive & Staff — 2,466 sq ft.

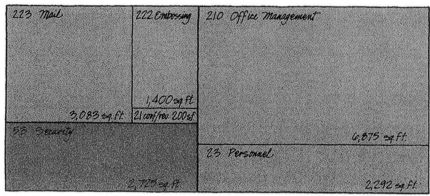

2nd Level   17,625 sq ft

410 Interchange — 2,025 sq ft.
520 Bank Services
423 Authorization — 4,992 sq ft.
422 Data Control — 1,983 sq ft.
220 Controller — 567 sq ft.
21 conf/rm 200 s.f.
221 Accounting — 1,142 sq ft.
510 User Project — 850 sq ft.
6,433 sq ft.
21 conf/rm 200 s.f.
420 Oper. mgt. 400 s.f.
421 Data Entry — 1,833 sq ft.

223 Mail — 3,083 sq ft.
222 Embossing — 1,400 sq ft.
21 conf/rm 200 s.f.
210 Office Management — 6,875 sq ft.
53 Security — 2,725 sq ft.
23 Personnel — 2,292 sq ft.

1st Level   16,575 sq ft

Executive — 1
Administration — 2
Marketing — 3
Operations — 4
Member Services — 5
Data Processing — 6
Systems Develop. — 7
Projects — 8

Plate Six Functional Adjacency and Work Flow Diagram (Figure CS2-1).

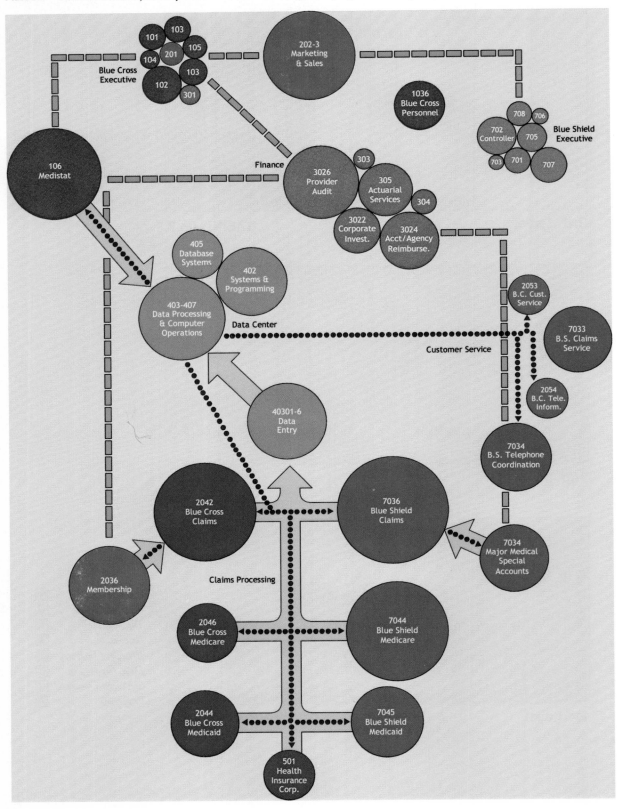

Plate Seven Isometric Stacking Diagram (Case Study Two: Blocking and Stacking- Development of an *ideal* building envelope for a Federal Agency).

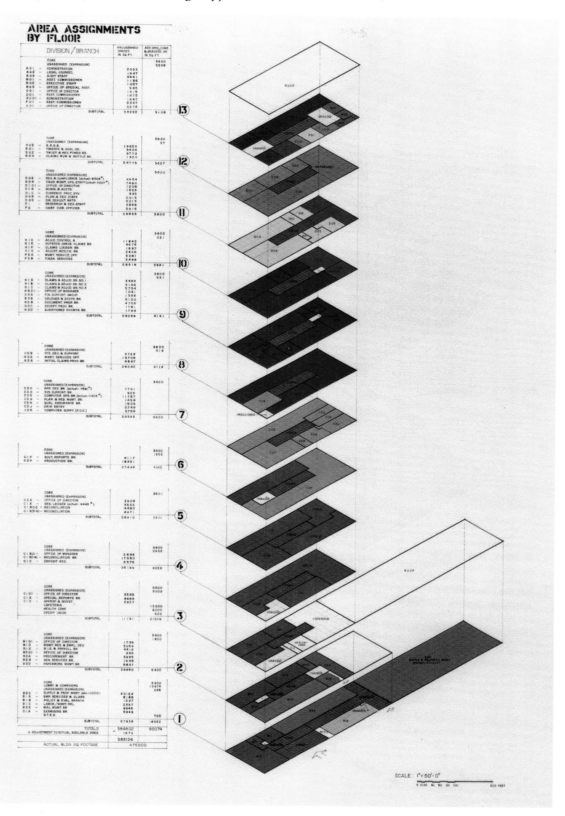

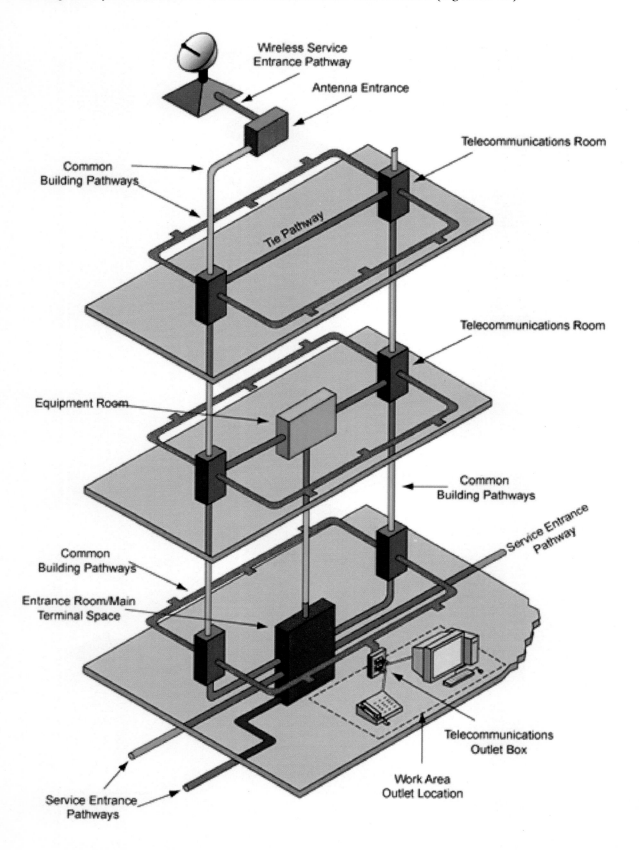

Plate Nine Drawing Data Center and Operations (Figure CS3-1 - Drawing Conversion Management Process).

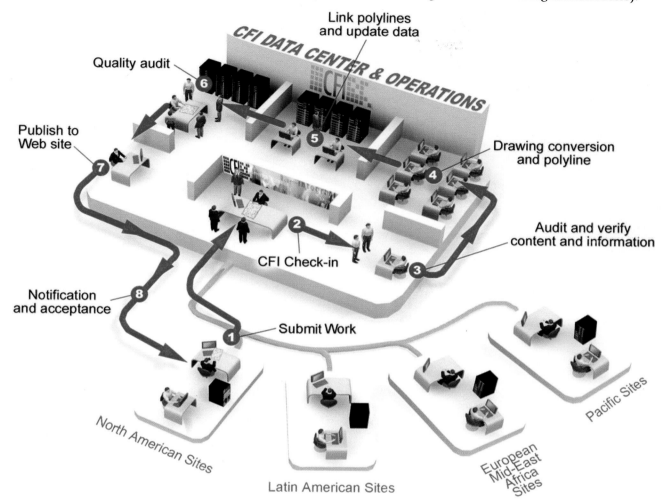

Plate Ten   Digital Café at a BP America Facility in the Midwestern U.S. (Figure CS5-3).

Plate Eleven   Space Types Defined on a Typical Office Floor Plan (Figure CS5-5).

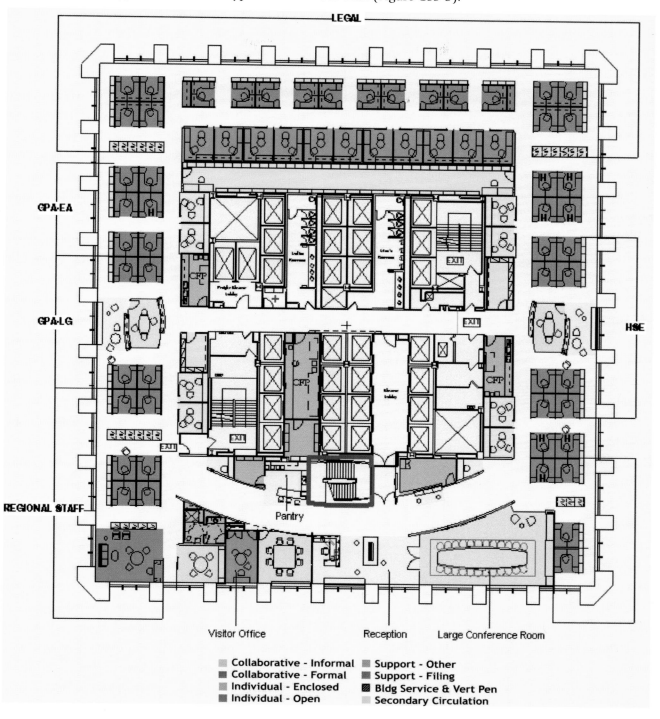

Collaborative - Informal
Collaborative - Formal
Individual - Enclosed
Individual - Open
Support - Other
Support - Filing
Bldg Service & Vert Pen
Secondary Circulation

This can cause some confusion if logical queries are not constructed carefully. If you have a room table with a column indicating whether rooms are occupied or not, and you search for all of the rooms that are *not* occupied, the results will not include any rooms with null in the occupancy column. *Null* means that the rooms may be occupied or not; their status is indeterminate.

The exception to this rule is that the *primary key*, the key value that uniquely identifies each record in a table, *must* have a value. Most database implementations provide for specifying whether or not *null* is permitted in certain columns.

C. J. Date has a basic issue with the inclusion of null on the ground that it undermines the entire relational concept by requiring the use of "three-valued logic" in which there are three truth values: *true, false,* and *unknown.*° He argues that, although "the problem of missing information is one that is encountered very frequently in the real world," the support of *null* only complicates matters. For example, if A > B and the value of A is unknown, then the result of the comparison is unknown regardless of the value of B. If A = B and A and B are both null, the comparison does not evaluate to *true* but the result is also *unknown.* Date recommends using default values instead of permitting *null* in order to stay within the binary realm of true and false.

Rule 4: Self-Descriptive

Metadata are data about data. Metadata describe the database itself, the objects it contains, and its structure. The structure of a database is often referred to as its *schema.* Both the user information and the schema must be stored in tables according to Codd's fourth rule, and they must be accessible in the same manner making the database self-descriptive.

In addition to the system tables, which are collectively called the *system catalog* or *data dictionary,* metadata include indexes, stored procedures, and other such objects that determine how the elements of the schema relate to one another. In theory, modification of the tables that contain user data should be possible at any time by changing the structure of the metadata, but often this is difficult to implement in practice (see Rule 9).

Rule 5: Comprehensive Data Sublanguage

Virtually all commercial RDBM systems run under the control of a variety of operating systems on different platforms. Relational systems may be written in and support different languages, input/output modes, and the creation of reports. Codd's fifth rule, however, specifies that there must be a single language for communicating with the database for data definition, modification, and administration. As we discussed earlier, SQL was first the de facto standard and later became the official standard for carrying out these functions.

SQL is a *declarative* or *nonprocedural* language in which the user says what she wants but not how to get it. SQL has both data-definition language (DDL) and data-manipulation language (DML) components, which the RDBMS uses to determine the best way of fulfilling the request. The system may take a brute-force approach and sequentially scan one or several tables for the requested data, or it may use an index to access the data directly. In either case, though, SQL operates at a "higher level of abstraction" than procedural languages.° By way of example, Date shows more than thirty lines required to perform a typical operation using procedural program code versus the single line of SQL necessary for the equivalent result.

Several basic categories of SQL operators and expressions are shown here:

- *Sessions, connections, and transactions*—CONNECT/DISCONNECT, COMMIT/ROLLBACK
- *Data definition* (TABLE, VIEW)—CREATE, ALTER, DROP
- *Data manipulation* (single row, CURSOR)—SELECT, INSERT, UPDATE, DELETE
- *Data control* (integrity and security)—GRANT/REVOKE, BACKUP/RESTORE
- *constraints* (KEY)—PRIMARY, FOREIGN
- *Table expressions* (JOIN)—INNER, LEFT, RIGHT, FULL, UNION
- *Conditional expressions*—AND, OR, NOT, LIKE, BETWEEN, IN, EXISTS, etc.
- *Scalar expressions*—NUMERIC, CHARACTER, etc.

The preceding words in full caps are reserved key words within the SQL language.° The general meaning of most of them should be discernible from the context, and many are discussed elsewhere in this chapter.

Rule 6: Views

Codd's sixth rule states that a relational database must not be limited to source tables. There are two kinds of tables in most RDBM systems, yet neither exists physically. Both are representations of the data within the system, as described by the metadata and presented to the user in the column and row format.

1. *Base tables* are what we usually refer to when we speak of relations. It is the base tables that we rigorously try to normalize in order to eliminate redundancy by making each attribute an atomic value.

2. *Views* are named virtual tables that are derived from the base tables. They are not copies of the data in the base tables but provide an alternative way of looking at selected data from one or several tables.

Views are created using the operators provided by the data sublanguage (SQL) to produce complex queries. Such queries may be saved and repeatedly invoked to restore commonly used views, simplifying data access. For example, a particular user's access to data may be restricted by defining a view that omits certain columns containing sensitive information.

Logical selections of data are presented to the user as views, which in theory should support the same methods for data manipulation as base tables. Changing data in a view should automatically change the underlying base table. In practice, however, this rule is seldom fully implemented. Often, the data presented in a view cannot be edited but only used to produce a report.

Base tables and views can be compared to model space and layouts in AutoCAD. A drawing is constructed of primitive objects defined as mathematical vectors in a 3-D coordinate space. The objects are presented to the user in model space as a dynamic, two-dimensional image that may be observed from different points of view and in different projections. AutoCAD layouts are specific views of the model defined using paper space. Layouts permit the user to combine different elements of the model or other models projected at different scales and from different points of view. Layouts in paper space may only be examined or printed. Changes to the drawing must be made in model space.

Rule 7: Set-Based Relational Operations

Flat files, even if they are represented as tables, are processed row by row. They can be processed sequentially, one row after the other, or randomly according to row (record) number. Each row also has a defined succession of fields. If the layout of the record is changed, the procedural application must be modified as well.

A relational database must support high-level set operations (intersection, union, difference, division) and basic relational algebra (selection, projection, join) that operate on rows as sets for data-manipulation purposes (SELECT, INSERT, UPDATE, DELETE). The relational algebra and set expressions operate on whole relations (tables) to produce other relations. This capability of operating on a base relation or a view *as a whole* applies not only to data retrieval (select) but also to the insertion, updating, and deletion of data.

Declarative SQL statements must ultimately be reduced to the equivalent of executable code using the following steps:

- *Parsing*—The statement is broken into individual words that are checked to make certain there is a valid verb, legal clauses, and no syntax errors.
- *Validation*—The statement is checked against the system catalog for correct table and column references, user authorization, and so on.
- *Optimization*—Alternative ways of executing the statement are evaluated. For instance, a statement may require that a table be searched and joined to another table.

Whether the search or the join is performed first may greatly affect the speed of execution, and it is up to the RDBMS to determine the best approach. Many different tactics may be evaluated by the optimizer, which is a CPU-intensive and time-consuming process. The assumption is that minimizing execution time will make up for the time spent analyzing the approaches.

- *Execution*—An execution plan is generated and carried out.

When all is said and done, optimized execution plans are comparable to procedural machine code, except that they are produced by the RDBMS. If they are saved as *stored procedures*, they can be used repetitively to execute queries without performing the first three steps each time.

Rule 8: Physical-Data Independence

An application that accesses data in the flat file we have described must be designed around the physical structure of the record. If a ten-character numeric field follows a thirty-character alphanumeric field, and an eighteen-character date field is added in between, the application must be changed accordingly.

The eighth rule states that relational applications must not be affected by changes in the way data are physically stored. Physical-data storage, access methods, and hardware can be changed without affecting the way the user perceives the data or how the application works with them. The logical structure consists only of tables, and each domain is simply defined according to data type and length.

Rule 9: Logical-Data Independence

Codd's ninth rule requires that the relationships among tables must also be subject to change without impairing the functionality of queries and applications. This applies to changes in the schema, table layouts, and logical relationships including the addition of new tables.

The logical-data-independence rule is very difficult to satisfy in all circumstances, and many systems do not. In most cases, there are many implicit connections between the actual table structures and the way the user sees the data. If a table is removed, for example, statements that depend on it will no longer function correctly, if at all. In some systems, it is necessary to depopulate the tables while their structure is being modified.

Rule 10: Integrity Independence

To ensure the accuracy and consistency of the data in an RDBMS, data integrity must be maintained. This must be managed through internal functions using the data sublanguage (SQL) rather than application programs. There are three basic kinds of data integrity:

- *Domain*—The set of permitted values in a column, on a basis of data type and format, permissible range of values, whether NULL is permitted, and so on. Domain integrity is also referred to as *attribute* integrity, which can also be maintained by inserting default values where explicit data are not available.
- *Entity*—The unique identification of individual rows in each table, usually through the use of a *primary key* in one or several columns (the latter being referred to as a *composite* key). This is a crucial concept in the relational-database model and ensures that there are no duplicate rows to introduce ambiguity. There can be only one primary key per table (which can be comprised of multiple columns), and attributes belonging to primary keys cannot contain *null* values.
- *Referential*—This is often the catch-all phrase to describe the mechanism for maintaining data integrity in general, because its principal purpose is to keep relationships among tables synchronized. Referential integrity is maintained through the use of a *foreign key*, which (like primary keys) can be comprised of multiple columns.

The relationship between one table and another is either that of *parent to child* or *child to parent*. Logical links are established between tables by matching the *foreign key* in a child table to the *primary key* in its parent. Only if *primary*-key values exist in the parent table can corresponding *foreign*-key values be entered into a child table. The intrinsic meaning of referential integrity lies in the fact that *orphans*, entries in a child table that do not have a parent, are not permitted.

Figure 7-2 illustrates the schema of the *facility-planning-access* database discussed in Chapter 7. This simple schema has two lines of *descendants* that define the organizational hierarchies pictured in Table 6-2.

1. The **management** hierarchy is fully implemented: *Group* is a parent of *Division*, *Division* is a parent of *Department*, and so on, down to the level of the *Detail* table, which is a child of *Subunit*.

2. Only the bottom two levels of the **location** hierarchy are implemented: *Building* is the parent of *Floor*, and *Floor* is the parent of *Detail*.

All the key field names are shown in boldface in Figure 7-2. If we follow the *location* hierarchy from top to bottom, we can see that *BldgID* is the primary key in the *Building* table. In the *Floor* and *Detail* tables, *BldgID* is a foreign key. *FloorID* is the primary key in the *Floor* table and is a foreign key in the *Detail* table.

Similar relationships can be followed through the six levels of the management hierarchy. The *Sequence* code (see Chapter 6) is the primary key for each record in the *Detail* table. The *Level* codes, as a group, plus the *Sequence* code, give each record its unique identity in the *Detail* table.

It is in this way that a hierarchy can be expressed through the use of multiple-linked tables. In this example, we have two separate hierarchies parenting (and grandparenting) the *Detail* table. Network relationships can also be expressed through different sets of keys. Primary-foreign-key relationships establish logical relationships among the data, but no single set of logical relationships can restrict other potential pathways. The foreign-key attribute in a child table cannot hold a value that does not exist as a primary key of its parent.

Because of these relationships, an RDBMS must be populated starting at the top of any hierarchy and working downward through the linked tables (INSERT). For the same reason, data must usually be *updated* or *deleted* from the bottom up if the changes affect the primary keys. (Please see the section titled "Populating the Database" in Chapter 7.)

Referential data integrity can be enforced in three ways:

1. Disallow data modifications that would violate any of the primary-foreign-key relationships. This can either be done declaratively, by placing restrictions on domains, or procedurally, using stored procedures or triggers.°

2. Implement cascading *updates* and *deletes* so that a change to one key field is automatically propagated throughout all related tables. Cascading updates are relatively predictable; but, with cascading deletes, deleting a key field in one table will cause all matching records in related tables to be deleted. In both cases, orphans are prevented; but, in the case of DELETE, unintentional loss of data can occur. (*Inserts* cannot be cascaded.)

3. Foreign key values can be set to null if there is no matching primary key after an INSERT, UPDATE, or DELETE operation. In most cases, however, none of these options is a good solution. In the *Facility-Planning* database discussed earlier, for example, workstations and support spaces could be created in the *Detail* table that are not assigned to any department, floor, or building.

Rule 11: Distribution Independence

Whether an RDBMS is maintained in one central location or distributed among multiple servers, logical data access must be user transparent. The data sublanguage must be capable of supporting distributed queries that can join data from tables on different servers, while maintaining data integrity and synchronization among multiple copies regardless of where the data actually reside. Logical accessibility should remain unaffected even when the physical configuration of the system is changed.

Rule 12: Nonsubversion

Modification of the data or database structure by any means other than the use of the data sublanguage (SQL) is not permitted. "Back doors" into the database, such as a low-level programming language might provide, cannot be used in ways that would bypass the rules and constraints allowing data or structural integrity to be violated. In practice, most of today's database systems support administrative tools allowing direct manipulation of the data structures.

RELATIONAL PROPERTIES AND NORMALIZATION

A *relation* is a collection of domains. Each domain has a heading, a fixed attribute identifying that column. In the tabular representation of a relation, the first row contains the domain headings and the body of the table consists of a set of *tuples* containing the data. Date identifies four essential properties common to all relations:°

- *There are no duplicate tuples.* This goes back to Codd's idea that the relational model is mathematics based and, by definition, mathematical sets do not contain duplicate elements.
- *Tuples are unordered from top to bottom.* Mathematical sets also are not ordered.
- *Attributes are unordered from left to right.* As in the case of tuples, this is where the table representation is somewhat misleading. Columns and rows in a physical table have a definite sequence; in a relation, they do not.
- *All attribute values are atomic.* In a relation, every cell contains a single value; there are no repeating groups.

Normalized relations are those that do not contain repeating groups; and, as Date puts it, "the unqualified term 'relation' is always taken to mean a 'normalized relation' in the context of the relational model." There are several levels of normalization to be considered, and each is more highly refined than the level that precedes it.

First Normal Form

Earlier we used the term *unit record* to convey the idea of atomic value. The eighty-column tab card represented *one* record of data input to a card reader or *one* check to be distributed to a particular payee. The unit-record configuration allowed small collections of data to be separated from one another and processed individually. Data constantly change, and the uniqueness of each such collection allowed one item to be processed with minimal effect upon the others.

The term *normalization* first appeared in our discussion of Codd's information rule. In the database, it is the field, or *attribute*, that constitutes the atomic value. Returning to our analogy of the chemical elements, the relatively small collection of attributes in a tupic can be compared to a molecule that has an explicit identity as a compound. Formed from two or more elements, tuples can interact with each other by normal "chemical" processes. They can be joined together, in their entirety or in part, into more complex substances as *views*.

In a procedure-based system, it was necessary to modify the program logic if either the data structure or the desired output changed. This introduced all the complexities of altering the elements' atomic structure. The principal goal of the database is to make information available to the user without his having to be concerned about its atomic structure, so the basic compounds must be reduced to their simplest forms.

The first normal form requires that each row be unique and that it have a *primary key*. The space-inventory record in Figure 10-2 satisfies neither of these requirements because

Figure 10-2   Space-Inventory Record Not in First Normal Form.

Department	Space1	Area1	Quantity1	Space2	Area2	Quantity2
3A1	Manager	150	1	Analyst	96	3

Figure 10-3   Space-Inventory Records in First Normal Form (1NF).

Department
3A1

Department	Space	Area	Quantity
3A1	Analyst	96	3
3A1	Manager	150	1

Figure 10-4   Space-Inventory Record with Sequence Codes and Buildings.

Department	Sequence	Building	Space	Area	Quantity
3A1	10	Headquarters	Manager	150	1
3A1	20	Headquarters	Analyst	96	3
4C3	10	Training Center	Director	225	1
4C3	20	Training Center	Clerk	64	4

there are multiple columns containing the same type of data. To reduce the elements of this table to first normal form, *repeating or multivalued elements must be unitized in a new-child table in which each domain contains atomic values.*

In Figure 10-3, each *Space* description has become the primary key in a new space-inventory-detail table. A new *Department* table has been created, and each *Department* can now have any number of *Spaces* related to it.

Second Normal Form

Second normal form assures that domains are assigned to the correct table *by relocating all nonkey attributes that do not depend upon the entire primary key.*

Figure 10-4 shows a little more of the space-inventory-detail table with some modifications. As it is likely that a space name may be used more than once within a *Department* (more than one analyst or clerk workstation, for example), *Sequence* codes have been added so that the space description is no longer used as a key. The *Department* and *Sequence* codes are now used as the primary key.

We also added the *Building* name, which we can see depends only on the *Department* and is therefore repeated. Moving the *Building* name to the *Department* table solves the immediate problem of normalizing the space-inventory-detail table. These modifications are illustrated in Figure 10-5.

The *Department* table, however, may need to be modified if some departments are divided among multiple buildings or for it to conform to third normal form.

Third Normal Form

Up to now, we have been directing our attention to attributes used as keys. The third level of normalization requires *the identification and removal of attributes that depend upon other nonkey attributes that are not alternative keys.* This condition implies that the remaining attributes are mutually autonomous and can be updated independently.

In Figure 10-6, we have added *Building* location information to the *Department* relation. We can immediately see that the *Address* and *City* domains do not depend upon the departments but on the *Building* identification, which is not the primary key. This has once again introduced redundancy, suggesting that the addresses belong in another table.

Figure 10-5   Space-Inventory Records in Second Normal Form (2NF).

Department	Building
3A1	Headquarters
4C3	Training Center

Department	Sequence	Space	Area	Quantity
3A1	10	Manager	150	1
3A1	20	Analyst	96	3
4C3	10	Director	225	1
4C3	20	Clerk	64	4

Figure 10-6   Department Records with Building Addresses.

Department	Building	Address	City
2B0	Headquarters	Randolph St.	Chicago
3A1	Headquarters	Randolph St.	Chicago
4C3	Training Center	Cantera	Hinsdale

Figure 10-7   Department Records with New Building Table (3NF).

Department	Building
2B0	Headquarters
3A1	Headquarters
4C3	Training Center

Building	Address	City
Headquarters	Randolph St.	Chicago
Training Center	Cantera	Hinsdale

What we have been doing at each stage of normalization is decomposing relations into smaller modules: projections specified by their join dependencies. *Projection* is the opposite of *join*, whereby we create subsets of the domains in a relation, discarding redundancies that existed in the larger set. When we normalize relations, we must be sure that no data are lost in the process, meaning that any decomposition must be reversible if the tables are later joined in a view.

Creating a separate *Building* table for the address information restores the form of the department relation (Figure 10-7) so that the *Address* and *City* data are not repeated.

The third normal form also disallows elements that can be derived from other attributes in a table. In the case of the space-inventory relation in Figure 10-5, if there had been an additional column for *Total Area* that was the calculated product of *Area* and *Quantity*, it would need to be deleted to restore the space inventory to third normal form.

Satisfying the first three forms is sufficient to normalize most tables under the majority of conditions. There are additional levels of normalization that can be performed, however, and three more are sufficient to guarantee that all relations will "be free of anomalies that can be eliminated by taking projections." °

Boyce-Codd Normal Form

Boyce-Codd normal form (BCNF) is a variation on the third normal form considering *all possible choices of candidate keys as primary key.* Date discusses *candidate* keys extensively. ° They are defined as one or more domains that *could* serve as primary keys, where no two tuples (rows) have the same value and no smaller set of domains is unique. If one domain happens to contain a unique identifier, it may not be immediately apparent that the data set contains more than one relation. In Figure 10-8, for example, if there were never more than two individuals with the same name in the *Person* domain, that column alone could be a primary key; but such is almost never the case.

Figure 10-8 Personnel with Rooms and Phones.

Room	Person	Phone
203	Bill	6689
203	Mary	6689
203	Teresa	4021
434	Bill	1338

Phone	Person	Room
1338	Bill	434
4021	Mary	203
6689	Teresa	203
6689	Bill	203

Figure 10-9 Room/Phone and Phone/Person Relations (BCNF).

Room	Phone
203	4021
203	6689
434	1338

Phone	Person
1338	Bill
4021	Teresa
6689	Bill
6689	Mary

Here, we have the following set of possibilities:

- A room can be occupied by many people.
- Each room can have multiple phones installed.
- Each phone is installed in only one room.
- A person can work in more than one room.
- Each person only has one phone in any given room.

In the top table in Figure 10-8, *Room* and *Person* serve as a composite primary key that determines the *Phone*. It is in third normal form. However, if we make *Phone* and *Person* a candidate key (also composite) in the bottom table, it is not in 3NF because the *Phone* number alone is sufficient to determine the *Room*. In this case, we find that *because the keys overlap*, more than one relation is contained in these three elements.

Figure 10-9 shows two tables projected from the three domains in Figure 10-8, which are both in Boyce-Codd normal form. In each decomposition, both columns are used as a composite primary key because none of the values of any single domain is unique. If we consider the *Room/Phone* table on the left as a child table, we can logically join it with the *Phone/Person* table on the right using *Phone* as a foreign key. Thus, views can be defined that *virtually* re-create either of the table layouts shown in Figure 10-8.

Fourth Normal Form

The four normal forms discussed up to this point all had one thing in common: they all involved *functional dependencies*. Dependencies refer to the many objects in the preceding tables that depend upon a singular key value within a relation. For example, in Figure 10-7, each building can only have one address; so any time that a building appears in a table, it must have the same address. This is the essence of the *integrity-constraint* concept: it establishes limits on the values that certain attributes can properly hold. The resolution of functional dependencies ultimately led us to the BCNF, in which overlapping keys and multilevel dependencies are considered.

Fourth normal form addresses *independent, multivalued* dependencies. In Figure 10-10, we have a table that contains two different kinds of unrelated data. There are a number of potential problems here even though this table is in BCNF because it is *all key* as constructed.

Ambiguity stems from the fact that the paintings in each room do not belong to the people in the rooms and that the number of either can vary independently. If another

Figure 10-10   Rooms with Personnel and Artwork.

Phone	Person	Artwork
203	Bill	Manet
203	Mary	
203	Teresa	Renoir
434	Rich	Constable

Figure 10-11   Room/Person and Room/Artwork Relations (4NF).

Room	Person
203	Bill
203	Mary
203	Teresa
434	Rich

Room	Artwork
203	Manet
203	Renoir
434	Constable

Figure 10-12   Personnel with Activities in Buildings.

Person	Activity	Building
Bill	Budget Meeting	Headquarters
Carl	Presentation	Headquarters
Don	Presentation	Annex
Mary	AutoCAD Class	Training Center
Rich	Excel Class	Training Center
Teresa	Budget Meeting	Headquarters

Figure 10-13   Person/Activity, Activity/Building, and Building/Person Relations (5NF).

Person	Activity
Bill	Budget Meeting
Carl	Presentation
Don	Presentation
Mary	AutoCAD Class
Teresa	Excel Class
Rich	Budget Meeting

Activity	Building
AutoCAD Class	Training Center
Budget Meeting	Headquarters
Excel Class	Training Center
Presentation	Annex
Presentation	Headquarters

Building	Person
Annex	Don
Headquarters	Bill
Headquarters	Carl
Headquarters	Teresa
Training Center	Mary
Training Center	Rich

person is assigned to Room 434, is another painting added as well? If a person can work in more than one room, Teresa could be the *Person* added to Room 434. Does she take the Renoir with her, requiring two transactions against the same table? If a new painting is added to Room 203, is a new row created or is the painting arbitrarily assigned to Mary?

Fourth normal form (Figure 10-11) is satisfied by splitting off the unrelated components of the primary key into new parent tables. The original table is only preserved if it contains other non-key attributes.

Fifth Normal Form

Finally, we come to what are sometimes known as *projection-join* dependencies. These are cases in which multiple elements have interdependencies that cannot be decomposed without loss by splitting them into two tables but can be resolved into three or more parent relations. At this level of normalization, though, the question is often to what extent such relations *should* be decomposed rather than whether further reduction is possible. Date maintains that the answer is usually "yes" because of data-integrity problems that can otherwise occur.°

In Figure 10-12, we have a set of relationships that arise in situations involving activity scheduling and the assignment of resources, such as room reservation or hoteling schemes. This table, like Figure 10-10, is *all key*, but it is not in 5NF because it contains what might be termed *pairwise-cyclic* dependencies. In this case, *Person* is linked to *Activity*, *Activity* is linked to *Building*, and *Building* is linked cyclically back to *Person*.

This relation has been decomposed into three parent tables in Figure 10-13, each of which is in 5NF because none any longer involves cyclic-join dependencies. If the three

Figure 10-14   Join Over Activity: Person/Activity and Activity/Building.

Person	Activity	Building
Bill	Budget Meeting	Headquarters
Carl	*Presentation*	*Annex*
Carl	Presentation	Headquarters
Don	Presentation	Annex
Don	*Presentation*	*Headquarters*
Mary	AutoCAD Class	Training Center
Rich	Excel Class	Training Center
Teresa	Budget Meeting	Headquarters

tables are recomposed into a view using a join query, the result is identical to the original table.

Projection-join dependencies can give rise to subtle update issues when insertions and deletions are made. Most seriously, *spurious tuples* can appear unpredictably when any two of the decomposed tables are joined. Figure 10-14 illustrates a join of only the *Person/Activity* and *Activity/Building* tables. Two bogus rows appear (underlined) that were not in the original relation, giving a total of eight instead of six rows. If we then join the *Building/Person* table to the first join, the false data are eliminated.

If it is not normalized, this type of dependency can be volatile with changes in the data. Using the data shown in Figure 10-13, joining the *Activity/Building* and *Building/Person* tables creates five spurious rows in a total of eleven. Curiously, however, the *third* possible join over *Person* gives the correct result, by chance, with *these particular data*.

As we have seen, the normalization process consists of breaking down, or decomposing, relations into simpler forms. If a set of relations is in the fifth normal form, we can be assured that there will be no anomalies in the schema that can be eliminated by further projections. In that sense, 5NF can be viewed as the final normal form so far as projections and joins are concerned.

One key point to be remembered is that the decomposition must be *reversible*. Tables created at any stage of reduction must be capable of being rejoined to the next higher level without loss of data. This applies even to spurious or extra tuples. If we have no way of telling the false rows from the real ones, data have indeed been lost!

The other key point is *functional dependence*. Functional dependence, as opposed to other types of dependence (such as multivalued), between tables can be thought of as relatively *permanent*. The association between a person and her social security number is permanent, while the fact that a painting happens to be in the same room with her is not. Interlocking relations on a basis of functional dependency provides the chief means by which integrity constraints are maintained.

THE SQL LANGUAGE

SQL is the standard data sublanguage discussed earlier under Codd's fifth rule where we listed several categories of words reserved to the language. We do not attempt an exhaustive explanation of SQL here (there is no shortage of excellent books on that subject), but let us now look at a few examples of how SQL syntax is constructed and how its declarative statements work.

SELECT

SELECT is one of the most frequently used words in SQL and is one of the four data-manipulation statements (SELECT, INSERT, UPDATE, and DELETE). Example 10-1 derives from the joining of the three tables pictured in Figure 10-13 into the relation from which they were decomposed. This example was constructed in Microsoft Access, which provides some very convenient tools for accomplishing this without actually having to write code.

As a matter of fact, SQL, being a rather verbose language, is often embedded into or generated by other procedural languages instead of being written directly.

```
SELECT ActivityBuilding.Activity, PersonActivity.Person,
        BuildingPerson.Building
FROM (ActivityBuilding
    INNER JOIN PersonActivity ON
        ActivityBuilding.Activity = PersonActivity.Activity)
    INNER JOIN BuildingPerson ON
        (ActivityBuilding.Building = BuildingPerson.Building)
        AND (PersonActivity.Person = BuildingPerson.Person);
```

Example 10-1. Selecting Domains from Multiple Tables

In this case, the Access design view is used to set up the statement largely by drag and drop. First, a new query is opened and the desired tables (or other views) are selected from a dialogue box. These objects, with the names of the domains they contain, are illustrated in the upper portion of Figure 10-15, along with the links establishing the primary- and foreign-key relationships. The grid in the lower portion of the view is used to select the fields to be displayed.

The resulting SQL statement (Example 10-1) is viewed in another window. The SELECT statement specifies the tables and fields to be utilized in the query. Access employs a *dot* notation for this purpose in the format: *Table.Field*. The FROM clause specifies how the tables are to be joined in the query, in this case using *inner* joins (the most ordinary type of join) on a basis of the equalities found between key fields.° The ON key word precedes each conditional expression making the necessary *Table.Field* comparisons.

INSERT

The INSERT statement is used to add data to an existing named table. Example 10-2 shows a single line entry from the script in the "Data Pipelining" section of Chapter 8. It creates a new tuple in a table called *rm*, which is one of the principal tables in *Archibus/FM*. This script is written in a dialect of SQL used by Sybase, Inc., the producer of one of several relational-database products that work with Archibus.

Figure 10-15   Join Over Activity/Building, Person/Activity, and Building/Person.

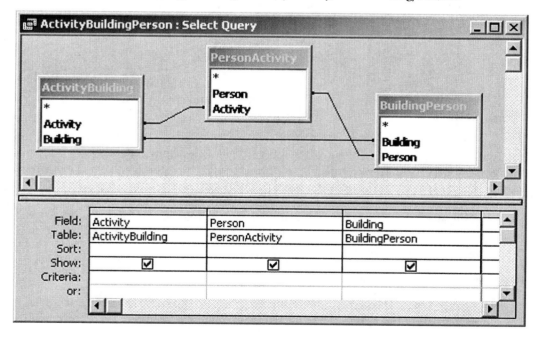

These data were generated from the spreadsheet records exported from the AutoCAD drawings described in the previous chapter (see Table 8-2).

```
INSERT INTO rm (bl_id, fl_id, rm_id, area,
        dp_id, dv_id, dwgname, ehandle, layer_name,
        name, rm_cat, rm_type, rm_std)
    VALUES ('A', '7', '7001', 598.8407,
        (Null), (Null), 'DwgNam', '10001', 'RM',
        'Senior Executive', 'PERS', 'EXEC-SR', 'A');
```

Example 10-2. Single Entry from an SQL Script

The statement is in two parts. First, INSERT INTO specifies the domain names of the columns into which data are to be inserted. These are written in the form of a comma-delimited list, which in this case includes thirteen out of the nearly thirty-five domains in the Archibus room (*rm*) table: *bl_id* (building ID), *fl_id* (floor ID), *rm_id* (room ID), and so on. The second part of the statement consists of the VALUEs to be inserted into the table, also in the form of a comma list.

Program Requirements

The program-requirements dialog box (Figure 7-20) described in Chapter 7 connects from AutoCAD to the Access database described in that chapter. There are a number of SQL statements embedded in the VBA (Visual Basic for Applications) code that drives the dialog box. °

DATABASE ADMINISTRATION

This chapter has been largely concerned with the origins and development of data-management concepts that led to the relational database as it is known and used today. Codd's twelve rules provide the basis for a sound and efficient structure in which to organize and manipulate information. The rules for normalization help us to anticipate and plan for conditions that would lead to impairment or corruption of constantly changing data.

One may make the observation, looking back at some of the principles of functional dependence and the normal forms, that they are largely applications of common sense. Studying Figure 10-2, for example, we sense intuitively that there may be a problem working with a space inventory in which every record contains descriptions of *two* rooms. Date makes exactly this point:

> All the formal ideas discussed . . . are nothing more or less than *formalized common sense*. The whole point of the theory underlying this area is to try to identify such commonsense principles and formalize them—which, of course, is not an easy thing to do! But if it can be done, then we can *mechanize* those principles; in other words, we can write a program and get the machine to do the work. °

The database administrator (DBA) is responsible for keeping everything running smoothly. It is the DBA, or a team led by the DBA, who is responsible for designing and implementing the database schema, but that simply establishes the structure. The machine may do a substantial part of the work; but the DBA must continually assess performance, maintain liaison with management and customers, and implement any necessary changes.

Ongoing administrative tasks also include the implementation of system integrity and security rules as data-backup and recovery procedures. The DBA must also coordinate closely with the group responsible for system-hardware maintenance.

Protection and Security

The fact that the logical database is inherently compartmentalized by its table structure is an asset to data-access control. Selected tables can be designated "read-only," and accessibility can be controlled on an individual table or even domain by domain. This would all be very simple if access were limited to a single user or if all users were limited to *reading* the database. One of the principal advantages of the relational database is its ability to share the same data among many users, but this brings with it the necessity of handling situations in which several users attempt to make changes simultaneously.

Data protection falls into three categories: *integrity, concurrency,* and *recovery,* which are all concerned with keeping the RDBMS operating smoothly and its data correct. *Security* is a fourth area of critical concern: protecting the database against misuse, corruption, or loss due to accident or by design.

Integrity

Data integrity means keeping the content of the database accurate at all times *to the extent possible.* Internal mechanisms for maintaining data integrity were discussed earlier under Codd's tenth rule.

Integrity refers to protecting the data from invalid, as opposed to illegal, input or alteration, recognizing that absolute accuracy cannot be ensured at all times. Internal checks and the use of validation tables can intercept such errors as invalid data types, field overflows, and even implausible values, but cannot trap errors such as misspelled names. In a room inventory, for example, with an area of 300 instead of 150 square feet, or three conference rooms where there should only be one, external data verification is required.

Concurrency

Concurrency control refers to the management of simultaneous access to the database. Loss of updates can occur if adequate measures are not provided to mediate between two or more users attempting to update the same record at the same time.

Record locks are among the most common methods of concurrency control. Individual tuples can be locked for the duration of a single transaction while all other records remain accessible. The entire database may be locked periodically while a backup copy is made. Less-frequent types of locking mechanisms include table locks, which are useful if many values in a single table are being updated, as well as locks on individual fields.

The term *deadlock* has a special meaning in database parlance, referring to processes competing for the same resources. For example, if two users each require access to two tables, and their queries are timed such that each user successfully locks one table, they may both wait indefinitely for the other user to unlock the other table.°

Recovery

When systems fail, as they inevitably do, database-recovery procedures must be available to quickly and accurately restore it to its last known correct state. Many disasters may befall; among them are hardware failure, software errors, incorrect or invalid data, viruses, and human error. There are four categories into which recovery facilities generally fall:

- *Backups*—Copies of part or all of the database. Backups are usually made on a regular schedule so that the database can be completely or partially restored from a recent copy. Depending on the volume of transactions issued to the database, backups may be performed weekly, daily, or even more frequently. Backup copies on tape or optical storage must of course be securely stored, often in a remote facility.
- *Journal*—An audit trail of all transactions and changes to the database along a time line. The transaction log retains the information essential to reproducing each transaction that has been made. The database-change log contains before and after images of every altered record. These images are copies of each record before and after it has been modified, including changes that may have been made to the schema itself.
- *Checkpoints*—Time-line markers established every few minutes. Checkpoints represent points in time when the database is considered stable: no transactions are currently in process, and transaction and database logs are synchronized. When it is necessary to restart the RDBMS using backups and journal data, checkpoints establish a baseline against which the database can be returned to a known state.

- *Recovery*—The defined process for restoring the database to an accurate state and resuming the regular processing of user requests. The three basic types of recovery procedure, depending upon the extent and nature of the failure, are:
    1. *Restore*—The most recent backup is reloaded, and all transactions processed since the backup checkpoint are rerun from the transaction log. This is the most time-consuming but least complex form of recovery.
    2. *Rollback*—Transactions are undone back to the most recent checkpoint, usually after a transaction has been abnormally terminated or a deadlock has occurred. The transaction log and *before* images are used to reprocess the transactions from the checkpoint correctly.
    3. *Roll-forward*—The most recent backup is reloaded and *after* images are applied up to the recent checkpoint. The transaction log and after images are used in conjunction with a backup, often following a system failure. Transactions are reprocessed from the last checkpoint.

Security

We are all aware of the general need for security for reasons that need not be restated here. We take for granted that the physical security of essential computing and telecommunications facilities is of critical concern. Our present interest, however, is with the internal security of the RDBMS.

As we mentioned earlier, the strategic importance of natural compartmentalization in the relational-database structure cannot be overstated. The ability to restrict each user's view of information is a substantial advantage in data management from a security perspective. Rights of access to specific tables, views, and domains can be individually controlled and monitored. Access privileges are managed using the data-control statements (GRANT, REVOKE) that SQL provides.

Authentication procedures positively identify persons attempting to gain access to the system. Password protection and magnetic-card readers are among many schemes used to identify prospective users and properly establish their level of authorization.

Once a user has gained access, authorization rules may restrict the operations he or she is permitted to carry out. The ROLE function in SQL associates specific applications with authorized users and establishes the extent to which any user is allowed to execute, administer, or modify an application. User-defined procedures allow system developers to implement additional security measures beyond the basic authorization rules.

Finally, encryption procedures may be employed to render the contents of the database unintelligible to would-be intruders. Encryption algorithms are most commonly employed in telecommunications, but they can also be used to thwart attempts to circumvent system protections that would violate Codd's twelfth rule.

User Interface

The logical database is built within a conceptual framework that is structured to reflect the meaning of the data and its intended use throughout the enterprise. There are many ways in which data can be viewed. Menus and forms of all kinds are used to present data to the user through logical and accessible interfaces.

Microsoft Access

Access was discussed in Chapter 7, and it supports a number of well-integrated methods for handling user input and output. In Figure 7-5, we showed an Access form used to select from among several different reports, each of which was described in some detail. The dialog box in the example provides for some data entry (titles, dates, circulation factor, etc.) and also serves as a switchboard for selecting the desired report. Customized reports, as well as special forms for data entry, can be designed using the Access interface.

Besides conventional reports, Access can display output using several different types of Web pages. Static HTML° pages can show snapshots of tables, queries, forms, or reports using a browser. Static pages are created from the database as needed, but have no permanent connection once they have been published to the Web server.

Data-access pages are dynamic HTML pages that you can publish to a Web server that maintains a direct link to your database through Open Database Connectivity (ODBC). ODBC is a driver protocol that allows a database and the server on which it resides to be accessed externally. Access can also serve as a *front end* to full-scale RDBM systems such as Microsoft's SQL Server. One method for achieving efficient, native-mode access to an SQL Server database is to use an Access project to create a client/server application. The connection is maintained using a component architecture known as *OLE DB*, using either conventional forms and reports or data-access pages in a Web environment.

Active server pages (ASPs) offer yet another level of flexibility for accessing databases through the Web. These are server-generated HTML files that possess most of the same functionality as Access forms when displaying and interacting with your database.

Several programming interfaces are available that can be used with Access as well. Data-access objects (DAO) are the basis for one of two object models that allow *Visual Basic (VB)* or *Visual Basic for Applications (VBA)* procedures to connect with your database. The program-requirements dialog box discussed in Chapter 7 (Figure 7-20) and Chapter 8 communicates with Access from AutoCAD using DAO and VBA in AutoCAD.

The second object model presents ActiveX Data Objects (ADOs), which allow the manipulation of database information through an OLE DB provider. OLE DB data-access technology can be accessed directly for greater performance and functionality, but this requires programming in C++, making maintenance more difficult and expensive. Microsoft's Visual Studio.NET development platform supports several programming languages as well as ADO, ADO.NET, and OLE DB. In fact, ADO.NET offers extensive scalability for Web-based applications and has an integrated-relational structure.

Large-Scale RDBM Systems

Large-scale RDBM systems are available from many different companies such as Oracle, Sybase, IBM, and many others. Several of the major dedicated CAFM systems have been designed around these high-performance database products. Archibus/FM, for example, supports Sybase, Oracle, and SQL Server databases. Facility Information Systems, Inc. (FIS), developed a comprehensive corporate real estate system designed around Oracle. FIS was acquired in 2003 by FAMIS Software, Inc., a leading provider of facility software for the public sector, creating what is purported to be the largest enterprise FM software company.

Several of these enterprise-level products are discussed in the next chapter. All of them are capable of integrating with leading enterprise-resource-planning (ERP) systems, many of which use the same relational core. These systems provide increased availability—because they can be backed up while in use, they have quick and automatic recoverability in the event of system failure, and the ability to set up and administer complex security schemes.

CLOSING WORDS

In this chapter, we have gained a basic understanding of the electronic card index known as the relational database. E. F. Codd's twelve rules for the characteristics of an ideal database capture the essence of the relational concept. Multiple tables are tightly linked together with minimal redundancy, allowing data to be input, queried, and extracted without the user having to be concerned with the underlying structure or the mechanics of information retrieval.

We enter and extract data from databases by making logical queries, most often using the structured query language called SQL. Most often, the details of the queries themselves are hidden beneath forms and dialog boxes that are part of a graphic user interface. The database administrator is responsible for maintaining the user interface and the integrity of the collected information, as well as behind-the-scenes functions such as security, backup, and recovery.

AutoCAD, the engineering design program to which we have made repeated reference throughout this text, is also a database program. Underlying its graphic interface is a hierarchy of lists within lists that define the objects that make up your drawings. Very soon after its first release, AutoCAD's designers provided access to the program's internal data structures through a programming language called AutoLISP. An SQL interface known as the AutoCAD SQL Extension made a brief appearance in the early 1990s, and later, when Auto-CAD was completely recast in object-oriented C++, a number of programming interfaces became widely used. ObjectARX, the AutoCAD Runtime Extension, provides a compiled-language environment in which add-on programs can interface directly with the application and its data. ActiveX Automation and Visual Basic for Applications (VBA) are vehicles for accessing AutoCAD data using the Component Object Model (COM). Most recently, the .NET framework has provided yet another method of access to AutoCAD's internal data structures.

It is hardly surprising that as CAFM applications began to appear in the 1980s, first using flat files and hierarchal databases, AutoCAD would be a natural choice for database applications requiring access to drawings. One thing that all CAFM systems have in common is that they integrate some kind of relational database with a CAD program, usually AutoCAD. The scope of such applications has continually grown in the related areas of facility and asset management, leading to such descriptions as IWMS, the Integrated Workplace Management System.

In the following chapter, we look at several CAFM systems belonging to this subclass of composite applications. These programs all consider the physical components of an organization's infrastructure as deployable assets having a measurable, useful life. The leading applications invariably contain modular components for tracking information relating to space, building operations, real property and leases, furniture and equipment, and the computing/telecommunications infrastructure.

CASE STUDY   International Chemical Company

This program involves the implementation and ongoing maintenance of a turnkey CAFM solution for one of the world's largest chemical companies. The company is a major science and technology enterprise that provides chemical, agricultural, and plastic products to numerous markets in 175 countries. With sales in excess of $40 billion, the company employs over forty thousand people. Although it is based in the United States, nearly two-thirds of the company's sales and half of its property investment are outside the United States.

In all, the company manages nearly two thousand buildings in some five hundred different locations, totaling approximately forty million square feet. In this case, the company opted to outsource its facility-management operation to its facility-management services provider.° The knowledge gained from a six-hundred-million-square-foot client portfolio enables the provider to offer comprehensive CAFM services on a plug-and-play basis. Facility-data-management services are provided through a hosted environment that includes all software applications, network hardware and infrastructure, drawing maintenance, training of personnel, and quality management.

Roughly 25 percent of the company's space inventory needed to be migrated from previous legacy systems into a new *TRIRIGA* database. The remainder of the space-utilization data was developed from electronic architectural drawings linked to the TRIRIGA database using space-defining polylines.

DATA-NETWORK HOSTING

As is often the case with specialized database applications, it can be cost-effective to have the system as well as the data hosted by an application service provider, or ASP. This places the responsibility for system operation and maintenance, as well as the preparation and input of data, on professional specialists, but requires special measures to ensure that the network infrastructure is reliable and secure.

Infrastructure Architecture

Any system capable of external communication uses routers and switches that must be physically as well as electronically protected. Automated battery backups are used to ensure a constant supply of electrical power to the network. RAID technology helps protect against data loss due to disk failure. Parts subject to malfunction, such as hard disks and network cards, must be kept on site to ensure that the system is kept up and running.

Figure CS4-1 shows a schematic diagram of the data center used to host this global corporation's CAFM network. The network provides standard Web browser access to all space-related data, drawings, and reports on the hosted CAFM Web site. Redundant application and database servers are maintained at two separate locations, with information being copied to backup servers

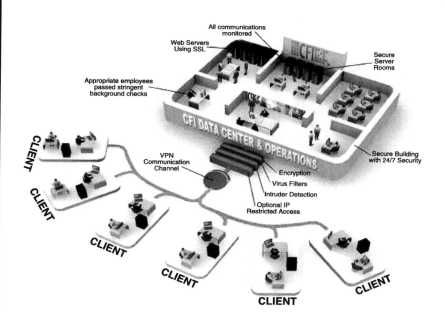

Figure CS4-1   Hosted-Data-Network Configuration.

(continued)

(continued)

nightly. In the case of failure or a lost connection to the primary servers, access can be switched to the other server within thirty minutes.

Both secure network facilities are monitored by security personnel on a 24-7 basis, with access limited to authorized employees who have passed stringent background checks. Server rooms are secured and continuously monitored by exterior and interior camera systems.

Data security is provided through a virtual private network (VPN) communication channel with several levels of protection. Industry-standard firewall and intrusion-detection hardware resist network attacks and ensure privacy. The latest antivirus technology is used, and virus definitions are automatically updated nightly. Secure Sockets Layer (SSL) technology is used to encrypt incoming and outgoing data.

Hosting Methods

Several different hosting methods are available depending upon different needs and budgets, ranging from dedicated solutions to a less-expensive shared-server environment. Dedicated Internet or leased-line connectivity, together with dedicated-server hardware, is the most flexible from the user's point of view. In addition to maximizing bandwidth, this option allows for IP (Internet Protocol) access restrictions to the server to be made based on the IP source.

In a second scenario, Internet connectivity is shared, but the server hardware is dedicated to each user. This configuration allows the user to make modifications to her server and its setup, also allowing for IP source restrictions. Finally, a shared-server environment allows servers and connectivity points to be shared among multiple users. In this environment, connectivity bandwidth and hardware loads are continuously monitored to ensure

that performance and required levels of security are maintained.

IMPLEMENTATION

TRIRIGA was chosen for this application because it is among the clear market leaders in terms of its technology, capability, and market share.° It provides a rich solution capable of future expansion.

The initial scope of this implementation was space and occupancy management, but TRIRIGA offers many additional modules including project management, property and lease management, telecommunications and cable management, preventive and demand maintenance, hoteling, and so on. Additional modules can be added by purchasing the appropriate module license, adding the appropriate data, and training the users. The system is fully capable of being expanded in place without the need for reimplementation. (See Chapter 11 for a summary of TRIRIGA'S basic capabilities.)

Figure CS4-2 illustrates the home page for TRIRIGA Facility Center 8i. This is the starting point through which the user navigates to manage space occupancy and utilization, which help assure optimal productivity levels from day to day. Alternative strategies can be effectively developed and compared for the distribution of space and equipment over time. Linked AutoCAD drawings generate reports that allow space relationships to be visualized and managed effectively, tracking utilization and vacancy rates, adjacency, and security information. Individual departments and cost centers are accountable for the space they use and the expense incurred. Space accounting is done according to recognized standards where it can be assigned to specific departments or shared among several business units, while common

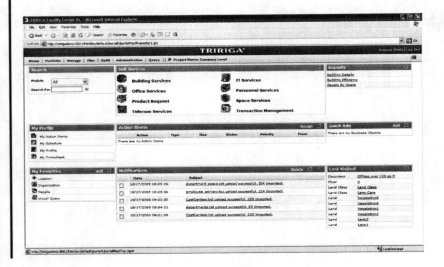

Figure CS4-2 TRIRIGA Home Page.

space is proportionately allocated among all the groups that have access to it.

With the facility-management module as the core, the groundwork has been established for implementing a comprehensive asset-management system. Effective inventory management of furnishings, movable and fixed equipment, consumables, and disposable items directly impacts the organization's bottom line while enabling sustainable elements of planning and design. A well-designed and -implemented asset-management system will ultimately track the procurement, installation, assignment, charge-backs, repair, and disposal of all assets throughout their useful lifecycles.

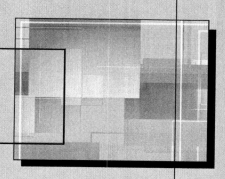

CHAPTER 11

Computer-Aided Facilities Management

It is not necessary to change. Survival is not mandatory.
—W. Edwards Deming

In Chapter 5, we discussed the close relationship of workplace programming and space planning to the management process. In an information-based global economy, the importance of facilities management as an integral part of enterprise resource planning (ERP), supply chain management (SCM), and systematized total quality management (TQM) cannot be underestimated. It naturally followed that database systems encompassing all aspects of facility planning and operations would evolve. In particular, it is the interoperability of computer-aided-design applications such as AutoCAD with the relational database that has led to the development of powerful tools for monitoring current resource utilization and predicting future needs, as well as increasing the focus on operations that can be improved.

The Sarbanes-Oxley Act of 2002, which was created by the U.S. Congress to raise investor confidence by improving corporate governance, has become a new motivating factor in the development of facility-related information systems. Although the chief executive and chief financial officers bear the primary responsibility for certifying financial statements, the chief information officer is accountable for the implementation of internal IT controls. In addition to the delivery of accurate financial statements, Sarbanes-Oxley requires that systems producing the underlying data related to operational details be properly managed and documented.

Tracking asset lifecycles from procurement to implementation, support, and disposal has therefore become one of the key areas in asset management. A new acronym, *IWMS*, for *integrated workplace management system*, appeared in the late 1990s; in the last several years, the development of such systems has been rapid. An IWMS consolidates CAFM with computerized maintenance management systems (CMMS), real estate management systems (REMS), and other such applications. There is little doubt that those in capital asset management, of which facilities are a significant element, will be held accountable by senior management for providing accurate information.

The subject of the next chapter, *sustainability*, exemplifies yet another area in which CAFM has become an essential tool: managing energy-related facility concerns. *Commissioning* of building components such as heating, ventilating, air conditioning, refrigeration systems and their controls, lighting controls, and domestic hot water, along with renewable energy systems such as solar and wind, requires that the performance of such systems is in accordance with the original program and sustainable-design parameters. Building heating and cooling loads, water consumption, irrigation systems, and runoff, as well as interior thermal comfort, are but a few of the parameters that must be optimized to qualify a facility for certification under the LEED initiative: Leadership in Energy and Environmental Design. Compliance with standards such as *ISO 9000* (quality) and *ISO 14000* (environment) necessitates close monitoring of operations and environmental conditions.

CAFM APPLICATIONS

Numerous applications have been developed that integrate table-driven and drawing-driven data that fill this need. They all address similar concerns: planning, space, property, and asset management, but each is scalable to a different degree. Some are far more complex than others to implement and administer, so choosing an appropriate hardware and software package is not a trivial matter. Often, the services of a specialized facilities consultant with qualified technical personnel are required.

In the remainder of this chapter, we look briefly at three leading CAFM products that illustrate the range of software available, from local-area-network-based solutions to global enterprise systems that are either part of or capable of being integrated with large-scale IWMS, ERP, SCM, or financial reporting infrastructure. We conclude with an in-depth look at a single package called *Archibus/FM* that falls in the upper-middle portion of this range.

Archibus commands a significant share of the CAFM user base because it can address itself to the needs of the smaller user as well as the large, Web-enabled enterprise. Archibus is comprised of several structurally independent modules that access a common relational database, as well as containing the essential functional elements common to this class of software. The structural clarity of Archibus/FM makes it an excellent vehicle with which to explore the use and functionality of a dedicated CAFM system, but this does not mean it is necessarily the default solution for all users. Every application package has its own unique strengths and design features, and the prospective user should carefully evaluate the attributes of an offering in terms of how it meets each particular set of requirements. Moreover, like all software in today's technology market, these applications continue to evolve at a rapid pace.

TRIRIGA

TRIRIGA° is a broad-scope, infrastructure-management system that addresses virtually all aspects of workplace design and operations at the enterprise level. Initially, its focus was on large-scale capital project management; operations and maintenance management, including that of the supply chain; and real estate. Operating on a component-based, Web services platform, TRIRIGA offers a range of integrated standard solutions that can be adapted to the specific processes and mission priorities of different organizations. The developer's objective is to offer an integrated solution that appeals to large organizations that want to minimize their concern with integration issues. Following are some of the different process areas that the software addresses, most facility related and directed toward controlling process efficiency to reduce the initial and continuing costs of asset ownership.

Projects

- *Program management*—Coordinate and integrate the structures of multiple projects.
- *Supply chain management*—All aspects of procurement: requests for quotations, contracts, purchase orders, bid awards, shipping notifications, and invoices.
- *Project administration and field management*—Manage project documentation, personnel, and resource accountability; monitor response times to keep critical tasks on track.
- *Cost, schedule, change management, and risk assessment*—Monitor and resolve project issues, maintain approval processes to ensure proper accountability, forecast costs in the context of historical budget data, maintain current scheduling parameters.

Operations

An integrated, computerized maintenance management system (CMMS) is accessed through Web portals tailored to the needs of maintenance supervisors and field technicians. Service level agreements (SLAs) are monitored according to specified criteria, together with preventive and on-demand maintenance, from service requests, approval, and dispatch to historical cost tracking and reporting.

Real Estate

Advanced portfolio- and transaction-management technology provides comprehensive re-porting at various levels of detail through customizable reports, summaries, and graphical dashboard displays. Once again, emphasis is placed upon the ability to customize the in-terface and outputs with a minimum of programming. Portfolio-, contract-, and lease-management functions interface with the space-process-management area, as well as with other disciplines such as finance, the human resource information system (HRIS), and customer resource management (CRM).

Facilities

TRIRIGA's facilities-management process area was largely developed through the incorpo-ration of a product that initially became available in the early 1990s known as *Span FM*. Span was one of the first applications to integrate facilities management operations through sharing of information among databases (Figure 11-1). Span was further devel-oped and refined in the client-server environment of the late 1990s after it was acquired by Peregrine Systems and rechristened *FacilityCenter*. The current TRIRIGA product is yet another incarnation of this application, integrated with Projects and the other process ar-eas described previously, using a service-oriented platform. At the present time, Facility-Center is also maintained in a client-server version for some clients who have not yet chosen to adopt the new Web server technology.

Another important feature of this enterprise application is its bidirectional CAD link that supports AutoCAD and Bentley Systems' *MicroStation*, both preeminent CAD programs. The integration of the CAD database effectively leverages the other elements of the TRIRIGA plat-form by presenting graphic representations of queries and objects from the relational side of the system. Finally, an internal document manager provides for document integrity through versioning control by maintaining a full document history and security access controls.

FAMIS

Most of the key components of *FAMIS*° are also deployed in a Web environment consist-ing of platform-independent, browser-based, and Java-enabled services provided by an Oracle application server. FAMIS is often referred to as a *native* Oracle application, mean-ing that it is built using Oracle development tools, forms, and report services. As such, it does not rely on protocols such as ODBC° to connect its interface components to the un-derlying relational database. FAMIS is sometimes described by systems integrators and other IT professionals as an Oracle *tool kit*, as it requires a trained database administrator to install and set up, but the result is an extremely efficient enterprise application with vir-tually unlimited scalability.

Figure 11-1   Diagram of Span FM Application Modules.

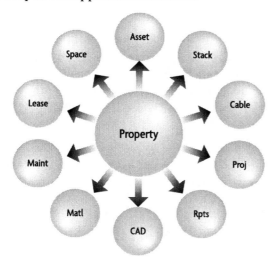

Today's FAMIS product line is the result of the acquisition of Facility Information Systems, Inc., by FAMIS Software, Inc., in 2003. FAMIS Software formerly was a leader in public sector facilities software while Facility Information Systems built its reputation in the corporate and real estate markets. FAMIS has also migrated its most current applications to a service-oriented, Web-based structure, for which it uses the term *adaptive architecture*. Supplementing its functionality are components that allow FAMIS to integrate with general ledger and other financial modules such as those from Oracle and PeopleSoft. Integration tools are also provided that allow the application to integrate with non-Oracle database applications using XML or ODBC connections. An interface to RS means estimating data is also available for maintenance management operations.

FAMIS can be used with any number of platforms, of which Windows, Unix, and Linux are considered the most desirable. The developer recommends using separate machines for the Oracle middle-tier application server and Oracle AS infrastructure server, for reasons of scalability and security. Multiple, middle-tier application servers can be clustered into a single environment supported by a single infrastructure host server. The application can also be hosted remotely by an application service provider (ASP), freeing the user from the task of maintaining the system.

The Famis **Drawing Coordinator** provides a link between the **Occupancy Management** module and spatial data stored externally in CAD and geographic information systems (GIS). The drawing coordinator interfaces with Autodesk's MapGuide and MicroStation.

The following brief overview of Famis business areas describes the kinds of information that are maintained in its comprehensive database.

Project Management

Capital Projects includes tools for budgeting, scheduling, and tracking projects and resources in real time. Some specific areas addressed include maintaining budget revisions and the reasons for them; managing scope changes, change orders, and contract change requests; and processing work orders to track the specific tasks being performed.

Maintenance and Operations

- *Maintenance Management*—All aspects of preventive and remedial maintenance; alterations and renovations are electronically tracked from task identification to completion.
- *Utility Management*—The utilization of heating and cooling energy, gas and electricity, water, and effluents are monitored to provide broad visibility of consumption patterns to facilitate conservation and cost control.
- *Inventory Control*—Performance of both cyclical and physical inventories to optimize on-hand stock levels of furniture, equipment, supplies, and replacement parts with the capability of procuring, locating, and properly charging back items.
- *Wireless Modules*—Maintenance, utility management, and inventory-control functions can all be conducted in the field using platform-independent devices from cellular phones to any type of portable computers.
- *Tool and Key Control*—The distribution and monitoring of tools, locks, and keys from request through issuance, return, transference, and maintenance.
- *Facility Assessments*—Comprehensive oversight of all business areas to identify, prioritize, fund, and remediate defects and insufficiencies. Integration with Capital Projects and Maintenance Management facilitates the generation, fulfillment, and closing out of work orders.

Real Estate

Real Estate Administration tracks owned and leased properties for both owners and tenants. Historical data, the terms of each lease, maintenance information, and data pertaining to the structures related to properties are tracked in detail. Lease components are linked to space and equipment inventories through the **Occupancy Management** module. The functions of the real estate administration business area can be integrated with financial and other such systems.

Facilities Management

- *Space Management*—Maintains a comprehensive inventory of space utilization including personnel assignments and departmental allocation.
- *Occupancy Management*—In conjunction with **Space Management,** provides a centralized tool for monitoring enterprise spatial resources, including ownership and availability of properties, work-space and support-space utilization, and personnel.
- *CAD Interface*—A two-directional interface between CAD drawings and the database that accurately represents the space inventory and area calculations.
- *Strategic Planning*—Modeling of current and projected resource needs by means of master planning scenarios. Resource data from individual planning units can be consolidated up to the enterprise level throughout the organization.
- *Stack and Block*—Produces stacking diagrams based upon adjacency algorithms, manually changing values, or drag-and-drop on-screen manipulation.
- *Graphical Report Server*—Allows automated production of accurate reports through the CAD software, using information directly exported from the facility database.

Performance Management

Performance Scorecard monitors employee and organization work performance to help improve effectiveness through the use of key performance indicators (KPIs); KPIs are measurable goals that are monitored over specific time periods.

SERVICE ORIENTATION

The term *service-oriented architecture* (SOA) represents a continuation of a long-term trend that has utilized progressively higher levels of abstraction to bring greater functionality to software. We remarked in Chapter 1 about the "appropriation" of the term *architect* by other professions, of which software architects, the creators of these multitiered structures, were but one example.

One of the main driving forces behind service orientation has been the need for interoperability among different types of applications. The ability to export data from an AutoCAD drawing to an Excel spreadsheet using a VBA macro provides a certain level of functionality, particularly when there is no CAFM system in place or when preparing data with which to populate one. However, the cost of changing or upgrading enterprise-level systems can be reduced if a business system (new or existing) can be packaged as a service and integrated with other services without the need for programming explicit links. Service orientation has been largely enabled by manifold increases in raw computing power over the past two decades, coupled with the delicate tension structure that we call the *World Wide Web*.

Web services have been implemented in most software products and platforms currently in use. Web services, although not intrinsically service oriented, communicate with clients through a set of standard protocols. This has made the Web a natural mechanism for implementing the kind of loose coupling that services provide. *Loose coupling* means that the client, or user, of a service is essentially independent of the service in the sense that it need not be concerned with the platform providing the service or the language in which it is coded.

Procedural languages were first organized into modules, which were superseded by objects. Every object is a specific instance of a general class that has specific properties, together with methods for interfacing with it. Objects *inherit* methods and properties from parent objects: a *line* in AutoCAD is a special case of an *entity* object. *Polymorphism* is another characteristic of objects; they can appear in different guises at different times. An AutoCAD *entity* can be a *line* or a *polyline*, or it can be a *circle*. The third defining characteristic of objects is *encapsulation*, which means that their internal operation is hidden from the user. Providing objects with multiple interfaces allowed them to be used as *components*, which was the next level of the building block.

Services are not dissimilar from components in that they both combine information and behavior. Objects have *attributes* (methods and properties), while services are adaptable through their *aspect* or the context in which they appear. Services, once published, can be utilized individually or as hierarchies to provide a flexible environment in which applications can evolve easily with changing needs.

ARCHIBUS/FM

Archibus/FM° has been a leader in the CAFM arena since the founding of the company in 1976. The core program provides a multiple-document interface allowing its user to look at data tables, reports, and drawings while working interactively with an underlying client-server database. Archibus/FM is capable of connecting with several major relational-database products such as Sybase, Oracle, and Microsoft SQL Server. Direct access to native-format AutoCAD drawings is provided by a drawing engine licensed from Autodesk.

The basic application consists of several modules, each of which deals with a specific area of facility and infrastructure management. Although it provides access to both database and drawing information, the primary orientation of the core program is toward the database side, principally its tables and the reports that can be produced by making queries and performing various other actions. Figure 11-2 illustrates the *Rooms* table with a linked room highlighted on the corresponding floor drawing, along with a dialog box for selecting the key value of the floor.

To balance this focus, Archibus/FM provides an application known as the *overlay* for AutoCAD. The Archibus/FM overlay provides a drawing-centric view of the database by

Figure 11-2   Archibus/FM Standard Interface Showing Tables and CAD Floor Plan.

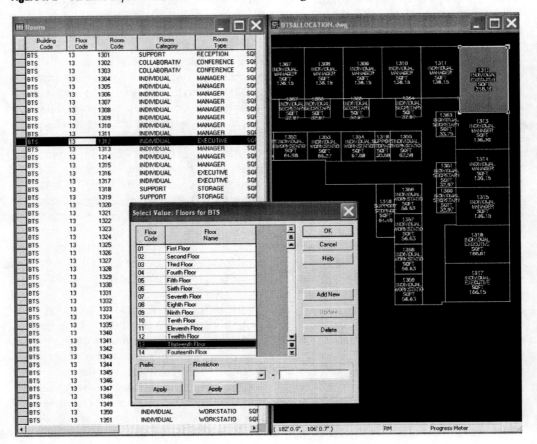

adding pull-down menus and toolbars to the AutoCAD interface. (A corresponding view of the overlay appears in Figure 11-10.) These tools allow the user to access the same database from the drawing side, both for data input and report production. The overlay allows the facility manager to draw information from the underlying database, placing text and business graphics directly onto the drawings.

In response to the growing use of the World Wide Web protocol for accessing information remotely, Archibus introduced a set of dynamic pages that allow a user to interact directly with the Archibus/FM database and its drawings using the Web. This separate product, known as *Archibus/FM Web Central*, uses an application development tool known as Cold Fusion that enables the rapid development of dynamic, interactive Web pages. *Cold Fusion* evolved from the Web server scripting standard known as the *common gateway interface* (*CGI*), which began as a means of executing procedures to make Web pages interactive. Operating as a service of the operating system, Cold Fusion contains language extensions to HTML that can read and write data to and from databases, populate forms with live data, process form submissions, and perform various other kinds of conditional processing. Web Central is thus able to access Archibus/FM data over the Web, also providing tools for creating dynamic custom pages.

Archibus provides yet another means of interacting with its database via scenarios constructed using its *Executive Information System* (*EIS*). The EIS very much resembles an on-screen slide presentation with hyperlinks to other images, procedures, and views of the facility database.

Throughout the rest of this chapter, we look at each of Archibus' seven application modules as presented by its *Navigator*, the primary means of interfacing with the core program (Figure 11-3). There is also an eighth "module" that provides the knowledgeable user with the ability to customize the database itself through the use of metatables that expose the database structure.

Real Property and Lease Management (1.0)

The real property and lease management module consists of six activity classes. The first manages necessary background information pertaining to organization structure and all properties that are owned or leased (item 1.1 in Figure 11-4). Special tables are provided in this module for recording areas that are negotiated between owner (*lessor*) and occupant (*lessee*), which may be different than those measured from CAD drawings. Either measured or negotiated areas may be used for property-management calculations and reports, as required by the various building area management tasks (1.2).

Owned and leased properties are separately managed by the next two activity classes, although the activities themselves, such as abstracts, budgets, and financials (accounting, charge-backs, and analysis) are similar. Abstracts contain summaries of key descriptive information regarding properties, such as leasehold improvements, responsibilities, options, and renewal dates in the case of leases; physical descriptions of amenities, pertinent regulations, and tax due dates in the case of owned property; and ownership and other key individuals in both instances (1.3 and 1.5).

One particular feature of the property-management module is the ability to track the amount of permeable area on the site, of particular interest in complying with LEED Sustainable Sites and other environmental requirements (see Chapter 12). Similarly, abstract activities are provided for tracking the inventory and allocation of parking spaces throughout a facility.

The management activities allow graphically tracking and representing leased areas as well as departmental space assignments; facilitating the maintenance of budgets, financial histories, and projections; and monitoring of cash flows. The depreciable costs of leasehold improvements can be effectively monitored along with rental income collected. Similar reports allow owned property income and operating expenses to be allocated to cost centers.

Specialized property-related activities are provided for managing compliance with regulations pertaining to codes and other performance guidelines relating to fire safety, accessibility, and the protection of the environment. Step-by-step plans for bringing non-compliant facilities into line can thus be developed with appropriate financial foresight.

Figure 11-3   Archibus/FM Explorer or *Navigator*.

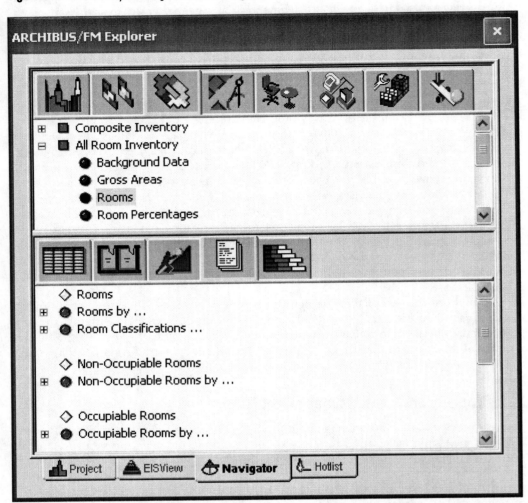

Likewise, property and other tax liabilities can be analyzed on a continual basis and reminders posted to ensure timely payments.

The property and lease management module also provides an activity class dedicated to special projects (1.4) relating to many of these same areas: lease responsibilities and options, land parcels, buildings and improvements, and regulation-compliance issues. Various reports can be produced, including contact lists, activity and communication logs, and activity costs.

Finally, executive and other quick-access reports provide summaries of decision-support information pertaining to budgets and cash flow, as well as selected property and lease data (1.6). These quick-access report formats are typical of most of the other Archibus/FM modules.

Strategic Master Planning (2.0)

The activities of the six classes in the strategic master planning module (Figure 11-5) address several of the workplace-programming concepts discussed in Chapter 6 and Chapter 7. Archibus/FM uses the term *space budgets* to refer to a "consistent description of space needs that enables development and use of space data across all activities of the . . . module."° This permits comparisons such as historical usage to current space utilization and business forecasts to programmed facility requirements.

Figure 11-4   Real Property and Lease Management Activities.

1.1	Manage Background Data
1.2	Manage Building Areas
1.3	Manage Properties
1.3.1	Manage Property Abstracts
1.3.2	Manage Property Budgets
1.3.3	Manage Taxes
1.3.4	Manage Regulations
1.3.5	Manage Property Assets
1.3.6	Manage Property Financials
1.4	Manage Leases
1.4.1	Manage Lease Abstracts
1.4.2	Manage Lease Budgets
1.4.3	Manage Suites and Dept. Spaces
1.4.4	Manage Lease Financials
1.5	Manage Projects
1.6	Quick Access & Executive Reports

Figure 11-5   Strategic Master Planning Activities.

2.1	Requirements Programming
2.2	Forecasting
2.3	Allocation
2.4	Layout
2.5	History
2.6	Quick Access

The ability to analyze historical trends and compare them with current forecasts provides an important reality check to strategic planning. Relationships can be established between changes in business activity, organizational structure, or other parameters and the necessary allocation (or *re*-allocation) of space. Changing trends along a time line can be graphically represented to compare relative efficiency as well as the effectiveness with which different kinds of space are being used.

Business forecasts are often based, at least in part, on expected performance extrapolated from historical data, combined with management's expectation of changes in the market, product or services themselves, or other factors integral to the business plan. A master occupancy plan may encompass several different models for the future allocation of facility resources, potentially affecting organizational groups at different levels. Ratios can often be established between work spaces that are a direct function of personnel and required support spaces that are not, and these factors can help to determine future allocations.

Space allocations are made by assigning various organizational units to buildings and floors. Earlier, we discussed several methods, such as the adjacency matrix, for prioritizing the relationships among organizational elements. Archibus refers to this process as *affinity mapping* and provides tools for using designated affinity priorities to generate graphics apace allocation diagrams. Stacking diagrams represent potential space allocation and use on a floor-by-floor basis. Dynamic bubble diagrams can also be produced to aid in developing block plans on a basis of organizational or functional groups. A point-scoring feature aids in making comparative evaluations of different potential solutions and adjusting "trial" layouts accordingly. This process helps to maximize the efficiency of space utilization while maintaining a balance with functional-adjacency relationships.

Quick-access reports include views of proposed stacking plans, departmental forecasts, and programmed areas together with historical data regarding past trends in space utilization.

Space Management (3.0)

As its name explicitly states, the space-management module provides for management and detailed administration of the space inventory (Figure 11-6). Its principal activity classes consist of tools for implementing two different inventory methods: what Archibus calls the *all-room* (3.1) and *composite* (3.2) inventories. Both types of space inventory clearly delineate how space is used on a floor. Rooms or functional areas are assigned identification

Figure 11-6   Space-Management Activities.

3.1	Composite Inventory
3.1.1	Background Data
3.1.2	Gross Areas
3.1.3	Vertical Penetration Areas
3.1.4	Service Areas
3.1.5	Group Areas
3.1.6	Room Areas
3.1.7	Room Percentages
3.2	All Room Inventory
3.2.1	Background Data
3.2.2	Gross Areas
3.2.3	Rooms
3.2.4	Room Percentages
3.3	Personnel
3.4	Hoteling
3.5	Room Reservations
3.6	Emergency Preparedness
3.7	Quick Access

keys that allow them to be described in database tables, which are directly linked to their corresponding polyline representations on an AutoCAD drawing, as shown in Figure 11-7.

All-Room Inventory (3.1)

The all-room inventory is potentially the most detailed of the two methods and is the simpler in terms of drawing layers, of which there are two. Gross floor areas are outlined using polylines on one layer while all other spaces are polylined on a second *room* layer. Spaces are designated according to room type and category in separate tables that are linked to the CAD drawings. Table 9-3 and Table 9-4 show such a classification system taking into account BOMA's distinction between office and common areas (see Chapter 9). *All* spaces from mechanical areas to circulation (both horizontal and vertical) are classified as rooms, which may be reclassified simply by changing table entries. The all-room methodology is thus completely flexible in its application.

Charge-backs are the result of accounting for space utilization in a way that allows assigned areas and areas used in common to be associated with the parts of the organization that use them. This makes it possible to charge each organizational unit (cost or profit center) for its proportionate share of the facility. Archibus/FM provides for calculating chargebacks using both inventory methods, and several variations are available.

The room charge-back method with the all-room inventory treats every space, enclosed or not, as a separate element. Each room is classified, as a whole or in part, by type and by individual if desired. Common support spaces can be prorated among the individually

Figure 11-7   Highlighted Rooms on Archibus/FM Table and Linked Drawing.

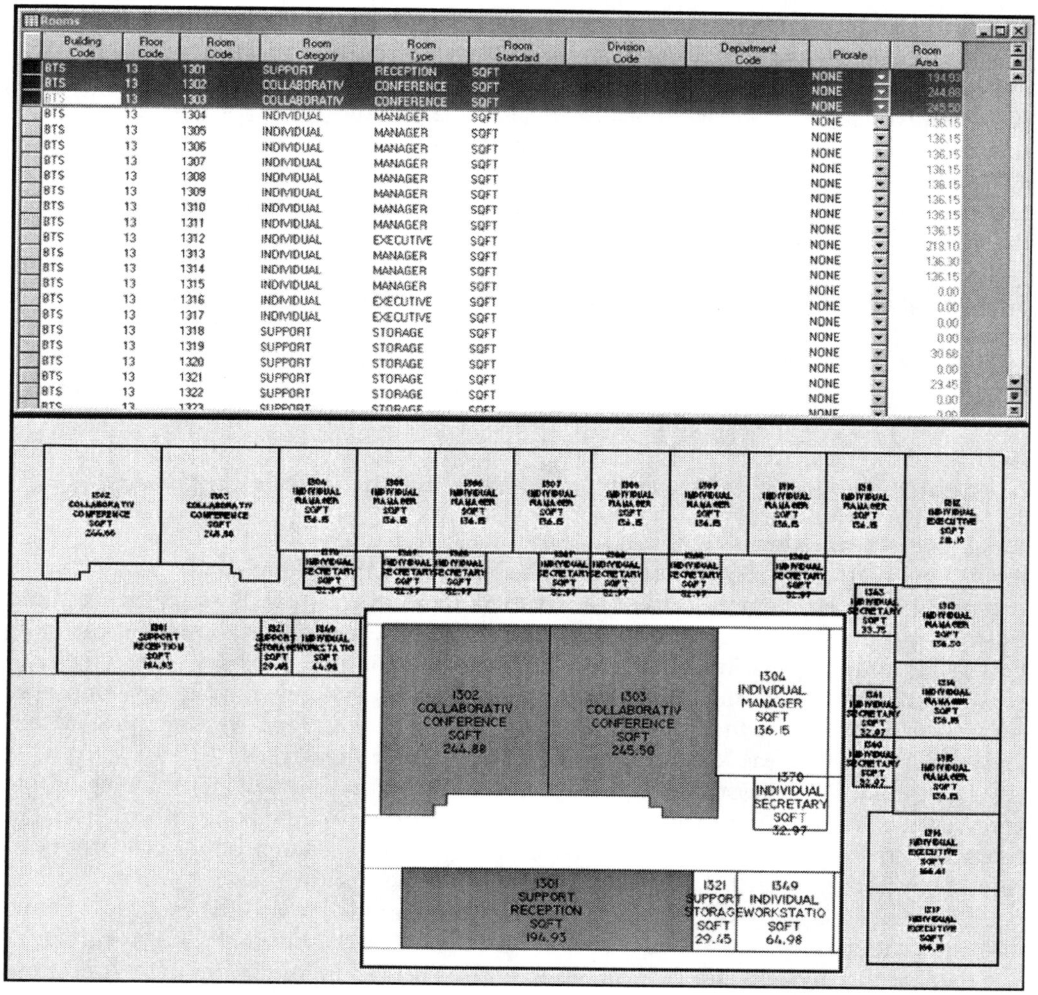

assigned rooms by floor, building, or a complex of buildings when summed according to departmental allocation.

Composite Inventory (3.2)

The composite inventory provides an effective method of classifying space and making area calculations without having to delineate individual rooms. Departmental boundaries, gross areas, vertical penetrations, common areas, and many of the other space categories discussed in Chapter 8 and Chapter 9 are inventoried using polylines on different drawing layers with their associated data stored in multiple tables.

Two variations on the BOMA method discussed in Chapter 9 derive from the changes in "new BOMA" published in 1996. Traditional BOMA charge-back allocates common areas among the organizational elements on a floor, while the "enhanced" method allows common spaces to be prorated among all the groups in a building.°

The *group* charge-back method prorates common areas among various defined groups throughout the facility. There is also a *room* charge-back variation within the composite method that prorates both rooms and service areas identified as common spaces. If rooms are identified in which space is shared among groups of employees, charge-backs can be made on a *percentage* basis, allocating area proportionately among the number of people from different departments. This kind of charge-back is possible with both the composite and all-room inventory methods.

Personnel, Hoteling, and Room Reservations (3.3–3.5)

The ability to maintain a table of employees as part of the facility database is one of the most useful features of the space-management module. In addition to allowing accurate charge-backs of individual work areas, rooms that are used on a part-time or intermittent basis can also be accurately tracked and scheduled. Figure 11-8 shows several representations of a query as to the location of and information regarding an employee.

Room reservations are typically handled on a short-term basis. Virtually all types of collaborative spaces may need to be reserved for specific purposes according to their functional capability. The scheduling activity involves locating an appropriate room, then creating and confirming reservations. The space tables can be searched for available rooms of a given type at a given time and location and a booking immediately confirmed with all the necessary parties.

A longer-term variation on the reservation activity class is often referred to as *hoteling*. Space must often be provided for the use of telecommuters and consultants on a regularly scheduled basis. Temporary space may also need to be assigned to special project teams or to facilitate a department relocation. These kinds of time-sensitive activities can be effectively coordinated through the CAFM system. If employee information is being maintained, room bookings can be inferred into the charge-back calculations so that departments can be charged for the space they use, even on a temporary basis. The historical information that is generated can then be used in forecasting and the preparation of space budgets.

Emergency Preparedness (3.6)

A substantial amount of the information needed for emergency-preparedness activities is contained in the space-management background data and drawings. Disaster-recovery and business-continuity planning relies on much of this same information. The roster of employees and organizational hierarchy can form the basis for emergency contact lists and situation recovery planning. Documentation of buildings, floors, rooms, and employee locations on CAD drawings can be used to generate egress plans, locate critical-system zones, and identify areas containing potentially hazardous materials.

Egress plans and posted directives produced from up-to-date CAD drawings show egress routes from all occupied areas and floors. Important-system zone information and hazardous-material locations depicted on drawings can be used to produce instructions to internal response personnel as well as fire and police authorities.

Ongoing emergency-preparedness activities also include monitoring essential systems and equipment using the available furnishings and equipment inventory. The same part of the database that tracks this information can be used to follow equipment condition,

Figure 11-8   Employee Table, Location Query, and Plan Designation.

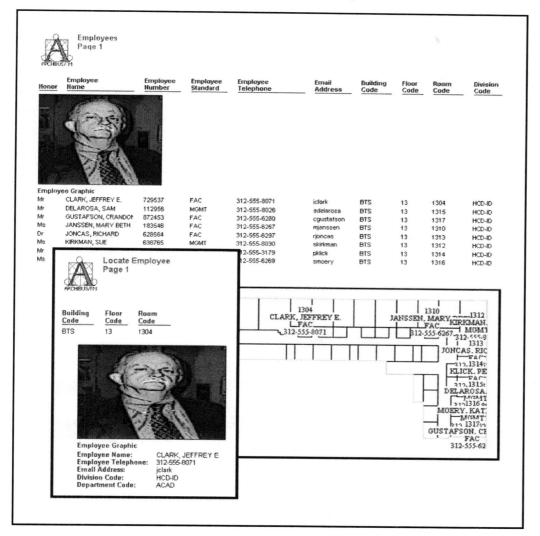

depreciation, and replacement costs relative to preparing insurance claims. Preparedness activities such as making periodic reviews of egress plans and updating emergency evacuation procedures also key into regulation assessments by the federal, state, and local authorities.

Design Management (4.0)

There are only three activity classes in the design-management module of Archibus/FM: *Doors, Windows,* and *Room Finishes* (Figure 11-9). Because most of its functionality is contained in an AutoCAD component, the Archibus/FM overlay, the module's basic activities focus on maintaining inventories of these components and producing schedules and reports.

The overlay is a package of AutoCAD enhancements consisting of menus, toolbars, and application procedures that allow you to access your facility-management database from within AutoCAD. The overlay supports the other Archibus modules as well, providing enhanced drafting tools that facilitate the creation and maintenance of the graphic side of the database. Whereas the primary emphasis of Archibus is upon tables and reports, allowing you limited ability to view and manipulate your drawings, the reverse is true with the overlay. You are looking at your drawings in AutoCAD, yet you can access many

Figure 11-9 Design-Management Activities.

4.1   Doors
4.2   Windows
4.3   Room Finishes

Archibus tables, query the database, and in fact migrate information from the database directly onto your drawings as text.

The overlay's design module provides drawing tools to create building components such as walls, doors, and column and ceiling grids. Block libraries of symbols with embedded data fields are used to populate your drawings with intelligence linked to the relational tables. Some of the important overlay functions follow:

- Drawing walls and architectural and ceiling grids
- Copying, moving, or erasing walls and cleaning up intersections
- Drawing and modifying doors and windows
- Inserting door and window blocks and asset data
- Creating and editing door, window, and finish tables and schedules

Archibus uses the term *asset symbol* to refer to the graphic representation of a facility component on an AutoCAD drawing. An asset symbol can be a polyline or a block together with its associated text. *Asset text* is information from relational-database tables (*asset tables*) displayed along with the symbol on a designated drawing layer. Asset symbols use *extended-entity data* to maintain table names and primary keys connecting the AutoCAD object to the database.°

Each domain, or column, of an asset table is either table driven or drawing driven, meaning that data are either entered directly into the table or generated by the linked drawing. Archibus tables containing drawing-driven asset data are defined with fields for the AutoCAD drawing name and entity handle.° Table-driven data may be entered directly, through a form, or by selection from validating tables elsewhere in the database. (Figure 11-10 shows a data-editing form and validating table being used to change a room category in the overlay.) Archibus uses drawing names and entity handles to locate and highlight asset symbols in the drawing.

Like all visible objects (entities) in AutoCAD, a polyline has properties such as the layer it is drawn on, its color, line type, and so forth. As we discussed earlier in Chapter 8, polylines have an *area* property that they share with certain other entities such as circles, ellipses, and regions. They also have the property of *length*, which lines also possess. The values of these properties are what are stored in the facility database so that if the configuration of a room changes in some manner, its area, as represented by the polyline, will be updated as well. Other characteristics of the room can be associated with the polyline such as room number and occupancy data.

Blocks are also used as asset symbols to represent people or physical objects within spaces, including employees assigned to rooms, furniture, and telecommunications equipment, faceplates, and jacks. This allows records in the database to be linked to their specific locations in the CAD drawing. Archibus/FM is capable of producing literally hundreds of standard and customized reports, and the AutoCAD overlay can place and highlight similar information onto your drawings. Tables and legends, as well as room annotations, can be created from custom-designed queries, as shown in Figure 11-11 and Figure 11-12.

Figure 11-10   Editing Room Data Using a Validating Table in the Overlay.

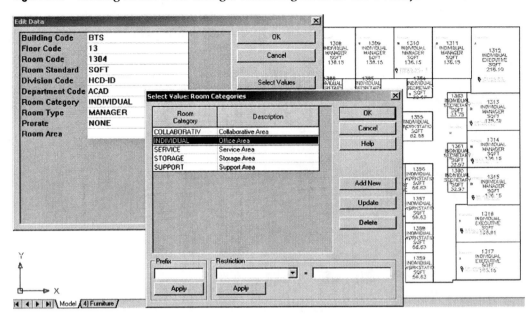

Figure 11-11   Occupancy Plan from Floor Layout Using the Overlay.

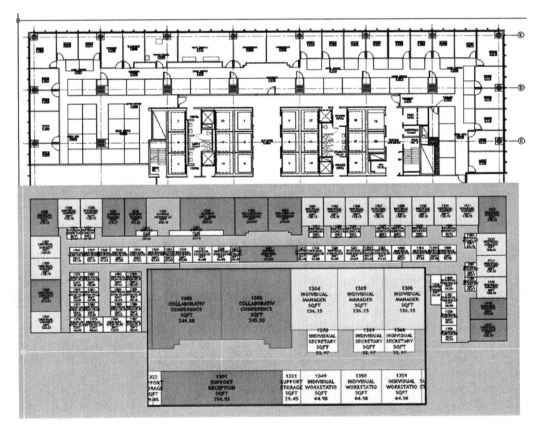

Cataloging and Uncataloging

Whenever an asset symbol is created or edited in a drawing, the overlay propagates the change to the corresponding database table. New records are created if necessary. This process is referred to as *cataloging*. Archibus provides two cataloging commands; one

Figure 11-12   Tabular Report Added to a Drawing Using the Overlay.

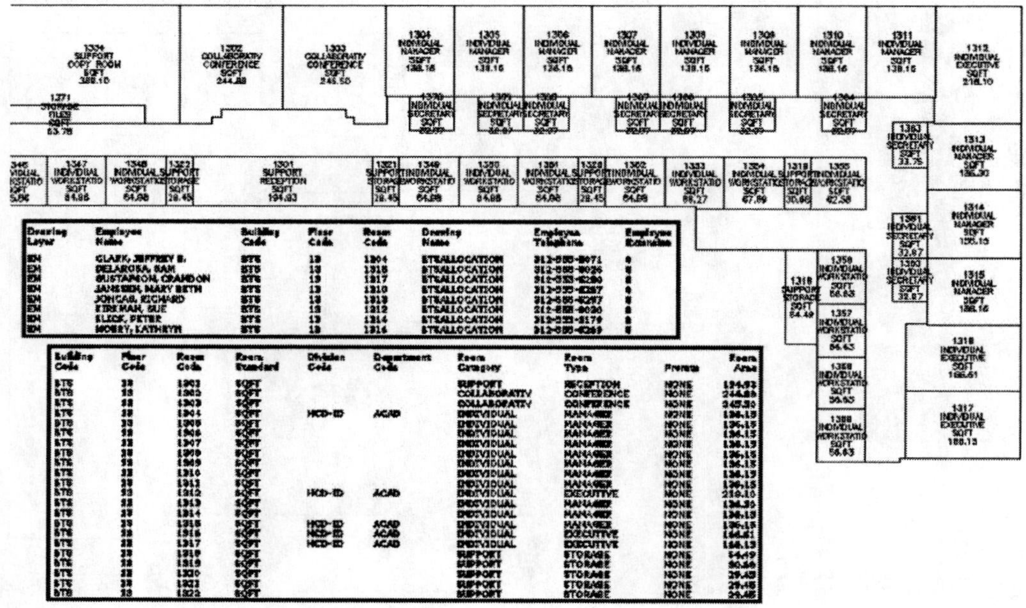

creates or updates links to a selected set of objects, and the other does the same for all objects on the current layer. Cataloging is usually required when new asset symbols have been created using the overlay but do not yet have corresponding database records.

Sometimes it is necessary to break the link between an asset symbol and the database, an action known as *uncataloging*. This deletes the drawing name, layer, and entity handle from the table on the database side, leaving the extended-entity data intact in the CAD drawing. The uncataloged drawing entity loses its asset text but can be relinked at a later time if desired.

Inference

Inference (Archibus calls its command *INFER*) effectively uses area polylines as selection boundaries to automatically modify the database values associated with asset symbols within a space. For example, a room's organizational and location codes can be applied to the furniture and equipment contained within automatically, by inference. When furniture is moved to a different space, the asset text immediately reflects its new owner and location.

Furniture and Equipment Management (5.0)

An important facility-management function is the documentation and tracking of movable assets such as furniture and equipment (Figure 11-13). Beyond just tagging and record keeping for insurance purposes, a physical inventory is essential for reconfiguring department areas and planning moves. The physical inventory needs to be more extensive than just an insurance inventory if it is to allow the planner to identify similar items in accordance with furniture standards, such as a description of each item indicating its condition and potential for reuse. The furniture standards themselves should be consistently documented as new items are added and old ones discarded. If a bar-code system is implemented, the electronic inventory can be kept current by periodically making field surveys simply by scanning the tags applied to individual furniture items.

Links between the database tables and CAD drawings simplify the production of move instructions, including furniture tags and plans clearly indicating source and destination locations. Archibus/FM provides trial layout drawing features that aid in the planning process. The furniture and equipment module works closely with space management in the use of common tables and drawings. The telecom and building-operations modules

Figure 11-13   Furniture and Equipment Management Activities.

5.1		Manage Background Data
	5.1.1	Facilities Background Data
	5.1.2	Manage Standards and Libraries
5.2		Track Tagged Furniture
	5.2.1	Track Locations and Ownership
	5.2.2	Perform Surveys
	5.2.3	Manage Insurance
	5.2.4	Manage Leases
	5.2.5	Manage Warranties
5.3		Track Equipment
	5.3.1	Track Locations and Ownership
	5.3.2	Perform Surveys
	5.3.3	Manage Insurance
	5.3.4	Manage Leases
	5.3.5	Manage Warranties
5.4		Track Furniture by Standards
	5.4.1	Manage from Surveys
	5.4.2	Manage from CAD Layout
	5.4.3	Manage Insurance
	5.4.4	Manage Leases
	5.4.5	Manage Warranties
5.5		Manage Moves and Layouts
	5.5.1	Plan Moves and Layouts
	5.5.2	Manage Move Orders
5.6		Create Management Reports
	5.6.1	Manage Churn
	5.6.2	Report on Costs and Depreciation
5.7		Quick Access

draw upon much of the same information to monitor equipment maintenance and the telecommunications infrastructure. Given the propensity of staff members to move furniture and equipment at will, the automated update capability can help ensure that the inventory reflects reality.

Asset Tracking
Data on all the assets of an organization are maintained by the finance and accounting departments, but the greatest accuracy can only be ensured if there is sufficient detail to back them up. The furniture and equipment module is capable of producing reports on maintenance costs and depreciation by any of several methods. The tagged furniture and equipment inventory tracks detailed depreciation statistics, while such data can be tracked on a more general level with the furniture-standards inventory. In either case, warranty, insurance, and lease data can be maintained as an integral part of the inventory system.

Move Coordination and Churn
Short of knocking down walls and doing major construction, most moves involve the relocation of people and equipment. Archibus provides for the development of pro forma configurations using trial copies of inventory drawing layers and database tables. Pro forma CAD layouts with their associated asset information are first used to develop a

Figure 11-14 Telecommunications and Cable Management Activities.

6.1		Manage Background Data
6.2		Develop Inventory
	6.2.1	Work Areas
	6.2.2	Horizontal Cabling
	6.2.3	Telecom Areas
	6.2.4	Backbone Cabling
6.3		Track Net Segments and Software
	6.3.1	Ilet Segments
	6.3.2	Software
6.4		Track Employees and Extensions
	6.4.1	Employees
	6.4.2	Telephone Extensions
6.5		Manage Moves
6.6		Run Help Desk
6.7		Quick Access

desired relocation strategy; then move orders based on the pro forma layouts can be used to coordinate multiple relocations. Every move from one specified room to another—whether it involves employees, tagged furniture, equipment, or some combination of these—can be phased as desired without duplicating items onto several move orders.

Churn or *churn rate* is calculated as the number of items in a specified group that change during a given time period, expressed as a percentage. The manage-churn activity produces a number of reports on churn rates of both employees and tagged furniture. Such reports can be used to analyze move patterns and frequency with the goal of maximizing the efficiency of move planning while minimizing costs.

Telecommunications and Cable Management (6.0)

Equally as important as furnishings and conventional office equipment is the voice and data communications infrastructure. In today's business organization, this includes information-network connections, both internal and external, as well as conventional voice communications. The telecommunications and cable management module provides tools for developing and maintaining an inventory of network devices and their interconnections (Figure 11-14). An interactive and graphic physical inventory of network components—from entrances to backbones to horizontal cabling, individual faceplates, and jacks—can be invaluable to IT technicians when maintenance issues arise. Finally, the same inventory can facilitate relocating people and equipment and managing the numerous details associated with terminating old connections and establishing new ones in different locations.

Telecommunications Standards
ANSI/TIA/EIA-568-B is the current edition of the *Commercial Building Telecommunications Cabling Standard*. Specifically, the Telecommunications Industry Association (TIA) divides the standard into three sections, of which the first, *568-B.1*, covers the design, installation, and field testing of a generic, structured cabling system. The other two sections contain reliability specifications for cables and interconnection hardware. The International Standards Organization (ISO) also publishes standard known as *ISO/IEC 11801:2002*, which covers several different classes of cabling and connecting hardware.

Figure 11-15   Physical Elements of the Telecommunications Infrastructure (see Plate Eight). °

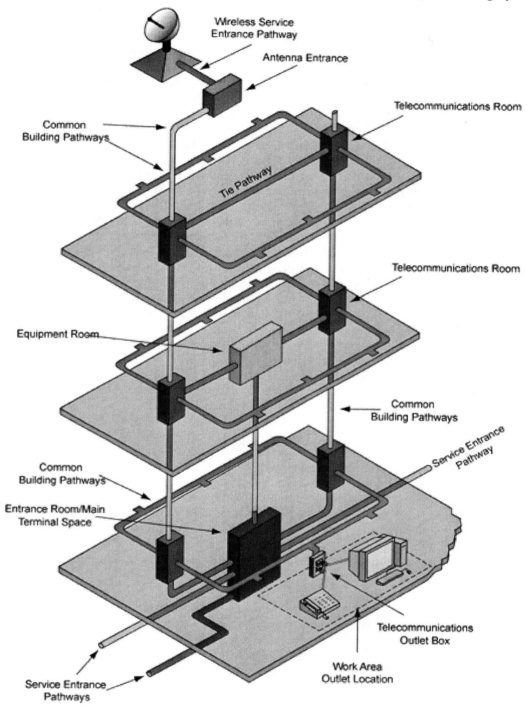

The TIA published a supplemental standard designated as *569-B* in 2004, which proposes to standardize design and construction practices for telecom pathways and spaces. *TIA-569-B* details six functional areas of design consideration for telecom systems, which include work area and horizontal cabling, telecommunications rooms, equipment rooms, building and backbone cabling, and the entrance facility (see Figure 11-15). Each of these areas generally include the following:

- *Work area*—The work area is the principal location in which people interact with telecom equipment. Outlets provide the interface to the cabling system through jacks, which are mounted in faceplates. Access is usually through the wall or floor but can be through the ceiling. Work-area equipment and connecting cables are within the scope of the ANSI and ISO specifications, and the supplemental standard provides that special consideration be given to fittings that transition between building and furniture system pathways. Major considerations are that furniture pathway capacity not be reduced by corners that will force installed cable to an intolerable bend radius (less than one inch) and that sufficient slack space be provided. It is expected that equipment cords and adapters are to have the same performance capability as the horizontal cabling to which they connect.

- *Horizontal cabling*—The horizontal-cabling structure extends from outlets in the work area to the telecommunications rooms containing horizontal cross-connects. The cross-connects consist of mechanical terminations (*punch-down blocks*) and patch cords (*jumpers*). The horizontal-cabling system also includes telecommunications outlets as well as consolidation points or transition-point connectors.

- *Telecommunications rooms*—These are dedicated rooms that provide access to the interconnections of horizontal pathways and the vertical backbone. At least one such space must be centrally located on each floor for telecom cable terminations, equipment, and cross-connects and can serve no other function. Multiple rooms on a floor must be interconnected by a conduit or equivalent pathway. At least two dedicated electrical duplex outlets on separate circuits are required in addition to convenience outlets at six-foot intervals in the perimeter walls. One wall must have three-quarter-inch plywood to a height of eight feet, access to an electrical ground, appropriate lighting, and no suspended ceiling.

- *Equipment rooms*—These are centrally located, dedicated rooms for telecom equipment serving specific occupants of the facility. This function may also be served by telecommunications rooms depending upon the size of the installation, as in the case of a multitenant floor. Equipment room locations should be accessible to the installation of large equipment, allow for expansion, and be protected against vibration and electromagnetic interference (EMI). They must have access to backbone pathways, 24-7 HVAC service, a dedicated electrical panel, electrical ground, and a ceiling/lighting configuration similar to telecom rooms.

- *Building and backbone cabling*—The backbone cabling system interconnects telecommunications rooms, equipment rooms, and entrance facilities. Backbone cables, main and intermediate cross-connects, punch-down blocks, and jumpers are included in the backbone. In a multibuilding campus environment, the backbone extends horizontally among the buildings.

Backbone cabling is most often arranged in a star configuration, meaning that each horizontal cross-connect is connected to an intermediate cross-connect that is then linked to a main cross-connect, or else the horizontal is linked directly to the main cross-connect. Permitted cable distances are limited by the specification at all levels of the system hierarchy, and the actual configurations of all pathways and interconnects must be documented for use by service providers.

- *Entrance facility*—This is the main point of service entry to a building and backbone pathways among buildings in a campus facility. The physical requirements are similar to those for telecom and equipment rooms. In addition, the entrance facility should be in a dry location close to the main electrical service room and other related utilities.

The relationships of these elements are illustrated in Figure 11-13. These are the physical components of the network, the maintenance of which is the responsibility of the facility manager, especially when it comes to changes in the network's physical configuration. Maintaining the essential information necessary to support the physical network is the function of the *develop-inventory* activity class (6.2). The module provides four activities for managing them according to their relative level in the hierarchy:

- Work Areas (6.2.1)
- Horizontal Cabling (6.2.2)
- Telecom Areas (6.2.3)
- Backbone Cabling (6.2.4)

The Logical Network
Management of the logical elements of the network, on the other hand, is usually done by information-systems personnel who may rely on the facilities manager for timely data.

The logical network consists of local area networks (LANs), subnets, individual users, and software. Archibus/FM also provides inventory activities for these elements, including both voice and data. Network segments, usually physical groupings of users in a single location, and subnets, groupings of LANs that may or may not be contiguous, are tracked by the *net segments and software* (6.3) activity class. The software available on the network can also be tracked along with the equipment on which it is installed, together with licenses in use and available.

The *employees and extensions* (6.4) activity class for voice and data networks maintains the identities of the users of the systems in user profiles, including e-mail addresses, log-in identification codes, passwords, telephone extensions, and cellular and fax numbers. On the data side, network user names and the Net segments to which personnel belong can also be tracked. Extensions, from a telecom point of view, may relate to people, rooms, or jacks in the inventory.

All of this information can be used as required to produce employee directories. Graphic and tabular reports document all the components of the telecom network, showing the flow of voice and data communications throughout the facility. As connections are made and changed, the database can be updated to reflect current conditions.

Background Data and Network Administration
In the process of developing background information, it is necessary to determine which devices need to be included in the inventory. The *background data* (6.1) is similar to that required for the other Archibus modules, including organizational structure; buildings, floors, and so on; and employees. Room types and categories tables are also used to designate service areas such as equipment rooms and telecom closets. The same basic background tables are used for telecom management as for other modules, so the background data may only need to be augmented with the special room categories and types needed.

Developing the telecom inventory can be accomplished in stages. Devices are represented by asset symbols in CAD drawings, such as the horizontal cable runs from work areas to telecommunications rooms. Backbone information can be added to the level of detail required on the one hand and wall plates and jacks in rooms and at individual workstations on the other.

Move management (6.5) of telecom equipment in Archibus is closely integrated with the furniture and equipment management module. Move orders produced through the furniture and equipment module allow the coordination of furniture and equipment relocations and tracking of move costs and include options for telecom circuit tracing and updating of connection data.

Ongoing maintenance of the telecommunications infrastructure is also facilitated by *help desk* activities (6.6) that enable personnel to report problems and track the resolution of service issues.

Building Operations Management (7.0)

Once again, much of the background information required for the activities in this module is likely to exist if other modules, such as space management, are being used. It is the database side of the CAFM system that is the primary focus of *building operations management* (Figure 11-16). Drawings may only be used incidentally to illustrate the locations of

Figure 11-16 Building Operations Management Activities.

7.1	Manage Background Data
7.2	Establish On-Demand Data
7.2.1	Create and Review Requests
7.2.2	Estimate and Schedule Requests
7.2.3	Generate On-Demand Work Orders
7.3	Plan Preventive Maintenance
7.3.1	Schedule Preventive Maintenance
7.3.2	Forecast Work and Resources
7.3.3	Generate PM Work Orders
7.4	Manage Active Work
7.4.1	Manage Active On-Demand work
7.4.2	Manage Active Preventive Work
7.4.3	Manage All Active Work
7.4.4	Update Work
7.5	Manage Equipment and Resources
7.5.1	Manage Equipment
7.5.2	Manage Labor
7.5.3	Manage Parts Inventory
7.5.4	Manage Tools
7.6	Analyze History and Finances
7.6.1	Review Work History
7.6.2	Create Management Reports
7.6.3	Analyze Finance
7.7	Quick Access

maintenance issues. Archibus/FM provides tables for data pertaining to the many aspects of preventive maintenance and work-order tracking. These include, in addition to the essential space, organization, and employee information, such items as equipment, parts inventory and vendor sources, tools, tradesmen or craftspersons, and a system of work classifications.

Establishing On-Demand Work (7.2)

Managing requests for work to be done is a key activity in building operations management. In addition to preventive maintenance, an essential function of a CAFM system such as Archibus involves the creation and scheduling of on-demand work requests and generating work orders to carry out the necessary activities. When a work request is approved, the required resources must be estimated and scheduled and time lines established. Work orders can be created using a form such as the one illustrated in Figure 11-17*a*.

One of the advantages of using an automated work-order system is the ability to plan and schedule work assignments to maximize the effectiveness of available resources. Tasks of a similar type can be scheduled to take advantage of manpower and equipment that may be in a particular location.

Planning Preventive Maintenance (7.3)

Preventive maintenance is carried out by the use of work orders as well but is scheduled proactively to forestall maintenance problems before they occur. Archibus classifies preventive-maintenance activities into two categories: equipment-related and housekeeping

Figure 11-17   Building Operations Management Work-Order Forms.

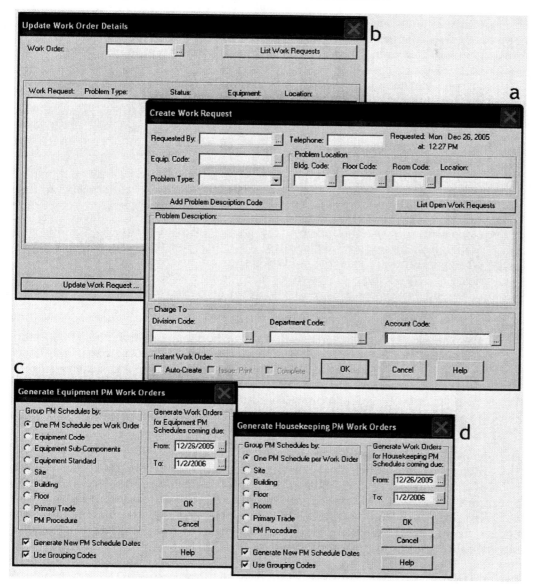

tasks. The latter are typically defined by location, such as window washing and painting. Typical Archibus dialog boxes are shown in Figure 11-17, *c* and *d*, that assist in generating preventive-maintenance work orders of both types.

Even more than with on-demand work orders, the scheduling of preventive work can be optimized by coordinating necessary activities so that they are carried out at the most propitious times. Intervals can be set as necessary for repetitive activities so that adequate notice can be given. Specific tasks can be scheduled on either a fixed or a floating basis, which means either that the activity must be carried out at the exact time scheduled or when a preceding activity has been completed. When a work order is actually generated for a given task, it is usually ensured that parts and tools are available along with the tradesperson to actually execute the work.

Active Work and Resource Management (7.4–7.5)
The management of active, on-demand work involves tracking outstanding but uncompleted work orders and their necessary resources. If an item is not completed on schedule,

it can be tracked within the system to determine whether the cause is a lack of resources or some other conflict that may require rescheduling (Figure 11-17*b*).

Preventive-maintenance activities are tracked in a similar manner to identify whether the work cycles are being disrupted for any reason. This may require reviewing why a particular item has been overlooked or whether the work cycle needs to be changed due to work-resource issues.

Often, on-demand and preventive work need to be managed together, and a CAFM system can accommodate that requirement as well. All outstanding work orders can be reviewed via consolidated reports showing the status of assigned work requests and resources. Floor plans can be highlighted to graphically depict the status of all work including uncompleted items. These can then be effectively monitored by building, assigned trade or craftsperson, or equipment required.

Of course, the availability of equipment and resources needs to be ensured as part of the building operations management function. Archibus breaks this class down into four activity areas:

- *Equipment (7.5.1)*—This involves the tracking of equipment use, location and availability, and necessary replacement parts, including work history, maintenance schedules, utilization analysis, and the printing of inventory bar-coding labels.
- *Labor (7.5.2)*—Manpower resources are tracked by trade and individual for scheduling purposes. Workloads can be analyzed on an ongoing basis, taking into account individual performance, available hours, and other factors related to the scheduling of work.
- *Parts Inventory (7.5.3)*—Parts needed and on hand are monitored to balance storage and purchasing costs. Vendor contact information is defined, parts received are entered into inventory, and links can be established with the accounting system to facilitate payments.
- *Tools (7.5.4)*—As with parts, monitoring tool activity helps to smooth out maintenance operations by ensuring that the status and location of all tools are known so that they can be effectively scheduled.

History and Financial Analysis (7.6)

This activity class ensures that the data accumulated through effective day-to-day management of work, equipment, and resources can be used in the future budgeting of building-operations costs. Numerous reports can be produced summarizing work histories and analyzing patterns of labor utilization and past expenses. In the same manner that space utilization can be charged back to departments or profit centers, maintenance budgets can be established and costs properly allocated.

Task Categories

All of the Archibus/FM modules are broken down into activity classes and activities. Activities in turn are subdivided into task categories, tasks, and subtasks, a total of six hierarchical levels. In examining each of the application's modules, we have followed the sequence of icons appearing at the top of its Windows-Explorer-like interface called the *navigator* (see Figure 11-3). When a specific activity has been selected, a second row of icons appears in the navigator identifying several task categories that are classified on a basis of various formats in which data can be viewed or manipulated. Five major task categories among the seven are used with (also) seven operational modules. These are discussed in the following section.

Another seven task categories are associated with an eighth module, which is used to provide for database administration and customization. The system module is the subject of the next section.

Tables

As discussed in the preceding chapter, tables are the basic database format. Each row represents a record containing defined fields, which are identified by column headings. Data may be directly entered into a field's cell, overwritten by double-clicking on the cell, or edited by placing the cursor at the desired position within the cell. Fields may contain text characters (fixed or variable length), they may be integer or real values with decimals, or they may be calendar dates and times. (Several examples of the table format in Archibus are shown in Figure 11-2 and Figure 11-7.)

Text data have several possible format restrictions. Such fields may contain any combination of characters up to the specified field length; they may be alphanumeric but forced to uppercase; or they may be restricted to only alphabetic, uppercase characters. Code fields are typically defined using the uppercase format. Numeric fields accept only numbers that are displayed in the defined format, as determined by the underlying relational-database engine (usually including integer, small integer, fixed and floating-point decimals of single or double precision). Any of these formats may be displayed as currency, with or without separation characters.

As an aid to data entry, Archibus also provides list fields, which are completed using a predetermined set of values presented in a drop-down box. Date fields can be completed by using a pop-up graphical calendar. The names of files containing graphic images can be entered into certain fields using the standard Windows Open dialog box and viewed as desired. Required key fields can be easily and accurately filled in by double-clicking on them, which activates a dialog box from which items in the key's validating table may be selected.

Reports

Archibus reports present tabular data in a printable format with headings and page numbers. Totals and subtotals at multiple levels can be specified as desired, and the report document can be previewed page by page on the screen. You can zoom in or out to look over the report as a whole or examine a particular portion. Many standard report views may be customized from within the application, including those that embed graphic images, AutoCAD drawings, and bar codes for inventory activities.

Like all views of the database, reports are real-time representations of your facility information. Reports that include graphic images often use color highlighting to indicate different classifications of space polylines or other asset symbols. These may be modified dynamically by changing color codes and hatch patterns. In addition, you can toggle between report and table views of the same data.

Actions

Views were discussed in the last chapter (Codd's sixth rule stating that relational databases must not be limited to source tables). Views are virtual tables, derived from combinations of base tables, that provide alternative ways of looking at selected data. In general database terminology, views result from performing queries upon the data. Virtually all of the task categories within Archibus perform actions of some kind, including the creation of tables and reports, using either view procedures (AVW files) or more elaborate programs written in a dialect of the BASIC language (ABS files). Any of the tasks or subtasks may initiate AVW or ABS procedures or a combination of both. Archibus distinguishes between *actions* that create views, however, and *queries*, on a basis of their relative complexity. Actions often perform specific procedures without necessarily producing displayed output.

Actions selected from the Archibus/FM navigator are performed immediately. These usually initiate updates to the database or make necessary calculations. Such actions are carried out by standard procedures supplied with the application or created by the user. Examples (from the lease module) include *update building area totals, update property tax rates,* and *perform prorations and update lease areas.* All of the standard navigator actions contained in the space-management module are outlined in Appendix D.

Archibus also performs certain actions when views are executed. Standard views are usually run from navigator tasks and subtasks, but they also may be invoked by name and saved using the menu bar. Views are also a type of procedure that may be extensively customized using dialog boxes that prompt for specific criteria. Information from multiple tables can be combined or special processing actions specified. Custom ABS procedures may also be written to perform specialized tasks.

Queries

Archibus queries are a type of view that usually performs extensive calculations or presents graphic data with highlighting to aid in the analysis of drawings and tabular data. Usually, queries are presented in report format, although they can be toggled to table view if desired. Many queries produce charts, including stacking diagrams for analysis and presentation.

Drawings

The drawings task category contains specialized views for working with tables and drawings in which linked data are involved. A one-to-one correspondence exists between specific kinds of drawing information and a record in its corresponding table. For the links to be established and maintained, each piece of drawing information resides on a particular layer in the drawing. Drawing views open the necessary tables, allow for the selection of the desired drawing, and set up the required layer configuration once the drawing is opened.

Background/Background Tables

Structural facts about an organization, its people, and the physical environment are all examples of background information. These tables are generally populated first when creating a CAFM database so that they can be used to validate entries and simplify data input to other tables. Like the list fields that are accessed using drop-down boxes, Archibus presents validating tables as dialog boxes when information is entered to limit the choices to legal values while minimizing the necessity of keyboarding. An example is shown in

Figure 11-2, where *Floor Code* values are selected from a background table to populate a record in the *Rooms* table. Background tables can also be updated using these same dialog boxes, as evidenced by the *Add New, Update,* and *Delete* buttons.

Executive Reports

The seventh and final task category included in the lease, space, furniture and equipment, and building operations modules is *executive reports*, which provides quick, direct access to many of the modules' most often used reports and queries. For instance, the lease module includes financial profiles and property and lease detail reports and summaries. The space module provides quick access to departmental stacking, allocation, and room standards reports, a charge-back report, plans highlighting rooms and groups by organizational unit, and an employee locator (shown in Figure 11-8). Archibus also maintains a hot list of other frequently used tasks.

System (8.0)

The Archibus/FM system module contains four classes of activities dedicated to system administration and customization (see Figure 11-18). The activities of the *schema* class (8.1) deal specifically with customizing the system under the following headings, each having its identifying icon:

- Define (8.1.1)
- Navigate (8.1.2)
- Control (8.1.3)
- Update (8.1.4)

Define

Included in the *define* category are tasks for adding tables and fields to the standard schema, copying them, and working with validating tables and the fields that reference them. To facilitate modifying the schema in this way, Archibus provides an "Archibus/FM Tables" table and an "Archibus/FM Fields" table that reflect the structure of the application and the

Figure 11-18   System Administration Activities.

8.1	Schema
8.2	Security
8.3	Step Up and Connectivity
8.4	Web Template Publishing

elements of its data tables. These two tables are, in effect, metadata tables that enable the structure of the other tables—hence, the schema—to be modified from within the application.

Changes in the schema are implemented by defining new tables as records in the "Archibus/FM Tables" table, new fields as rows in the "Archibus/FM Fields" table, and editing or deleting records in the "Archibus/FM Fields" table. Figure 11-19 shows these metatables. In the example, the *Rooms* table (which is used by all modules) is highlighted in "Archibus/FM Tables," which causes the data structure of *Rooms* to be displayed in "Archibus/FM Fields." Each record in the "Archibus/FM Fields" table defines the format in which *Rooms* data are stored. Only a few of the format specifications are shown in the figure;

Figure 11-19 Archibus/FM Tables and Archibus/FM Fields Metatables.

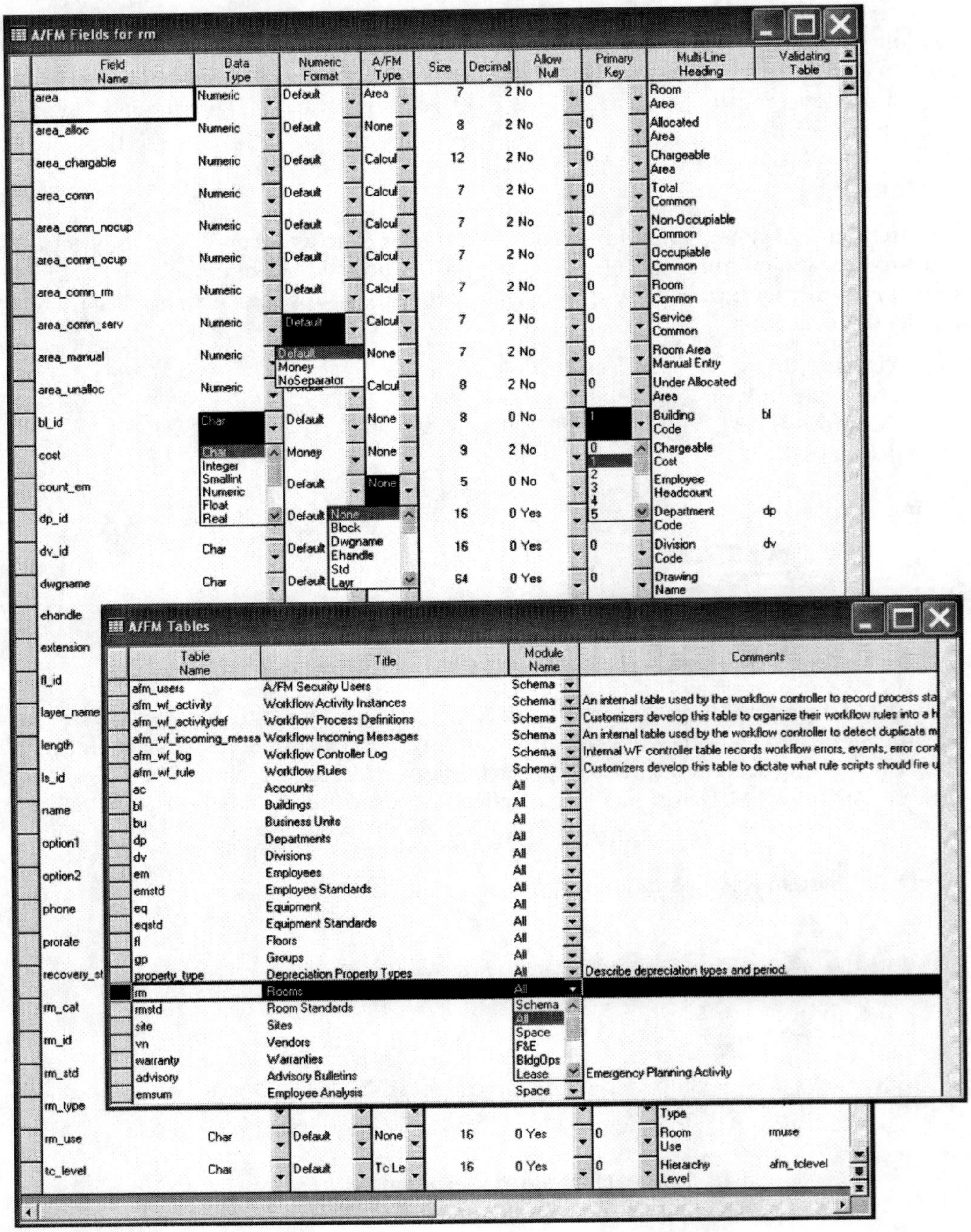

there are about two dozen altogether. If we look at the *bl_id* record, we can see that it defines the *Building Code* in the *Rooms* table, which has the *Data Type* "Char" (alphanumeric), that the *Size* of the field is eight characters, and that this field is the first element of the *Primary Key*, the *Validating Table* for which is *Buildings* (*bl*).

Navigate

The *navigate* task category performs a similar function for the Archibus/FM navigator. By adding records to the tables that define each of the six levels of the navigator outline, functions can be added or removed. The order in which items are displayed in the navigator hierarchy can be altered as well. Changes are implemented when the navigator window is closed and reopened.

Control

This task category allows table-driven properties of the schema to be modified, such as elements of the drawing list, asset types as represented by AutoCAD polylines or blocks, and asset layers. A table of connection information maintained using the telecom module is also accessible through the control function, which assists in tracing duplicate or broken connections.

Layer characteristics such as color, line type, and the asset type of objects placed on the layer can be modified using the "Archibus/FM Layers" table. Renaming the standard layers is generally not recommended because it is then necessary to modify the schema itself as well as many views and other procedures that refer to the standard names.

Update

The last of the four task categories provided for modifying the schema is *update*. The update tasks either force a re-read of the schema definitions (*Reset Schema*) or update the physical structure of the schema following changes to schema table definitions (*Update Schema*). The schema must be reset when changes are made in Archibus/FM asset types, layers, preferences, tables, and fields so long as the changes do not involve added or deleted records. The schema must be *updated* when new tables or fields are defined; when the definitions of data types are modified; or when names, validating tables, or fields are changed.

These tools can make customizing the Archibus database convenient, in view of the various complexities of relational databases that were described in Chapter 10. Structural changes should only be attempted on a duplicate of your production database, however, in many cases, it is necessary to depopulate the database tables before making such changes, and only when the new database has been thoroughly tested and everything works correctly should it be implemented.

Figure 11-20 Navigator Icons for the System Module.

Security (8.2)

Enable and *secure* control user identification and password security over access to the facility database (icons are shown in Figure 11-20, *a* and *b*). *Enable* simply turns security on and off and permits changing passwords.

The *secure* task establishes named user groups that may be assigned different levels of access to specific fields. Archibus/FM provides four levels of security:

- *System Manager*—Unrestricted access to all tables and commands, including those that control system security and changes to the schema.
- *Expert*—Access to features requiring technical knowledge of the application such as importing and deleting data or creating new views.
- *Edit*—Standard user access with the ability to edit data in tables and drawings, and modify views (sort, change field visibility, etc.).
- *Review*—Restricted access: read only, but users can change the format of tables and reports, change layer visibility of drawings, and save view changes as new views.

When security is activated, the default setting for all fields is *edit*, except for users with a *review-access* setting.

Step Up and Connectivity (8.3)

This is an activity class providing utility functions for importing and exporting data from the application. *Step up* (Figure 11-20*c*) consists of utilities for converting tabular data from legacy dBASE versions to the current Archibus release, as well as asset drawings created in earlier AutoCAD releases.

Connect (Figure 11-20*a*) provides similar utilities for importing human resource data from dBASE and Oracle formats, importing and exporting data from/to different file formats such as text and Microsoft Access and Excel.

Drawing Publishing (Figure 11-20*d*) facilitates outputting customized sets of AutoCAD drawings in both DWG and DWF° formats. Using these functions, rules can be established for publishing and the display of desired asset types.

Web Template Publishing (8.4)

The Web template activity class contains tasks for publishing reports and tables in the form of *ColdFusion* templates. ColdFusion is a separate application that produces templates for reports that can be viewed on the World Wide Web without the need for writing HTML or SQL code. Archibus views, sorted as desired and formatted in either grid or report layouts, can produce entire groups of ColdFusion template files for viewing over the Web.

CLOSING WORDS

As relational-database systems became widely available during the 1980s, they gained almost universal favor as the engines driving CAFM applications, replacing flat-file data managers such as dBASE. CAFM systems typically integrate the facility database with a CAD program and provide methods of linking drawing objects with various kinds of information stored in the database.

Different applications emphasize diverse aspects of asset management, many now being Web-enabled components of an *integrated workplace management system*. Six major classes of data are common to nearly all enterprise-level products: Strategic Planning, Space Management, Furniture and Equipment, Telecommunications and Cabling, Building Operations, and Real Estate and Property Management. We have looked at three such applications—TRIRIGA, FAMIS, and Archibus/FM—which are among today's leaders in CAFM software.

Chapter 12 through Chapter 16, the final part of the book, address several aspects of the framework in which all facility planning must be carried out: environmental guidelines, accessibility regulations, and building codes. The concern of the next chapter is principles of sustainability, including the LEED initiatives of the U.S. Green Building Council. LEED (Leadership in Energy and Environmental Design) falls squarely into the interest and responsibility area of architects, space planners, and facility managers.

CASE STUDY BP America

The 1998 merger of British Petroleum (BP) with Amoco Corporation created one of the world's largest petrochemical companies.° At the time of the merger, Amoco had major office and refinery facilities in the Chicago area, its headquarters, as well as in Cleveland, Tulsa, Houston, and Atlanta. Very shortly thereafter, BP added Arco oil company to the mix, with operations in Los Angeles and Dallas. BP's principal U.S. operations were in Cleveland and Houston. The operations of three different corporate cultures needed to be integrated in a short period of time.

Under the direction of Amoco's Engineering Support Services group, a CAFM initiative had already been under way for several years at the time of the mergers. An Archibus system using an Oracle database was already in place at Amoco's Midwest research center and several other locations, capable of tracking space utilization by region, state, city, campus, individual buildings, and floors. In addition to the standard and custom reports developed by the company's integration consultant, the common Oracle database system allowed key data to be passed to corporate management's ERP system for financial reporting.

During 1998 and 1999, most of Amoco's other U.S. facilities were either added to the system or converted from separate databases previously maintained in each locale. In the case of a Midwest refinery facility, which encompassed nearly a dozen office and laboratory buildings, floor plans needed to be verified onsite and CAD drawings created. An independent facility database was constructed; then, after thorough testing, it was consolidated into the corporate database. The corporate general offices, which occupied nearly forty floors of an office tower in downtown Chicago, needed to have their plans converted to AutoCAD from a legacy system that had been in use there for many years. Later, this building would be sold and the functions there moved to other locations. Simultaneously, several Amoco facilities in other cities were documented and the information updated in the new format.

DRAWING, DATABASE, AND APPLICATION ISSUES

In the process of building the consolidated facility database, major changes were made to the standard schema consisting of nearly 250 tables. Multiple sites needed to be maintained in a single database while restricting views and reports to a single site at any given time. In addition, the standard Archibus/FM Building-Floor-Room space hierarchy had to be modified to allow multiple complexes to be defined for individual sites. *Complexes* allow for the proration of rent charge-backs for shared spaces among multiple linked buildings (see Table 6.2).

The complex provides an intermediate level of detail between allocating space by building or to the entire site.

All of the AutoCAD drawings linked to the Archibus system were either set up or converted to a uniform format so that space utilization could be calculated according to BOMA standards. This required a complete audit of the drawings for every building and floor to ensure that it was formatted according to corporate standards and had the correct information on each drawing layer. This was of particular importance with drawings of buildings such as the Amoco general office tower, which had been originally produced by a legacy CAD system.

While these tasks were in process, the groundwork was also being put in place for the merger implementation by the creation of a common operating environment for all corporatewide information systems. This entailed the development of installation scripts for all software applications used individually or in combination. Authorized software modules were installed on central servers in order to set up or reconfigure individual workstations by downloading appropriate packages where needed. This provided effective control over security as well as software licensing. In the case of CAFM workstations, scripts for the automatic installation of Archibus and AutoCAD as an integrated system were developed and tested. As the data and drawings were completed at various locations, they were migrated to test servers, after which they could be consolidated into the Oracle production database.

Customized Archibus/AutoCAD Procedures

A number of custom views and reports were added to the standard Archibus space management module (see earlier section in Chapter 11 and Appendix D) to support the expanded BP space and organizational hierarchies. The *all-room* inventory format was used in conjunction with a highly detailed set of room Categories, Types, and Standards necessary to allocate occupied space by cost center and identify vacant areas wherever they existed. Special reports included utilization by room, room standard, floor, and building complex, as well as cost center and company code. These reports, using appropriate rental rate options in effect at each location, facilitated accurate space charge-backs. Corporate standards were implemented for equipment utilization as well as compliance with disability guidelines.

MERGER-INTEGRATION METRICS

Prior to the merger, the three companies maintained major facilities in seven locations totaling in excess of ten million square feet with 17,500 employees. Each company had its separate and distinct corporate culture. As one of the effects of the merger would be to consolidate the combined operation's office and research facilities over a three-year period, success would depend upon how quickly and effectively the three cultures could be integrated. By 2002, the company's nine thousand employees would occupy 3.9 million square feet (see Figure CS5-1). At the end of the consolidation, overall space utilization on a per-person basis would be reduced to three-quarters of what it had been.

As a result of its ten-year history with CAFM systems, BP's facilities organization had a sizable store of information related to the past *efficiency* of its space utilization. These data included such metrics as area per person, cost per workstation, and vacancy rates for underutilized space. Efficiency, however, measures only one dimension of functionality. The other main component needed to manage inevitable culture shifts within the organization was perceived to be the *effectiveness* with which the work environment supported the integration of people, technology, and space.

The recognition of efficiency and effectiveness as key performance indicators (KPIs) derived from a facilities-change program implemented by BP Global Property Management in 2000. Measurements were drawn from the work environments at its U.S. facilities consisting of nearly four million square feet, much of the data provided by the CAFM systems then in place. Relevant criteria for KPIs were established that could be clearly defined, were measurable, could be benchmarked, and that would encourage desired behaviors.

Collaborative Space

One of the most important criteria was determined to be the percentage of collaborative space within each facility. Collaborative spaces promote two or more people having informal conversations and sharing ideas. As defined by BP, such spaces are of several varieties. They can be conventional meeting places such as conference rooms or informal meeting areas such as lounge seating near a window, as well as other support areas such as break areas, cafeterias, and mail-drop stations where people can bump into one another and generate ideas.

The configurations of such areas are highly variable according to function. According to BP's Midwest regional property manager, "You need a certain amount of collaborative space to make a facility effective. It depends on the type of work group and the type of space. You don't need much for areas devoted to heads-down activities like legal, finance, or computer programming, [but] for a trading operation or customer service function or any activity where there's a lot of banter going on, you need collaborative space to promote that interaction."

Criteria were established by researching facilities considered to be particularly effective both inside and outside of the company. Different work environments at various locations in the United States and the United Kingdom were toured, including several in Silicon Valley. Internal tenant improvement work was surveyed globally to determine what types of space were considered by BP employees to function most effectively. All these considerations went into developing work environments at BP that would have a consistent look and feel and support brand values while meeting these requirements. The guidelines that were arrived at for collaborative space were described as follows: "Best-in-class projects at BP fell into a very defined range of having between 20 and 35 percent collaborative space. With more than 35 percent . . . occupants tend to feel it's wasteful. Under 20 percent they feel the space is not productive." Overall, the goal was to plan spaces that would drive positive culture change and reinforce the feeling that this is a great place to work.

Key Performance Indicators for the New Work Environment

Through its extensive research, BP determined several measurements that would provide planning guidelines for both renovated and newly constructed spaces. Spaces are evaluated according to the following criteria, given the need to maximize both efficiency and effectiveness:

- Efficiency KPIs
  - Cost per square foot
  - Dollars per workstation
  - Square feet per person (utilization rate)
  - Square feet per workstation
  - Percentage of vacant space

Figure CS5-1 BP America's Major Facilities Before and After Consolidation.

(continued)

(continued)

Figure CS5-2   New Work Environment KPIs.

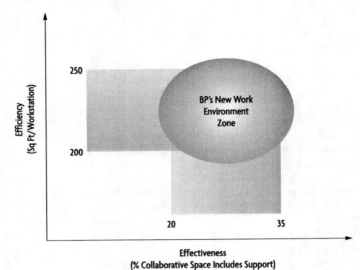

Figure CS5-3   Digital Café at a BP America Facility in the Midwestern U.S. (see Plate Ten).

- Effectiveness KPIs
  - Percentage of collaborative space
  - Ratio of workstations to employees

In terms of utilization efficiency, BP determined that critical efficiency is reached at between 200 and 250 square feet per person (see Figure CS5-2). When a business unit indicates a need to develop a new space, BP's facility staff performs a thorough programming analysis of how the group works. The middle of the optimal range for collaborative space (approximately 25 percent) is the target for traditional work groups. As planning proceeds and specific needs are determined, workstations developed, and conference rooms added

or modified, the planning staff endeavors to keep the utilization rate and collaborative space percentage within what is termed "the zone."

BP's property executive observes that "we have a lot of spaces at 350 square feet per person with very little collaborative space. We can present these business units with an opportunity to reduce their costs and improve the productivity of their work group. We do that by putting in some of these innovative meeting areas like small cafés so that during the day people can have a collision with another person and generate ideas." The digital café pictured in Figure CS5-3 is equipped with collaboration amenities such as readily available network and telecommunications connections.

MANAGING CHANGE

Since the KPI measurement standards were implemented, BP America has constructed well over thirty thousand square feet and currently has in excess of three hundred thousand square feet being planned or under construction. KPIs together with the capability of the CAFM system make effective space management a daily reality. The KPIs can be calculated by business unit and by location (floor, building, etc.), which gives planners a powerful tool for prioritizing their work. Because the business units are responsible for funding new construction in their areas, work may be deferred until they have the necessary funds in their individual budgets.

A pilot implementation of the system was recently completed in the United Kingdom as part of an overall plan for global implementation. Some of the other principal benefits of BP's system are described here:

- *Visual drill down for facility and project managers*—BP's system integration consultants built a Web interface to the CAFM system that offers the user a dynamic view of his facilities. Interactive Web pages allow queries to be made of the database by drilling down through the space hierarchy to find specific information regarding individual properties or buildings. Figure CS5-4 illustrates several levels of the drill-down process,

from the *country* to *region* to *site* to *floor* level. Using the reports produced, planners can easily determine the ratios of individual, collaborative, and support spaces, allowing for vertical penetrations and other loss factors.

- *Better local/regional space planning by minimizing vacancies*—The majority of planning work is generated by moves and restacking efforts. The CAFM system allows managers to be proactive in seeking out available space. If a business unit needs to relocate from Houston to Chicago, for example, vacant space in existing BP properties can be easily located rather than necessarily having to lease additional space. Even if the vacant space is scattered in small pockets, those pockets can be consolidated through re-stacking, making a contiguous block of area available.

- *Interactive tool for service level agreement discussions concerning rent*—The amount of detail that the system is capable of producing lets business unit managers know precisely what they have, which is particularly helpful when annual rent discussions take place and service level agreements (SLAs) are negotiated. Figure CS5-5 shows the level of detail with which a typical office floor is documented in the Archibus/FM system.

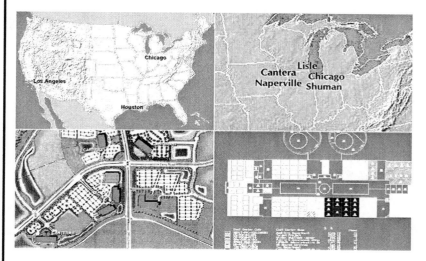

Figure CS5-4   Drill-Down Web Page Views of Company Facilities.

(continued)

(*continued*)

Figure CS5-5 Space Types Defined on a Typical Office Floor Plan (see Plate Eleven).

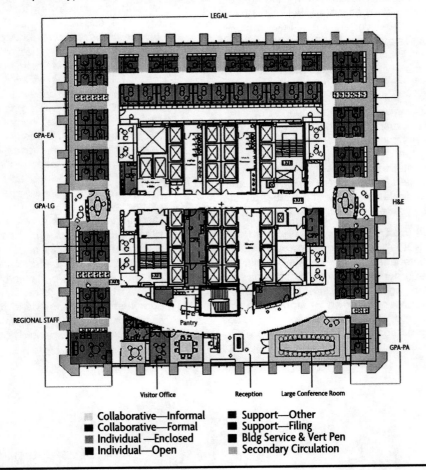

Collaborative—Informal
Collaborative—Formal
Individual —Enclosed
Individual—Open
Support—Other
Support—Filing
Bldg Service & Vert Pen
Secondary Circulation

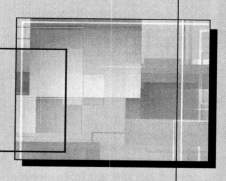

CHAPTER 12

Sustainable Design

The "control of nature" is a phrase conceived in arrogance, born of the Neanderthal age of biology and philosophy, when it was supposed that nature exists for the convenience of man. . . . It is our alarming misfortune that so primitive a science armed itself with the most modern and terrible weapons, and that in turning them against the insects it has also turned them against the earth.
—Rachel Carson°

In 1969, in an attempt to put forward an interdisciplinary response to what was then known as the "ecology crisis," ecologist and philosopher Paul Shephard wrote: "Where now there is man-centeredness, even a pathology of isolation and fear . . . ecology as applied to man faces the task of renewing a balanced view. . . . The ideological status of ecology is that of a resistance movement." Arguing for the necessity of a wholesale reversal of the Western view of natural science, Shephard characterized Rachel Carson and other writers of the time as "subversives," noting that the crisis could only be averted through the admission and acceptance of "an element of humility which is foreign to our thought, which moves us to silent wonder and glad affirmation."°

When these words were written, it had been seven years since the appearance of Rachel Carson's *Silent Spring*, which challenged the notion that the mechanical sciences and technology had the ability to control nature. The *control* mentality originated in the Industrial Revolution; here is Jacob Bronowski's description of which we quoted earlier: "The machines changed the organization of society, and shifted the centre of a man's life from his cottage home to the daily factory." This fundamental change marked the beginning of man's isolation from the natural environment in which he had been completely integrated since the dawn of history. He became a part of what came to be called the *industrial complex*, which was built almost entirely on scientific advances and the convergence of technologies. One of the early consequences was large-scale neglect of the delicate connections to the natural environment.

This disconnect was the result of arrogance of the kind that Dr. Bronowski passionately argued should be supplanted by the principle of tolerance. As he put it, "When people believe that they have absolute knowledge, with no test in reality . . . they aspire to the knowledge of gods."° Bronowski was referring specifically to the scientific theories of racial purity that powered the eugenics movement at the beginning of the twentieth century, ultimately leading to genocide on an unprecedented scale during the Second World War. In terms of basic human needs, however, it is hard to imagine even now, with fifty years' hindsight, how we believed that housing projects such as Robert Taylor Homes could resolve any problems of urban living. Was it completely by chance that the rise of environmentalism in the United States coincided with the decline of institutionalized racism during the final quarter of the last century? It was during this period that minority inhabitants of degraded urban neighborhoods began to speak with a common voice for environmental justice in regard to factories and incinerators that polluted the water and the air.

We are only a few years into the current century, but there are some hopeful signs that humility is becoming less foreign to our thought process and that the spiritual ideals of sustainability will help us connect once again with our earth.

ENVIRONMENTAL AWARENESS IN THE UNITED STATES

Not long after the first moon landings produced the unforgettable picture of Earth as a small blue and white sphere suspended in the cosmos, dawn of the first Earth Day was celebrated on a clear morning in April 1970.

There has always been an appreciation of the value of wilderness in the American consciousness. Thoreau's Walden, published in the mid-nineteenth century, emphasized the dualist faces of nature: power and tranquility. At the very beginning of the twentieth century, Theodore Roosevelt led a popular campaign for conservation that resulted in the formation of the National Park system. Franklin Roosevelt's New Deal in the 1930s instrumentalized the formation of such agencies as the *Soil Conservation Service* and the *Tennessee Valley Authority (TVA)*. In the first instance, scientific methods were brought to bear on controlling the erosion of agricultural land, which in conjunction with a prolonged drought had produced the dust bowl. The TVA successfully exploited the natural environment as a clean source of electric power.

After World War II and through the 1950s, explosive economic growth and the need for housing pushed suburban construction far beyond city boundaries. Some of us have vivid memories of driving through the sulphurous black and orange clouds of airborne effluent from the lakefront steel mills south of Chicago. The deterioration in air quality and urban water supplies together with newspaper stories about nuclear fallout led to a growing public awareness of the need to protect the environment.

Within months following Richard Nixon's election in the acrimonious 1968 presidential contest, the new president established a cabinet-level *Environmental Quality Council*. In his 1970 State of the Union Address, Nixon outlined to Congress a multifaceted program to create national air-quality standards, improve water-treatment facilities, end dumping of untreated waste, and create effective standards for reducing pollution from motor vehicles. By the end of that same year, the *U.S. Environmental Protection Agency* had been formed as an autonomous regulatory body that would enforce environmental policy, conduct research on pollution and the methods for controlling it, and provide grants to assist in implementing cleanup efforts.

Another impetus toward *resource-efficient design* came in the form of the OPEC oil crisis of the early 1970s. One effect of worldwide oil shortages and escalating prices was a generation of smaller and more energy-efficient automobiles, which in the end gave way to the SUVs that became popular by the millennium.

Attempts at reducing energy losses in buildings resulted in the use of more and better insulation as well as enclosures being made virtually airtight to reduce air-infiltration losses. This had the unfortunate consequence of causing interior air quality to deteriorate, which ran counter to increasing consciousness of occupant safety and comfort. Stakeholders began to realize that occupant health issues were a potential source of significant liability. In recent years, however, increased productivity as a result of attention to the comfort and health of workers has come to be viewed as a sought-after asset by building owners and facility managers.

Unforeseen Consequences

Partially as a result of Carson's landmark book, which appeared at the beginning of a period of great social unrest, environmentalism became a national political movement that demanded that government act to preserve the earth by regulating pollution-causing industries and their products. The specific thrust of Carson's book was against the pesticide DDT, the use of which has been all but completely discontinued around the globe.

Underscoring the need for a thoughtful and balanced approach when dealing with nature, however, a recent editorial in *Forbes* magazine pointed out that an estimated two million people die annually from malaria, most of them children under five and 90 percent of them in Africa. In 1996, when South Africa stopped using DDT, its malaria cases increased by a factor of ten. When the country reversed this policy four years later, new malaria cases quickly dropped by nearly 80 percent. The piece went on to say that deaths from malaria could be greatly reduced by using minute quantities of DDT to control the mosquito population, yet officials insist on using alternatives that are significantly less effective. If nearly a half-million people can be protected from malaria through the indoor use of the same quantity of DDT sprayed during the 1960s on a single American farm, one must question the morality of not doing so.°

GLOBAL ENVIRONMENTALISM

In Chapter 5, we described a number of influences on business organization and culture: process reengineering and systems integration to name but a couple. The principles of total quality management, which W. Edwards Deming introduced to Japan in 1950, became the foundation of international standards for quality control and, later, environmental management, under the leadership of the *International Standards Organization*.

The ISO's quality standard, known as *ISO 9000*, established a certification process for quality management beginning in 1987. It established quality management system standards that could be applied to any organization, regardless of size and whatever its product or service. Management system standards provide a model to follow, which, in the case of ISO 9000, was based on an international consensus as to state-of-the-art practices.

ISO 9000 gained wide acceptance during the early 1990s by companies that were part of the global supply chain or wished to demonstrate that they met the quality standards required to enter that marketplace. Certification is based on factors such as a customer-driven focus, a process-based approach to manufacturing activities and resource management, a systems approach to management—integrating and aligning processes for optimal results; establishing mutually beneficial supplier relationships; and, perhaps most important, continual overall performance improvement as a permanent objective of the organization. All of these factors require an ongoing commitment to the total quality management cycle.

A similar cyclical approach can also be seen in the construction lifecycle management process. Whereas in TQM the four parts of the cycle are *program* → *implement* → *evaluate* → *refine* (Figure 5-12), in the construction process they are often labeled *predesign* → *design* → *construction* → *occupancy*. In this process, the final task in the construction phase involves the *commissioning* of the facility to ensure that all of its performance standards have been met and verified. The occupancy phase has three stages that complete the cycle:

1. *Startup*—The building systems are fine-tuned and operations and maintenance staff are trained.

2. *Operation and maintenance*—Ongoing facility management in which maintenance plans are developed and implemented, periodic inspections and post-occupancy evaluations are conducted, and ongoing training and education are provided.

3. *Next use*—At the end of the building's (or tenant space's) lifecycle, options for reuse, recycling, and salvage are identified and a plan for implementation is developed, leading back into the next predesign phase.

Following the rapid and widespread acceptance of ISO 9000, the International Standards Organization formed the *Strategic Advisory Group on the Environment* (*SAGE*) to determine whether international environmental standards were needed. Many countries around the world had already launched major initiatives for environmental management, eco-labeling, and other standards.

In 1992, an ISO committee was created, made up of representatives from standards and environmental organizations, together with industry and government representing numerous countries, to draft a new series of standards based on the earlier model. The *ISO 14000* family of standards, first published in the fall of 1996, are comprehensive, including environmental management systems, auditing, performance evaluation, labeling, and lifecycle assessment. Environmental aspects of product standards are also addressed. The principal sections of the ISO 14000 standards series are listed in Table 12-1. Its goals are threefold:

- Promote a unified global approach to environmental management.
- Develop the ability of organizations to attain an acceptable level of environmental performance and measure improvement.
- Promote international trade and facilitate the reduction in nontariff trade barriers.

Since the publication of ISO 9000, and ISO 14000 several years thereafter, there has been wide and growing acceptance of both sets of standards worldwide. In the case of ISO 14000, the existence of a set of international standards has given a global perspective to environmental concerns and the need for a cleaner and healthier Earth. Many countries and regional authorities have promoted requirements addressing environmental issues that vary widely. As acceptance grows, this single standard can help keep a focus on the need for a positive global environmental strategy while minimizing conflicts between regional interpretations.

The certification process, through which companies gain recognition for their compliance to ISO standards, may well eclipse the basic ethical motivations for sound environmental management. Numerous organizations sought ISO 9000 registration as a result of increasing demands from their customers, making certification a business necessity in many regions and industries. So far, ISO 14000 registration seems to be following a similar pattern.

At the end of 1993, there were fewer than fifty thousand organizations certified to the original ISO 9000 standards. By the turn of the century, the number had increased to more than four-hundred thousand. Certifications to the newest version of the quality standard (ISO 9001:2000) were slightly over a half million with total registration approaching six-hundred thousand by the end of 2003.

Table 12-1   ISO 14000 Series of Environmental Standards

Standard	Title / Description
14000	Environmental Management Principles Guide to Systems and Supporting Techniques
14001	Environmental Management Systems Specification with Guidance for Use
14010	General Principles of Environmental Auditing
14011	Auditing of Environmental Management Systems
14012	Qualification Criteria for Environmental Auditors
14013/15	Auditing Guidelines, Programs, Reviews and Assessments
14020/23	Environmental Labeling
14024	Labeling: Practitioner Programs Guiding Principles, Practices, and Certification Procedures
14031/32	Guidelines on Environmental Performance Evaluation
14040/43	Lifecycle Assessment General Principles and Practices
14060	Guide for the Inclusion of Environmental Aspects in Product Standards

ISO 14000 effectively began in 1995, with 257 certifications in 19 countries. By the millennium, there were nearly 23,000 certifications in 98 countries. At the end of 2003, the number of certifications had increased to over 66,000 globally in 113 countries, with the largest increase yet (34 percent) over the preceding year.°

PRINCIPLES OF SUSTAINABILITY

At its annual convention in 1993, the *American Institute of Architects* (*AIA*), together with members of other design professional organizations, announced a *Declaration of Interdependence for a Sustainable Future*, which stated an ambitious set of goals for establishing sustainable design practices. In brief, these goals included the following:

- Social and environmental sustainability must become a fundamental principle of professional responsibility and practice.
- Policies, procedures, practices, products, services, and standards for sustainable design must be developed and improved on a continual basis.
- Sustainable design policies and standards must be implemented in new built environments and existing facilities brought up to such standards.

To those ends, there was a general consensus that leaders in business and government, the building industry, and the general public needed to be educated so that sustainable design policies and practices would become the norm and be fully integrated into our lifestyles. People who occupy buildings, including their owners and operators, must learn to recognize that developing more energy-efficient facilities that promote conservation and recycling of natural resources will benefit global society, both on a short-term basis and in the long run.

Most sources of information on the subject of sustainable, or *green* design, agree that the following principles form the core of a sustainable philosophy.°

1. People are a part of nature and must exist within its bounds as part of a healthy and mutually sustainable system. Planning and development must promote the beneficial coexistence of human enterprise and natural systems.

2. Human design decisions affect the natural world and this inherent interdependence means there are consequences at every level. Design parameters need to recognize far-reaching as well as immediate effects.

3. There are integral relationships between material and spiritual considerations that affect each other. The human sense of place must be considered on all scales from individual dwelling and community to global trade and industry.

4. Long-term value must be a primary consideration in the creation of all designed material objects. Future generations should not have to bear the cost or confront the dangers of ill-considered processes and products.

5. The full lifecycle of products and processes must be understood and solutions implemented in a manner that emulates natural systems in which there is no waste.

6. As an integral part of the natural world, human designs must maximize the safe and responsible use of natural energy sources.

7. All man-made material objects have a lifecycle. Human design decisions cannot solve all problems; hence, planning has inherent limitations. Nature must be considered with humility and should be looked at as a model to be emulated, not controlled or, worse yet, ignored.

8. The relationship between human enterprise and natural processes must be continually improved by the sharing of knowledge. All human participants, be they planners, governmental authorities, manufacturers, suppliers of services, or end users, must exercise ethical responsibility in the pursuit and maintenance of long-term sustainable conditions.

SIX MAJOR AREAS OF SUSTAINABILITY AWARENESS

There are six principal areas into which the awareness of sustainable design considerations can be classified, one being the interior environment, which is the main focus of all facility planning and facility-management endeavors. The other areas—site planning, water supply, energy management, use of materials, and waste disposal—all come into play to a greater or lesser extent according to the nature of a particular set of program requirements.

Site Considerations

A common real estate maxim consists of one word echoed three times: *location*. As originally used, this statement was made strictly in reference to present and future economic value. It has always been prudent to look beyond questions of short-term economics, but questions of sustainability require us to incorporate an additional level of awareness into our thinking.

Mankind is part of an interdependent natural system, and we must develop a respect for the landscape so that the negative consequences of our actions are minimized. The elimination of industrial pollution such as acid rain and other careless side effects of our presence, including soil erosion and groundwater contamination, need to be an integral part of our planning methodologies at all scales.

During the twentieth century, it became almost a habit to abandon existing sites in highly industrialized areas in favor of new development on unspoiled land. This trend needs to be reversed if the ever-decreasing supply of biodiverse natural habitats is to be preserved. Many plant and animal species are becoming extinct on a global scale, and incorporating restoration strategies into our planning efforts becomes more critical as more local ecosystems are increasingly disturbed.

No individual site should be treated without regard to its context. Planning and design must properly address the questions of how to reestablish the connection between developed areas where all traces of nature have been destroyed and the natural organic processes that will help to restore them. Incentives need to be found to make restoration and habitation of previously disturbed areas in the landscape fabric, especially in urban areas, primary goals of new development.

Water Supply

Freshwater is an essential but limited resource, given that over 95 percent of the earth's supply is seawater. Our freshwater supply is entirely dependent on the atmospheric water cycle, powered by the sun. Evaporation purifies water raised from the oceans, which returns to the land as rain and snow. Freshwater reserves are held in natural underground aquifers that are often a great distance from where water is needed. To this extent, the water can be costly and energy intensive to extract while equally slow to replenish.

Globalization of industrialized society brings with it increased population growth, necessary food production, and other types of consumption that increasingly stress the finite sources of natural freshwater. Ironically, cost considerations tend naturally to promote the wise use of water in dry areas. It is in areas where water is relatively plentiful that conscious effort is necessary to promote sustainable use.

As with other areas of sustainable design consciousness, water conservation requires user awareness and education. The average person is unaware, for example, that the flush toilet is the largest single user of water within buildings. Water-conserving flush mechanisms are widely available that limit consumption to a low volume per flush cycle for solids (less than 1.6 gallons) while providing for a partial flush of liquid wastes using even less water.

In like fashion, urinals and lavatories should be automatically operated and have flow rates of from one to two gallons per minute. The flow rate of shower fixtures should not exceed 2.5 gallons per minute. Garbage-disposal units waste water and put excessive loads on waste-treatment facilities. Appliances used in kitchen and laundry areas should be water-saving models that meet sustainable energy standards.

Rainwater from roof surfaces and paved ground surfaces such as parking lots can often be collected to serve secondary uses such as toilet flushing and irrigation. Wells and springs (i.e., groundwater) used for drinking, cooking, and bathing need to be carefully protected from contamination and used only for purposes requiring freshwater to minimize the costs of purification.

Energy Management

One of the lasting consequences of the Industrial Revolution is the ever-expanding use of fossil fuels with the habit of thinking of them as inexhaustible. There seems to be a mental disconnect in regard to acts such as turning on an electric light switch, air conditioning a sealed building, commuting alone by automobile, the coal mining that destroys forests and grazing land, periodic oil spills that foul marine habitats and groundwater, and air pollution from numerous sources.

Although improved technology to some extent has mitigated the increased economic cost of exploration and production of progressively harder-to-obtain oil and natural gas, the environmental cost continues unabated. Unrestrained consumption of fossil fuels continues to produce acid rain, smog, and climate changes at an accelerating rate. A suburban office building can have the most efficient internal energy systems but can hardly be considered *green* when six times its internal area is devoted to asphalt parking filled with cars that arrived with only their driver.

Within an office building, as much as half the total energy load can be attributable to electric lighting. Curiously, demand for artificial lighting traditionally occurs during normal business hours when there is daylight. The efficient use of daylight in conjunction with effective lighting controls has been shown to be capable of reducing consumption by up to 80 percent.° Control is important because the occupant's ability to adjust the local illumination level can enhance her sense of well-being while reducing demand for both lighting and cooling. As anyone who has worked for an extended period of time at a computer monitor can attest, a lower lighting level without glare is highly desirable. Indirect ceiling lighting and task-ambient fixtures in furniture systems can contribute to energy-efficient illumination with a high degree of control.

Walls of glass can lead to extreme overheating unless effective measures provide for shading during the summer months. These include attention to building orientation and the placement of glazing during design, the use of overhangs on south-facing elevations, and skylights with adjustable louvers.

Technologies exist for the effective use of alternative, renewable, primary energy sources such as sun and wind; but their potential availability and the feasibility of use in a particular area must be considered early in the development or redevelopment process. Solar technologies can be as simple (and as passive) as orienting buildings to maximize the positive effects of seasonal variations in intensity and the angle of incidence of sunlight. Sun is usually an asset in cold climates to provide passive heating, but such loads must be actively controlled in warm environments. There are numerous measures that can be taken to mitigate the need for energy inputs if only they are considered as an integral part of the design process. For example:

- Building orientation is a critical parameter, especially in areas where there is intense western exposure late in the day.
- Site features, trellises, and natural vegetation all can be used to provide shading without impeding air circulation and natural light.
- The colors of wall and roofing materials have a substantial effect on the reflection or absorption of solar radiation, while overhangs effectively shade wall surfaces and openings, controlling heat buildup while providing changing visual patterns.
- Shutter systems, louvers, and screens can all provide for active management of solar loads on glass surfaces.

Features that promote wind and natural air movement are assets in warm and humid climates. In cold climates, wind contributes to the rate of heat loss. This can be minimized in such locations by the proper orientation of openings, barriers to prevent infiltration, and the effective use of insulation and building mass.

Materials Utilization

The choice of building materials should be made in the context of a lifecycle analysis so that all the environmental implications are known and can be comparatively evaluated. Estimates of the proportion of municipal waste resulting from demolition and construction activities range as high as 40 percent in some locations. Finding ways to recycle building materials is rapidly becoming as much a matter of economics as a result of environmental concerns as the cost of disposal in landfills continues to rise.

Beginning with each material's source and method of extraction, including necessary energy consumption through all the stages of fabrication, transportation, application, and ultimate disposal or reuse, measurable environmental effects need to be a part of the cost/value equation. Positive attributes of sustainability are that raw materials be renewable, nontoxic, and locally available; that they not require excessive energy in use or generate waste; and that they have the potential for being recycled.

As Vitruvius fittingly observed, the architect should avoid the use of materials "that are not prepared and produced on the spot," because bringing desired materials from a great distance involves great trouble and expense. Assuming their source is sustainable, locally produced materials incur less energy cost and are less polluting due to reduced need of transport in accord with the principle of parsimony. Natural materials require less energy and are cleaner to produce while being less likely to contaminate the indoor environment. Durable materials reduce the need for the fabrication and installation of replacement products while saving on energy costs associated with maintenance.

There are essentially three classes of building materials that can be obtained from renewable sources:

- **Natural materials such as wood, earth, and stone and natural fabrics such as cotton, wool, hemp, and jute**—New lumber should be obtained from sustainably managed sources; and products used in fabrication or installation, such as adhesives, should not contain toxic chemicals or emit volatile organic compounds (VOCs).
- **New materials fabricated from recycled products**—Aluminum from recycled material uses less than one-fourth the energy to produce than virgin metal. Cellulose insulation is more efficient per unit thickness than fiberglass. Care must be taken, however, to avoid energy-intensive salvage processes, toxic components in recycled products, and waste.
- **Man-made materials such as plywood and engineered lumber**—Products using or manufactured with ozone-depleting chlorofluorocarbons (CFCs) and related compounds should be avoided, along with VOCs and other materials such as new aluminum that is environmentally disruptive to obtain and requires great amounts of energy to produce.

Interior Environment

Attention to the question of sustainability in the interior environment is a relatively recent phenomenon. As we have seen, many companies, particularly those operating in environmentally conscious markets, devote considerable effort to the exercise of environmental responsibility under ISO 14000. They have developed the in-house resources and skills to manage sustainable process development guidelines together with strong commitments in this area out of concern for their public image.

In the past, the principal active sectors have included the automotive, communications, electronics, and information technology industries. Corporate responsibility can be a strong driver, especially when combined with the potential for cost savings on energy and materials in manufacture. One source of continuing difficulty, though, even where large companies have attempted to incorporate environmental considerations into their processes, has been the development of markets for remanufactured goods—recycled materials—together with an infrastructure for recycling and recovery. Such issues—together with significant gaps in technology relating to substitute materials and processes, inadequacies in decision-making data support, lack of awareness and training, and the lack of resources in small and medium-size enterprises—continue to present significant obstacles.

Within the energy sector itself, design considerations are largely focused on minimizing the environmental consequences of product plants and processes. In transport, the focus is largely the environmental impact of operations, equipment, and infrastructure. Transport operators are major purchasers of materials and equipment but do not customarily design vehicles. They have, however, a natural incentive to purchase vehicles that are energy efficient, economical to operate, and environmentally friendly. Some financial companies are beginning to actively apply environmental considerations in their investment policies and, like other large users of office space, are bringing an increasing awareness of sustainable design to the development of their offices and other facilities.

Attention has more recently been focused on the construction sector because construction projects and buildings are major users of energy. Building owners and managers are recognizing that active management of facility assets can produce a substantial return on investment, not only during construction but throughout the operating life of buildings as well. The growth of facility management as a discipline and the application of technology to track operating costs and environmental factors are producing an ever-increasing amount of information relating to the true environmental impact of construction procedures and building products. These factors are more frequently being viewed as cost-saving opportunities during and after initial construction as well as at the end of a facility's lifecycle. Included are advance consideration of non-waste-producing construction methods and products together with alternative systems and procedures that promote recycling and adaptive reuse.

Waste Management

It has been estimated that on a per capita basis, U.S. citizens generate more garbage than any other nation on Earth. At the rate of up to five pounds of solid waste per person per day, this adds up to over 200 million tons per year. All methods of disposal affect local ecologies. Incinerating garbage can pollute the air; landfills often contaminate the water; and, in general, there are no methods of waste disposal that are completely without environmental risk.

Many household and commercial products commonly contain toxic materials that can pose serious health risks, such as mercury, cadmium, and lead from batteries; cleaning products; and organic solvents of all kinds. The only sure way to eliminate the environmental effects of waste is to prevent its generation. This can be accomplished in three ways:

- Use products that are nontoxic and durable.
- Reuse materials and products onsite and collect items that are suitable for recycling elsewhere.
- Compost or otherwise process biodegradable wastes.

Eliminating disposable products and their packaging requires policy foresight in purchasing. Buying in bulk and choosing durable, reusable items can greatly reduce solid waste. More than one-fifth of all solid waste consists of plastics that contribute to waste volume and release toxic gases when incinerated. Locally produced goods require less transport and storage and should need less packaging.

When planning for a facility, the availability of potential markets for recycled materials should be determined along with the feasibility of alternative means of dealing with organic waste. On-site systems that utilize captured rainwater or treated wastewater can reduce the need for freshwater input to waste-disposal systems. Composting and anaerobic digestion systems each offer distinct advantages for waste treatment. Composting is a familiar method of processing yard waste and even sewage sludge, although it requires sufficient open space with a permeable surface. Anaerobic digestion is slower in operation but can handle food, animal, and solid human waste. In addition to producing high-quality organic fertilizer, anaerobic systems produce a gas stream similar to natural gas that is rich in energy and can be used as fuel.

Minimizing the use of water; reducing the volume of nonbiodegradable waste products; and implementing simple, reliable, and passive (non-energy-consumptive) technologies for processing waste all can contribute to a high level of sustainability on a continuing basis. This is an area in which innovation is needed; and, of course, conventional backup systems must be provided unless the system is sufficiently redundant that large-scale failure is unlikely.

LEADERSHIP IN ENERGY AND ENVIRONMENTAL DESIGN—LEED

One of the most significant new sources of sustainability guidelines in the United States is a product of the *U.S. Green Building Council (USGBC)* called *LEED°*, for *Leadership in Energy and Environmental Design*. LEED offers an expanding set of consensus-based standards developed by its membership. The LEED mission statement is as follows:

> LEED encourages . . . global adoption of sustainable green building and development practices through the creation and implementation of universally understood and accepted standards, tools and performance criteria.°

The vision of LEED is to bring about a market transformation in the building industry, which is viewed as being one of the principal sources of environmental influence contributing to the exhaustion of natural resources. The USGBC's position is that buildings should be created in harmony with the natural environment and that not to do so represents a continuing threat to our environmental, social, and therefore economic sustainability.

The USGBC actively advocates the refocusing of emphasis by real estate markets, the construction industry, and the design professions toward sustainable design in the interest of promoting economic prosperity, environmental restoration, and health and social welfare. The council believes that on a basis of the green building movement's advances in building technology, integrated design and operating practices, all of which fall within the realm of *facility management,* buildings that endure will be those that prove most flexible in terms of their utilization and compatibility with their surroundings. The LEED green building rating system is intended to provide a mechanism by which sustainable design becomes a default standard in the marketplace, recognizing that green building design embodies *best practices* for the real estate industry on a long-term basis.

LEED Green Building Rating System

The core of LEED operations is directed by a management committee and a steering committee responsible to the USGBC board of directors. Within this structure are several categories of subcommittees for horizontal and vertical markets as well as technical scientific advisory, education, and training.

LEED horizontal-market products address various building types and the stages of building lifecycles. All horizontal-market products are built on a system of credits for meeting specific performance guidelines within the major areas of sustainable design awareness described in the previous section, plus a set of bonus categories for process and design innovation. Each is developed and maintained by its own product committee, working with the technical advisory groups.

To provide recognition for buildings or projects meeting the LEED performance guidelines, a uniform system of certification levels applies to the different standards. This system is based upon the percentage of credits achieved under the cumulative performance guidelines, as follows:

- *Platinum*—80 percent
- *Gold*—60 percent
- *Silver*—50 percent
- *Certified*—40 percent

LEED vertical-market products are developed to address special technical considerations applicable to certain building types in specific use and occupancy classifications. These include educational, factory and industrial, institutional, and mercantile facilities. For example, in health-care facilities, functional process loads from laboratories, specialized diagnostic and treatment equipment, laundries, and waste-handling systems require special consideration. In general, process loads are excluded from the LEED vertical-market products, which address such systems as HVAC, refrigeration, and lighting, assuming there are no extraordinary operational loads or industrial processes that should be optimized.

Several of the LEED horizontal-market products are the following:

- *CI*—Commercial interiors
- *CS*—Core and shell
- *EB*—Existing buildings (operations and maintenance)
- *H*—Homes
- *NC*—New construction
- *ND*—New developments

LEED-CI integrates with LEED-CS to establish sustainable design parameters for developers and tenants. LEED-EB is specific to ongoing facility management as it addresses building operations, maintenance, and retrofitting of in-place building systems. It is currently the one rating system that is intended for ongoing certification of building performance after design and initial construction have been completed. New buildings are generally certified under LEED-NC and must be at least two years old before EB can be applied if they have not previously been certified.

LEED-H for homes and LEED-ND for neighborhood developments principally address the residential aspects of sustainable design. ND emphasizes the smart-growth aspects of site selection and development, focusing on such considerations as density, optimal mixtures of uses and housing types, proximity to public transit, and planning for pedestrian and bicycle access.

In a similar vein, LEED-H is directed toward the home-building industry, promoting the effective use of land, water, energy, and building construction resources, together with materials and practices designed to protect the health of workers and occupants. LEED-H is designed to further the perception by builders that green homes are more profitable, by consumers that they are better built, and by both groups that green upgrades are seen to be beneficial and representative of good value. The incentive is for builders of high-quality homes to differentiate themselves in the marketplace.

LEED for Commercial Interiors—LEED-CI

LEED-CI° applies principally to tenant improvements in new and existing office space. Certification offers opportunities for community recognition, independent validation of environmental achievements, assistance through a variety of government programs, and marketing exposure through USGBC media.

As with the other standards, an applicant project must have its compliance with a sufficient number of points on the LEED-CI specification documented in addition to all of the necessary prerequisites. Registering for LEED certification in the early phases of a project is recommended, as consideration of design options with these specific criteria in mind will maximize the project's potential for certification.

Point values for certification in the various LEED-CI categories are summarized in Table 12-2. Specific category requirements are described in the following sections.

Sustainable Sites
Site Selection
The purpose of this section is to provide incentives to tenants to choose buildings that have implemented sustainable design strategies and systems. Maximum credit toward certification is given if a facility or tenant space is located in a LEED certified building. Otherwise, incremental credit is given if the building meets at least two or more of the following requirements.

Table 12-2 LEED for Commercial Interiors—Categories and Point Values

LEED-CI Categories and Point Values			Target	Points	Maximum Points
Sustainable Sites	Site Selection	LEED Certified Building		3.0	7 12%
		or			
		Brown field Redevelopment		0.5	
		Stormwater Rate and quantity		0.5	
		Stormwater Treatment		0.5	
		Non-Roof Heat Absorption	50%	0.5	
		Roof Heat Absorption	100%	0.5	
		Light Pollution	50%	0.5	
		Water Efficient Irrigation	20%	1.0	
		Wastewater Technology	5%	0.5	
		Water Use Reduction	10%	0.5	
		On Site Renewable Energy		1.0	
		Other Measurable Attributes		0.5	
	Density and Community Connectivity			1.0	
	Public Transportation			1.0	
	Changing Rooms and Bicycle Storage			1.0	
	Automobile Transportation			1.0	
Water Efficiency (Use Reduction)			20%	1.0	2:4%
			30%	2.0	
Energy and Atmosphere	Energy Related System Commissioning				12 21%
	Minimum Energy Efficiency				
	CFC based Refrigerants				
	Lighting Power		15%	1.0	
			25%	2.0	
			35%	3.0	
	Lighting Controls			1.0	
	HVAC		15%	1.0	
			30%	2.0	
	Equipment and Appliances		70%	1.0	
			90%	2.0	
	Enhanced Commissioning			1.0	
	Measurement and Payment Accountability			2.0	
	Green Power		50%	1.0	
Materials and Resources	Storage and Collection of Recyclables				14 25%
	Long-Term Commitment			1.0	
	Building Reuse—Interior		40%	1.0	
			60%	2.0	
	Construction Waste Management		50%	1.0	
			75%	2.0	
	Resource Reuse		5%	1.0	
			10%	2.0	
	Furniture and Furnishings Reuse		30%	1.0	
	Recycled Content Materials		10%	1.0	
			20%	2.0	
	Materials Manufactured Regionally		20%	1.0	
	Materials Extracted and Manufactured Regionally		10%	1.0	

Table 12-2 (*continued*)

	Rapidly Renewable Materials	5%	1.0	
	Certified Wood Products	50%	1.0	
Indoor Environmental Quality	Minimum Indoor Air Quality Performance			
	Environmental Tobacco Smoke			
	Outdoor Air Delivery Monitoring		1.0	
	Increased Ventilation	30%	1.0	
	Construction Indoor Air Quality—During Construction		1.0	
	Construction Indoor Air Quality—Before Occupancy		1.0	
	Low Emitting Materials — Adhesives and Sealants		1.0	
	Low Emitting Materials — Paints and Coatings		1.0	
	Low Emitting Materials — Carpet Systems		1.0	
	Low Emitting Materials — Composite Wood and Laminating Adhesives		1.0	
	Low Emitting Materials — Systems Furniture and Seating	1.0		17 30%
	Indoor Chemicals and Pollutants		1.0	
	Controllability of Environmental Systems—Lighting	90%	1.0	
	Controllability of Systems—Temperature and Ventilation	50%	1.0	
	Thermal Comfort—Compliance		1.0	
	Thermal Comfort—Monitoring		1.0	
	Daylight and Views	75%	1.0	
	Daylight and Views	90%	2.0	
	Daylight and Views—Seated	90%	1.0	
Innovation and Design Process		4.0		5:9%
	LEED Accredited Professional		1.0	
				57:100%

1. *Brownfield redevelopment*—A documented *brownfield* redevelopment site from which previous environmental contamination has been removed.°

2. *Stormwater rate and quantity*—After development, the rate and quantity of stormwater discharge from the site must be maintained or improved through the use of retention ponds, collection of rainwater for productive use, or other means.

3. *Stormwater treatment*—Systems are in place that remove 80 percent of suspended solids and 40 percent of phosphorus on average.

4. *Nonroof heat absorption*—Light-colored building materials or shaded areas from landscaping produce a solar-reflectance index (SRI) of 30 or more, and open-grid pavement constitutes at least 30 percent of all parking areas and other impervious paved surfaces. Alternatively, half of all parking is underground or covered by structure, or open-grid paving is used for half of the exterior parking area.

5. *Roof heat absorption*—Either a high SRI rating for at least three-quarters of the roof surface (78 for low-sloped roofs less than or equal to 2:12, 29 otherwise) or a *vegetated* (green) roof constituting half the total roof surface. The two methods may be combined if the area relationships meet a defined standard.

6. *Light pollution*—Exterior and interior light levels are controlled so that they are below minimum standard recommendations,° and lighting is shielded so that minimal light crosses the property boundaries.

7. *Water-efficient irrigation*—High-efficiency irrigation technology is provided, or a combination of captured rain and recycled site water is used to achieve a 50 percent reduction in the use of drinking water for irrigation. Additional credit is given to buildings that use *no* potable water for ongoing site irrigation or have no permanent landscape watering systems.

8. *Wastewater technology*—Municipal drinking water used for building sewage conveyance is reduced by at least half, or all wastewater is treated on site before entering the municipal sewage system.

9. *Water-use reduction*—Overall water use is reduced by 20 percent with an ongoing plan to ensure compliance by future occupants.

10. *On-site renewable energy*—Between 5 and 10 percent of the building's total energy use is supplied by on-site renewable energy systems.

Additional consideration may be given if the building can demonstrate other quantifiable environmental performance attributes that are recognized by other LEED rating systems.

Density and Community Connectivity
These standards are intended to preserve natural resources and habitats and promote development in urban areas having existing infrastructure. Qualifying buildings will be located in an established community with a minimum density of sixty thousand square feet per acre, typically a two-story business district, or one within a half mile of a residential area with ten units per acre on average. In the latter case, the facility must have access to at least ten basic community or commercial services, such as retail stores, a grocery, bank, post office, school, park, or other such amenities.

Public Transportation
In order to reduce the developmental effects of automobiles and associated pollution, qualifying buildings will be within one-half mile of commuter rail transportation or rapid transit or have immediate access to public bus lines that can be used by building occupants.

Changing Rooms and Bicycle Storage
Qualifying buildings will provide convenient changing rooms, shower facilities, and secure storage for bicycles that can accommodate at least 5 percent of facility occupants.

Automobile Parking
Tenant parking cannot exceed the minimum number required by local zoning regulations, and priority parking for car or van pools must be made available to at least 5 percent of the tenants. Other provisions are that no subsidized or priority parking will be provided for building occupants and that no new parking will be added in conjunction with renovations.

Water Efficiency
The goal of these standards is to reduce the load on water supply and wastewater systems by maximizing the efficiency of their use. Baseline drinking water needs are to be established and demand reduced through the use of high-efficiency fixtures, dry-fixture technologies, and occupant sensors. The *1992 Energy Policy Act* fixture-performance requirements must first be met.

Two levels of compliance are specified, 20 percent and 30 percent, with greater credit given commensurate with the amount of reduction.

Energy and Atmosphere
Prerequisites
There are three prerequisites that must be met before a facility can qualify for credits in this category.

1. *Energy-related systems* must be installed and calibrated, and their performance to specification verified through a *commissioning* process. A designated commissioning authority, not directly responsible for project design or construction management, must document the owner's program and the design of all energy-related systems. A commissioning plan must be developed, its requirements incorporated into the

construction documents, and a report submitted documenting the adherence of energy-consuming systems to the original program and all sustainable design parameters. Minimum systems that must be included are heating, ventilating, air conditioning, and refrigeration (HVAC&R) and their controls, lighting controls, and domestic hot water, as well as renewable energy systems such as solar, wind, and so on.

2. *Minimum energy-efficiency* levels must be established for all tenant space systems and be in compliance with the most stringent energy codes currently in effect.°

3. *CFC-based refrigerants* in HVAC and refrigeration systems must be eliminated, either through the replacement or retrofitting of existing systems or through the installation of new equipment that uses no such refrigerants.

Once these basic prerequisites have been met, credit is given for the optimization of energy performance in several categories as listed here.

Lighting Power
Increasing levels of credit are given for reducing lighting power density 15, 25, or 35 percent below the established industry standards.°

Lighting Controls
Lighting controls must be designed to be daylight responsive, maximizing energy performance in all occupied spaces within a fifteen-foot perimeter adjacent to windows and under skylights.

HVAC
HVAC systems must be state of the art in accordance with the New Building Institute, Inc.'s, *E-Benchmark* criteria and be provided with appropriate zoning and controls (separate zones for different solar exposures, private offices, and special support and interior spaces). Alternatively, increasing levels of credit are given for 15 and 30 percent better performance criteria than the applicable *ASHRAE/IESNA*° standard.

Equipment and Appliances
Office, other electronic equipment, and commercial food-service equipment that is *Energy Star* eligible is given proportionately higher levels of credit for 70 and 90 percent utilization of *Energy Star*–rated appliances.

Enhanced Commissioning
Prior to the development of bid documents (ideally before the start of design), an independent commissioning authority must be designated that will review energy-related design activities and contractor submittals. A manual must be developed that covers the information necessary to recommission energy systems in tenant spaces. Operators and maintenance personnel must be properly trained, and procedures established to resolve commissioning issues following occupancy.

Measurement and Payment Accountability
For tenants occupying less than 75 percent of the building area, negotiate leases whereby tenants pay their own energy costs and provide submetering equipment to record their energy use. For major tenants, continuous-metering equipment must be installed that will facilitate measuring energy use in such areas as lighting systems and controls, cooling loads and heat-recovery cycles, boiler efficiency, indoor water supply, and outdoor irrigation systems. Credit is given on a basis of actual usage measured during building operation that can be compared with predicted energy savings.

Green Power
Credit is given if at least half of the tenant's energy is provided from renewable sources such as low-impact water, wind, solar, or geothermal sources. This may be accomplished by means of a green-power contract with a local utility, procurement from a supplier meeting Green-e renewable power requirements, or through Green-e certified Tradable Renewable Certificates (TRCs).°

Materials and Resources

Storage and collection of recyclables is a prerequisite for gaining credit for material and resource management. A program must be in place and a designated area provided for collecting and processing materials to be recycled, including paper, corrugated cardboard, aluminum cans, glass, plastics, and the like.

Long-Term Commitment
On the premise that remaining in the same location reduces the environmental consequences of tenancy resulting from wasted materials and the need for transport, credit is given for occupants who either own their space or have committed to at least a ten-year lease.

Building Reuse—Interior
Increasing levels of credit are given for 40 and 60 percent levels of reuse of nonstructural interior components. These considerations are typical of this section in that they are aimed at maximizing the useful life of buildings and cultural resources, conserving resources, reducing waste, and reducing the environmental effects of new buildings as a consequence of materials use, manufacturing, and transport.

Construction Waste Management
Differential credit is given for attaining 50 and 75 percent levels in the recycling of demolition, construction, and packaging waste. The objective is to keep such materials out of landfills by finding recyclable uses for cardboard, wood, masonry, concrete, metal, gypsum board, and glass. Construction recyclers and haulers must be designated to handle such materials along with a place on the construction site to carry out necessary staging activities.

Resource Reuse
Differential credit may be obtained by incorporating salvaged materials into the project design, including posts and beams, brick, doors and frames, paneling and cabinetry, as well as decorative items. The intent is to reduce waste and the demand for new materials and must be demonstrably applicable to more than 5 or 10 percent of the building.

Furniture and Furnishings Reuse
With the same intent as the previous section, credit is given for the identification of opportunities to reuse furniture in the project design to the level of at least 30 percent of the budget. Furniture and furnishings include casework, seating, and furniture and filing systems, as well as decorative lighting and accessories.

Recycled Content Materials
To increase demand for building products incorporating material having recycled content, which reduces the environmental costs of extracting and processing new material, credit is given at the 10 and 20 percent levels for ensuring that specified recycled materials are installed and quantities verified. To this end, a project goal must be established and suppliers utilized who can achieve it.

Materials Manufactured Regionally
In order to promote demand for materials extracted and manufactured regionally, that is, within a five-hundred-mile radius, credit is given for a minimum of 20 percent of the combined value of interior construction, furniture, and furnishings. In this case, the distance is determined from the point of assembly, even if the material point of origin is outside the five-hundred-mile radius.

Materials Extracted and Manufactured Regionally
In conjunction with the preceding section, additional credit is given for at least 10 percent of construction, furniture, and furnishings products being extracted, recovered, harvested, *and manufactured* within five hundred miles of the project. The intent of both of these provisions is to support the regional economy and lessen environmental impact due to transportation.

Rapidly Renewable Materials
Rapidly renewable interior construction, furniture, and furnishings products receive credit if they constitute at least 5 percent of all project materials and products. This provides an incentive to diminish the use of raw materials in limited supply or only renewable on a long-term basis. Examples of such products include wool carpeting, linoleum flooring, and cotton batt insulation. A project goal must be established for the use of these materials, suppliers identified who can meet it, and quantities tracked during installation.

Certified Wood Products
Credit is given if at least 50 percent of all new wood-based materials and products are certified° for environmentally responsible forest management, including furniture and furnishings. Quantities must meet a preestablished project goal and be installed by qualified suppliers, and documentation must be maintained during installation.

Indoor Environmental Quality

Prerequisites
Two prerequisites must be met for a facility to qualify for credits in this category:

1. ***Minimum indoor air quality performance*** must be established to ensure the comfort and well-being of all occupants. HVAC systems must be designed to meet the applicable ventilation standards,° and potential site issues must be identified including system constraints that may prevent the outside air requirement of the referenced standard being met.

2. ***Environmental tobacco smoke*** must be controlled and the exposure to interior space occupants, as well as building systems and surfaces, minimized or prevented. This may be accomplished by prohibiting smoking in all spaces except designated areas no closer than twenty-five feet from entrances, operating windows, or ventilation intakes. Alternatively, if interior smoking areas are provided, they must have exhaust air systems directly to the exterior of the building and be negatively pressurized to prevent infiltration to adjoining areas. If residential dwelling units are present, very tight construction is required and core penetrations must be sealed to prevent infiltration of tobacco smoke into adjacent spaces and public areas.

Once these basic prerequisites have been met, credit is given for the monitoring and optimization of air quality performance in several categories. The rationale is that providing such performance incentives will help ensure that systems are designed and maintained for the long-term comfort and well-being of all occupants.

Outdoor Air Delivery Monitoring
Monitoring and alarm systems must be provided to give timely feedback on air quality status so that appropriate operational adjustments can be made. Carbon-dioxide sensors must be present on mechanical ventilation systems for spaces where the occupancy density exceeds forty square feet per person, or within each naturally ventilated space where natural ventilation systems are utilized. Outdoor airflow measuring devices must be provided on mechanical ventilation systems in all other circumstances that are effective at expected operating conditions within 15 percent of the design minimum.

Increased Ventilation
Credit is given for mechanically ventilated spaces if the outdoor air rates to occupied spaces are maintained at a level at least 30 percent above the minimum standard required under the first prerequisite described earlier. Naturally ventilated spaces must meet defined industry recommendations° or be designed on the basis of an analytical computer model that will ensure that at least 90 percent of all occupied spaces are effectively ventilated.

Construction Indoor Air Quality Management
During construction and prior to occupancy, a management plan must be in effect that protects the HVAC system and minimizes the exposure of absorptive materials such as gypsum drywall, insulation, ceiling tile, and carpeting to airborne contaminants and moisture.° If air-handling equipment is used during construction, filtration devices must be placed at each return-air grille.

After construction is complete and all filtration media replaced immediately prior to occupancy, the building air must be flushed out over a two-week period with a prescribed volume of outside air. Alternatively, indoor air quality testing must be carried out using U.S. EPA protocols° for contaminants such as particulates, carbon monoxide, VOCs, and formaldehyde.

Low-Emitting Materials

Incremental credit is given for managing each of several sources of indoor air contaminants that give off strong or irritating odors or are potentially harmful to installers and occupants. These fall principally into five categories:

1. *Adhesives and sealants°*—Low-VOC materials must be specified in construction documents and manufacturer data attesting to emission limits reviewed to ensure that specified limits are met.

2. *Paints and coatings°*—Same as number 1.

3. *Carpet systems°*—Low-VOC carpet products and systems must be specified in construction documents and manufacturer documentation provided to ensure that the specified standards are met.

4. *Composite wood and laminate adhesives*—No urea-formaldehyde resins may be used in composite products, including core materials or field- and shop-applied adhesives. Manufacturer data must attest to the fact that products used meet the specified standards.

5. *Systems furniture and seating*—Credit is given based upon whether the systems used are *Greenguard* air quality certified or, alternatively, whether they have been tested by an independent laboratory and found to be in compliance with the U.S. EPA testing protocols regarding VOCs, formaldehyde, total aldehydes, and 4-phenylcyclohexane.

Reused furniture that is older than one year at the time of occupancy is excluded from these requirements. Systems furniture includes workstations configured using interconnecting modular panels with panel-hung work surfaces and storage components. Certain freestanding types of furniture may be included in this category if they have been designed to function as part of an office system, as are task and guest chairs used with the systems. Other freestanding occasional furniture is excluded from this category.

Indoor Chemicals and Pollutants

Special-use areas where potential contaminants are used must be designed with separate plumbing and exhaust systems so that they are effectively isolated from the remainder of the building. Credit is given where permanent architectural entryway systems are used to inhibit the introduction of contaminants by occupants.

Grilles or grates must be used at exterior entryways to prevent the entry of dirt and other particulates. Hazardous gases must be controlled by maintaining negative air pressure relative to surrounding spaces, with no air recirculation, in physically segregated spaces with partitions to the underside of the structural slab above. Containment drains suitable for hazardous liquid wastes must be provided where such use is anticipated, such as maintenance and laboratory facilities.

Controllability of Environmental Systems

To further productivity and promote the comfort and well-being of building occupants, a high degree of control must be provided for individual occupants and specific kinds of groups in multifunction spaces such as conference rooms and classrooms, specifically as they pertain to the following:

1. *Lighting*—In general, 90 percent of all occupants must be able to adjust for their individual preferences and task needs, as well as those using shared, multi-occupant spaces.

2. *Temperature and ventilation*—Credit is given if 50 percent of all occupants can adjust thermal and ventilation to individual preferences, as well as all shared, multi-occupant spaces. The availability of operable windows satisfies these requirements if they are located within twenty feet of occupants and the windows meet applicable standards.°

Thermal Comfort

Credit is given for compliance with thermal standards that ensure a high degree of comfort for all occupants through the design of HVAC systems and the tenant space envelope.°
Additional credit is provided if a permanent monitoring system is in place and there are means of adjustment necessary to maintain the desired comfort level in conformance with the standard.

Daylight and Views

Increasing levels of credit are given for providing daylight to 75 and 90 percent of spaces. Introduction of daylight and exterior views provides tenant occupants with a sense of connection to the outdoors. Qualifying areas must provide a minimum daylight factor of at least 2 percent excluding direct sunlight penetration, or a minimum of twenty-five foot-candles, demonstrable using a computer simulation or physical model.

Additional credit is given if the tenant space is developed to maximize the opportunities for exterior views by lowering partition heights and using interior glass. Direct line of sight must be provided for 90 percent of the people in all regularly occupied areas.

Innovation and Design Process

Several credits may be given to projects in which exceptional or innovative solutions are demonstrated that perform substantially above the LEED Green Building Rating System standards or in ways not specifically covered by the system. These may include any of the categories we have listed or other measures, such as lifecycle performance of specific materials or assemblies, and the education of tenants and other occupants regarding environmental issues.

Credit is also given when at least one LEED-accredited professional is a principal member of the project team. In this way, the required design integration and certification process can be optimally managed.

LEED for Existing Buildings—LEED-EB

LEED-EB° emphasizes the operations and maintenance of existing buildings to ensure optimal performance from a sustainability standpoint. In some respects, LEED-EB has the most stringent requirements of all the LEED rating systems. Whereas the maximum possible point score for LEED-CI is 57, the top of the LEED-EB scale is at 85 points. Whereas LEED-CI requires a minimum score of 21 to qualify for the lowest level of certification, LEED-NC requires 32 points for a comparable rating. In both cases, the percentage of credits achieved under the cumulative performance guidelines is nominally 40 percent, actually 37 to 38 percent as shown in Table 12-3. The values shown for each level of certification are the minimum necessary to qualify.

Table 12-3   Comparative Point Values of LEED Rating Scales

	CI-Commercial Interiors		NC-New Construction		EB-Existing Buildings	
Sustainable Sites	7	12%	14	20%	14	16%
Water, Efficiency	2	4%	5	7%	6	6%
Energy & Atmosphere	12	21%	17	25%	23	27%
Materials and Resources	14	25%	13	19%	16	19%
Indoor Environmental Quality	17	30%	15	22%	22	26%
Innovation & Design Process	5	9%	5	7%	6	6%
	57	100%	69	100%	87	100%
Platinum	42	74%	62	75%	64	75%
Gold	32	56%	39	57%	48	56%
Silver	27	47%	33	48%	40	47%
Certified	21	37%	26	38%	32	38%

Given that LEED-EB is the only one of the LEED rating systems intended to be used to certify the performance of older buildings, systems retrofits and upgrades may be required in order to qualify for certification. Additionally, many of the rating categories have to do with site and other considerations over which there may be limited control in a strictly interior buildout. Whereas an individual occupant may seek certification within an existing building under LEED-CI, the building as a whole must meet a wider range of qualifications under LEED-EB.

The requirements that follow are partially summarized from the LEED-EB rating system. These criteria are largely unique to existing buildings, and the full EB list includes many of the same green rating factors that were included in the CI standard. Where specific criteria are noted as *required*, they are prerequisites that have no point value but must be satisfied before any points can otherwise be earned in that category.

Sustainable Sites

- *Erosion and sedimentation control (required)*—Prevention of soil loss due to stormwater runoff or wind erosion during construction, protection of topsoil, and prevention of storm sewer and downstream sedimentation. Operation and maintenance activity must be documented to ensure ongoing compliance.
- *Age of building (required)*—Buildings must be at least two years old to be eligible for certification under LEED-EB.
- *Plan for green site and building exterior management*—A low-impact site and green building exterior management plan must be in place that addresses issues such as the following: landscape waste, fertilizer use, irrigation management, and paints and sealants used on the building exterior.
- *High-development-density building and area*—Encourages high-density buildings in high-development-density areas. A qualifying building will have at least sixty thousand square feet per acre within an area having that same average density, typically a two-story urban development.
- *Alternative transportation*—Options include commuter rail or rapid transit within a half mile of the building, two or more public bus lines within a quarter-mile, or an available conveyance system linking the building with public transportation. (This is an increased emphasis over the CI requirement.) Consideration is also given to programs that promote the use of alternative fuel or hybrid vehicles and car pooling, including preferred parking and telecommuting.
- *Reduced site disturbance*—Protection or restoration of open space on the site through the use of native or adapted plants not requiring mowing or substantial amounts of irrigation water. Program must provide for documentation of quarterly inspections.
- *Stormwater management*—Mitigation of stormwater runoff by capturing rainwater for functional use or other viable measures such as increasing the absorption capacity of the soil, providing retention basins or other biologically based and innovative stormwater-management programs.

Water Efficiency

- *Minimum efficiency (required)*—Reduction of potable water usage to within 20 percent of the level that would be attained if all building plumbing fixtures were to meet the performance requirements for automatic water control systems as specified in the 1992 Energy Policy Act (baseline value). Documentation must be provided as to water use per occupant and on a square footage basis.
- *Discharge water compliance (required)*—If subject to the EPA National Pollution Elimination System Clean Water Act, the facility must demonstrate compliance with NPDES requirements for grease traps, oil separators, and other filtration devices to ensure the proper disposal of all building discharges.
- *Use reduction*—Achieve a 10 to 20 percent reduction of potable water usage below the baseline value required under minimum efficiency, as described earlier.

Energy and Atmosphere

- *Optimize energy performance*—From one to ten points may be given on a basis of the EPA Energy Star score achieved by the building. Starting with an Energy Star score of 63, an additional LEED-EB point is earned for each four points added to the EPA score up to 99.

- *On/Off-site renewable energy*—From one to four points may be given according to the rate of green power utilization, which may be produced on site or obtained from a Green-e-certified power marketer or utility program or in the form of Green-e Tradable Renewable Certificates.
- *Operations and Maintenance: Education, Maintenance, Monitoring*—Educational programs must provide at least twenty-four hours of training per year for each person engaged in building systems operation and maintenance. A documented preventive maintenance program must be in place to deliver post-warranty equipment maintenance. Continuous monitoring of systems that regulate indoor comfort and air quality must be provided, including alerts for conditions requiring adjustment or repair.
- *Additional ozone protection*—Base building HVAC, refrigeration, and fire-suppression systems using *HCFCs* or *Halon* must be phased out or emissions reduced to levels conforming to the EPA Clean Air Act.
- *Performance measurement: metering and reduction reporting*—From one to three points may be earned for performance measurement that will demonstrate continuing optimization of building energy and water consumption, such as electric and natural gas metering, tracking of heating and cooling loads, irrigation systems, ventilation distribution, and air volumes. Another point may be gained from additional measures taken to reduce significant pollutants.
- *Document sustainable building costs*—Maintain records of aggregate building operating costs over the previous five years (or the life of the building) to document the financial impact of implementing LEED-EB on a continuing basis.

Materials and Resources

- *Waste stream audit (required)*—Develop a program for reducing the amount of waste produced by the building, beginning with an analysis of the current waste stream to identify reduction options through reuse, recycling, management, and procurement policies.
- *Toxic material source reduction (required)*—Reduce the amount of mercury introduced into the building (less than one hundred picograms per lumen hour) through the selective purchasing of lightbulbs.
- *Alternative materials*—Purchase supplies and equipment that meet specified sustainability criteria. From one to five points may be given for each 10 percent of total purchases that exceed criteria such as containing at least 70 percent salvaged material from on or off site, 50 percent rapidly renewable or materials processed within five hundred miles of the site.
- *IAQ-compliant products*—Up to two points may be earned for each 45 percent of annual purchases made under a documented program to optimize the use of air quality compliant materials within the building and elsewhere on site. Categories included are adhesives, sealants, paints, and coatings with low VOC content, and carpet and cushioning meeting CRI Green Label Testing Program requirements.
- *Sustainable cleaning products and materials*—Up to three points may be earned for each 30 percent of annual cleaning and disposable janitorial product purchases that can be documented as meeting applicable sustainability guidelines.
- *Occupant recycling*—Up to three points may be earned for the diversion from landfill disposal or recycling of 30 to 50 percent of the building's total waste stream. Included are materials such as paper, cardboard, glass, plastics, fluorescent lamps, and batteries. Architectural wall panels may be included as recycled components of the waste stream each time they are relocated and reinstalled.
- *Additional toxic material source reduction*—Minimize the amount of mercury introduced into the building (less than eighty picograms per lumen hour) by all light bulbs containing the element.

Indoor Environmental Quality

- *Asbestos abatement: removal or encapsulation (required)*—The building must have an asbestos-management program in place and all applicable regulatory requirements must be identified. A current survey locating all asbestos-containing materials must be maintained as a basis for carrying out necessary abatement activities on an ongoing basis.

- *PCB removal (required)*—The building must have a documented management program in place for abating PCBs and combustion by-products in the event of fire.
- *Documenting productivity impact: absenteeism, health care, other*—Maintain records of absenteeism and health care expenditures for building occupants over the previous five years (or the life of the building) to document the financial impact of implementing LEED-EB on a continuing basis.
- *Indoor chemical and pollutant source control: particulates, high-volume printing/copying*—Provide and maintain outside air intake and return filters to ensure a specified level of particle-removal effectiveness. Isolate high-volume reproduction rooms with slab-to-slab partitioning and separate ventilation systems.
- *Green cleaning: entryways, janitorial closets, policy, pests, equipment*—Reduce the exposure of occupants and maintenance staff to hazardous contaminants: dirt, dust, pollen, and particulates at entryways (one point); isolation of janitorial facilities (one point); use of sustainable cleaning systems, equipment, and products (two points); environmentally low-impact pest management (two points).

Innovation in Upgrades, Operations, and Maintenance

As LEED-CI provides credits for exceptional or innovative sustainable design solutions, LEED-EB specifies comparable incentive points for documented actions resulting in the delivery of additional environmental benefits over and above those established by the LEED rating system. Up to four points may be awarded in this category with an additional point available for having a LEED-accredited professional as a principal member of the building's project team.

LEED for New Construction—LEED-NC

LEED-NC° was the first in the growing series of LEED rating systems, first published in 1999. A second version was issued in June 2001. LEED-EB was initially published in 2002, with Version 2 following in October 2004. The first edition of LEED-CI appeared in November 2004.

Many of the categories listed in the LEED-NC Table of Contents are the same as or very similar to those in the other two rating systems, in some cases with variations in the number of points given for compliance or in the level of compliance required to gain a specific number of points. The final category of LEED-NC, for instance, provides the same four points for innovation in design, plus a fifth for having an accredited LEED professional integrally involved in the project team. The following sections highlight those items particularly applicable to new building construction, which obviously offers opportunities for implementing green technologies that existing facilities and interior buildouts may not.

Sustainable Sites

- *Site selection*—New buildings may not be developed on prime farmland, land that provides a habitat of any endangered species, land within one hundred feet of any wetland, where the elevation is less than five feet above one hundred-year flood levels, or where such development results in the loss of public parkland.
- *Urban redevelopment and brownfields*—Preference is given to developing in urban areas with existing infrastructure, thereby preserving natural resources and habitats. Brownfield redevelopment is particularly encouraged with remediation of toxic conditions in accordance with EPA sustainable redevelopment requirements.
- *Alternative transportation*—Up to four points may be earned if the facility is located within specified distances of public transportation, secured bicycle storage and changing facilities are provided for at least 5 percent of building occupants, alternative-fuel refueling stations are provided for at least 3 percent of parking capacity, and preferred parking for car and van pools are provided for 5 percent of the building occupants.

- *Reduced site disturbance*—The building must be designed with a minimal site footprint to minimize disruption. Earthwork and clearing of vegetation may only be done within forty feet of the building perimeter, five feet of primary roadways, and twenty-five feet beyond other paved areas. Alternatively, credit may be given if at least half of the remaining open area is restored with natural vegetation. Additional credit is given if the total development footprint exceeds local zoning open space requirements by 25 percent.
- *Stormwater management*—The development must produce no increase in the rate and quantity of stormwater runoff to adjacent areas. If permeable land areas are less than 50 percent of the site, stormwater must be managed in such a way as to produce a 25 percent reduction of runoff. Additional credit may be given for implementing treatment systems that will remove at least 80 percent of all suspended solids and 40 percent of the average annual amount of phosphorus.

Water Efficiency

- *Water-efficient landscaping*—Implement high-efficiency irrigation technology, captured rainwater, or recycled site water to reduce consumption of drinking water by half or, for additional credit, entirely eliminate the use of potable water for irrigation.
- *Innovative wastewater technologies*—Reduce the consumption of drinking water from the municipal supply by at least half through the use of alternative waste disposal systems, or treat all wastewater on site to tertiary standards.

Energy and Atmosphere

- *Optimize energy performance*—From two to ten points may be earned by using regulated energy components° to reduce design energy cost by from 20 to 60 percent in new buildings (10 to 50 percent in existing buildings).
- *Renewable energy*—From one to three points may be earned through the implementation of self-supplied renewable technologies such as solar, wind, hydro, geothermal, biomass, and biogas expressed as a percentage of annual energy cost in the range of 5 to 20 percent.
- *Ozone depletion*—Base building HVAC, refrigeration, and fire-suppression systems may not utilize HCFCs or Halon.

Materials and Resources

- *Building reuse*—One to two points may be given on a basis of the amount of existing building structure and shell reused during renovation or redevelopment: one point for 75 percent, two points for 100 percent. A third point may be earned if at least 50 percent of nonshell components are also reused (walls, ceiling systems, and floor coverings).
- *Recycled content*—In order to increase demand for building products incorporating recycled materials, one to two points may be earned for specifying 25 to 50 percent of building materials containing an average of 20 percent postconsumer or 40 percent postindustrial recycled content material.

CLOSING WORDS

Chapter 12 and the next four chapters address several codes and guidelines currently in effect. These are key components of the framework in which all space planning must be undertaken.

Modern building, zoning, and fire codes have been developed over the course of the last century primarily for the purpose of ensuring public safety and establishing local authority over real estate usage. *Accessibility* has always been a focus of fire codes; but in more recent years the word has taken on a new level of meaning, emphasizing the usability of facilities by people with disabilities in everyday situations.

This chapter began by tracing some of the significant events in the rise of global environmentalism during this same period. It is not a coincidence that the same cyclical model that describes *total quality management* (see Figure 5-12) underlies the picture of sustainable design and construction. Operating on a global scale, it was the International Standards Organization that established defined quality benchmarks in 1987 (ISO 9000), followed by a set of environmental standards in 1996 (ISO 14000). Following a discussion of the principles of sustainability, this chapter considered the developing LEED guidelines. All of these are the concern of architects, space planners, and facility managers.

The subject of building codes covers a large area. From the time of Hammurabi to the great fire of London in 1666, to Chicago's Iroquois Theater fire of 1903, structural integrity and fire protection have been the principal concerns of building codes and regulations. It would be impossible to cover, even in a volume entirely devoted to the subject, all the important building and fire codes currently in effect. Thus, we have chosen to use the recently published *International Building Code* (2003) as a model.

The *IBC* consolidates three major codes, described in Chapter 13, that establish, on a "prescriptive and performance-related" basis, a uniform set of building standards intended to be adopted and used by local jurisdictions on a worldwide basis. Beginning with a section briefly describing the overall structure of the model code, the next chapter covers use and occupancy classifications, types of construction, fire-resistive construction, and active fire-protection systems.

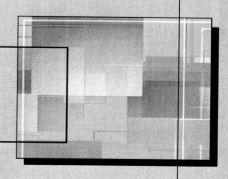

CHAPTER 13

Building Codes

*If a builder constructed a house for a seignior, but did not make his work strong,
with the result that the house which he built collapsed and so has caused the
death of the owner of the house, that builder shall be put to death. . . .*
—*The Code of Hammurabi (c.1700 BCE)*°

As the epigraph suggests, building codes have been with us for some time. Generally regarded as the first to have been put in writing, the preceding example is nearly four thousand years old and has, without doubt, the steepest penalty for noncompliance!

Some things do not change a great deal, however, and the primary concerns of all building codes today are the soundness of the structure and the safety of its occupants. In addition to the general building codes, there are many specialized standards governing mechanical, plumbing, and electrical systems; fire safety; and accessibility.

Most contemporary codes specify levels of performance to be achieved rather than the specific materials and construction techniques to be employed. Fire resistance, for instance, is generally stated in terms of the length of time a wall must stand without failing under certain conditions rather than that such and such a material of such and such a thickness must be used. There are many detailed requirements to be sure, but the intent is to leave it to the ingenuity of the architect, designer, and builder to meet the required performance levels in a functional, economical, and well-designed structure.

The space planner and facilities professional must have a working knowledge of the codes governing the construction and use of buildings in the jurisdiction within which they work. Although many codes are national in scope, they are almost always administered locally; so compliance necessitates dealing with local officials to get drawings approved, secure the necessary permits, and pass required inspections. Many large cities have their own code documents, which often have more stringent requirements than the national codes. In those cases, the levels of performance specified by the local regulations need to be thoroughly understood and followed.

This chapter gives you an overview of the typical building code structure, using the recently compiled *International Building Code* as a model. Virtually all codes today follow a uniform format developed by the Council of American Building Officials (CABO). The *IBC*, which was first drafted in 1997 by the International Code Council (ICC), Inc., is largely a consolidation of three major code documents currently in use throughout most of the United States. The Building Officials and Code Administrators (BOCA) International, Inc., code was (and still is) used extensively in the northern part of the country east of the Mississippi River. The Southern Building Code Congress International (SBCCI), Inc., published the *Standard Building Code* (*SBC*), used largely in the southeast. The third, the *Uniform Building Code* (*UBC*), published by the International Conference of Building Officials (ICBO), was used in the west.

The *IBC*, first published in its present form in 2000, "establishes minimum regulations for building systems using prescriptive and performance-related provisions."° Through its various committees, the International Code Council provides a forum for an ongoing model code development process. New editions of the *IBC* will be published every three years, as was the case with its predecessors, which will ultimately be phased out. As the new model code is adopted and used by local jurisdictions on an international basis, it is intended that the public health, safety, and welfare will be better protected through the consistent application of its provisions.

Although specific requirements differ, the content of most current codes is similar, covering the following subject areas of particular interest to the space planner and designer:

- Classification by Intended Use and/or Occupancy (300–400)
- Classification by Construction Type, Height/Area Limitations (500–600)
- Fire-Resistance Ratings and Fire-Protection Systems (700 and 900)
- Interior Finishes and Environment (800 and 1200)
- Means of Egress and Accessibility (1000–1100)

The numbers in parentheses indicate major sections of the *IBC*. Most of these are discussed throughout the remainder of this chapter. Sections 1000 and 1100, on Egress and Accessibility, are covered in Chapter 14 and Chapter 15. The *Americans with Disabilities Act (ADA)* of 1990 is the subject of the final chapter.

The first two sections of the *IBC* deal with how the *Code* is to be administered (100) and some necessary definitions (200). Beginning with Section 13 and continuing through the remaining 60 percent of the volume, the *IBC* spells out numerous detailed requirements regarding permitted materials, design parameters, and other special considerations:

- Energy Efficiency (1300)
- Exterior Walls, Roof Assemblies, and Rooftop Structures (1400–1500)
- Structural Design, Tests, and Special Inspections (1600–1700)
- Soils and Foundations (1800)
- Concrete, Aluminum, Masonry, Steel, Wood (1900–2300)
- Glass and Glazing (2400)
- Gypsum Board and Plaster, Plastic (2500–2600)
- Electrical, Mechanical, and Plumbing Systems (2700–2900)
- Elevators and Conveying Systems (3000)
- Special Construction (3100)
- Encroachments into the Public Right-of-Way (3200)
- Safeguards during Construction (3300)
- Existing Structures (3400)
- Referenced Standards: Appendices and Indices (3500)

Most of the material covered in these later sections is outside the scope of this book and is not discussed in detail here. Required levels of performance for permitted materials are spelled out along with guidelines for the correct preparation and installation. Seismic design requirements may be applicable depending on the building's location and are covered in sections 1600 through 2300. Emergency and standby power systems are included under electrical requirements, and plumbing facilities are detailed to the extent of the number of fixtures for each sex, location, and travel distance. Otherwise, reference is made to the ICC *Electrical Code, International Mechanical Code*, and *International Plumbing Code*.

Once again, the organization of the *IBC* is representative of the other codes; however, the emphasis on many details differs from one to another. You need to be aware of how the specifics of the local codes in your city or municipality may affect your planning, particularly if your facility has features or conditions that may be viewed as unusual.

USE AND OCCUPANCY CLASSIFICATION

The first way that most building codes classify structures or portions of structures is use and occupancy. Standard classifications are determined by fire safety considerations and the relative hazards of the intended use. If a particular building type does not fit into one of the standard classifications, the nearest, most stringent classification is usually required.

There is some variation in terminology from one code to another in the specific designations; but, in general, there are ten classifications:

- Assembly (303)
- Business (304)
- Educational (305)
- Factory and Industrial (306)
- High Hazard (307)
- Institutional (308)
- Mercantile (309)
- Residential (310)
- Storage (311)
- Utility and Miscellaneous (312)

Many buildings have multiple uses and, therefore, multiple occupancy classifications. Codes generally permit this, with the notable exception of certain *high-hazard* uses, to the degree that the different use spaces are separated by appropriate fire barriers. These may consist of appropriately rated fire walls, fire partitions, smoke barriers, or horizontal assemblies.

- *Fire walls* must have adequate structural stability to withstand the collapse of construction on either side, without itself collapsing, for the length of time specified by its fire-resistance rating (705).
- *Fire partitions* are wall assemblies having a one-hour fire-resistance rating, constructed of materials permitted by the particular building type (708).
- *Smoke barriers* provide a one-hour fire-rated, continuous membrane separating spaces (709).
- *Horizontal assemblies* are floor and roof assemblies separating occupancies, having an appropriate fire-resistance rating for the types of occupancies separated (711).

In some circumstances, different *nonseparated* uses may be permitted. The classification of each portion of the building must be determined; and in such cases, the entire building's construction must meet the requirements of the most restrictive category. Other requirements may apply to various portions of the building based on use, except that the most restrictive fire-protection system and high-rise building requirements apply (302).

Table 13-1 shows the required separations between different occupancy classifications, in hours, for buildings without an automatic sprinkler system. The separations may be reduced by one hour (but not to less than one hour) for buildings equipped with sprinklers, except for *Group H* and category *I-2* areas. Note particularly that *H-1* occupancy is not permitted to coexist with any other occupancy within the same structure.

Numerous special circumstances and exceptions are spelled out in different codes. For example, in private residences, the *IBC* specifies that a garage must be separated from the residence and its attic by a minimum of one layer of $\frac{1}{2}$-inch gypsum board on the garage side. Doors into the residence must be solid core wood or honeycomb core steel, 1 and $\frac{3}{8}$ inches thick. In no case can openings from the garage open directly into a sleeping area (302.3.3).

Following is an inventory of the ten groups and subcategories with examples of each. The examples listed here are by no means exhaustive but are intended to give a sense of the kinds of spaces included. Applicable codes for your locale should be consulted when analyzing and classifying occupancies, at least until the ICC codes are universally adopted.

Assembly

The *Assembly* classification includes buildings or portions of buildings where people gather together. Its common denominator is a high density of people relative to floor area. Rooms used for assembly purposes by fewer than fifty people and adjunct to another use classification normally are not considered Assembly but are part of that occupancy.

Table 13-1   Required Occupancy Separation in Hours°

Category	Code	A-1	A-2	A-3	A-4	A-5	B	E	F-1	F-2	H-1	H-2	H-3	H-4	H-5	I-1	I-2	I-3	I-4	M	R-1	R-2	R-3,4	S-1	S-2	U
Assembly	A-1		2	2	2	2	2	2	2	3	2	2	1	1	4	4	2	1	2	2	2	2	2	3	2	1
Assembly	A-2			2	2	2	3	3	N	N	1	N	2	4	4	4	3	2	2	2	2	3	2	2	1	
Assembly	A-3				2	3	2	2	4	2	2	1	N	2	4	3	1	2	2	2	3	2	1	1		
Assembly	A-4					2	N	N	3	3	1	1	2	3	2	N	1	2	4	3	2	1				
Assembly	A-5						N	4	2	2	2	3	3	2	3	2	1	4	3	3	2					
Business	B							4	4	4	2	2	2	2	2	N	1	3	1	2						
Educational	E								3	2	2	2	2	3	3	2	1	1								
Factory & Industrial	F-1									2	2	2	2	2	2	N	1									
Factory & Industrial	F-2										2	2	2	2	2	2	N									
High Hazard	H-1											2	2	2	2	2										
High Hazard	H-2												2	2	2	2										
High Hazard	H-3													3	3	3										
High Hazard	H-4														2	2										
High Hazard	H-5															1										
Institutional	I-1																									
Institutional	I-2																									
Institutional	I-3																									
Institutional	I-4																									
Mercantile	M																									
Residential	R-1																									
Residential	R-2																									
Residential	R-3,4																									
Storage	S-1																									
Storage	S-2																									
Utility & Misc.	U																									

A-1 *Performing arts facilities*, usually with fixed seating, including theaters, movie theaters, and television and radio studios where there are live audiences.

A-2 *Food and drink establishments*: Restaurants, banquet halls, night clubs, taverns, and bars.

A-3 *Amusement, recreation, and worship facilities* not elsewhere classified in *Group A*. These include art galleries and exhibition halls, churches, courtrooms, indoor swimming pools and tennis courts, libraries and museums, transportation waiting areas, and so on.

A-4   Indoor sports facilities with spectator seating, such as arenas, skating rinks, swimming pools, and tennis courts.

A-5   Outdoor facilities for viewing or participating in activities such as amusement park structures, bleachers, grandstands, and stadiums.

Business

B   The *Business* group includes structures used, in whole or in part, for office, professional, or other service-oriented activities, including record storage and computer facilities. Examples include, in addition to corporate, professional, and governmental offices, banks, fire and police stations, research and testing laboratories, post offices, and television stations. The Business classification also includes airport traffic control towers as well as postsecondary educational facilities (above the twelfth grade).

Educational

E   The *Educational* classification includes any structure or portion thereof used for educational purposes through the twelfth grade by six or more people at any given time. This group also includes day-care centers for six or more children older than $2\frac{1}{2}$ years of age.

Factory and Industrial

The *Factory and Industrial* group includes facilities for assembling or disassembling, fabricating, manufacturing, finishing, packaging, repair or processing operations not included in *Group H.*

F-1   Moderate-Hazard: Aircraft, motor vehicles, business machines, dry cleaning and laundries, electric power plants, food processing, millwork and woodworking shops, printing and publishing, refuse incineration, textiles, and so on.

F-2   Low-Hazard occupancies are those involving the manufacture or fabrication of noncombustible materials presenting a minimal fire hazard. Examples are nonalcoholic beverages and ice, along with ceramic, glass, and metal products.

High Hazard

The *High-Hazard* classification includes facilities used for the fabrication, generation, processing, or storage of materials constituting a physical or health hazard in significant quantities. Extensive definitions and tables showing allowable quantities of specific types of materials are spelled out in detail by the *IBC.* Sections dealing with mass-detonating explosives and materials qualifying for the *H-1* designation were added or expanded in the 2003 edition of the *Code.*

H-1   Buildings and structures containing *materials that pose a detonation hazard:* Explosives, oxidizers, unstable reactive materials, and detonable pyrophorics (materials that are spontaneously combustible when exposed to air). This is the only classification that cannot be part of a mixed-occupancy building.

H-2   Materials that present a deflagration hazard, or a hazard from accelerated burning, such as flammable or combustible liquids and gases, combustible dusts, nondetonable pyrophoric and unstable substances, as well as water-reactive materials.

H-3   Materials that readily support combustion or pose a physical hazard. Examples include combustible fibers, flammable solids, oxidizing cryogenic liquids, oxidizing gases, and other oxidizers.

H-4   Materials that are intrinsically health hazards, including corrosives and materials classed as toxic or highly toxic.

H-5   Semiconductor fabrication plants and similar research and development facilities where hazardous production materials are used or stored in significant quantities.

Where hazards exist in multiple groups within the same structure, the facility must conform to the code requirements pertaining to each of the occupancies so classified.

There are situations in which certain hazardous materials in limited, specified quantities may be utilized in buildings or portions thereof, and they are not classified as *Group H*. In such cases, particular care must be taken to check applicable codes in a particular jurisdiction, such as the *IBC*, the *International Mechanical Code*, the *International Fire Code*, and the National Fire Protection Association (NFPA) codes.

Institutional

The *Institutional* group applies to uses wherein the occupants have physical limitations due to age or health or wherein they are detained for correctional or penal purposes and their liberties are therefore restricted.

I-1   *Supervised residential environments* housing more than sixteen people on a twenty-four-hour basis due to age or disability. The occupants are assumed to be familiar with their surroundings and capable of responding to an emergency without physical assistance. This group includes assisted-living facilities, group homes, alcohol and drug centers, and social rehabilitation and convalescent facilities.

I-2   *Medical, surgical, psychiatric, or other custodial care* where there are six or more people on a twenty-four-hour basis who are not capable of self-preservation. Included in this group are hospitals, nursing homes, mental health centers, detoxification facilities, and full-time child-care centers housing children younger than 2½ years. Such facilities housing fewer than six people are classed as *Group R-3*.

I-3   *Secured facilities* housing six or more individuals. Occupants are presumed to be incapable of self-preservation due to security measures or restraints not under their control. This group includes prisons, reformatories, and detention and correctional centers.

I-4   *Day-care facilities* for people of any age receiving custodial care for less than twenty-four hours, provided by persons other than family members not in the person's own home. This group also includes part-time child-care facilities for six or more children under 2½ years of age, unless all rooms exit directly to the exterior. In this special case, a day-care center accommodating no more than one hundred children may be classified as *Group E*.

Mercantile

M   The *Mercantile* classification includes buildings, or portions thereof, that are accessible to the public for the display and sale of merchandise. Storage space for stocks of goods and merchandise incidental to these activities are included in this group. Examples include wholesale or retail stores and sales rooms, department and drug stores, and automotive service stations.

Residential

The *Residential* group consists of buildings or portions of structures in which sleeping accommodations are provided, not otherwise classed as *Group I*.

R-1   *Transient occupancies* (less than thirty days) such as hotels, motels, and transient boarding houses.

R-2   *Permanent occupancies* in which there are more than two dwelling units. Included are apartment buildings, dormitories, and similar structures.

R-3   *Permanent single or duplex* occupancies including adult and child-care facilities housing fewer than six people for less than twenty-four hours.

R-4   *Residential care and assisted-living facilities* housing six to sixteen occupants excluding staff.

Storage

Storage occupancy is intended for facilities in which nonhazardous materials are stored.

S-1   *Moderate-Hazard:* Aircraft, books, paper, cardboard and cardboard boxes, clothing, furniture, lumber, soaps, sugar, tobacco products, and wax candles.

S-2   *Low-Hazard* storage includes noncombustible materials such as beer and wine, electric motors, food products, glass, gypsum board, meats, metals and metal parts, parking garages, and washers and dryers.

Utility and Miscellaneous

U   Utility structures not belonging to any specific occupancy classification, generally of an accessory nature. Examples include barns, private garages, carports, greenhouses, fences higher than six feet, and retaining walls. The *IBC* specifies that they be constructed to meet code requirements appropriate to the fire and life-safety hazards they present.

Special Building Types

In addition to the basic occupancy and construction requirements specified by the *International Building Code*, there are numerous other special detailed requirements for certain types of uses and occupancies. The following examples are not intended to be an exhaustive list. Code documents applicable to your particular jurisdiction and project type should be consulted for pertinent details.

High-Rise (403)

High-rise buildings are considered to be those in excess of seventy-five feet, which is the maximum height usually accessible by fire department equipment. Construction must be of noncombustible materials and (new) structures equipped with an automatic sprinkler system having a secondary water supply (Section 903).

Exceptions are allowed to the sprinkler requirement in areas used for telecommunications equipment and open parking, provided that such areas are equipped with an automatic fire-detection system. Such facilities must be separated from other areas of the building by one-hour fire-resistive walls and two-hour-rated floor/ceiling assemblies.

Many other requirements are set forth in the *IBC* to maximize life safety in the event of emergency. Included are the following:

- Automatic fire and smoke-detection system
- Emergency voice/alarm communication system
- Fire command center
- Fire department communication system
- Standby power, light, and emergency systems
- Automatic stairway door operation and stairway communications

Atriums (404)

Atriums (or *atria*) are openings through two or more floor levels other than enclosed stairways, elevators, escalators, or other vertical penetrations that are closed at the top and not classified as a mall. (A *mall* is a specific type of atrium within a covered mall building that provides pedestrian access for two or more tenants and does not exceed three floor levels.)

Atrium floors may only be used for nonhazardous activities. Materials and decorations used must meet all applicable fire codes. In general, the *Code* requires that a building containing an atrium be sprinklered throughout its entirety, with certain exceptions. Areas adjacent to an atrium need not be sprinklered if separated by a two-hour fire barrier.

Automatic fire-detection systems including smoke detectors are required. Standby power is also required for any necessary smoke-control equipment.

Other than in the lowest level where the atrium provides the means of egress, exit travel distance within the atrium is limited to two hundred feet. The specifics of egress and exit access requirements are discussed more fully in the next two chapters.

Covered Mall Buildings (402)

A *covered mall building*, by definition, cannot exceed three floor levels at any point nor have more than three stories above grade. It is a single building housing multiple tenants and other occupants including retail stores, restaurants, entertainment facilities, and offices. Many of the tenants have a main entrance into one or more interior malls.

Anchor buildings are exterior perimeter structures that have direct access to the covered mall. They are also required to have means of egress independent of the mall. *Food courts* are public seating areas within the mall that support immediately adjacent food service tenants.

Occupant loads determine the required means of egress from the mall and are based on the mall's *gross leasable area*, exclusive of the anchor buildings. Gross leasable area is the total floor area designed for exclusive tenant occupancy and use, measured from the outside of the tenant walls to the centerlines of joint (demising) partitions. These measurements are represented on a *lease plan*, which the building owner must provide to the building and fire departments. The lease plan delineates the location of each occupant and its required exits.

Automatic sprinkler, standpipe, and smoke-control systems are required throughout the mall and all tenant spaces, occupied or not. In addition, sprinkler protection for the mall is required to be separate from that protecting tenant spaces or anchors. Standby power and emergency voice/alarm communication systems must be present in mall facilities exceeding fifty thousand square feet.

The *IBC* specifies numerous special requirements related to means of egress and exit distances. For example, if the travel distance to the mall from a tenant space accessible to people (other than employees) exceeds seventy-five feet, or the occupant load is greater than fifty, at least two means of egress must be provided. An assembly occupancy having an occupant load greater than five hundred must be located immediately adjacent to a principal mall entrance so that at least one-half its occupants exit directly out of the building. Required egress for anchor buildings must be independent of the mall; and, conversely, mall egress is not permitted through anchor buildings.

The minimum width of a mall must be at least twenty feet—wider if necessary to adequately serve its occupant load. The minimum required width of exit passageways or corridors leading out of a mall is five feet six inches. The maximum travel distance from any point within a tenant space to an outside exit or to the mall cannot exceed two hundred feet. The maximum travel distance to an exit from within the mall also cannot exceed two hundred feet.

Institutional

Medical, Surgical, Psychiatric, or other Custodial Care (407)

Group *I-2* facilities have special corridor requirements. Corridors must be continuous to exits and generally separated from other areas.

Smoke barriers must be provided that divide every floor used for patient sleeping or treatment facilities, or other floors with an occupant load of at least fifty people, into a minimum of two smoke compartments. Smoke compartments are permitted a maximum area of 22,500 square feet, and the travel distance from any point within the compartment to a smoke barrier door cannot exceed two hundred feet.

Smoke compartments containing patient sleeping rooms must be equipped with automatic sprinkler systems throughout. Independent means of egress must be provided from each smoke compartment without a person having to return through the compartment from which he started. Adequate space must be provided for all occupants to take refuge on either side of each smoke barrier in the event of emergency: thirty square feet per person on floors where patients are confined to beds or litters, and six square feet per person otherwise.

Waiting rooms and similar *unlimited* areas may be open to a corridor except when the areas are used for patient sleeping or treatment rooms, hazardous or other incidental uses having specific separation requirements. Such open spaces must be protected by an automatic fire-detection system. The corridor into which the spaces open must be protected as well, either by an automatic fire-detection system or quick-response sprinklers within the same smoke compartment. The spaces must be planned to ensure that access to the required exits is not obstructed.

Nursing stations and related clerical areas may not be open to corridors except where they are constructed according to corridor requirements. Mental health treatment areas housing people not capable of self-preservation cannot open to the corridor except as follows:

- The area is located to permit supervision by facility staff.
- The area is planned so as not to obstruct exits.
- The area does not exceed 1,500 square feet.
- The area is equipped with an automatic fire-detection system.
- There is not more than one such space in any smoke compartment.
- Walls and ceilings of the space are constructed according to corridor requirements.

Corridor walls must be constructed of materials consistent with the overall building's construction and must extend continuously from the floor to the underside of the floor, roof deck, or ceiling membrane above in order to form a barrier limiting the transfer of smoke. Doors must provide an effective barrier to smoke and be equipped with positive latching. Locking devices must not restrict egress from patient rooms except in mental health facilities.

Secured Detention and Correctional Facilities (408)
Group I-3 facilities also have special requirements with respect to egress. Provisions need to be made for the release of occupants in the event of emergency. In mixed-occupancy buildings, exiting can be through areas with classifications other than *I-3*, so long as those areas meet code requirements for their classifications and are not high-hazard areas.

Floors used for sleeping, or where there is an occupant load of at least fifty people, must have smoke barriers dividing them into at least two smoke compartments. No more than two hundred people can be housed in a single smoke compartment. Exception is made to this requirement if an exit from the space opens directly into another building separated by a two-hour fire assembly or fifty feet of open space, a secured yard, a holding area providing at least six square feet for every occupant of the space, or a public way.

The travel distance from any room door required for exit access within the compartment to an exit through the smoke barrier cannot exceed 150 feet. No point in any room within a smoke compartment can be more than 200 feet from a smoke barrier exit door. Six square feet of readily available space on either side of a smoke barrier must be provided for residents from the other side in the event of an emergency. Each smoke compartment must have its own means of egress so that no one need return through the area in which the emergency exists.

Sleeping areas must be separated from common group activity spaces, corridors, and adjacent sleeping areas by smoke-tight partitions. The exit access corridor must be similarly separated from common spaces. Doors in these partitions must also be of substantial construction to resist the passage of smoke and have a clear width not less than twenty-eight inches.

In the *I-3* occupancy group, the difference between the lowest and highest finished floor levels may not exceed twenty-three feet. At least one-half the occupants of any story must be able to exit directly without having to pass through another floor within the connected areas. Unenclosed openings between floor levels are permitted so long as the normally occupied areas are visually open to observation by supervisory personnel and there is sufficient egress capacity for all occupants to exit simultaneously from all levels.

Vertical exit enclosures may be constructed with glazing in doors and interior walls, subject to the following limitations:

- They may not serve more than four floors.
- Exit doors must have a ³/₄-hour fire rating.
- The total area of glazing may not exceed 5,000 square inches, with individual panels not larger than 1,296 square inches (nine square feet).
- Glazing must be protected on both sides by sprinklers and installed in stress-relieving gaskets to prevent breakage before the sprinklers have time to operate.
- Obstructions such as window treatments must not interfere with automatic sprinkler operation.

Egress doors may be locked in accordance with the type of use; however, there are several requirements in the *IBC* governing how this may be done. Keys that unlock exterior doors must be available at all times and operate the locks from both sides of the door. Remote locking mechanisms must be configured so that they are capable of being released by minimum staff in less than two minutes. Provisions for remote locking are not required so long as not more than ten locks need to be unlocked in order for occupants to be moved to a place of refuge in less than three minutes. Power-operated doors and locks must have redundant, mechanically operated release mechanisms at each door.

Performing Arts Facilities

Motion Picture Projection Rooms (409)

The principal concern with spaces in this classification is the presence of motion picture film in conjunction with projection equipment that produces heat, hazardous gases, dust, or radiation.

Projection rooms must have a minimum floor area of eighty square feet for a single machine, plus at least forty square feet for each additional projector. The ceiling height must be at least seven feet six inches and a clearance of not less than two feet six inches must be provided beside, behind, and between projectors. Openings into the auditorium area cannot exceed 25 percent of the dividing wall's area, and they must be completely closed with glass or other approved material.

Where cellulose acetate or other safety film only is used, signs must be posted at the door and inside the room, in one-inch block letters, saying: SAFETY FILM ONLY PERMITTED IN THIS ROOM. Where flammable cellulose nitrate film is used, additional NFPA fire regulations apply.

Air supply and exhaust ventilation are also important concerns. The air supply may be drawn from adjacent spaces as long as the amount of air is equivalent to that exhausted by the projection equipment and there is sufficient air when other mechanical systems are not operating. On the subject of return air, however, the *IBC* states that each projection machine must be provided with an exhaust duct that draws air from each lamp and exhausts it directly to the outside of the building. This system may also be used to exhaust projection room air, but it cannot be connected into any other type of exhaust or air-return system.

Stages and Platforms (410)

The area of a *stage*, for code purposes, includes the backstage and support areas, in addition to the performance area itself, that are contained within a fire-rated enclosure separating it from the remainder of the building. The *proscenium* is that part of the stage in front of the curtain (or drop), usually with an enclosing arch. Where the stage height (measured from the lowest point on the stage to the roof or floor deck above) exceeds fifty feet, a two-hour, fire-rated wall is required. This proscenium wall must extend continuously from the foundation to the roof of the structure.

In order to close off the stage from the auditorium in the event of fire, a curtain of a fire-approved material (or a water curtain) must be provided. This fire curtain must be capable of preventing the glow from a severe fire onstage from becoming visible to the audience for twenty minutes, while intercepting smoke, flame, and gases from the stage. It also must be capable of making a controlled descent from the full-open position in less than thirty seconds, with the last eight feet of closure taking not less than five seconds.

Contained within the two-hour fire enclosure is the *fly gallery*, a platform above the stage from which stage operations such as the movement of scenery are controlled. Also in this space is the *gridiron*, which supports hanging or flying scenery, lighting, and other stage effects. *Pin rails* provide for the attachment of lines using belaying pins, and catwalks allow stage personnel to move about in this space. These facilities are not considered to be additional floors or mezzanines for the purposes of the *Code*, nor is an additional fire rating required. However, the *IBC* does require that these areas be constructed of approved materials consistent with the type of building. Smoke and fire tests are required of the materials proposed before the theater can be certified for occupancy.

Backstage support spaces such as dressing and property rooms, scene docks, workshops, and storerooms must also be separated from the performance area by a two-hour wall. These spaces must be separated from each other by one-hour walls and protected by appropriate fire-door assemblies. Openings other than the necessary doorways at stage level are not permitted to connect these spaces with the stage.

At least one approved means of egress is required from each side of the stage, from each side of the area below the stage, from each fly gallery, and from the gridiron. The gridiron may be exited by alternative means such as a steel ladder or spiral staircase to a roof scuttle.

Stages must be protected by an automatic fire-extinguishing system, with sprinklers installed under the roof and gridiron, in the fly galleries, behind the proscenium wall, and in the backstage support spaces such as dressing rooms. Sprinklers are not required in certain instances (stages less than thousand square feet in area or fifty feet in height), but then there are restrictions on the extent of combustible materials allowed. Curtains may not retract vertically, and combustible hangings are generally limited to a single backdrop.

A *platform* is a raised area within a building used for the presentation of music, drama, or other entertainment, or for worship. The platform has no hanging curtains, scenery, or stage effects other than sound and lighting. Permanent platforms must be constructed of materials that are in keeping with the overall building construction. Where the space under a platform is used for any purpose other than equipment, wiring, or plumbing, the platform must be of one-hour fire-rated construction.

Motor-Vehicle-Related Occupancies (406)

Private garages belonging to *Group U* may have an area of up to three thousand square feet if they are used for the storage of private or pleasure-type motor vehicles where no fuel is dispensed or major repair work is done. If the occupancy is mixed, the *Code* requirements for the main portion of the building pertain to the *Group U* portion as well.

Carports must be open on at least two sides and have a floor of approved noncombustible material. Asphalt flooring is permitted at ground level. The area of floor used for parking automobiles or other vehicles must be pitched toward a drain or the main vehicle entry door opening to direct the movement of liquids.

Open parking garage requirements chiefly pertain to vehicular safety. Guards at least three feet six inches high must be provided at interior and exterior openings on floors or roofs where vehicles are parked where the vertical distance to ground exceeds two feet six inches. Vehicle barriers two feet high and capable of resisting a single six-thousand-pound horizontal load are required at the ends of drive lanes and parking spaces where the adjacent floor elevations differ by more than one foot.

Vehicular ramps are not considered as required exits. Publicly accessible stairs and exits must meet all code requirements for egress. Where stairs are used only by parking attendants, there must be a minimum of two, each not less than three feet wide. Mechanical lifts may be provided for the use of garage employees only, and they must be enclosed by noncombustible construction.

The exterior of a parking garage structure must have uniformly distributed openings on two or more sides to ensure natural ventilation. Such openings must constitute at least 20 percent of the total perimeter wall area of each tier and 40 percent of the perimeter length. Interior walls must be at least 20 percent open with the openings uniformly distributed.

Open parking garages may be contained within mixed-use buildings, in which case the separation requirements in Table 13-1 apply. When the building is used exclusively for the parking or storage of private motor vehicles, office, waiting, and toilet facilities may be provided on the ground-level tier without separation, up to a maximum combined area of one thousand square feet.

Enclosed parking garages must have a mechanical ventilation system that complies with applicable mechanical codes.

Motor fuel-dispensing facilities must comply with all applicable fire codes (the *IBC* refers specifically to the *International Fire Code* here as well as in the section on repair garages). Canopies over pumps must be constructed of noncombustible materials or materials having a one-hour fire-resistance rating. Such canopies must have an unobstructed height of thirteen feet six inches. In addition, there are special requirements with respect to combustible materials used within or on a canopy. For example, panels of light-transmitting plastic are permitted but may not be placed closer than ten feet from any building on the same property. The total area of such plastics is limited to one thousand square feet with no individual panel exceeding one hundred square feet.

Repair garages are buildings or portions thereof in which major repairs are made such as engine overhauling, painting, or body and fender work. Facilities provided for offices or salesrooms operating in direct connection with the garage must be separated from the garage as provided in Table 13-1. Any door openings between them must be equipped with self-closing, fire-rated doors.

The floors of repair garages must be concrete or similar noncombustible and nonabsorbent materials. Mechanical ventilation systems are required, with controls located at the entrance to the garage. Facilities where vehicles fueled by gases such as nonodorized LNG or hydrogen are serviced must be equipped with an approved flammable gas-detection system.

The foregoing examples deal with only the most common special building types treated by the *International Building Code*. Others include underground buildings, special amusement buildings, aircraft-related occupancies, combustible storage, and hazardous materials. The highly specialized requirements dealing with buildings in the *Group H* classification are some twenty pages in length.

TYPES OF CONSTRUCTION (600)

Structures are also classified on a basis of construction type, and the *IBC* has specific requirements as to the construction types permitted for many of the use and occupancy classifications. There are five construction types (typical of most current codes), and all except *Type IV* are subclassified *A* or *B* on a basis of the degree to which they are resistant to combustion. The subclassification depends upon whether a particular building element is *protected*, that is, chemically treated or covered in order to gain a fire rating of one or more additional hours.

Construction-type classifications are as follows:

Types I and II—All building elements listed in Table 13-2 must be of noncombustible materials that will not ignite or burn or release flammable vapors when subjected to heat or fire.

Table 13-2 Fire-Resistance-Rating Requirements for Building Elements°

Building Element	Type I A	Type I B	Type II A	Type II B	Type III A	Type III B	Type IV HT	Type V A	Type V B
Sturctural Frame Including Columns, Girders, Trusses	3	2	1	0	1	0	HT	1	0
Bearing Walls Exterior	3	2	1	0	2	2	2	1	0
Interior	3	2	1	0	1	0	1/HT	1	0
Nonbearing Walls and Partitions Exterior	*See Fire Separation Distance Table*								
Nonbearing Walls and Partitions Interior	0	0	0	0	0	0	1	0	0
Floor Construction Including Supporting Beams, Joists	2	2	1	0	1	0	HT	1	0
Roof Construction Including Supporting Beams, Joists	15	1	1	0	1	0	HT	1	0

Type III—Exterior walls must be of noncombustible materials, but interior elements may be constructed of any materials permitted by the *Code*. Fire-retardant-treated wood framing meeting code quality standards is permitted within exterior wall assemblies having a two-hour rating or less.

Type IV (Heavy Timber, HT)—Exterior walls are of noncombustible materials, and the interior building elements are constructed of laminated or solid wood with no concealed spaces. As with *Type III*, fire-retardant-treated wood framing meeting code quality standards is permitted within exterior wall assemblies having a two-hour rating or less.

Wood framing must be constructed of sawn or glued laminated timber; floors and roofs of sawn or glued laminated planks, either splined or tongue-and-groove. Nominal column dimensions must be a minimum of eight by eight inches when supporting floor loads and six by six inches when supporting only roof and ceiling loads. Wood girders and beams supporting floors must be at least six by ten inches. Timber arches and framed timber trusses must be minimally eight by eight inches. Flooring planks are required to be nominally three inches thick and covered by one-inch-thick flooring laid diagonally or crosswise. Roof planks must not be less than two inches thick. Engineered structural panels or wood planks may alternatively be used if set on edge and closely spaced, four inches wide for floors and three inches wide for roofs. Other types of decking with similar structural and fire-resistive properties may be used for roofs.

Type V Structural elements and exterior and interior walls may be constructed of any materials permitted by the *Code*.

The structural frame, as used in Table 13-2, consists of the columns and girders, beams, trusses, and spandrels directly connected to the columns and bracing members that carry gravity loads. The elements of floor and roof panel assemblies having no connection to the columns are not considered part of the structural framework.

Exterior bearing walls must meet the minimum fire-resistance rating based upon fire-separation distance as shown in Table 13-3, as well as complying with the hourly ratings in Table 13-2.

In *Type I* construction, the fire-resistance ratings of the structural frame or bearing walls may be reduced by one hour when supporting a roof only.

In *Type II-A*, *Type III-A*, and *Type V-A* (*protected*) construction, an approved automatic sprinkler system may be substituted for the one-hour fire-resistance-rated construction, provided such a system is not required by the *Code* for other reasons or used for an allowable height or area increase. This substitution is not permitted for exterior walls.

In *Type IV* construction, interior nonbearing partitions must consist of at least two layers of one-inch-thick matched boards, laminated construction at least four inches thick, or one-hour fire-rated construction.

Use Classifications and Construction Types

There are no hard and fast rules specifying fixed relationships between use/occupancy classifications and construction types so long as the required area separations and

Table 13-3 Fire-Resistance-Rating Requirements Based on Separation Distance°

	Fire Separation Distance (Feet)						
	< 5	5 to < 10		10 to < 30			> 30
Type of Construction	All	LA	Others	LA, LB	ILB, V.B	Others	All
Group H	3	3	2	2	1	1	0
Groups F-1, M, S-1	2	2	1	1	0	1	0
Groups A, B, E, F-2, I, R, S-2, U	1	1	1	1	0	1	0

fire-resistance ratings are met. As mentioned earlier, the *International Building Code* permits certain trade-offs in required construction types where fire protection systems are provided. In fact, the concept of *protected* construction is recognized in the *A* and *B* subclassification of all but one category.

The *IBC* does give specific requirements in certain cases, however, often offering alternatives intended to create incentives to upgrade fire-protection measures. For example, in high-rise buildings, fire-resistance-rating reductions are allowed where sprinkler control valves with supervisory and water-flow-initiating devices for each floor are provided (403.3). In this case, *Type I-A* construction may be reduced to *Type I-B*. In groups other than *F-1*, *M*, and *S-1*, construction may be further reduced to *Type II-A*. Fire barriers enclosing vertical shafts other than exit stair and elevator shaft enclosures may be reduced to a one-hour rating where sprinklers are provided within the shafts at the top and at alternative floor levels.

Covered malls, including anchor buildings, may be built using any construction other than *Type V* without any area limitation. In this case, the *IBC* requires that the mall, along with attached anchors and parking structures, have a minimum of sixty feet of permanent open space on all sides (402.6).

Habitable underground structures must be of *Type I* construction (405.2). The *IBC* defines as an underground building any building having occupiable spaces more than thirty feet below the level of exit discharge. In addition, automatic sprinkler systems are required at the highest level of exit discharge and all levels below.

Table 13-4 Allowable Building Heights and Floor Areas°

		Type I A	Type I B	Type II A	Type II B	Type III A	Type III B	Type IV HT	Type V A	Type V B
Height in Feet		UL	160	65	65	65	55	65	60	40
Group	**Height in Stories**									
A-1	S A	UL UL	5 UL	3 15,500	2 8,500	3 14,000	2 8,500	3 15,000	2 11,500	1 6,500
A-2	S A	UL UL	11 UL	3 15,500	2 9,500	3 14,000	2 9,500	3 15,000	2 11,500	1 6,000
A-3	S A	UL UL	11 UL	3 15,500	2 9,500	3 14,000	2 9,500	3 15,000	2 11,500	1 6,000
A-4	S A	UL UL	11 UL	3 15,500	2 9,500	3 14,000	2 9,500	3 15,000	2 11,500	1 6,000
A-5	S A	UL UL	UL UL	UL UL	UL UL	UL UL	UL UL	UL UL	UL UL	UL UL
B	S A	UL UL	11 UL	5 37,500	4 23,000	5 28,500	4 19,000	5 36,000	3 18,000	2 9,000
E	S A	UL UL	5 UL	3 28,500	2 14,500	3 23,500	2 14,500	3 25,500	1 18,500	1 9,500
F-1	S A	UL UL	11 UL	4 25,000	2 15,500	3 19,000	2 12,000	4 33,500	2 14,000	1 8,500
F-2	S A	UL UL	11 UL	5 37,500	3 23,000	4 28,500	3 18,000	5 50,500	3 21,000	2 13,000
H-1	S A	1 21,000	1 16,500	1 11,000	1 7,000	1 9,500	1 7,000	1 10,500	1 7,500	NP NP
H-2	S A	UL 21,000	3 16,500	2 11,000	1 7,000	2 9,500	1 7,000	2 10,500	1 7,500	1 3,000
H-3	S A	UL UL	6 60,000	4 26,500	2 14,000	4 17,500	2 13,000	4 25,500	2 10,000	1 5,000

Table 13-4   *(continued)*

H-4	S	UL	7	5	3	5	3	5	3	2
	A	UL	UL	37,500	17,500	28,500	17,500	36,000	18,000	6,500
H-5	S	3	3	3	3	3	3	3	3	2
	A	UL	UL	37,500	23,000	28,500	19,000	36,000	18,000	9,000
I-1	S	UL	9	4	3	4	3	4	3	2
	A	UL	55,000	19,000	10,000	16,500	10,000	18,000	10,500	4,500
I-2	S	UL	4	2	1	1	NP	1	1	NP
	A	UL	UL	15,000	11,000	12,000	NP	12,000	9,500	NP
I-3	S	UL	4	2	1	2	1	2	2	1
	A	UL	UL	15,000	11,000	10,500	7,500	12,000	7,500	5,000
I-4	S	UL	5	3	2	3	2	3	1	1
	A	UL	60,500	26,500	13,000	23,500	13,000	25,500	18,500	9,000
M	S	UL	11	4	4	4	4	4	3	1
	A	UL	UL	21,500	12,500	18,500	12,500	20,500	14,000	9,000
R-1	S	UL	11	4	4	4	4	4	3	2
	A	UL	UL	24,000	16,000	24,000	15,000	20,500	12,000	7,000
R-2	S	UL	11	4	4	4	4	4	3	2
	A	UL	UL	24,000	16,000	24,000	16,000	20,500	12,000	7,000
R-3	S	UL	11	4	4	4	4	4	3	3
	A	UL	UL	UL	UL	UL	UL	UL	UL	UL
R-4	S	UL	11	4	4	4	4	4	3	2
	A	UL	UL	24,000	18,000	24,000	16,000	20,500	12,000	7,000
S-1	S	UL	11	4	3	3	3	4	3	1
	A	UL	48,000	26,000	17,500	26,000	17,500	25,500	14,000	9,000
S-2	S	UL	11	5	4	4	4	5	4	2
	A	UL	79,000	39,000	26,000	39,000	26,000	38,500	21,000	13,500
U	S	UL	5	4	2	3	2	4	2	1
	A	UL	35,500	19,000	8,500	14,000	8,500	18,000	9,000	5,500

Stages must normally be constructed of materials in keeping with the requirements for floor construction in the building in which it is housed, except that *Type II-B* or *Type IV* is permitted if the stage is entirely compartmentalized behind a proscenium curtain as described earlier. In buildings of *Type II-A*, *Type III-A*, and *Type V-A* construction, a fire-resistant stage floor is not required so long as the space below is equipped with an automatic fire-control system (410.3).

Of course, there are many special requirements for buildings in the hazardous-use category, but there the emphasis is largely on the nature and quantity of materials handled or stored, specific fire and other kinds of detection systems needed, and necessary separation from other structures. These factors underscore the need to carefully check all applicable codes whenever there are extraordinary conditions or specialized facility requirements to be satisfied.

Building Height and Area Limitations (500)

There is one set of general constraints that relate both to the use/occupancy classifications and building types, however, and that pertains to height and area limitations. These are summarized in Table 13-4.

For the purposes of the *IBC*, building area includes the area within the exterior walls (or fire walls) excluding interior vent shafts and courts. Building height is measured from the grade plane to the average height of the highest roof surface. The *grade plane* is a reference elevation that represents the average ground level at the building's exterior walls. Story height is measured between finished floor surfaces or from finished floor surface to the top of the roof rafters. Basements are generally considered to be those portions of the building that are partially or entirely below the grade plane.

Special note should be made of the phrase "or fire walls" as used in the preceding description. For the purpose of determining allowable building area, the *Code* treats parts of a building that are separated by fire walls as separate buildings. *Fire wall*, in this context, means a "fire-resistance-rated wall having protected openings, which restricts the spread of fire and extends continuously from the foundation to or through the roof, with sufficient structural stability under fire conditions to allow collapse of construction on either side without collapse of the wall."° In densely populated areas where buildings may share a party wall along property lines, such walls must be constructed as fire walls without openings in order to create separate buildings (503.2).

There are a number of exceptions permitted to the basic limitations set forth in Table 13-4. Some acknowledge practical considerations related to building use. For example, one-story aircraft hangars and buildings where aircraft are manufactured are exempted from height limits. This exemption carries with it the requirement that the building be provided with an automatic fire-extinguishing system and that it be surrounded on all sides by open space at least $1\frac{1}{2}$ times as wide as the building's height (504.1).

Some less arcane special provisions relate to mixed-usage situations such as office buildings (*Group B*), shopping centers (*Group M*), or apartment buildings (*Group R*) constructed above enclosed parking garages (*Group S-2*). The *Code* allows these structures to be treated as separate and distinct buildings when certain conditions are met. In the preceding situation, for example, the basement and/or first story above the grade plane must be of *Type I-A* construction and be separated from the building above by a horizontal assembly providing a minimum three-hour fire-resistance rating. Vertical shafts, stairways, ramps, and so on must be protected by two-hour enclosures extending above and below any such openings. Necessary openings into such enclosures must also be fire rated (508).

Providing sufficient open space around one- and two-story buildings can also yield an allowable increase in area. Typically, the building must be surrounded by public ways or yards not less than sixty feet wide. Thus situated, one-story, *unsprinklered* buildings in *Group F-2* or *Group S-2* may have unlimited area. *Sprinklered*, one-story buildings in *Groups A-4, B, F, M,* or *S* may also have unlimited area. Two-story buildings in *Groups B, F, M,* and *F* qualify for the same area exemption if sprinklered. One-story motion picture theaters qualify if they are of *Type I* or *Type II* construction (507).

The *IBC* provides a formula for calculating allowable area increases where there is a desirable combination of special factors present. One such factor is having more than 25 percent of the building's perimeter adjacent to a public way or open space twenty feet or more in width. The other and most important factor is the presence of an automatic sprinkler system. Except for high-hazard usages, area limitations (Table 13-4) are generally permitted to be increased by 200 percent for multistory buildings and 300 percent for single-story buildings when an approved sprinkler system is provided. Building heights may be increased by twenty feet (in addition to the area increase) up to a limit of sixty feet, except for high-hazard uses and certain institutional uses of construction *Type II-B* or below (504, 6).

FIRE-RESISTANCE-RATED CONSTRUCTION (700)

In reading the foregoing sections, you should have gotten the sense that many—if not most—code provisions are concerned with protecting a building's occupants in the event of fire or other emergency. Beyond basic structural integrity, fire safety is the overriding consideration in all space planning and building construction.

There are several levels to effective planning for fire safety. Use and occupancy classifications take into account the kinds of activities that may reasonably be expected to take place within a building or portion of a building. Where there are dissimilar occupancies adjacent, the required fire separations are greater, particularly where hazardous activities or materials are involved.

Another major consideration is the familiarity of the individual occupant with his or her surroundings together with the amount of control a person is able to exercise over her ability to leave a given area. In theory at least, people are not as much at risk in their homes, with which they are assumed to be completely familiar, as they might be in a crowded theater or club that they have only occasionally, if ever, visited before. Even so, we witness news reports of fire-related disasters in both private and public facilities almost daily.

Smoke is often more dangerous than fire. It spreads rapidly and, depending upon the kind of materials burning, can release any number of toxic gases. Poorly maintained or malfunctioning household heating systems, for example, can produce carbon monoxide gas as a result of incomplete combustion without there even being an open fire.

To lessen these risks, the codes specify performance levels for the smoke and fire resistance of the materials used in constructing individual parts of each type of building. In addition, there are specific requirements for active safety devices such as fire alarms, smoke detectors, and other warning systems. Other requirements are to be found in the sections of the *Code* related to egress as well, and we examine each of these in turn.

Fire-Resistance Ratings and Fire Tests (703)

We have already looked at the definition of fire wall. Now we need to distinguish among several other terms used when speaking of fire-resistance-rated construction.

- *Fire-Resistance Rating*—The length of time a building material, assembly, or component can maintain its structural integrity and its ability to confine a fire, as determined by specific tests prescribed by the *Code*.
- *Fire-Rated Assembly*—A combination of building components, including all construction materials and hardware, that together comprise a building element that has passed various tests under fire conditions and has been given a fire rating.
- *Fire-Protection Rating*—Length of time, in hours or minutes, that an opening assembly can maintain its ability to restrict the spread of fire through an opening that is protected.
- *Fire Separation Distance*—Distance from the building face to the closest lot line; centerline of a street, alley, or public way; or a line midway between two buildings on the same property, measured at right angles to the line.
- *Fire Barrier*—A rated horizontal or vertical assembly designed to impede the spread of fire, in which any openings must be protected.
- *Fire Partition*—A vertical, fire-rated assembly (typically one-hour) in which openings are protected.
- *Fire Blocking*—Building materials used to impede the passage of fire through concealed spaces.
- *Fire Door Assembly*—A combination of a rated fire door, frame, and other hardware that provide a specified degree of protection to an opening.
- *Fire (Smoke) Damper*—Automatic, rated devices installed in ventilation ducts and air-transfer openings designed to resist the movement of fire and/or smoke. Such devices are controlled by heat and smoke-detection systems and often can be positioned from a remote location.
- *Smoke Barrier*—A continuous, horizontal or vertical membrane (wall, ceiling, or floor assembly) designed to resist the movement of smoke.

The *International Building Code* refers extensively to a specific testing methodology formulated by the American Society for Testing and Materials: *Test Methods for Fire Tests of Building Construction and Materials* (*ASTM E119*). These tests are principally of three types:

1. Materials or assemblies are rated according to their ability to maintain their structural stability under specific fire conditions;
2. The degree to which flames, smoke, and gases penetrate the assembly; and
3. The degree to which the measured temperature increases on the protected side of the assembly.

The *IBC* discusses alternative methods for determining fire resistance according to calculations spelled out elsewhere in the *Code* or engineering analysis based on similar designs, but these procedures also refer to ASTM E119 together with other ASTM, Underwriters Laboratories (UL), and National Fire Protection Association (NFPA) tests.

Exterior Walls (704)

Exterior walls must be fire rated as set forth in Table 13-2 and Table 13-3. Specific requirements extend to projections such as cornices, balconies, and overhangs. In general, projections from walls of *Type I* or *Type II* construction must be of noncombustible materials unless the building is three stories or less in height. Where the construction is *Type III*, *Type IV*, or *Type V*, any approved material may be used. As we have seen, each of the building types has its own height limitation based on use.

Where parapets are required, they must have the same fire-resistance rating required for the supporting wall. Parapets must extend a minimum of thirty inches above the intersection of the roof and wall surfaces, and the upper eighteen inches must have noncombustible facings including flashing and coping materials. Parapets are not required where the wall is not required to be rated due to fire-separation distance (typically in the residential-use category), or where the roof is rated for not less than two hours or is entirely constructed of noncombustible materials including its deck. Buildings of less than one thousand square feet per floor are also exempted.

Other than those in the hazardous-use classification, buildings are permitted to have unlimited unprotected openings in the exterior walls at the first story, so long as they face a street and have a fire-separation distance of more than fifteen feet. The permitted areas of windows and other openings above the first story are as shown in Table 13-5. The values shown in the table are stated in terms of the percentage of the area of the exterior wall. When both protected and unprotected openings occur in any story, the *Code* provides a formula for computing the allowable area.

In determining fire-separation distances and wall, opening, and roof-protection requirements, multiple buildings on the same property are treated as if they had a property line between them. If their aggregate areas do not exceed the limits specified under the general height and area limitations for a single building, they may be treated as one building. If the buildings are of different construction types or use classifications, the most restrictive category must be used.

Fire Walls (705)

Each portion of a building separated by one or more fire walls is considered by the *Code* to be a separate building. Where fire walls separate groups required to be separated by a fire barrier, the most restrictive requirements for each separation must be used. Such walls must have sufficient structural stability under fire conditions so that the wall will remain standing for the length of time for which it is rated following the collapse of the construction on either side. They are required to have a fire-resistance rating, by classification group, not less than indicated in Table 13-6.

The *IBC* states that in the horizontal direction, fire walls must be continuous from exterior wall to exterior wall and extend at least eighteen inches beyond the surface of the exterior walls. In most circumstances, however, this requirement is mitigated by having the exterior walls, extending at least four feet on either side of the fire wall, rated for at least one hour. Fire walls may also terminate at the interior surface of noncombustible exterior sheathing, siding, or other exterior finish extending at least four feet on either side of the fire wall.

Table 13-5   Maximum Area of Exterior Wall Openings°

Classification of Opening	Fire Separation Distance (feet)							
	0 to 3	> 3 to 5	> 5 to 10	> 10 to 15	> 15 to 20	> 20 to 25	> 25 to 30	> 30
Unprotected	Not Permitted	Not Permitted	10%	15%	25%	45%	70%	No Limit
Protected	Not Permitted	15%	25%	45%	75%	No Limit	No Limit	No Limit

Table 13-6   Fire Wall Fire-Resistance Ratings°

Group	Fire-Resistance Rating (hours)
A, B, E, H-4, I-R-1, R-2, U	3
F-1, H-3, H-5, M, S-1	3
H-1, H-2	4
F-2, S-2, R-3, R-4	2

Table 13-7   Minimum Roof-Covering Classification by Construction Type°

Type I		Type II		Type III		Type IV	Type V	
A	B	A	B	A	B	HT	A	B
B	B	B	C	B	C	B	B	C

Table 13-8   Fire-Resistance-Rating Requirements for Fire-Barrier Assemblies°

Occupancy Group	Fire-Resistance Rating (hours)
H-1, H-2	4
F-1, H-3, S-1	3
A, B, E, F-2, H-4, H-5 I, M, R, S-2	2
U	1

Where the building on both sides of the fire wall is protected by an automatic sprinkler system, the fire wall may terminate at the interior surface of noncombustible sheathing.

The *Code* says that fire walls must extend vertically from the foundation to a point thirty inches above the higher of the adjacent roof surfaces, but there are a number of exceptions to this requirement. Walls rated for two hours or longer may terminate at the underside of the roof sheathing or slab, provided the lower roof assembly and all supporting elements are rated for one hour or more, there are no openings in the roof closer than four feet to the fire wall, and the building has no less than a Class B roof covering on either side. The *IBC* permits certain other exceptions to this requirement for specific building construction types and use groups.

Roof assemblies are classified as Class A, Class B, and Class C, based upon their effectiveness against severe, moderate, or light fire-test exposure, respectively (Section 1505). Acceptable roof-covering classifications by construction type are given in Table 13-7.

Fire Barriers (706)

Fire barriers are used to separate areas having different uses or to separate a single occupancy into different fire compartments. Typically, they are walls extending from the floor to the underside of the floor or roof slab above. Such walls must extend continuously through concealed spaces such as those above suspended ceilings. No single opening is permitted to be more than 120 square feet in area, and the total widths of all openings may not exceed 25 percent of a wall's length.

Separation requirements between different occupancies are shown in Table 13-1, and similar requirements apply to incidental-use areas. Table 13-8 gives the requirements for fire-barrier assemblies separating areas having the same occupancy.

For specific fire-resistance ratings governing the separations between vertical-exit enclosures, horizontal exits, and exit passageways, the *IBC* refers to a detailed set of exit requirements in the section on egress. For example, all interior stairways must be enclosed with two-hour walls in a building four stories or greater or one-hour walls otherwise. (Egress and accessibility are the subjects of Chapter 14 and Chapter 15.)

Shaft Enclosures (707)

The basic requirements for shafts used as exit enclosures are similar to those described earlier for interior stairways. Enclosures are not required for escalators or stairways that are not essential means of egress if the building has an automatic sprinkler system throughout.

Elevator lobbies must be provided on every floor on which elevators open into a fire-rated corridor. They must have at least one means of egress and completely separate the elevators from the corridor by fire barriers and required opening protection. Street-level elevator lobbies in office buildings need not be separated if the entire street floor is sprinklered. Lobby separation is also not required in buildings where automatic sprinkler systems are installed, except more than four stories above the lowest level of fire department vehicle access or for buildings in *Groups I-2* and *I-3*.

Shaft enclosures are intended to serve as fire barriers and are constructed in the same manner: extending from slab to slab. Openings and penetrations must be protected and made only for purposes related to the use of the shaft. For instance, ducts may never penetrate exit shaft enclosures. Shafts that do not extend all the way from the top to the bottom of the building must be fully enclosed at the floor level at which they terminate. The shaft enclosure must maintain the same fire-resistance rating as the floor at which it terminates but not less than that required for the shaft enclosure. If the bottom of a shaft terminates in a room related to the purpose of the shaft, the room must maintain the same fire rating as the shaft itself (or be protected by fire dampers).

Fire Partitions (708)

Fire partitions are typically one-hour-rated walls extending from slab to slab, often used to separate areas of similar occupancy in the same building. Examples include residential dwelling units, tenant spaces in covered malls, guest rooms in hotels, and corridor walls. The *IBC* permits demising walls separating tenants to be fire partitions, but this is an area where local code requirements should be thoroughly checked. The precise requirement may depend upon the specific occupancy involved, and local codes frequently require two-hour walls rather than one.

Smoke Barriers and Smoke Partitions (709–710)

Smoke barriers also require a one-hour fire-resistance rating. They may consist of wall assemblies or be a fully enclosed compartment consisting of wall, ceiling, and floor assemblies in which only limited openings are allowed to make them smoke resistant. In *Group I-2* occupancies, described earlier in this chapter, where floors greater than a specified area must be divided into multiple smoke compartments, pairs of close-fitting, opposite-swinging doors with automatic-closing devices are required. Otherwise, opening protectives must have a minimum fire-protection rating of twenty minutes.

Smoke partitions are wall assemblies constructed to limit the transfer of smoke. They are not required to have a fire-resistance rating unless specified elsewhere in the *Code*. This is a new classification in the 2003 edition of the *IBC*.

Horizontal Assemblies (711)

The *IBC* requires floor and roof assemblies to have a fire-resistance rating not less than that required for a building's construction type. Such assemblies must be continuous without openings, joints, or penetrations except as explicitly permitted. They are rated on a basis of the entire floor/ceiling or roof/ceiling assembly, from the surface of the ceiling material below to the floor or roof surface above.

Suspended ceilings may consist of gypsum board or lay-in acoustical tile, and the latter must be installed in such a manner as to resist a moderate upward force. Access panels are only permitted in such ceilings if rated for the purpose. Penetrations such as skylights may be made through a fire-rated roof deck so long as the structural integrity of the roof construction is maintained, except in certain specified circumstances.

Penetrations (712)

Penetrations through fire-rated walls, horizontal assemblies, and membranes must be made in a manner that will maintain the required level of protection for the building element involved. Where sleeves are used, such as for pipes or conduits, they must be securely fastened to the assembly penetrated. In concrete or masonry walls where the penetration is up to one square foot and the penetrating item is no larger than six inches in diameter, concrete, grout, or mortar may be used to maintain the fire-resistance rating. Alternative filler materials, fire-stop, and draft-stop systems may be used that meet ASTM testing standards.

Fire-Resistant Joint Systems (713)

Joint systems designed to protect walls and horizontal assemblies are required where these building elements intersect. Approved fire-resistant joint systems must impede the passage of fire for a period of time at least equal to the rating of the floor, wall, or roof in or between which it is installed. Such joint systems must be tested according to ASTM E 1966 and Underwriters Laboratories specifications (UL 2079). They must be securely installed in or on the entire length of joint where required and possess the ability to withstand normal building movement while resisting the passage of fire and hot gases.

Where exterior curtain walls intersect fire-rated floor/ceiling assemblies, any voids created must be sealed with an approved material tested according to ASTM E119.

Fire-resistant joint systems are not required within single-dwelling units, floors within malls or open parking structures, mezzanine floors, walls in which unprotected openings are permitted, or roofs where openings are permitted.

Fire-Resistance Rating of Structural Members (714)

The fire-resistance ratings of structural building elements must comply with all requirements for the type of construction used and must be equal to or greater than the ratings of all assemblies that they support.

Individual structural members such as columns, girders, beams, trusses, and lintels that support more than two floors, one floor and a roof, a load-bearing wall, or a non-load-bearing wall over two stories must be individually protected in their entirety by materials with the required rating. Other structural members requiring a rating must be individually encased or protected by a membrane or ceiling as described earlier under horizontal assemblies.

Rated columns must be entirely protected, including connections to beams and girders. Fire resistance of columns must be continuous from the surface of the floor to the top of the column wherever columns penetrate a suspended ceiling. The construction of fire-rated assemblies enclosing trusses must be individually evaluated to demonstrate the necessary fire resistance using full-scale tests or approved calculations based on such tests.

Rivets, bolt heads, and other attachments to structural members must be covered by at least one inch of the fire-protection material. Stirrups and reinforcement ties for concrete or masonry reinforcement may project not more than $1/2$ inch into the fire protection. No pipes, conduits, wires, ducts, or other services may be embedded in the protective covering of any structural member requiring individual encasement.

Exterior structural members subject to external fire exposure must be protected as indicated in Table 13-2, according to the type of construction for exterior load-bearing walls. Where impact damage to the fire-protective covering of a structural member is at risk due to moving vehicles or other activities, it must be protected by corner guards or an enclosure of metal or other noncombustible material.

Opening Protectives (715)

The subject of opening protectives is one on which the *IBC* refers to numerous testing standards, notably NFPA and UL. These include fire doors, shutter assemblies, and fire-protection-rated glazing.

Table 13-9 Opening Protective Fire-Protection Ratings°

Type of Assembly	Required Assembly Rating (Hours)	Minimum Fire Door and Shutter Assembly Rating (Hours)
Fire Walls and Fire Barriers	4	3
Having a Required	3	3
Fire Resistance Rating	2	1.5
Greater Than One Hour	1.5	1.5
Fire Barriers Having a Required		
Fire Resistance Rating of One Hour		
Shaft Exit Enclosure and Exit		
Passageway Walls	1	1
Other Fire Barriers	1	0.75
Fire Partitions		
Corridor Wall	1	0.33
	0.5	0.33
Other Fire Partitions	1	0.75
Exterior Walls	3	1.5
	2	1.5
	1	0.75

Fire doors and fire-shutter assemblies must be installed in accordance with NFPA 80 and provide an appropriate fire-protection rating for the assembly in which it is installed, according to Table 13-9. Specific rating requirements for different types of doors, including side-hinged or pivoted swinging doors, doors in corridors and smoke barriers, and doors in exit enclosures refer to NFPA 252, UL 10B, UL 10C, and UL 1784.

This section also regulates fire-protection-rated glazing in doors and fire-window assemblies. Specific area limitations are established for glazing in doors in relation to the use and fire-protection rating of the door, subject to NFPA 80 and NFPA 252. Glazing in fire-window assemblies is regulated by NFPA 80 and NFPA 257. Typically, glazing in fire partitions and fire barriers must provide a forty-five-minute rating.

Fire doors and fire-protection-rated glazing must be permanently factory-labeled by the manufacturer with the manufacturer's name and other pertinent information such as the test standard, the third-party inspection organization, and the fire-protection rating granted. In certain cases, where required for fire doors in exit enclosures, the maximum transmitted temperature end point is shown on the door label. Fire doors and fire shutters must be equipped with fire-exit hardware including automatic-closing devices.

Ducts and Air-Transfer Openings (716)

Other types of opening protectives that are of concern in maintaining the integrity of fire-resistive construction are those necessary to accommodate mechanical ductwork. Here the *Code* is concerned with fire, smoke, and ceiling dampers located within air-distribution and smoke-control systems. There are also combination fire/smoke dampers that perform multiple functions. Dampers automatically interrupt the flow of air through a duct system, restricting the passage of heat, smoke, and fire during an emergency. Fire and smoke dampers are often also used to control the volume of air flow within the HVAC system during normal use.

Dampers are of two types: static and dynamic. Static dampers automatically shut down during a fire and may only be used in static HVAC systems. Dynamic dampers are used where fans continue to operate during an emergency situation and must be fire-resistance rated. Like fire doors, fire and smoke dampers must bear the label of an approved testing agency (UL 555 and UL 555S). The minimum fire-damper rating for less than three-hour fire-rated assemblies is 1½ hours; otherwise it is three hours.

Ceiling dampers are used in suspended ceilings that are part of a rated ceiling and floor/roof assembly. They are designed to prevent heat from traveling through the duct system or entering into the cavity between the ceiling and floor or roof above. Ceiling radiation dampers close when heat above a specified level rises into the duct; they are labeled under UL 555C.

Other Fire-Resistance Requirements (717–721)

The remaining several parts of section 700 are concerned with many details concerning the establishment and maintenance of desired levels of fire resistance. The treatment of concealed spaces with fire blocking and draft stopping creates barriers to the spread of fire and smoke through floor/ceiling and attic spaces, connections between horizontal and vertical spaces, and stairways (717). Performance levels for plaster as well as thermal and sound-insulating materials are specified (718–19).

Prescriptive Fire Resistance (720) details requirements for fire-resistance-rated building elements. It consists principally of nearly twenty pages of tables listing various materials and construction assemblies, in three parts:

1. Minimum protection of structural parts (in hours) for various noncombustible insulating materials;
2. Fire-resistance periods (hours) for various walls and partitions
3. Minimum protection for floor and roof systems (in hours).

Finally, *Calculated Fire Resistance* (721) sets forth procedures by which the fire-resistance of specific materials or assemblies may be determined by calculations. This section contains numerous formulas and tables dealing with concrete assemblies, concrete masonry, clay brick and tile masonry, steel assemblies, and wood assemblies.

FIRE-PROTECTION SYSTEMS

We have seen in the previous sections of this chapter how codes, as exemplified by the *International Building Code*, approach the issue of minimizing safety threats to building occupants. First, by establishing use and occupancy classifications, the *IBC* sets up different requirements that derive from the nature of different kinds of buildings and the physical features likely to be found in them. The requirements take into account how familiar the occupants are likely to be with their surroundings and how much control they may or may not have over their own well-being in the event of an emergency.

Second, the codes categorize building types on a basis of the materials used in their construction. This enables relationships to be established and specific requirements set up between the anticipated functions within a building and the fire resistance offered by its many elements, structural and otherwise. These requirements take the form of fire-resistance ratings given in hours, together with height and separation distances.

Third, detailed performance specifications are given for the building elements themselves, beginning with parameters for testing them to determine the level of safety each provides. These are organized according to the physical parts of the building such as exterior walls, interior fire walls and barriers, vertical penetrations, ducts, and other openings.

Next, we come to the systems required for the *active* protection of building occupants: devices that are relied upon to inhibit the spread of fire and smoke, and alarm systems that let the building's inhabitants and other responsible parties know an emergency exists.

Automatic Sprinkler Systems (903)

Automatic sprinkler systems are dedicated systems of water pipes, with a suitable water supply, designed to meet specific fire-protection-engineering standards. The part of the system placed within habitable spaces is usually a network of pipes connected to overhead sprinklers arranged in a regular pattern.

Approved automatic sprinkler systems are required by the _IBC_ in many locations classified according to occupancy and use. The term _fire areas_ refers to the portions of buildings separated by fire barriers (Section 706), which defines the separation requirements in hours as set forth in Table 13-8.

Use Classifications

In the _Assembly_ group, sprinklers are required not only in the area so classified but on all floors between the assembly area and the level of exit discharge (with the exception of _Group A-5_). In _Group A-1_, _Performing Arts Facilities_, sprinklers are required if the fire area exceeds twelve thousand square feet, the occupant load is three hundred people or more, the fire area is located on a floor other than the level of exit discharge, or the fire area contains a multitheater complex. In _Group A-2_, _Food and Drink Establishments_, the requirements are five thousand square feet, occupant load of at least three hundred, or a fire area not on the level of exit discharge. The requirements for _Group A-3_, _Amusement, Recreation, and Worship Facilities_, and _Group A-4_, _Indoor Sports Facilities_, are similar to _Group A-1_. In _Group A-5_, _Outdoor Facilities_, sprinklers are required in accessory spaces such as retail areas, press boxes, and concession stands that occupy more than one thousand square feet.

Curiously, the _Group B, Business_, classification has no section of its own in the _IBC_. However, _all_ buildings except those (in most instances) in _Group R-3_, _Permanent Single or Duplex_ and _Group U, Utility_, having a floor level with an occupancy load of at least thirty people, located fifty-five feet or more above the lowest level of fire department vehicular access, must be sprinklered. This covers most office facilities, especially those in high-rise construction.

There are other general requirements for sprinkler systems as well. Stories and basements without openings to provide exterior access with a minimum dimension of thirty inches must have a sprinkler system. The same is true where any portion of a basement is more than seventy-five feet from such openings.

Buildings in _Group E, Educational_, require sprinklers in any fire areas exceeding twenty thousand square feet or any portion below the level of exit discharge, unless every classroom has its own exterior door at ground level.

In the _Factory and Industrial_ group, the criteria for _Group F-1, Moderate-Hazard_, are at least twelve thousand square feet of floor area, a fire area located more than three stories above grade, or a total of all fire areas exceeding twenty-four thousand square feet (including mezzanines). Woodworking operations in the _F-1_ group cannot exceed 2,500 square feet without being sprinklered, because the type of use involves finely divided combustible materials and waste.

As might be expected, sprinkler systems are required in all _Group H, High Hazard_, facilities. Similarly, buildings containing a _Group I, Institutional_, fire area must be sprinklered.

Sprinkler requirements for _Group M, Mercantile_, and _Group S-1, Storage_ (_Moderate Hazard_), occupancy are the same as for _Group F-1_. Repair garages, tire-storage areas, and commercial parking garages exceeding specific area requirements must be sprinklered.

Standards

Most codes, including the _IBC_, refer to NFPA 13 (Standard for Installation of Sprinkler Systems) for the specifics of sprinkler system design and installation. Specific requirements for sprinkler-head coverage depend upon the type of occupancy, but the typical spacing of heads is from twelve to fifteen feet with each having a coverage in the range of ninety to two hundred square feet.

There are several types of sprinkler systems covered by NFPA 13. In addition to the basic standard, the _IBC_ and the _IFC_ now allow the use of less expensive NFPA 13R systems in Group R buildings up to four stories high. There is also an NFPA 13D variant for single and duplex residential units.

A secondary water supply is generally required for systems in high-rise buildings that is capable of maintaining a calculated pressure for at least thirty minutes. Residential systems such as NFPA 13R may use a single-combination water supply so long as it is capable of supplying the sprinklers and normal domestic demand simultaneously.

Except for automatic sprinkler systems in one- and two-family dwellings, limited systems with fewer than twenty sprinklers, and certain other special circumstances, all critical components of NFPA 13 systems must be electrically monitored and alarmed. Alarms indicating trouble in the system and those indicating an actual fire condition must be distinctly different.

Alternative Automatic Fire-Extinguishing Systems (904)

There are numerous alternatives to water-based fire-protection systems. These are regulated by other NFPA standards and include wet and dry chemical systems (NFPA 17), foam (NFPA 11 and 16), carbon dioxide (NFPA 12), and halon° systems (NFPA 12A). Such systems are used in areas that contain electrical, computer, and telecommunications equipment or where valuable items susceptible to water damage may be located. Special commercial cooking systems must be used in food-service areas and kitchens where grease fires may occur.

Standpipe Systems (905)

Standpipes and fire hoses comprise manual fire-protection systems that are usually installed when a building is constructed. Standpipe installations are governed by NFPA 14 (Installation of Standpipes, Private Hydrants, and Hose Systems). The *IBC* defines five types of standpipe systems:

1. *Automatic dry:* Normally filled with pressurized air, the dry-standpipe system admits water into the pipe automatically when a hose valve is opened.
2. *Automatic wet:* Piping system is normally filled with water and connected to a supply capable of automatically providing sufficient water upon demand.
3. *Manual dry:* Piping system is normally empty and requires water pumped into the system from a fire department pump truck.
4. *Manual wet:* Piping system is connected to a water supply sufficient to maintain water within the system but must be connected to a fire department pump truck to meet system demand.
5. *Semiautomatic dry:* Piping system is connected to a supply capable of meeting system demand upon the automatic activation of a device, such as a deluge valve, to admit water into the system.

The code also refers to three *classes* of standpipe systems:

1. *Class I:* For use by fire departments; has 2.5-inch hose connections.
2. *Class II:* For use by building occupants or fire department personnel; has 1.5-inch hose stations (such as fire hose cabinets).
3. *Class III:* System has both 1.5- and 2.5-inch hose connections for use by building occupants and fire department personnel.

Class III standpipe systems are required in buildings where the floor level of the highest or lowest story is located more than thirty feet above or below the nearest level of fire department vehicle access. If the building is equipped with an automatic sprinkler system throughout, *Class I* standpipes are sufficient. *Class I* manual standpipes are also permitted in open parking garages where the highest floor is not more than 150 feet from fire department access.

In *Assembly (Group A)* occupancies, automatic wet standpipes are required in non-sprinklered buildings having an occupant load of more than one thousand people, except for open-air-seating structures without enclosed spaces. *Class I* automatic or semiautomatic dry standpipes or manual wet standpipes may be used where the highest floor used for human occupancy is not more than seventy-five feet above the lowest level of fire department vehicle access.

A *covered mall* must have a standpipe system throughout if it meets the conditions outlined earlier. Otherwise, the mall must be equipped with *Class I* hose connections capable of delivering 250 gallons per minute from the most remote outlet. Hose connections are required within the mall at the entrance to each exit passageway, at each floor landing within enclosed stairways, and at all exterior public entrances.

Class III wet standpipes are required for *stages* larger than one thousand square feet, with both sizes of hose connection at either side of the stage. The hose connections alone are sufficient if the entire fire area is covered by an automatic sprinkler system, and the hoses may be connected to the same water supply so long as it has the same flow rate as *Class III* standpipes. In that case, hoses attached to the 1.5-inch connections must be long enough to reach all parts of the stage area.

Locations of Hose Connections and Cabinets
Class I and *Class III* standpipe systems must have hose connections at specific locations:

- in all required stairways, at every floor level above and below grade, at the landing between floors;
- on each side of the wall adjacent to all horizontal exits and where exit passageways join other areas of the building;
- adjacent to all public entrances in covered mall buildings and where exit passageways empty into the mall area;
- on the roof or at the highest landing of stairs having roof access, where the roof has less than a 33 percent slope.

In addition, the *IBC* specifies that building officials may require additional connections where the most remote point of a floor is more than 150 feet from a hose connection in a nonsprinklered building (200 feet if the building is sprinklered). In buildings with more than one standpipe, NFPA 14 governs how they may be interconnected.

Class II standpipe connections must be accessible and placed so that a hundred-foot hose can reach within thirty feet of any point in the building. Fire-hose cabinets must be identified by appropriate signage or have a clear glass panel providing visual identification.

Portable Fire Extinguishers (906)

The *IBC* refers to the *International Fire Code* for requirements relating to portable fire extinguishers. In general, fire extinguishers may be surface mounted or placed in clearly marked cabinets in readily visible and accessible locations. They must bear approved labels and be regularly tested.

Fire extinguishers of specific types are often required in kitchen areas and break rooms. Usually they must be placed within seventy-five feet of the location where they may be required.

Fire Alarm and Detection Systems (907)

The *International Building Code* lists the specific elements of fire alarm and detection systems that must be included on construction documents and approved before installation. These include, at minimum, the following:

- A floor plan indicating the intended uses of all rooms together with details showing ceiling height and construction
- Planned locations of notification and alarm-initiating devices along with alarm control and warning devices to indicate system malfunctions
- *Annunciator:* A central control unit indicating the status of the system using alphanumeric displays and indicator lamps
- Main power and battery backup connections, conductor types and sizes, and voltage-drop calculations
- Manufacturers and model numbers of all equipment and devices together with their listed ratings

Here, once again, the *IBC* lists specific requirements by occupancy type, referring frequently to the *International Fire Code* and NFPA 72 (the *National Fire Alarm Code*).

A—A manual fire alarm system is required for occupant loads greater than three hundred people. Where the occupant load is one thousand or more, activation of the system must trigger an emergency, voice-alarm communications system, which must have an approved emergency power source.

B—A manual fire alarm system is required for occupant loads greater than five hundred people, or more than one hundred people above or below the lowest level of exit discharge.

E—A manual system is required for occupancies greater than fifty. If sprinklers or smoke detectors are installed, they must be connected to the building fire alarm system. Manual fire alarm boxes are not required if an extensive automatic warning system is installed, including *all* of the following:

- Special areas such as interior corridors, auditoriums, cafeterias, gymnasiums, shops, and laboratories are provided with smoke or heat detectors, or other approved devices;

- Off-premises monitoring together with the ability to activate an evacuation signal from a central point; and

- The building has a two-way communications system in all normally occupied spaces with a constantly attended central office from which a general evacuation alarm can be initiated.

F—A manual system is required for buildings two or more stories high that have an occupant load of five hundred people above or below the level of exit discharge.

In the preceding occupancies and in *Group M*, which follows the requirement for manual fire alarm boxes is *waived* if the building is equipped throughout with an automatic sprinkler system and notification devices that will activate the alarm system when water flow is detected.

H—The *IBC* requires manual fire alarm systems for *Group H-5* occupancies and specifies smoke-detection systems for highly toxic gases, organic peroxides, and oxidizers, referring to several pertinent chapters of the *International Fire Code*.

I—A manual alarm system together with an automatic fire-detection system is required in all institutional occupancies. Automatic, electrically supervised smoke-detection systems are also mandated in waiting areas open to corridors.

- In *Group I-1* occupancies, smoke alarms are required in sleeping areas unless the building is equipped throughout with an automatic fire-detection system.

- Corridors in nursing homes and detoxification facilities (*I-2*) and spaces open to corridors are required to have automatic fire-detection systems. Corridor smoke detection is not required so long as patient sleeping units are equipped with smoke detectors with a visual indicator in the corridor and an audible and visual alarm at the attending nursing station.

- *Group I-3* occupancies must have manual and automatic fire alarm systems for alerting the staff. Key-operated manual fire alarm boxes may be locked in areas occupied by detainees so long as supervisory staff are present and have keys readily available. Approved automatic smoke-detection systems must be installed in housing and sleeping areas, day rooms, group-activity spaces, and other contiguous common areas used by residents. Detectors may be placed in exhaust ducts from cells or behind approved protected guards to prevent tampering or damage.

M—In *Mercantile* occupancies other than *Covered Mall Buildings* (Section 402), a manual fire alarm system is required for occupant loads greater than five hundred people, or more than one hundred above or below the lowest level of exit discharge.

R-1—Manual fire alarm systems are required except when the building is two stories or less in height, all individual guest rooms and contiguous spaces are separated by one-hour fire partitions, and each guest room has its own exit. Manual alarm boxes are not required if the building has an automatic sprinkler system throughout that will activate the alarm upon water flow.

Automatic fire alarm systems are required in all interior corridors serving guest rooms, unless each room has an egress door leading directly to an exit. In buildings not equipped with an automatic sprinkler system, smoke alarms in guest rooms must have a backup electrical supply and be annunciated by guest room at a central, staff location at which the alarm system may also be manually activated.

R-2, R-3, and R-4—Regardless of occupant load, single- or multiple-station smoke alarms are required in each sleeping room, on the wall or ceiling immediately adjacent to each sleeping area, and in each habitable story within a dwelling unit.

In addition to the requirements by occupancy group, the *IBC* spells out numerous specific alarm features such as power source and backup provisions, system response and fire department communication features, mounting locations for fire alarm boxes, and the like.

For example, in *Special Amusement Buildings*, when two smoke detectors or a single detector with alarm verification activate an automatic sprinkler system, the system must perform three actions:

1. Illuminate the means of egress with a specified amount of light.
2. Stop any conflicting alarm sounds and visual distractions.
3. Activate a directional exit marking system along with prerecorded audible instructions directing occupants to the nearest exit.

High-Rise Buildings (with certain exceptions) must be provided with automatic fire alarm and emergency voice/alarm communications systems. Smoke detectors are required as well and must be connected to the automatic fire alarm system. Smoke-detection devices must be placed in all mechanical, electrical, and telecommunications rooms and in elevator machine rooms and lobbies that do not have sprinkler protection. They must also be provided in the main-air return or exhaust-air plenums and at each connection to a vertical riser serving at least two stories from a return-air duct or plenum of an air-conditioning system.

Emergency voice/alarm communications systems in high-rises must sound an alarm and be followed by voice instructions directing occupants to approved terminal areas on the alarmed floor or the floors immediately above or below. Such designated areas may include corridors, elevator lobbies, rooms or tenant spaces exceeding one thousand square feet, dwelling or sleeping units in *R-1* and *R-2* occupancies, or other designated refuge areas.

Two-way fire department communications systems are also required in high-rises and must be designed in accordance with NFPA 72. Such systems must enable communications between a central fire command center and other critical areas: elevators and elevator lobbies, enclosed exit stairways, areas of refuge, fire-pump rooms, and emergency-power rooms.

Atriums connecting two or more stories must have a fire alarm system. In *Groups A, E,* or *M*, such occupancies must be provided with an emergency voice/alarm communication system. Other special building types have specific alarm requirements, such as *Covered Mall Buildings* in excess of fifty thousand square feet. These buildings, like atriums, must have emergency voice/alarm communications similar to those described for high-rises that must be accessible to the fire department.

Visible Alarms

In new construction or where existing fire alarm systems are upgraded or replaced, visible alarm notification is required by the *2003 IBC* in all public and common areas. In employee work areas as well as dwelling and sleeping units in *Group R-2* occupancies, wiring systems must be designed to accommodate visible alarms where necessary. In *Group I-1* and *Group R-1* occupancies, visible alarm notification, activated by both in-room smoke alarms and the building fire alarm system, must be provided.

Emergency Alarm Systems (908)

Emergency alarms are required in all *Group H* occupancies for the detection and notification of an emergency condition. Continuous-gas-detection systems must be provided, for example, in *Group H-5* facilities where hazardous production materials (HPMs) are used. Gas-detection systems are also required where toxic and highly toxic gases are used or stored.

Gas-detection systems must be designed with both audible and visible alarms, provide warning both within and outside of the area where an emergency may be expected to occur, and be separate from all other alarm systems. A gas-detection system must automatically close gas-supply pipes and tubes at the source, for whatever type of gas is detected, by means of a shutoff valve. This requirement may be waived where the facility is constantly attended, and has readily accessible shutoff valves that are operated at low pressure (less than fifteen pounds per square inch).

Machinery rooms must contain a refrigerant detector with an audible and visual alarm. Ammonia-system machinery rooms must be equipped with a vapor detector. Detectors and alarms must be provided in approved locations as specified by the *International Mechanical Code*.

Smoke-Control Systems (909)

This section of the *International Building Code* establishes minimum engineering requirements for mechanical and passive smoke-control systems when they are required by provisions specified elsewhere in the *Code*. It establishes "minimum requirements for the design, installation and acceptance testing of smoke control systems that are intended to provide a tenable environment for the evacuation or relocation of occupants."° Once again, the *IBC* refers specifically to the construction documents including "sufficient information and detail to adequately describe the elements of the design necessary for the proper implementation of the smoke control systems."°

The bulk of the section specifies detailed levels of performance and provides engineering formulas for calculating design factors, determining control methods, and establishing control-system standards. Included are such factors as smoke-barrier construction and opening protection, pressurization methods and pressure differentials across smoke barriers, mechanical smoke-exhaust systems for large enclosed volumes such as atriums and malls, detection and control systems, and acceptance testing.

Smoke and Heat Vents (910)

Approved smoke and heat vents must be installed in the roofs of single-story buildings in several use and occupancy classifications to facilitate the removal of smoke and heat from the building. Such vents must be labeled and capable of being operated by automatic and manual means. In sprinklered buildings, smoke and heat vents must operate automatically. In nonsprinklered buildings, vents must be automatically actuated by means of a heat-responsive device. Gravity-operated drop-out vents are also permitted in certain circumstances. This type of vent contains heat-sensitive glazing designed to shrink and drop out of the vent opening when the air temperature reaches a certain level in a defined time period.

Smoke and heat vents are required in the following occupancy classifications:

- buildings or portions thereof, in *Group F-1* and *Group S-1*, having more than fifty thousand square feet of undivided area, or where maximum exit access travel distance is increased as allowed in the *Means of Egress* section of the *Code*;
- *Group H-2* or *H-3* buildings exceeding fifteen thousand square feet of single-floor area;
- areas of buildings in *Group H* used for storing any of several classes of liquid and solid oxidizers, detonable organic peroxides, unstable or water-reactive materials as required for a high-hazard classification; and
- buildings or portions thereof used for high-piled combustible rack or stock storage in any occupancy group.

Vent spacing and vent area to floor area ratios are specified in the *IBC* according to occupancy group and commodity classification. Maximum vent spacing varies from seventy five to 120 feet and area ratios from 1:30 to 1:100. Smoke and heat vents must be located twenty feet or more from adjacent fire walls and lot lines, and ten feet or more from fire barrier walls. They must be uniformly located within roof areas above high-piled storage, coordinated with structural members, roof pitch, sprinkler, and draft-curtain locations. Draft curtains, where required, must be constructed of lath and plaster, gypsum board, sheet metal, or other approved materials that resist the passage of smoke, and have a minimum depth of from four to six feet.

Engineered, mechanical smoke-exhaust systems are permitted as an alternative to smoke and heat vents, where approved. When exhaust fans are used, they must have a maximum individual capacity of thirty thousand cubic feet per minute with a maximum distance between them of one hundred feet. Exhaust fans must be uniformly spaced within the draft-curtained area and be automatically operated by the sprinkler system or by heat detectors. Individual manual controls must also be provided for each fan. Supply air for an exhaust-fan system must be uniformly distributed around the perimeter of the area served, at or near the floor level, and provide at least half of the necessary exhaust volume.

Fire Command Center (911)

A fire command center must be provided in *high-rise buildings* for fire department operations. It must be an accessible room having a minimum of ninety-six square feet and a minimum width of eight feet, separated from the remainder of the building by at least a one-hour fire-resistance-rated fire barrier. A layout of the command center must comply with NFPA 72 and be approved prior to installation. The features to be provided are listed in detail by the *IBC* and include all communications systems, annunciator units, status-indicator displays and controls for emergency power, air-handling systems, sprinkler valve and water-flow detectors, and fire pumps.

The command center must also have an adequate work surface at which are available schematic building plans detailing typical floor layouts, building-core configurations and means of egress, fire-protection systems, fire-fighting equipment, and fire department access.

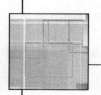

Closing Words

This chapter addressed those sections of the *IBC* that are most applicable to the interests of facility-planning professionals including designers, architects, and facility managers. Beginning with a section briefly describing the overall structure of the model code, we looked at the following subject areas:

- Use and occupancy classification
- Types of construction
- Fire-resistance-rated construction
- Fire-protection systems

The next chapter explores the *IBC* standards as they relate to means of egress. General requirements covering occupancy loads and the necessary widths of all passageways used for exiting are spelled out here. Areas of refuge for people who may be temporarily unable to reach an exit are discussed. Aisles, corridors, doors, changes in level, illumination, and signage are addressed in conjunction with exit access, exits, and exit discharge. These factors are all integral parts of the planning process as they relate to wayfinding.

All current codes are periodically updated; therefore, the current documents adopted by state and local authorities must always be consulted regarding specific code issues.

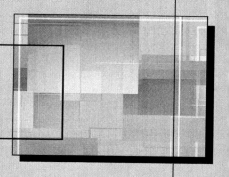

CHAPTER 14

Egress

But, Mousie, thou art not alone, In proving foresight may be vain; The best-laid schemes o' mice an' men Gang aft agley, An' leave us nought but grief an' pain, For promis'd joy!
—Robert Burns (c. 1780) °

$\mathbf{A}$s we pointed out at the beginning of the preceding chapter, the very existence of the *International Building Code* is the result of efforts by many organizations, governmental and otherwise, to agree upon consistent requirements to protect and accommodate the public. In few areas has there been greater success than in the promotion of accessibility to buildings of all types by individuals with disabilities. By no means have all the barriers been eliminated, but we have succeeded in making accessibility considerations a part of our national design consciousness.

This chapter continues where Chapter 13 left off, following the basic outline of the *International Building Code*. The final two chapters cover the *IBC*'s Accessibility requirements (Section 1100) and the related provisions of the Americans with Disabilities Act (ADA), which was passed into federal law in 1990.

The technical guidelines in the major section (Section 4) of the ADA *Standards for Accessible Design* (28 C.F.R. Part 36) are the same as those in the American National Standard Institute's document *A177.1*. The second paragraph of Section 1100 of the *IBC* incorporates *ANSI A177.1* by reference, referring to it as *ICC A117.1*, making it part and parcel of the *IBC*. The *ICC/ANSI A117.1-1998* version was extensively rewritten to bring it into line with federal accessibility requirements to the extent possible.

The NMHC/NAA, a major association representing the housing industry, stated in a 2000 memorandum that it considered the fact that the ICC codes have been designed to comply with both the *ADAAG* (*Americans with Disability Act Accessibility Guidelines*) and *FHAG* (*Fair Housing Accessibility Guidelines*) a "very critical victory for apartment/seniors housing developers." ° The memorandum goes on to say that "[c]onflicting accessibility requirements found in the existing codes have been removed and the accessibility provisions have been 'main streamed' into the body of the code. For example, accessibility provisions related to means of egress are in the means of egress chapter and not in a separate section."

One important provision included in the *IBC*, derived from this recent version of *A117.1*, is the distinction between two different types of dwelling or sleeping units. Each provides a different level of accessibility, type *A* and type *B*. The type *A* dwelling unit is fully compliant with the *ADAAG* standards maintained by the U.S. Departments of Justice (DOJ) and Transportation (DOT). The type *B* unit also meets the *FHAG* standards of the U.S. Department of Housing and Urban Development (HUD). Both of these are discussed ° presently.

Lest one acquire the impression that the path to enlightenment is followed without difficulty, however, this same NMHC document criticizes several organizations, notably the NFPA, for promoting a set of fire and life-safety codes that compete with the ICC code family. ° The Building Owners and Managers Association (BOMA) International, an NMHC counterpart representing commercial real estate interests, also published a position paper encouraging "the ICC and the NFPA to resolve their differences and collectively put forth one single set of comprehensive model building codes for the built environment."° Once again we are reminded of the necessity of checking which codes govern in your jurisdictional location. It is critically important, though, that we understand the intent and *meaning* of the codes in order to plan spaces that are not only accessible but as safe as we can make them.

FIRE HAS AN ATTITUDE°

Almost at the hour that I finished writing the first draft of this chapter in a Chicago suburb, there was a deadly fire in a forty-year-old office building downtown. Originally designed as a corporate headquarters by a major architectural firm,° the building was purchased by Cook County in the mid-1990s for its administrative offices. Originally built without an automatic sprinkler system, sprinklers were not a part of the extensive renovation that was performed nearly ten years ago.

Sprinklers are not currently required in Chicago buildings constructed prior to 1975, although several recently renovated buildings have been equipped with such systems. One state-owned building, constructed in 1920 and refurbished at about the same time as the county building, is now fully sprinklered.° In addition, its stairwell doors were enhanced with electronic hardware that unlocks them automatically in case of emergency, as well as a system that allows firefighters to manually pressurize individual stairwells against smoke infiltration.

In the case of the Cook County building, it is likely that sprinklers would have at least slowed the development of the intense fire on the twelfth floor. According to the NFPA, the chances of dying in a fire, as well as the average property loss per fire, are cut by one-half to two-thirds when a sprinkler system is present.° There were several other system and procedural failures, however, that are believed to have contributed to the six deaths in this fire.

A plan of the Cook County building shows a smoke tower adjacent to the southeast stairwell where the fatalities occurred (Figure 14-1). A mechanically operated vent in the smoke tower was designed to pull smoke out of the stairwell vestibule, preventing its intrusion into the stair tower. In this case, a combination of factors contributed to smoke filling the stairwell. The smoke tower venting system may not have operated properly to begin with, and firefighters using the stairwell door adjacent to the fire area could not prevent smoke from entering the stair once the door was forcibly opened. People attempting to use this stairwell to evacuate areas above the fire floor were turned back due to the activity blocking the stair enclosure at the twelfth floor. Many were then trapped in the stairwell by egress doors that were locked from the stairwell side. This is a permissible practice under the *IBC* (Sections 1008.1.8.7 and 403.12) and Chicago codes, frequently done for security reasons. Subsequent to this event, however, the local chapter of BOMA sent a letter to its members strongly supporting retrofitting older buildings with electronic devices to facilitate remote unlocking of stairwell doors in case of emergency.°

Yet another factor in this complex chain of events was a series of evacuation orders that were apparently issued by building management personnel. Fire department procedures for high-rise fire events call for fire officials to assume control of the fire command center, which the codes require for that purpose, including all communications and annunciator systems (*IBC*, Section 911). In this case, there were at least three orders given to evacuate the building. The first announcement was made shortly after 5 p.m., when the fire was discovered, advising people to evacuate the twelfth floor. This was followed shortly by an announcement that the entire building should be evacuated. One such announcement was heard as high as the thirty-second floor, sending additional people into both stairwells. It was reported that office workers were never advised to avoid the southeast stairwell. According to 911 tapes in which announcements can be heard in the background, the final announcement was made twelve minutes after firemen arrived but not by fire officials.

Figure 14-1 Stairwell and Smoke Tower.

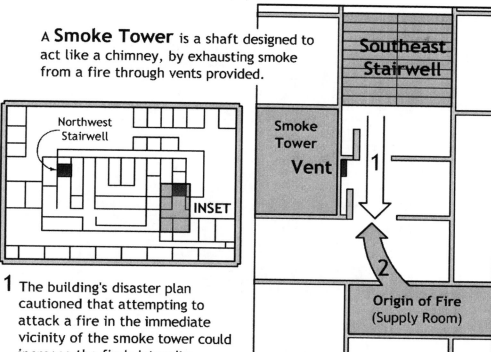

A **Smoke Tower** is a shaft designed to act like a chimney, by exhausting smoke from a fire through vents provided.

1 The building's disaster plan cautioned that attempting to attack a fire in the immediate vicinity of the smoke tower could increase the fire's intensity. Firefighters opened the southeast stairwell next to the smoke tower, only a few feet from where the fire started.

2 Intense heat and smoke then rushed into the stairwell toward the firefighters.

The origin of the fire was a supply room on the 12th floor of the Cook County Adminstration Building.

The Chicago fire commissioner was quoted as saying that the fire department did not order the building evacuated and that people on the upper floors probably would have been safer had they stayed in their offices and waited for help.° He further observed that "people are still running scared from September 11."° As it was, a number of people attempted to get out of the building using the blocked stairwell and were trapped for an extended period of time between the fourteenth and twenty-seventh floors, the topmost of these being the only floor on which a stairwell door was found to be unlocked (Figure 14-2).

During a press conference on October 25 at the city's 911 center, Chicago's fire commissioner told reporters: "It was one of those things that if you picked that one day where a lot of little things could go wrong, they did. And they lined up."°

What could have made a difference? Any *one* of the following measures might have prevented loss of lives in this fire:

- An automatic sprinkler system to suppress the fire at its origin
- Pressurized exit enclosures to prevent the intrusion of smoke and toxic gases
- Automatic sensors to deactivate the locking mechanisms on all stairway exit doors
- Properly executed building emergency evacuation procedures

All of the items listed are standard fire-prevention and detection methods advocated by the codes. Moreover, pressurizing stairwells and unlocking doors can be implemented

Figure 14-2   Locations of Fire-Related Events.

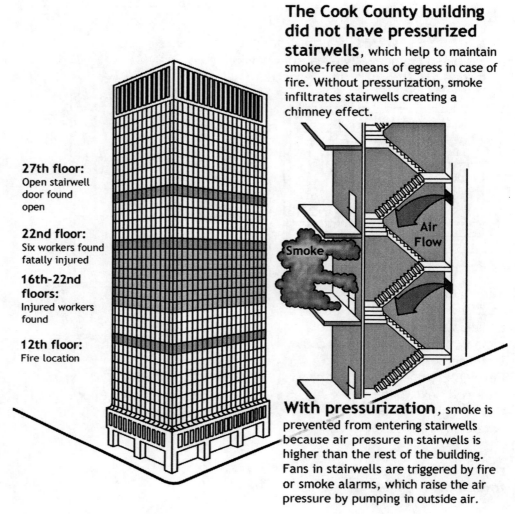

The Cook County building did not have pressurized stairwells, which help to maintain smoke-free means of egress in case of fire. Without pressurization, smoke infiltrates stairwells creating a chimney effect.

27th floor:
Open stairwell door found open

22nd floor:
Six workers found fatally injured

16th-22nd floors:
Injured workers found

12th floor:
Fire location

Smoke

Air Flow

With pressurization, smoke is prevented from entering stairwells because air pressure in stairwells is higher than the rest of the building. Fans in stairwells are triggered by fire or smoke alarms, which raise the air pressure by pumping in outside air.

in a relatively fail-safe manner. It is easier to determine when a fan that is supposed to run all the time has stopped than it is to test a smoke tower vent fan that may not have been needed since the building was built. If the door-latching mechanisms are designed so that electrical power is required to keep the doors secured, then the doors will unlock automatically if the power and, therefore, the pressurizing fans fail.

In the final analysis, there is no substitute for a well-conceived and executed plan for dealing with an emergency situation. Chapter 13 referred to numerous NFPA regulations governing fire-resistant construction, alarm systems, and fire-suppression systems including sprinklers. The *Life Safety Code* (NFPA 101) also calls for evacuation/relocation plans and regularly scheduled drills for certain occupancy types. Building owners and facility managers are mandated by NFPA 101 to hold regular emergency drills for occupants of workplaces, educational institutions, and health-care facilities. It is incumbent upon responsible building owners and operators to implement and effectively communicate emergency procedures. The key components of emergency preparedness may be summarized as follows:

- Occupant familiarity with emergency procedures through knowledge and practice
- Early warning through an alarm or voice communication system
- Adequate means of egress

MEANS OF EGRESS (1000)

Let us begin with some essential definitions:

- *Means of egress*—A continuous and unobstructed path of vertical and horizontal egress travel from any occupied portion of a building or structure to a public way. A means of egress system consists of three separate and distinct parts:
  - The exit access
  - The exit
  - The exit discharge°
- *Accessibility*—A site, building, facility, or portion thereof that complies with *this chapter*° (or *these guidelines*, in the case of the ADA°).

The latter definition, although self-referential, means that people with disabilities must have equal access to the means of egress. An *Accessible Route* is a "continuous unobstructed path connecting all accessible elements and spaces of a building or facility. Interior accessible routes may include corridors, floors, ramps, elevators, lifts, and clear floor space at fixtures. . . ."° Incorporating this definition, the *IBC* arrives at the following:

- *Accessible means of egress*—A continuous and unobstructed way of egress travel from any point in a building or facility that provides an accessible route to an area of refuge, a horizontal exit, or a public way.°

The preceding chapter began with a discussion of how building codes mandate the design and construction of all kinds of structures in such a way as to minimize the chances of life-threatening situations arising. Then we saw how fire-protection systems are intended to provide building occupants with knowledge of a threat as well as the means to take action against it. This chapter describes how the codes help ensure that occupants are able to get out of a building safely under all circumstances, particularly when all other safeguards have failed. The accessibility provisions are intended to ensure that design foresight includes people with disabilities.

The components of a means of egress system are as follows:

- *Exit access*—A path that leads from any occupied portion of a building or structure to an exit.
- *Exit*—The portion of the egress system that is separated from other interior spaces by fire-rated construction and opening protectives. The exit protects the egress path at the point of transition between the exit access and the exit discharge. Included are exit passageways and enclosures, doors at ground level, horizontal exits, exterior exit ramps, and stairs. Stairs, of course, are not accessible to many people with disabilities.

   A *horizontal exit* is a path, on approximately the same level, from one building to another or to a protected compartment beyond a fire or smoke barrier. Remember that, from a code standpoint, areas separated by a fire wall are considered to be separate buildings.
- *Exit discharge*—A path that leads from the exit to a public way. *Public way* means a street, alley, or open parcel of public land that is available on a permanent basis.

General Requirements

The egress system may not be interrupted by building elements other than those that are components of the means of egress system, nor may the required capacity of a path be reduced by projections other than those specifically permitted.

General Means of Egress (1003)

All three elements of a means of egress system must have a ceiling height of at least seven feet. Objects may protrude below the ceiling so long as minimum headroom of eighty inches is maintained over at least 50 percent of the ceiling surface. Barriers at least twenty seven inches high must be provided where the vertical clearance is less than eighty inches. Door closers and stops may reduce headroom to not less than seventy eight inches.

Horizontal projections may not protrude more than four inches into an egress path if they are located more than twenty seven inches or less than eighty inches above the walking surface, except stair and ramp handrails, which may project 4.5 inches. The same height requirements apply to freestanding objects mounted on posts or pylons, and such objects may not overhang the post or pylon by more than twelve inches. Objects of this kind may not reduce the clear minimum clear width of accessible routes.

Where elevation changes of less than twelve inches occur along a path of egress, the floor surface must slope. Where the change in elevation is six inches or less, ramps must either have handrails or floor finish materials that contrast with the adjacent flooring. Special provisions apply if the slope is greater than 5 percent (Section 1010). One or two steps are permitted in certain building types and at locations not required to be accessible. Walking surface materials in general must be securely attached and slip resistant.

Occupant Load and Egress Width

The *IBC* specifies that occupancy loads be determined on a basis of the average area allocated per person for the planned occupancy type. Minimum required egress width is measured in inches, based upon the calculated occupant load.

Occupant Load (1004)

Occupant load is calculated using the actual planned density for which a space is designed or using the occupancy type as listed in Table 14-1. Where the path of egress from one major area leads through another, the total calculated occupant load must include the number of people needing to exit both spaces. If the calculated density for a planned occupancy exceeds the value given in Table 14-1, the *IBC* permits the higher value to be used so long as the load does not exceed five square feet per person and all other code requirements.

In such cases, a plan showing approved aisles, seating, and fixed equipment must be provided and, if required by the building official, posted. Any space in the *Assembly* group must have its planned occupancy load conspicuously posted near the main exit.

Egress capacity may not decrease in the direction of travel toward an exit. Where exits serve multiple floors, the occupancy load of each individual floor is used to determine the exit capacity of that floor. Where an egress path converges from floors above and below, its capacity from the point of convergence must be at least equal to the sum of the two floors. The occupancy of a mezzanine exiting into an area below must be included in the occupancy density of the space into which the path of egress flows.

The occupancy load of spaces with fixed seating and aisles is determined by the actual number of seats planned. The width of seating booths is measured along the backrest, allowing twenty four inches per person. Fixed seating without arms is calculated on a basis of eighteen inches per person over the seating length.

Outdoor areas used only for building service may have only one means of egress. Courtyards and similar outdoor spaces usable by building occupants are treated similarly to enclosed areas for code purposes. The required egress capacity of an outdoor area that exits through a building must be added to the occupant load of the building. As in most other situations, in which there are multiple occupancy classifications to be considered, the most stringent requirements apply.

Egress Width (1005)

Minimum egress width in inches is determined by multiplying the planned occupancy density by the factors in Table 14-2, subject to other more specific requirements elsewhere in the *Code* (such as Assembly occupancies in Section 1024). If more than one egress path is required from a given location, the loss of one in an emergency cannot reduce the overall capacity by more than half. Maximum capacity must be maintained continuously from any floor level to the ultimate point of exit discharge.

Table 14-1   Maximum Occupancy Density (Square Feet per Occupant) °

Group	Occupancy	Floor Area per Occupant (Sq. Ft.)
A-1	Assembly with Fixed Seats	Determined by Number of Seats

	Assembly without Fixed Seats
	    Standing Space
	    Concentrated (Chairs)
	    Unconcentrated (Tables & Chairs)
		5 Net
		7 Net
		15 Net

	Stages and Platforms	15 Net
A-2	Kitchens (Commercial)	200 Gross

A-3	Airport Terminal
	    Waiting Areas
	    Baggage Claim
	    Concourse
	    Baggage Handling
		15 Gross
		20 Gross
		100 Gross
		300 Gross

	Assembly Gaming Floors (Blackjack, Slots etc.)	11 Gross
	Bowling Centers	7 Net
	Courtrooms (Other Than Fixed Seating Areas)	40 Net
	Exercise Rooms	50 Gross

	Library
	    Reading Rooms
	    Stack Area
		50 Net
		100 Gross

	Locker Rooms	50 Gross
A-4	Skating Rinks & Swimming Pools Decks Rink and Pool	15 Gross 50 Gross
B	Business	100 Gross
E	Educational Classrooms Shops and Vocational Rooms	20 Net 50 Net
F	Industrial	100 Gross
H-5	Fabrication and Manufacturing	200 Gross

I	Institutional
	    Outpatient Areas
	    Sleeping Areas
	    Inpatient Treatment Areas
		100 Gross
		120 Gross
		240 Gross

M	Mercantile
	    Basement and Grade Floor Areas
	    Areas on Other Floors
	    Storage Stock Shipping Areas
		30 Gross
		60 Gross
		300 Gross

R	Residential	200 Gross
R-2	Dormitories	50 Gross
S	Accessory Storage Areas, and Mechanical Equipment Room	300 Gross
	Warehouses	500 Gross
S-2	Aircraft Hangars	500 Gross
	Parking Garages	200 Gross
U	Agricultural Building	300 Gross

Table 14-2   Egress Width per Occupant Served (Inches per Occupant)°

Occupancy	Without Sprinkler System		With Sprinkler System	
	Stairways	Other Egress Components	Stairways	Other Egress Components
Occupancies Other Than Those Listed Below	0.3	0.2	0.2	0.15
Hazardous H-1, H-2, H-3, & H-4	0.7	0.4	0.3	0.2
Institutional I-2	NA	NA	0.3	0.2

Doors that swing into the path of egress may not obstruct more than 50 percent of the required width, nor project more than seven inches into the required width when fully open. This requirement does not apply to dwelling and sleeping units in groups *R-2* and *R-3*.

Using Table 14-1 and Table 14-2 together as an example, a thirty-thousand-square-foot floor of a sprinklered office building would require a minimum of sixty inches of egress stairway width. A waiting area in an airport terminal, on the other hand, requires the same exit capacity for only three thousand square feet of gross floor areas.

Accessible Means of Egress (1007)

If a space is required to be accessible, at least one accessible means of egress must be provided. If the *Code* otherwise calls for multiple exits, at least two accessible egress paths are required. Accessible means of egress must lead continuously to a public way, consisting of routes, stairways within exit enclosures, elevators and platform lifts, horizontal exits, and smoke barriers that meet specific accessibility criteria specified by the *Code*.

Areas of Refuge

In certain situations, such as those in which an exit discharge is not accessible, provision must be made for assisted rescue. The same applies to exit stairways open to the exterior, although here a code-compliant area of refuge is an acceptable alternative. An area of refuge is a protected space where disabled persons can wait for assistance with a reasonable expectation of safety.

Accessible exit stairways must be at least forty eight inches wide, clear of its handrails, and either incorporate or be accessed from an area of refuge or a horizontal exit. An elevator may be considered part of an accessible egress system if it is equipped with backup power and specified signaling devices and is accessed from an area of refuge. These requirements are mitigated by the presence of an automatic sprinkler system. The *IBC* generally does not permit platform lifts for wheelchairs as part of an accessible means of egress, except in specified circumstances.

Buildings having at least four stories above or below the level of exit discharge must have an elevator in at least one path of egress. If the building has an automatic sprinkler system, this requirement is waived on floors where there is a horizontal exit or an accessible ramp at or above the exit discharge level.

The permitted travel distance to an area of refuge may not exceed the equivalent maximum distance to an exit for the occupancy group; and, of course, it must be able to be reached from the space it serves by an accessible means of egress. Required refuge areas must have direct access to an accessible exit stairway or elevator as already described. Where an elevator lobby is designated as an area of refuge, it must meet the *Code* requirements for smokeproof enclosures unless the elevators themselves are within a refuge area formed by a horizontal exit or smoke barrier. Areas of refuge, in general, must be separated from the remainder of the floor by a smoke barrier and designed to minimize the intrusion of smoke, unless the space itself and the area it serves are fully sprinklered.

Each area of refuge must provide space for one thirty- by forty-eight-inch wheelchair per two hundred occupants without compromising the required egress path width. No wheelchair space may be obstructed by more than one adjoining wheelchair space. Two-way communications including both audible and visible signals must be provided between any refuge area and a central control point. If the command center is not always manned, the area of refuge must also have access to a public telephone. All refuge areas must have identification signage stating: AREA OF REFUGE, along with the international accessibility symbol, meeting the requirements of *ICC A117.1*. Areas of refuge must also contain posted information describing how the space and communications equipment are to be used in an emergency situation. Where the *Code* requires illuminated exit signage, the area of refuge signage must also be illuminated and tactile signage compliant with *A117.1* must be provided at each door.

If exits and elevators serving an accessible space do not provide accessible means of egress, appropriate signage must be provided directing occupants to an approved egress system.

Exterior exit stairways and areas for assisted rescue must meet the same size and identification requirements already described. Such exterior areas must be at least 50 percent open, and separation walls must meet the *Code* criteria for exterior walls (Section 704). Open areas above must be configured to minimize the accumulation of toxic gases or smoke.

Doors, Gates, and Turnstiles (1008)

Means of egress doors must be unambiguously recognizable as doors. They cannot be concealed by curtains or other decorative elements such as mirrors or other reflective materials. The minimum door opening width must be at least thirty two inches measured along a perpendicular from the door stop to the face of the door and, in any case, must be wide enough to handle the occupant load of the space served. The maximum width of a swinging door leaf is forty eight inches. If a door opening contains two door leaves, one must provide the minimum thirty two-inch clear opening width. Two doors in series must have space between them equal to forty eight inches plus the width of any door swinging into the space. Doors in series must either swing in the same direction or away from the space between them. Door heights may not be less than six feet eight inches.

There are several exceptions to the door width and height requirements, based upon occupancy. For example, doors used for the movement of beds in occupancy group *I-2* (medical facilities) must provide a clear width of $41^1/_2$ inches. Door openings to resident sleeping units in group *I-3* (secured facilities) may be as narrow as twenty eight inches. Storm and screen units and entrance doors in *R-2* and *R-3* dwelling units do not need to be four feet apart.

In general, egress doors must be swinging and side-hinged, but the *IBC* specifies exceptions here, too. These include single-dwelling units in groups *R-2* and *R-3* and detention facilities under group *I-3*. Revolving and horizontal sliding doors that meet specific code criteria may also be used on occupancies other than group *H* (hazardous).

In group *H* occupancies, or where the occupancy load consists of at least fifty people, doors must swing in the direction of egress travel. The force necessary to open an interior, side-swinging door without a closer may not exceed five pounds. Other side-swinging, folding, and sliding doors must not require more than a thirty-pound force to move them when the force is applied to the latch side. Door latches must release with a force of fifteen pounds.

Power-operated and power-assisted doors must be designed to be opened manually in the event of a power failure so that the egress path is not compromised. The force necessary to set the door in motion may not exceed fifty pounds applied to the side from which egress is made.

Revolving Doors
Revolving door leaves must collapse when a force of not more than 130 pounds is applied within three inches of an outer edge. When collapsed, they must provide parallel egress paths with a minimum combined width of thirty six inches.

Wherever a revolving door is used, a code-compliant, side-hinged swinging door must also be provided within ten feet and in the same wall as the revolving door. No revolving door may handle an egress capacity of more than fifty people, nor may it account for more than 50 percent of the egress capacity at that point.

The speed of a revolving door must also be limited in accordance with its diameter. The *IBC* specifies eleven to twelve revolutions per minute (rpm) for a six-foot-six-inch door, decreasing to seven to eight rpm for a ten-foot door. Revolving doors also may not be placed within ten feet of a stairway or escalator.

Horizontal Sliding Doors

When horizontal sliding doors are used within an egress system, they must be power operated and also capable of being operated manually should the power fail. When such doors are placed in a series, they must be at least forty eight inches apart. In addition, the door assembly must have a standby power supply that is electrically supervised and be capable of opening the door to the minimum required width within ten seconds of activation.

In the event that manual operation is required, the doors must be easily operated from both sides. The force required to open the door along its normal path cannot exceed thirty pounds initially or fifteen pounds to either close it or open it to its maximum width. When a force of 250 pounds is applied against the side of the door adjacent to the operating device, the door must be operable using a force not exceeding fifteen pounds.

When a sliding door assembly is required to have a fire-protection rating, it must be automatic or self-closing by smoke detection and be installed according to the requirements for opening protectives (Section 715).

Access-Controlled Doors

Building entrance doors or doors to tenant spaces in occupancy groups *A*, *B*, *E*, *M*, *R-1*, and *R-2* may be equipped with an entrance and egress access-control system. A sensor on the egress side must automatically unlock the doors when an occupant approaches. Activation of the building fire alarm system, automatic sprinklers, or fire-detection system must also unlock the doors, which must remain unlocked until the fire alarm system has been reset.

Access-controlled doors must also have a manual unlocking device within five feet of the secured doors, located between forty and forty eight inches above the floor. The device must be clearly labeled PUSH TO EXIT and, when activated, cause the doors to be unlocked for a minimum of thirty seconds. In the event power is lost to the access-control system, the doors must automatically be unlocked.

Entrance doors of buildings in group *A*, *B*, *E*, or *M* may not be secured on the egress side when the building is normally open to the public.

Security Grilles

Horizontal sliding or vertical security grilles may be used to secure some exits in groups *B*, *F*, *M*, and *S*. Such grilles must be operable from the inside without the need of a key, special skill, or effort when the space is occupied and must be secured in the full-open position when the building is open to the public. Where multiple means of egress are required, only 50 percent of the available exits may be equipped with security grilles.

Landings and Thresholds

The *IBC* specifies that there must be a floor or landing at the same elevation on each side of a door. Interior landings must be level, but exterior landings may have a slope of up to one-fourth-inch per foot (2 percent). Elevation differences due to a change in materials are permitted, so long as the difference does not exceed one-half inch. In residential applications (group *R*), more latitude is given. For example, a door may open at the top step of an interior stair so long as the door does not swing over the step.

The width of a stairway landing must be at least as great as the stairway or door, whichever is greater. Fully open doors may not reduce required dimensions by more than seven inches. Landings must be at least forty four inches long in the direction of travel. If a landing serves an occupant load of at least fifty people, doors in any position may not reduce the landing to less than one-half its required width. Landing length in certain residential applications may be reduced to thirty six inches.

Thresholds of sliding doors serving dwelling units may not exceed three-fourths inch. Other thresholds are limited to one-half inch in height. Where floor level changes are greater than one-fourth inch at a doorway or threshold, the transition must be beveled at a slope not greater than 1:2 (50 percent).

Door Operations, Locks, and Latches

Egress doors in general must require no key, special knowledge, or effort to operate unless specifically permitted by the *IBC*. If the doors are required to be accessible, door handles, locks, and other operating devices may not require tight grasping, pinching, or twisting of the wrist. All such devices must be installed between thirty four and forty eight inches above the finished floor.

Obviously, locks and latches are required in many common uses. Buildings in group *A*, where the occupancy load is less than three hundred, and in groups *B*, *F*, *M*, and *S* may have their entrances equipped with key-operated locks operable from the inside. The *Code* stipulates, however, that a sign with one-inch-high letters be placed on or adjacent to the door on the egress side, stating THIS DOOR IS TO REMAIN UNLOCKED WHEN BUILDING IS OCCUPIED. In addition, it must be easy to determine that the locking device is locked. Discretion is given to building officials to revoke the use of such locking devices for cause.

Automatic flush bolts may be used where egress doors are installed in pairs, so long as the normally inoperable leaf has no surface-mounted hardware or door handle. In group *R*, where the occupancy load is ten or less, individual dwelling or sleeping units may be equipped with a dead bolt, night latch, and security chain, so long as all are operable from the inside without using a key or tool.

Manually operated flush or surface bolts are not permitted, except on the inactive leaf of pairs of doors serving equipment or storage rooms or doors not required for egress in dwelling or sleeping units. The unlatching of any leaf cannot require more than one operation except in places of restraint or detention or where the use of bolts is specifically permitted as already described.

Delayed egress locks may be used on doors serving occupancies *other than* groups *A*, *E*, and *H*, if the building is equipped throughout with an automatic sprinkler system or an approved automatic heat- or smoke-detection system. Emergency lighting must be provided at the doors, which must unlock automatically when any of the fire-suppression or warning systems are activated. Doors must unlock automatically if power is lost or by a signal issued from the fire command center. A pressure of not more than fifteen pounds against the door must cause it to unlatch automatically in not more than fifteen seconds (or thirty seconds, if approved). The doors must be posted, above and within twelve inches of the release mechanism: PUSH UNTIL ALARM SOUNDS. DOOR CAN BE OPENED IN 15 (or 30) SECONDS.

Doors in stairways may be locked on the side opposite the egress side, provided they are *always* operable in the direction of egress travel without a key or special effort. Where fire exit and panic hardware is installed, the device that actuates the release mechanism must be at least one-half as wide as the door. No more than fifteen pounds of force against the device may be required to open the door.

Egress doors from occupancy groups *A* or *E*, where the occupancy load is one hundred or more, and all hazardous occupancies except *H-4* may not have a latch or lock unless the mechanism has panic hardware. If balanced doors are used where panic hardware is required, the hardware must be of the push-pad type. The push pad may not be wider than half the width of the door, measured from the latch side.

Gates and Turnstiles

Gates used in a means of egress must meet all code requirements for doors. Horizontal sliding or swinging gates in fences or walls surrounding a stadium may exceed the four-foot-width limit. If stadium gates are constantly supervised when the public is present, panic hardware is not required. Safe dispersal areas are required within fifty feet of the enclosed space, however, at the rate of three square feet per occupant between the fence and enclosed area.

Turnstiles may not be placed where they will obstruct any required means of egress. Each device must turn freely in the direction of egress if power is lost or when manually released. Each device must be at least $16\frac{1}{2}$ inches wide below a height of thirty nine inches, and at least twenty two inches wide above that height. Turnstiles may not comprise more that 50 percent of the total required egress capacity, nor may any single device handle more than fifty people, where all of these provisions are met.

If a turnstile is part of an *accessible* route, it must not be a revolving device. It must have at least thirty six inches clear width up to a height of thirty four inches, and at least thirty two inches from thirty four up to eighty inches. Turnstiles more than thirty nine inches high must meet the requirements for revolving doors. Nonportable turnstiles serving an occupant load greater than three hundred must have a code-compliant, side-hinged swinging door within fifty feet.

Changes in Level or Floor

Stairways and Handrails (1009)

All stairways must be constructed of materials consistent with the construction type of the building; however, wood handrails are generally permitted. Surfaces of stair treads and landings must be solid, they must be level in all directions within a tolerance of 2 percent (one inch in four feet), and finish surfaces must be securely attached. Structures in groups *F*, *H*, and *S* (other than publicly accessible parking garages) may have openings in landings and treads, but the openings may not exceed $1^1/_8$ inches in any direction.

The required width of stairways is determined by occupancy density (Table 14-2) but typically must be at least forty four inches. A stair serving an occupancy load of fifty or fewer must have a minimum width of thirty six inches. Where a stairway lift is installed in group *R-2* or *R-3* occupancies, a clear passage width of at least twenty inches must be provided (measured from the folded position of the seat and platform fold when not in use). Aisle stairs in Assembly occupancies have special requirements (Section 1024).

Spiral stairways may be used as egress system components only in dwelling units, spaces smaller than 250 square feet, or performing arts stage areas. Treads must have a minimum width of twenty six inches and a depth of $7^1/_2$ inches at a point one foot from the narrow edge. Maximum riser height is $9^1/_2$ inches, and a minimum headroom of seventy eight inches must be maintained.

Standard headroom clearance in other than spiral stairways must be at least eighty inches measured perpendicular to a line connecting the edges of the nosings. This clearance must be maintained across the full width of the stairway and its landings.

All treads and risers within the same stair must be of uniform size and shape, although winders of a consistent shape may be used in conjunction with rectangular treads in the same flight. Stair riser height must be between four and seven inches. Tread depth must be a minimum of eleven inches. Riser height and tread depth may not vary by more than $^3/_8$ inch within any flight of stairs, as measured at right angles between leading edges of adjacent treads.

Winder treads must have a minimum depth of ten inches, eleven inches at a walk line one foot from the narrowest edge. The variance of tread depth at the twelve-inch walk line must not exceed $^3/_8$ inch. Winder treads are only permitted in means of egress stairways within dwelling units.

In group *R-2* and *R-3* dwelling units and in group *U* occupancies that are accessories to *R-3* buildings, maximum riser height may be $7^3/_4$ inches. The minimum tread depth is ten inches for rectangular and winder treads (at the walk line), and the minimum winder tread depth is six inches. Nosings of between $^3/_4$ and $1^1/_4$ inches must be provided where the tread depth is less than eleven inches on stairways with solid risers.

Circular stairways must meet the same tread depth and riser height requirements as straight runs, and the smaller radius must equal or exceed double the width of the stairway. Minimum tread depth at the narrow end must be at least ten inches, eleven inches measured one foot from the narrow end. Individual dwelling units in group *R-2* and *R-3* occupancies may be excepted, per the *International Residential Code*.

As a means of egress, the use of alternating tread devices is limited to mezzanines of not more than 250 square feet in occupancy groups *F*, *H*, and *S*, from guard or control stations in group *I-3*, and access to unoccupied roofs. Both sides of such devices must be provided with handrails. Minimum tread dimensions must be seven inches wide by 8.5 inches deep, with a maximum riser height of 9.5 inches. (Minimum dimensions are to 8.5 inches wide by 10.5 inches deep treads with eight-inch risers where there are no more than five occupants.)

The vertical rise of a single flight of stairs may not exceed twelve feet between floors or landings. Outdoor stairways and approaches must be designed so their walking surfaces will not accumulate water. Where snow and ice are an issue, exterior stairs must be protected to prevent their accumulation, except in group R-3 and related group U occupancies.

Where a sloping walkway, driveway, or public way meets a stair, possibly serving as a landing, the top or bottom riser may be reduced along the slope. The slope may not exceed one inch in twelve; however, the riser must have a distinctive marking stripe, and the tread must have a nonslip surface.

Landings are required at the top and bottom of every stairway. The smaller dimension of any landing may not be less than the width of the stair but need not be larger than forty eight inches if the stairway has a straight run. Doors opening into a landing may not reduce its required width by more than half, nor may they project more than seven inches into the landing when fully open.

Handrails of adequate strength are required on both sides of most stairs. A single riser at an entrance in R-3 occupancies does not require a handrail; nor do decks, patios, and walkways where the areas of walking surfaces on both sides of the riser are greater than would be required for a landing. The same applies to single-riser elevation changes in residential dwelling and sleeping units. Stairways within dwelling units, spiral stairs, and aisle stairs serving seating on one side may have a single handrail. Aisle stairs meeting all other assembly occupancy requirements also may have a single handrail at the center.

Handrail heights must be between thirty four and thirty six inches. Wide stairs must have intermediate rails so that all parts of the stair required for egress capacity are within thirty inches of a rail. Monumental stairs must have the handrail located along the main traffic path. Except in dwellings, where newel posts are permitted, handrail gripping surfaces must be continuous and have no obstructions. In general, handrails should be continuous from one adjacent stair flight to another or return to a wall or the walking surface. Where handrails are not continuous, they must extend at least one foot beyond the top and bottom risers, horizontally at the top and continuing the slope at the bottom. In dwellings, the lowest tread may be wider than the stair and have a turnout or volute. Handrails in dwellings, in stairs not required to be accessible, need only extend from the top to the bottom riser.

In buildings taller than four stories, one stair must extend to the roof through a penthouse unless the roof has a slope greater than 33 percent. If the roof is considered unoccupied, an alternating tread device may be used with a trap door.

Ramps (1010)

Ramps have the same basic construction and surface finish requirements as stairs. Design considerations pertaining to outdoor conditions are also the same as for stairs.

The minimum clear width of a ramp that is part of a path of egress is the same as that of a corridor in the same egress system (see Corridors later in the chapter) but not less than thirty six inches. The width may not decrease in the direction of egress, and there may be no projections into the required ramp or landing width. The maximum slope of an egress ramp is one in twelve (8 percent), but other ramps may have a slope of up to one in eight ($12^1/_2$ percent). The rise over the length of any ramp may not exceed thirty inches. No portion of a ramp may have less than eighty inches of headroom.

Ramps with a rise greater than six inches must have handrails on both sides that meet the same requirements as those for stair rails. Intermediate rails must also be provided seventeen to nineteen inches above the walking surface of the ramp or landing. Curbs or other barriers must be used to provide additional edge protection within four inches of the walking surface, except where handrails are not required or the vertical drop-off within ten inches of a landing is less than one-half of an inch.

Landings are required at the top and bottom of ramps as well as at entrances, exits, doors, and turning points. The width of a landing may not be narrower than the widest ramp leading to or from it, nor may the slope exceed 2 percent in any direction. The landing length must be at least sixty inches, and changes in level (risers) are prohibited. Doors that open onto a ramp may not reduce its clear width to less than forty two inches.

Where changes in the direction of travel occur at landings between ramp runs, the landing must be at least five feet square. Landings in nonaccessible dwelling units in groups *R-2* and *R-3* may be thirty six inches by thirty six inches, minimum.

Guards (1012)

As additional protection against falls, guardrails must be provided along stairs, ramps, landings, open-sided walking surfaces, mezzanines, and industrial equipment platforms located more than thirty inches above the floor below. Where glazing is used to enclose stairways, ramps, and landing but does not meet the *IBC*'s strength and attachment requirements, guards are also required. Guards are not required where their presence would interfere with the activity for which a space is designed, principally in performance-related occupancies.

Guards must be at least forty two inches high, except in group *R-3* occupancies and individual dwelling units in group *R-2*. There, where the guard also serves as a handrail, the height must be between thirty four and thirty eight inches above the stair tread nosing. The height in assembly seating areas is treated under Special Requirements (1024).

Ornamental guards may have openings, but below the thirty four-inch height they must be small enough that a four-inch spherical object will not pass through. From thirty four to forty two inches, the openings may be up to double that size. In assembly seating areas, the transition point between the four- and eight-opening requirement is at twenty six inches.

At the open side of a stairway, the triangular openings formed by the tread, riser, and bottom rail must have a six-inch maximum dimension measured in the same manner. Elevated walking surfaces in mechanical equipment areas or other nonpublic areas in *I-3*, *F*, *H*, or *S* occupancies may have up to 21-inch openings between intermediate rails, balusters, and the like. The same requirement pertains to roof-mounted equipment less than ten feet from a roof edge or open walking surface that is elevated more than thirty inches. Screened porches and decks must also have guards in similar locations.

Means of Egress Illumination and Signage

Illumination (1006)

Means of egress systems must be illuminated to a level of one foot-candle at the floor, except for aisles in group *A*; sleeping units in group *I*; dwelling units in *R-1*, *R-2*, and *R-3*; and group *U*. Performance areas in the assembly group may have the light level at the floor reduced to 0.2 foot-candle provided the full illumination is restored when a fire alarm system is present and activated.

Emergency electrical power must automatically illuminate all critical elements of the egress path. These include exit access aisles, corridors, and passageways in spaces requiring at least two means of egress, plus stairways and exterior egress components in buildings requiring two or more exits. Interior exit discharge elements (1023), exterior egress components on levels other than the level of exit discharge, and areas immediately adjacent to exterior exit discharge doorways in buildings required to have two or more exits must have emergency illumination.

Emergency power systems consisting of storage batteries or an on-site generator must function for at least ninety minutes. The initial illumination level must average one foot-candle along the egress path at floor level, declining to not less than 0.6 foot-candle (on average) at the end of the emergency lighting time duration.

Signage (1011)

Exit access must be indicated by easily visible exit signage where the path of egress travel is not readily visible or obvious to the occupants of a space or building. Exits and exit access doors must be marked by approved exit signs visible from all directions of egress travel.

Exit signs are not mandated in rooms or spaces requiring only one exit or exit access; sleeping areas in group *I-3*; individual sleeping or dwelling units in groups *R-1*, *R-2*, or *R-3*; or group *U* occupancies. Main exterior entrances that are clearly identifiable and obviously exits do not require exit signs, subject to the approval of local building officials. In group *A-4* and *A-5* occupancies, exit signs are not required on the seating side of vomitories where signage is clearly visible through the openings leading from the seating area. Emergency egress illumination is required at each vomitory or opening within the seating area.

Tactile exit signs (raised lettering or Braille) must be mounted adjacent to egress stairways, passageways, and exit discharges. All visual exit signage must be internally or externally illuminated. Exit signs illuminated internally must be labeled and illuminated at all times.

Externally illuminated exit signs must have legible lettering at least six inches high and two inches wide (excepting the letter *I*), with a stroke width of at least three-fourths of an inch. The word EXIT must be displayed against a contrasting background, easily readable without special illumination; and, if a directional arrow is provided, the arrow should not be easily alterable. External illumination must provide an intensity of at least five footcandles and be connected to an emergency power system as already described.

Aisles, Corridors, and Exit Access (1013)

"A continuous and unobstructed path of vertical and horizontal egress travel from any occupied portion of a building or structure to a public way" is the *IBC*'s definition of *egress*. This is further qualified in Section 1013 to limit egress passage through intervening spaces. In general, the layout of spaces must "provide a discernible path of egress travel to an exit."° Except in dwelling or sleeping units where egress is permitted through adjoining rooms such as a kitchen, the path of egress may not pass through adjoining spaces that are not functionally related. Exit access may not be through kitchens, closets, storage rooms or similar spaces, or any room that can be locked to prevent egress. Even in residential occupancies, the egress path may not lead through sleeping areas, toilet rooms, or bathrooms. On residential multitenant floors, each tenant space must have access to required exits without passing through other tenant spaces.

A notable exception to the preceding restrictions is in hazardous (group *H*) occupancies, where egress may be provided through adjoining or intervening areas so long as the spaces through which exit traffic flows are of the same or less hazardous occupancy groups.

In group *I-2* (medical and custodial care) occupancies, habitable rooms or suites must have an exit door leading directly to an exit access corridor (or out of the building at ground level). One intervening room is permitted under certain circumstances:

- It is not used as an exit access for more than eight people.
- The room layout provides for direct supervision by nursing personnel.
- It is part of a suite that is not used for sleeping purposes, and the travel distance to the exit access door is less than one hundred feet.

In the case of suites of nonsleeping rooms, *two* intervening rooms are permitted if the travel distance to an exit door is less than fifty feet.

Suites containing sleeping rooms may not exceed one thousand square feet without having at least two exit doors remote from one another or five thousand square feet in total. For suites not containing sleeping rooms, the maximum permitted area is 2,500 square feet with one exit and ten thousand square feet with two. Travel distance from any point in a group *I-2* occupancy to a room exit may not exceed fifty feet. The distance to an exit access door from any point in a suite of sleeping rooms may not exceed one hundred feet.

- *Common path of egress travel*—The portion of an exit access system that occupants must traverse before two separate and readily discernible paths to separate exits are available. Common paths of egress travel, including paths that merge, must be included when calculating permitted travel distance.

The common path of egress travel may not exceed twenty five feet in group *H-1*, *H-2*, and *H-3* occupancies. One hundred feet are permitted in group *I* and groups *B*, *F*, and *S* if the building is equipped with an automatic sprinkler system and in groups *B*, *S*, and *U* if the occupant load is less than thirty people. In other occupancies, the common egress path must not exceed seventy five feet.

Aisles must be provided within any occupied portion of an egress system that contains tables; seats; displays; and similar furniture, fixtures, or equipment (FF&E). Required aisle widths must be unobstructed except for open doors and handrails, which may encroach a maximum of seven inches. Aisles serving group *A* occupancies, including bleachers and grandstands, have special requirements noted in Section 1024.

In groups *B* and *M*, clear aisle width must be a minimum of thirty six inches or as determined by the occupant load (see Table 14-2). Nonpublic aisles that are not required to be accessible and serve fewer than fifty people may have a minimum width of twenty eight inches.

Where table or counter seating is adjacent to an aisle, nineteen inches must be allowed for the seating itself, without encroaching upon the required width of the aisle. The nineteen-inch allowance must be measured perpendicular to the edge, establishing a boundary parallel to the counter or table. Where the seating is fixed, the back of the seat determines the aisle boundary. Other side boundaries of aisles are measured to walls, edges of stair treads, or furnishings, allowing for the encroachment of handrails.

Aisle accessways, or secondary accessways leading to an aisle, must satisfy the pertinent occupancy load requirement (Table 14-2) but cannot be less than the minimum aisle width as already specified. If an aisle accessway is twelve feet or longer, twelve inches of width must be provided plus one-half inch for each additional foot or fraction thereof, up to the permitted maximum of thirty feet (measured to the centerline of the seat farthest from the aisle). The path of egress travel from any seat may not exceed thirty feet to the point where a person has a choice of two or more paths.

Balconies serving as means of egress must satisfy the same requirements as those for corridors as to width, headroom, projections, dead ends, and separation from the building interior by walls and opening protectives. Wall separation is not required where the balcony is served by at least two stairs and a dead-end condition does not require a person to travel past an unprotected building opening to reach a stair. Exterior balconies must be designed so that the accumulation of snow and ice does not impede egress (except in outdoor stadiums). The long side of an egress balcony must be at least 50 percent, and the area above the guardrails must be configured to minimize the accumulation of toxic gases or smoke.

Exit and Exit Access Doorways (1014)

Table 14-3 shows the maximum occupancies allowed for spaces with only one means of egress. If the planned occupancy exceeds these levels, two—in some cases, three—exits are required. At least two exits are required if the common path of egress travel exceeds the distances described in the previous section, and for certain kinds of mechanical equipment rooms. If the occupancy level exceeds the values shown in Table 14-6, three or more exits are required.

Exits and exit access doorways must be unobstructed at all times and located so that their availability is made obvious. Where two exits are required, the *IBC* states that "the exit doors or exit access doorways shall be placed a distance apart equal to not less than one-half of the length of the maximum overall diagonal dimension of the building or area to be served measured in a straight line between" them.° Exceptions may be allowed where exit enclosures are provided and connected by a one-hour fire-rated corridor, in which case the exit separation is measured along the shortest route within the corridor. If a sprinkler system is present, the separation distance may be reduced to one-third of the overall diagonal rather than one-half.

Where three or more exits are required, the preceding rule applies to at least two of them. The remaining exits must then be spaced in such a manner that if one is blocked in an emergency the others will be available. As with two exits, the one-third diagonal rule applies if the building is sprinklered throughout.

Table 14-3   Spaces with One Means of Egress°

Occupancy Group	Maximum Occupant Load
A, B, E, F, M, U	50
H-1, H-2, H-3	3
H-4, H-5, I-1, I-3, I-4, R	10
S	30

Boiler, furnace, and incinerator rooms larger than five hundred square feet and containing fuel-fired equipment exceeding 400,000-BTU input capacity require two exits, as do refrigeration machinery rooms larger than one thousand square feet. In both cases, the exit access doorways must be separated by a horizontal distance equaling at least half the maximum dimension of the room and there must be an exit within 150 feet of any point in the room. Tight-fitting, self-closing doors must swing in the direction of egress for all occupancy loads. One of the exits may be via a fixed ladder or alternating tread device.

Refrigerated rooms larger than one thousand square feet that are maintained at a temperature less than 68°F (20°Celsius) must have at least two exits. There must be an exit no farther than 150 feet from any point in the room; otherwise, travel distances are as specified under Exit Access Travel Distance (1015). Egress is permitted through adjacent refrigerated spaces.

Where two exits are required for stages based on area or occupancy load, they must be located on either side of the stage. Stairways, as well as lighting and access catwalks, are permitted a minimum width of twenty two inches. Ladders and spiral stairways may be utilized in the means of egress, and stairways need not be enclosed. A second means of egress from the fly gallery area is not required if a means of escape to a floor or roof above is provided. In this case, alternating tread devices or ladders are permitted.

Exit Access Travel Distance (1015)

Table 14-4 shows the maximum permitted exit travel distance, along the "natural and unobstructed path of egress travel"° from the most remote point on a floor to the entrance to an exit. Unenclosed stairs and ramps along the exit access are included in the distance measurement. Stairs are measured at the center of the treads along a plane defined by the tread nosings.

Travel distance in open parking structures may be measured to the nearest open stair riser, and the same applies to outdoor facilities with open stairs, ramps, and exit access elements.

The maximum travel distance can be increased to four hundred feet in one-story buildings in occupancy group *F-1* or *S* that are equipped with sprinklers along with automatic heat and smoke roof vents. If the last element of the exit access is a code-compliant egress balcony, the travel distance may be increased up to the greater of the length of the balcony or one hundred feet.

Corridors (1016)

Corridors must be fire-resistance rated according to occupancy as shown in Table 14-5. In most cases, the enclosing walls must be *fire partitions* as defined in Chapter 13.

Group *E* occupancies are exempted from this requirement if rooms used for classes have at least one exit door leading directly out of the building at ground level. Assembly rooms in this occupancy group are also exempt if at least half of the required egress

Table 14-4   Exit Access Travel Distance°

Occupancy Group	Without Sprinkler System (Feet)	With Sprinkler System (Feet)
A, E, F-1, I-1, M, R, S-1	200	250
B	200	300
F-2, S-2, U	300	400
H-1	Not Permitted	75
H-2	Not Permitted	100
H-3	Not Permitted	150
H-4	Not Permitted	175
H-5	Not Permitted	200
I-2, I-3, I-4	150	200

Table 14-5   Corridor Fire-Resistance Rating°

Occupancy Group	Occupant Load Served by Corridor	Fire-Resistance Rating (Hours)	
		Without Sprinkler System	With Sprinkler System
A, B, E, F, M, S, U	> 30	1	Not Required
H-1, H-2, H-3	All	Not Permitted	1
H-4, H-5	> 30	Not Permitted	1
I-1, I-3	All	Not Permitted	1
I-2, I-4	All	Not Permitted	Not Required
R	> 10	1	0.5

doors lead outside. Business occupancies requiring only one means of egress (Section 1014) do not need fire-rated corridors, nor do residential occupancies and open parking garages.

Fire-rated corridors may not be interrupted by intervening rooms and must provide continuous protection from each point of entry to an exit. Reception areas, lobbies, and foyers that are functionally part of the circulation system are permitted so long as they meet corridor construction requirements.

Corridor widths are determined by the egress width requirements set forth in Table 14-2 (Section 1005) but in general may not be less than forty four inches. Specific exceptions are the following:

- *Twenty-four inches*—Access to mechanical, electrical, and plumbing systems
- *Thirty-six inches*—Occupancy of fifty or less or within a dwelling unit
- *Seventy-two inches*—Required capacity of one hundred or more in educational occupancies and in health-care facilities where patients are not capable of self-preservation
- *Ninety-six inches*—Health-care facilities where bed movement is required

Dead ends may not exceed twenty feet in length where more than one exit is required. This limitation is extended to fifty feet in certain group *I-3* occupancies and in groups *B* and *F* where there is an automatic sprinkler system throughout the building.

Corridors are not permitted to be used as principal elements of the building ventilation system. They can supply makeup air to directly adjacent spaces such as toilet rooms and janitor closets, so long as an equivalent volume of outside air is supplied to the corridor. This restriction does not apply to dwelling units or tenants spaces of less than one thousand square feet.

The space above a corridor ceiling may be used as a return-air plenum under certain conditions:

- The corridor is either not required to be fire rated or is separated from the plenum by fire-resistance rated construction.
- The air-handling system serving the corridor is automatically shut down by the detection of smoke or water flow in the building's sprinkler system.
- The space above the ceiling is a component of an approved, engineered smoke-control system.

Exits and Exit Discharge

Exits (1017)

Exits that are part of an egress system may not be used for any other purpose. The *IBC* specifies that once a given level of protection is attained along a path of egress, it may not be reduced until the exit discharge has been reached. Structures designed for human occupancy must have at least one exit door that meets the *Code* requirements for egress doors (Section 1008).

Table 14-6   Minimum Number of Exits for Occupant Load°

Occupant Load	Maximum Number of Exits
1–500	2
501–1,000	3
> 1,000	4

Table 14-7   Buildings with One Exit°

Occupancy Group	Maximum Height of Building Above Grade Plane	Maximum Occupants (or Dwelling Units) per Floor and Travel Distance
A, B, E, F, M, U	1 Story	50 and 75 Feet
H-2, H-3	1 Story	3 and 25 Feet
H-4, H-5, I, R	1 Story	10 and 75 Feet
S	1 Story	30 and 100 Feet
B, F, M, S	2 Stories	30 and 75 Feet
R-2	2 Stories	4 Dwelling Units and 50-Feet Travel Distance

Number of Exits and Continuity (1018)

The minimum number of exits required is based upon the occupancy load as listed in Table 14-6, except as follows. The overall egress system must be configured in such a way that the required number of exits from all individual spaces and floors (including basements) is maintained in a direct path from the point of entry to the exit discharge or public way.

Table 14-7 describes the conditions under which buildings may have only one exit so long as the building has only one level below grade. Also included are buildings in group R-3 and one-story buildings occupied at the level of exit discharge (subject to the density requirements of Section 1014).

Maximum travel distance in group B may be increased if the building has an automatic sprinkler system. Buildings in group R-2 may have three stories if they are sprinklered. Open parking structures must have two exits unless vehicles are mechanically parked. Unenclosed vehicle ramps cannot serve as required exits unless designed for pedestrian access. Heliports are referenced by the *Code* as well and require two exits. If the touchdown area is smaller than two thousand square feet or shorter than sixty feet, the second means of egress may be a ladder or fire escape.

Vertical Exit Enclosures (1019)

Like exits, vertical exit enclosures cannot serve any function other than egress. Interior exit stairways and ramps must be enclosed by fire barriers with at least a two-hour rating if the enclosure connects four or more floors. A minimum one-hour rating must be provided if the enclosure is less than four stories, counting basements but not mezzanines.

There are a number of occupancy situations, however, in which vertical egress paths do not need to be enclosed. The *IBC* enumerates these as follows:

- Stairways not required as a means of egress, in open parking structures, group I-3 correctional facilities (408), and performance stages or platforms (410)
- Group A-5 buildings where the egress system is outside the building
- Stairways contained within sleeping units in group R-1 and residential dwelling units in R-2 and R-3
- Stairways with an occupant load of less than ten people not more than one floor above the exit discharge level, except in groups H and I

- Up to half of the egress stairways serving one adjacent floor, so long as there are two means of egress from both floors
- Interior egress stairways serving only the first two floors of a sprinklered building, so long as there are two means of egress from both floors

In the last two cases, neither of the interconnected floors may be open to other floors and the exceptions do not apply to groups *H* and *I*.

All opening protectives must meet the requirements of Section 715, as well as related passageways and fire door assemblies. With the exception of covered mall buildings (402), most openings in exit enclosures must either be for exit access into or egress from the enclosure. The *IBC* prohibits penetrations into exit enclosures except for equipment and services directly related to egress and fire safety, such as sprinkler piping and standpipes; fire department communication and electrical raceways; pressurization equipment and ductwork; and, of course, necessary exit doors.

Exit enclosure ventilation systems must be independent of other building mechanical systems. Openings into the enclosure must be protected by self-closing, fire-resistance rated devices. The enclosure ventilation must be designed in one of the three following ways:

1. Equipment and ductwork are inside of but separated from the remainder of the building by construction as required for shafts.
2. Equipment and ductwork are external to the building and directly connected to the exit enclosure by ductwork protected by construction as required for shafts.
3. Equipment and ductwork are inside the exit enclosure with air taken from and exhausted directly to the outdoors through ducts protected by construction as required for shafts.

Walls of a vertical exit enclosure that are also exterior walls must meet the applicable requirements for exterior walls (Section 704). Adjacent openings within ten feet must have a minimum one-hour rating with one-quarter-hour opening protectives, extending to the lower of the roofline or ten feet above the uppermost landing.

Enclosures under stairways (including unenclosed stairs) must have a fire-resistance rating of one hour or be equal to that of the stair if greater. This space may not be accessed from within the stair enclosure. The open space under exterior stairways may be fully enclosed by one-hour rated construction but may not be used for any purpose. Such spaces within single residential units in groups *R-2* and *R-3* are exempted from these requirements.

Vertical exit enclosures connecting more than three floors must have signs at the top, bottom, and at each landing indicating the floor, the direction of exit access, and the availability of roof access for the fire department. Signage must be located five feet above the floor where it will be visible regardless of the position of the doors. An exit stairway may not continue below the level of exit discharge without an approved barrier at that level. Directional exit signs must also be provided to prevent people from bypassing the exit.

Vertical enclosures in high-rise buildings (403) and underground structures (405) serving floors located more than seventy five feet above the lowest level of fire department access or thirty feet below the exit discharge level must be smokeproof or pressurized enclosures (*1019.1.8*). Pressurized stairways and smokeproof enclosures must open into a public way or into a two-hour fire-rated exit passageway or open space directly accessing a public way. Such exit passageways may have no other openings unless they are pressurized and protected in the same way as the smokeproof enclosure or pressurized stairway. A stairway within a smokeproof enclosure may be accessed through a vestibule or exterior balcony.

Exit Passageways (1020)

Exit passageway widths are determined by the egress width requirements set forth in Table 14-2 (Section 1005) but, in general, may not be less than forty four inches. Such passageways may not be used for any purpose other than egress. Doors that swing into the egress path (or handrails) may not obstruct more than 50 percent of the required width nor project more than seven inches into the required width when fully open.

Walls, ceilings, and floors must be constructed as fire barriers (706) having a fire-resistance rating that is equal to or greater than that of any connecting exit enclosure, or a minimum rating of one hour. Passageway openings must be limited to those required for entry from occupiable spaces and for egress. Where a passageway is used to extend an interior exit enclosure to the exterior of the building, a rated fire door must provide opening protection. Elevators may not open into exit passageways, nor may there be penetrations other than those directly related to egress and fire-safety systems.

Horizontal Exits (1021)

Buildings or areas connected by a horizontal exit must have at least a two-hour fire separation in the form of a fire wall (705) or a fire barrier (706). Opening protectives are, of course, required; and the vertical separation must be continuous through all floors unless floor assemblies are two-hour rated and have no unprotected openings. Fire-barrier walls must continuously divide the floor served by the horizontal exit. Opening protectives must be consistent with the fire rating of the wall, and fire doors must be self-closing or close automatically when activated by a smoke detector.

The space into which a horizontal exit leads must have adequate exits to meet its own occupancy requirements but need not include the additional occupancy that may be a consequence of persons entering through horizontal exits from other areas. At least one of the space's exits must lead out of the building or to an exit enclosure, and occupants must not be required to return through the area from which they came.

Where two or more exits are required, only half of the necessary exit width or number of exits may be provided by horizontal exits and they may not constitute the only means of egress. In group I-2, however, they may provide up to two-thirds of the exit capacity, and in group I-3 100 percent. In the latter case of secured facilities, at least six square feet of accessible refuge space must be provided for each occupant on both sides of the horizontal exit.

The refuge area of a horizontal exit must be large enough to accommodate its own occupants plus the anticipated load from the adjoining compartment, and it must either be public space or belong to the same tenant. In general, the capacity of areas of refuge is calculated on a basis of three square feet per person net, exclusive of stairways, elevators, other shafts, or courts. Group I-2 occupancies must allow fifteen square feet per ambulatory patient and thirty square feet per nonambulatory patient. Group I-3 occupancies require six net square feet per person.

Exterior Exit Ramps and Stairways (1022)

Exterior exit ramps and stairways may serve as components of a required means of egress in all occupancies other than group I-2 for buildings not exceeding seventy five feet or six stories in height. They must be open on at least one side and have at least thirty five square feet of open space adjacent to the level of each floor and intermediate landing, which must be located at least forty two inches above the adjacent floor or landing level.

Ramps and stairways must be located in accordance with the requirements for exit discharge. The adjoining open areas must be yards, courts, or public ways; but the remaining sides may be enclosed by exterior building walls. Separation from the interior of the building must meet the requirements of Vertical Exit Enclosures (Section 1019), and openings must be limited to those essential for egress from occupiable spaces.

Exterior exit ramps and stairways do not have to have a fire separation from the interior of the building in the following circumstances:

- Fire separation is not required in buildings up to two stories above grade where the level of exit discharge is the first story above grade, except for groups R-1 and R-2.
- The ramp or stairway is served by a balcony or another ramp connecting two remote exterior stairways or other approved exits. In this case, the perimeter opening must account for at least 50 percent of the enclosing wall height with the top of the openings at least seven feet above the top of the balcony.
- The structure in which the ramp or stair is located is not required to have enclosed interior stairways (1019).

- Exterior ramps or stairways are connected to open-ended corridors if the building is sprinklered; the corridors meet the requirements of Section 1016; and, where a direction change exceeding forty five degrees occurs, a clear opening of at least thirty five square feet (or another exterior ramp or stairway) is provided. Clear openings must be located so as to prevent the accumulation of toxic gases and smoke.

Exit Discharge (1023)

Exits must discharge to the exterior of and may not reenter a building. An exit discharge must be at or have direct access to grade. Up to one-half the number and capacity of a building's exit enclosures may lead to a path of egress through spaces on the discharge level, so long as the way out is immediately visible and unobstructed. The entire area of the discharge level must also be separated from areas below by construction with the same fire rating as the exit enclosure, and the path of egress must be protected by a sprinkler system. The entire egress level must either be sprinklered or separated from the egress path as required for exit enclosures.

Up to one-half of the exit enclosures may lead to egress through a vestibule, provided that the area serves only this purpose and discharges directly outside. The depth of a vestibule may not exceed ten feet from the exterior wall, nor may it be longer than thirty feet. Vestibules must also be separated from areas below by construction rated the same as the exit enclosure and from the remainder of the exit discharge level by wire glass in steel frames or the equivalent.

An exit discharge must have at least as much capacity as the exits it serves, and its components must be configured to minimize the accumulation of toxic gases and smoke. Exterior balconies, ramps, and stairways must be at least ten feet from adjacent lot lines or other buildings unless the exterior walls and openings are protected in accordance with fire-separation requirements (Section 704).

Egress courts must generally maintain a minimum width of forty four inches with the exception of group *R-3* and group *U* occupancies, which may be thirty six inches wide. Handrails and fully opened doors may not reduce the required width by more than seven inches, doors in any position by more than half, or other projections such as trim by 1.5 inches. The required width must be maintained to a height of seven feet. If the width of an egress court decreases along the path of exit travel, the decrease must be gradual with a guard provided that is at least thirty six inches high.

The *IBC* specifies that an exit discharge must, if possible, provide direct and unobstructed access to a public way. Where this is not possible, the *Code* allows for a dispersal area meeting the following conditions:

- The area must be located on the same property at least fifty feet from the subject building.
- It must have sufficient area to provide five square feet per person when in use.
- It must have a safe and unobstructed path of access from the building.
- It must be permanently allocated as a safe dispersal area and so identified.

Special Requirements

We conclude our discussion of egress considerations by looking at some special requirements for high-density (group *A*) and residential (groups *I-1* and *R*) occupancies.

Assembly (1024)

The special requirements for group *A* occupancies all focus on ensuring the safe egress of large numbers of people from a densely populated building or area. In principle, spaces having an occupancy load greater than three hundred people are required to have a defined main exit able to accommodate at least one-half of the actual load. The width of the main exit must be at least equal to the total width of all the egress paths leading to it. The main exit must discharge into a street or a dedicated space at least ten feet wide adjoining a street or public way.

In circumstances that do not permit a well-defined main exit or in which there may be multiple main exits, the necessary exits may be distributed so long as the total required egress width is maintained. All secondary exits must be capable of providing for at least one-half of the total occupancy load and must meet the same egress width requirements as pertain to the main exits.

Waiting areas in foyers and lobbies must not encroach upon the required clear width of the defined egress paths. Such waiting areas must be separated from the means of egress by fixed railings at least forty two inches high or by permanent walls. An unobstructed pathway of the necessary width must be maintained from a lobby or foyer to *all* of the main exits.

At least two means of egress are required from all balconies or galleries, one from each side, and at least one must lead directly to an exit. Interior exit stairs must meet all the requirements of vertical exit enclosures (1019), except that in auditoriums, theaters, and churches, stairways may be open between the balcony and the main floor. If a balcony or gallery contains accessible seating, there must be at least one accessible means of egress from it (1007).

Means of egress width is of particular importance in assembly occupancies, and the *IBC* treats them accordingly. There are separate sets of detailed, clear width capacities specified depending upon whether the facility has smoke protection or not. For example, if the assembly area is not smoke protected, at least 0.3 inch must be allowed per occupant on stairs with risers up to seven inches high and treads greater than eleven inches deep. If risers are greater than seven inches, the necessary width is increased by 0.005 inch per occupant. Where egress is in a downward direction, the width per occupant must be increased by 0.075 inch if there is no handrail within thirty inches horizontally. Ramped means of egress have similarly detailed requirements based upon the slope of the ramp.

Where the seating is smoke protected, the *IBC* provides a table of unit exit width requirements based on the occupancy load and the steepness of the slope.° The smoke-control system must be capable of keeping accumulated smoke at least six feet above the floor. If the facility has a roof, the lowest portion of the roof deck must be at least fifteen feet above the highest aisle or stairway. The width requirements in the table range between 0.05 and 0.25 inch per person based upon occupancy, whether there are handrails within thirty inches, and whether or not the slope is steeper than 10 percent. The *Code* requires that a life-safety evaluation, compliant with NFPA 101, be done if the table values are used.

Automatic sprinklers must also be provided within enclosed assembly seating areas, unless roof construction is more than fifty feet above the floor, press boxes and storage facilities total less than one thousand square feet, or the area is essentially open to the outside. The required clear egress width for outdoor, smoke-protected assembly areas may be determined using the table method or by multiplying the occupant load by 0.08 for aisles and stairs or 0.06 for corridors, ramps, tunnels, or vomitories, whichever is less.

Travel distance to an exit door must be less than two hundred feet in nonsprinklered buildings. In sprinklered buildings, the limit is 250 feet. Where there are aisles provided, the exit distance must be measured along the aisles and not over the seats. In smoke-protected assembly areas, two hundred feet is allowed from each seat to a vomitory or concourse entrance; another two hundred feet is allowed from that point to a ramp, stair, or walk on the building's exterior. In open-air seating, maximum travel distance from each seat to the exterior may not exceed four hundred feet.

The common path of egress travel normally cannot exceed thirty feet from any seat to a point where two paths are available, seventy-five feet if the assembly area serves fewer than fifty occupants. Fifty feet is the limit in smoke-protected assembly seating. Where one of the two paths of egress is through a row of seats to another aisle, there may be no more than twenty four seats in the row. The minimum clear walking space between rows must be twelve inches, plus 0.6 inch for every seat over seven in the row. If the area is smoke protected, the limits are raised to forty seats and an additional 0.3 inch over the required twelve for each additional seat.

All occupied portions of assembly areas with seats, tables, displays, and similar furniture and equipment must have aisles leading to exits or exit access doorways. Access aisles must comply with general requirements for aisles, corridors, and exit access (Section 1013); and the *Code* does not allow obstructions within the required aisle width, except for handrails. Specific group *A* aisle requirements are the following:

- Aisle stairs with seating on both sides—Forty-eight inches (thirty-six inches if the aisle serves fewer than fifty seats)
- Level or ramped aisles with seating on both sides—Forty-two inches (thirty-six inches if aisle serves fewer than fifty seats, thirty if fewer than fourteen seats)
- Aisle stairs with seating on one side only—Thirty-six inches
- Aisle subdivided by a handrail—Twenty-three inches on each side
- Aisle stair serving up to five rows—Twenty-three inches between seating and handrail

Aisle widths must be sufficient to provide egress for the occupancy capacity of the *catchment area* served by the aisle, which is the portion of the total assembly space served by the particular aisle. The *IBC* states that "in establishing *catchment* areas, the assumption shall be made that there is a balanced use of all means of egress, with the number of persons in proportion to egress capacity."°

Converging aisles forming a single egress path must maintain the same required capacity along the combined aisle. The width of those portions of aisles where egress is possible in either direction must be uniform. Aisles must terminate at a cross aisle, doorway, vomitory, concourse, or foyer that has exit access. Dead-end aisles may not be longer than twenty feet unless the seats beyond are no more than twenty four seats from another aisle. If the assembly area is smoke protected, dead-end vertical aisles may not be longer than twenty one rows unless they are fewer than forty seats from another aisle.

Rows of seating served by doorways or aisles at both ends may not have more than one hundred seats. Starting with a minimum clear width of twelve inches, the distance between rows must be increased by 0.03 inch for every seat beyond fourteen but need not be wider than twenty two inches. If there is an aisle or doorway at only one end, the spacing must increase by 0.6 inch from the twelve-inch minimum for each seat over seven, up to twenty two inches. If the assembly seating is smoke protected, the number of seats per row allowed is based on occupancy and set forth in a table showing the requirements for both single and dual access.° Bench-seating occupancy is calculated at the rate of one person per eighteen inches of bench.

Aisles with a slope of up to 12.5 percent must be ramped and provided with a slip-resistant walking surface. Aisles sloping at a greater rate must have steps extending their full width, with uniform treads at least eleven inches deep. Where the amount of slope follows that of adjacent seating areas, the riser height must be between four and eight inches and kept uniform within each flight. To the extent that it is necessary to maintain adequate sightlines, however, riser heights may vary; but nonuniformities must be indicated by a distinctive, one- to two-inch-wide marking stripe at each tread nosing. The contrasting marking stripe must be readily visible while descending, and in such circumstances the riser height may extend up to nine inches.

If floors are level and seating accommodates fewer than two hundred people or is at tables, seats do not need to be fastened to the floor. If the number of seats is greater than two hundred and the floor is level, seating must either be fastened to the floor or together in groups of three. If seating is on tiered levels and flexibility is a necessary design requirement, up to two hundred movable seats may be provided; however, plans showing proposed configurations must be approved by the building official. Seats for musicians or other performers, as well as small groups of seats of up to fourteen on level floors and separated by guards or partial-height walls, need not be fastened to the floor. All other seating in assembly occupancies or portions thereof must be securely fastened to the floor.

Handrails are required for aisles having a slope greater than 6.7 percent (one in fifteen), as are aisle stairs. Handrails are not required for ramped aisles with a slope less than 12.5 percent (one in eight) if there is seating on both sides or if there is a guard at one side that may be grasped in the same manner as a handrail. Discontinuous handrails are required to facilitate access to seating on both sides of an aisle. Such breaks must be between twenty two and thirty six inches measured horizontally, the handrails must have rounded corners and ends, and the intervals between breaks may not exceed five rows. An intermediate handrail below the main rail must be provided on handrails in the center of aisle stairs.

Where cross aisles are more than thirty inches above another floor, or where an elevation change occurs between a cross aisle and the adjacent floor or grade below that is thirty inches or less, guards must be provided that are at least twenty six inches high, relative to the floor of the aisle. (See also Section 1012.) Guards are not required if the seat backs along the cross aisle project at least twenty four inches above the aisle floor.

If a normal-height fascia or railing would interfere with the sightlines of adjacent seating, guards may be provided having a minimum height of twenty six inches. At the ends of aisles where the elevation is more than thirty inches above the floor below, a fascia or railing at least thirty six inches high is required, which must provide at least forty two inches between the nearest tread nosing and the top of the rail.

Emergency Escape and Rescue (1025)

Residential occupancies in group I-1 and group R must have, in addition to the normal means of egress, provisions for emergency escape and rescue. Sleeping rooms below the fourth story above the grade plane and in basements must have at least one exterior emergency escape and rescue opening into a public street, alley, yard, or court. High-rise buildings and occupancies other than group R-3 are exempted from this requirement if they are equipped with sprinklers or if sleeping rooms open into a fire-resistance-rated corridor with access to two exits in opposite directions. Emergency exits may open onto a balcony within an atrium if the balcony has access to an exit and there is another means of egress from the room. Basements with a ceiling height less than eighty inches, no more than two hundred square feet of floor area, or that are without habitable space are exempted from this requirement. If a basement has an exit door or exit access door leading directly to a public exterior accessway, additional emergency openings are not required.

Emergency escape and rescue openings must have a minimum clear height of twenty four inches and a minimum clear width of twenty inches, and the opening must be at least 5.7 square feet (five feet for floor openings). The bottom of such openings must not be higher than forty four inches above the floor.

Escape openings must be operational from the inside without the use of keys or special tools. Security devices such as bars or grilles are permitted so long as they also require no special tools or keys and do not reduce the required net clear opening. Where bars or other physical security devices are added, smoke alarms must be provided.

If the sill of an emergency rescue opening is below ground level, a window well must be provided having a minimum area of nine square feet and no dimension smaller than thirty six inches. If the window well is more than forty four inches deep, it must have a permanent ladder or steps at least twelve inches wide. The ladder may not project into the well by more than six inches and must not obstruct or be obstructed by the opening.

CLOSING WORDS

The main considerations of this chapter were means of egress and wayfinding. General requirements as to occupancy loads and the necessary widths of passageways used for exiting were discussed, together with areas of refuge for people who may be temporarily unable to reach an exit. Aisles, corridors, doors, changes in level, illumination, and signage were addressed in conjunction with exit access, exits, and exit discharge.

An American National Standard titled *Accessible and Usable Buildings and Facilities* is also published by the International Code Council. Also known as *ICC/ANSI A117.1*, this second document not only is largely the basis of the accessibility portion of the *IBC* but also is closely coordinated with the *ADA Accessibility Guidelines* (*ADAAG*) and the *Fair Housing Accessibility Guidelines* (*FHAG*). Paragraph 2 of the Accessibility section (1100) of the *IBC* incorporates *ANSI A117.1* by reference.

Chapter 15 covers the following related subjects:

- Accessible route and entrances
- Parking and passenger loading
- Dwelling and sleeping units
- Special occupancies and other features
- Signage

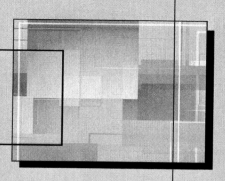

CHAPTER 15

Accessibility

> *[S]ociety's accumulated myths and fears about disability and disease are as handicapping as are the physical limitations that flow from actual impairment.*
> —*William J Brennan*°

So far in our discussion of building codes, we have looked at those sections of the *IBC* that are primarily concerned with ensuring the safety of building occupants. First, there are two different sets of overlapping classifications, one based upon occupancy and the other on the type of construction. In this way, potential dangers that may result from extraordinary events are considered from dual perspectives:

- What kinds of materials are to be used in the building's structure, and how resistant are they to fire, smoke, and other potential damage?
- What are the planned functional requirements of the people who will inhabit the structure?

Next, we considered those sections of the *Code* that specify fire-prevention and fire-protection measures:

- How can the building components be assembled most effectively to resist fire and smoke?
- What can be done to actively protect the inhabitants and warn them of an emergency?

In the preceding chapter, we looked at *egress*. Within the physical envelope that the building provides, there must be clearly defined (wayfinding) pathways arranged to allow people to get out of the facility safely if all else fails. Many of these may be the same pathways that are used as entrances and exits under normal circumstances, but alternate routes are required if a large number of people need to evacuate simultaneously or if a customary route is blocked by fire or smoke. As we have seen, a number of accessibility concerns were addressed in the foregoing sections.

Yet another set of requirements that has become a part of our design consciousness over the last twenty years or so is the federal mandate for *Accessibility*. Safety considerations do not drive the desire for accessibility as much as the recognition of the need for fairness and equality of opportunity, which were codified into the *Americans with Disabilities Act* in 1990. The United Kingdom followed suit with its *Disability Discrimination Act (DDA)* of 1995, and the European Union has since adopted another similar document.

In Section 1100, the *IBC* addresses a number of specific issues related to the application of *ICC/ANSI A117.1*, incorporating them into its very fabric. We examine these before turning to the *ADA Accessibility Guidelines (ADAAG)*° themselves in our final chapter.

ICC/ANSI A117.1-2003

There are ten major sections in the *ICC/ANSI* document: they are listed in Table 15-1 in the order that *A117.1* presents them. Several sections are new since the previous version of 1998.°

In the right-hand column of Table 15-1 are the numbers of the corresponding sections of the *ADAAG* (1991). Most of the technical specifications in *A117.1*, with the notable exception of Chapter 10, "Dwelling Units," are contained in Sections 4.2 through 4.35, and they are organized quite differently. Although *dwelling units* are defined by the *ADAAG*, their definition refers to "transient lodging" as opposed to private residential units, which *A117.1* does address.

The new federal ADA guidelines of 2004 will ultimately supersede the 1991 document, but as of this writing, the earlier law (as amended through 2002) remains in effect. Happily, the new ADA guidelines have been reorganized to conform with A117.1. We discuss the new federal accessibility guidelines in the next and final chapter and have noted significant changes where applicable. Sources of information on the current status of updated federal guidelines are also discussed at the beginning of Chapter 16.

The *ADA Accessibility Guidelines* document states at the outset that the technical specifications referred to earlier are substantially the same as *ICC/ANSI A117.1*, "except as noted in [the] text by italics."° Section 4.1, which sets forth minimum requirements, and Sections 5 through 10 are entirely different, dealing with specific types of public and commercial facilities.

The remainder of this chapter addresses the specific provisions of the *IBC* controlling "the design and construction of facilities for accessibility to physically disabled persons"° in accordance with *ICC/ANSI 117.1*.

Scoping Requirements (1103)

The *IBC* generally mandates that "[b]uildings and structures, temporary or permanent, including their associated sites and facilities, shall be accessible to persons with disabilities."° This section of the *IBC* deals largely with exceptions, however: situations in which accessibility is not required. (Requirements pertaining to existing buildings are treated separately: Section 3409.)

Employee work areas are generally required to comply with the *IBC* requirements for fire alarm and detection systems (Section 907), accessible means of egress (1007), and accessible routes (1104). However, work areas less than 150 square feet that are functionally required to be elevated at least seven inches above the finished floor are exempt from these requirements. Raised observation areas used primarily for security or life- or fire-safety purposes, such as galleries and prison guard towers, do not need to be accessible nor served by an accessible route. This also applies to common-use areas in detention facilities that do not serve cells required to be accessible.

Single and duplex residential facilities and the sites on which they are located are not required to be accessible, nor are buildings in group *R-1* with not more than five sleeping units for rent in which the owner or proprietor also resides. If part of a dwelling unit is used as a day-care facility, the portion not used for day care need not be accessible. Group *U* occupancies are exempt, except for paved work areas and areas open to the public in agricultural buildings and private garages containing accessible parking.

Single-occupant structures accessed above or below grade such as toll booths accessed by tunnels need not be accessible. The same applies to spaces used to maintain or repair equipment such as mechanical, electrical, and telecommunications rooms; elevator penthouses; and the like. Nonoccupiable spaces accessed only by ladders, catwalks, and such are exempt, as are construction sites.

Accessible Route and Entrances (1104–1105)

Wherever a building or an area within a building is required to be accessible, an accessible route must connect those areas with accessible entrances leading to accessible pedestrian

Table 15-1   *ICC/ANSI A117.1-2003* Compared with ADA

ICCA/ASTM *117.1-2003* and ADA ABA Accessibility Guidelines 2004			ADA Standards (1991–2002)
	100	Application & Administration	Purpose, Compliance, Clarification
	200	Scoping Requirements	Scope
Building Blocks	301	General	Scope
	302	Floor [or Ground] Surfaces	4.5
	303	Changes in Level	4.3 & 4.5.2
	304	Turning Space	4.2
	305	Clear Floor [or Ground] Space	4.2, 16, 18–20, 22–25, 27, 32, 34–35
	306	Knee and Toe Clearance	4, 19-2 4.2.4 4.15.5 4.17.4 4.2.4.3 4.32.3
	307	Protruding Objects	4.4
	308	Reach Ranges	4.2
	309	Operable Parts	4.27
Accessible Routes	401	General	Scope
	402	Accessible Routes	4.3
	403	Walking Surfaces	4.5.2 4.8.8 4.9.6 4.29.2 4.29.5
	404	Doors and Doorways [and Gates]	4, 13
	405	Ramps	4.8
	406	Curb Ramps	4.7
	407	Elevators	4.10
	408	Limited-Use/Application Elevators	Not Covered
	409	Private Residence Elevators	Not Covered
	410	Platform Lifts	4.11
General Site and Building Elements	501	General	Scope
	502	Parking Spaces	4.6
	503	Passenger Loading Zones	4.6
	504	Stairways	4.9
	505	Handrails	4.9
	506	Windows [not in ADA/ABA]	4.12
Plumbing Elements and Facilities	601	General	Scope
	602	Drinking Fountains	4.15
	603	Toilet and Bathing Rooms	4.22
	604	Water Closets and Toilet Compartments	4.16
	605	Urinals	4.18
	606	Lavatories and Sinks	4.19
	607	Bathtubs	4.20
	608	Shower Compartments	4.21
	609	Grab Bars	4.26
	610	Seats	4.26
	611	Washing Machines and Clothes Dryers	Not Covered
	612	[Stores and Steam Rooms-ADA/ABA]	Not Covered

(continued)

Table 15-1   *ICC/ANSI A117.1-2003* Compared with ADA (*continued*)

			Scope
Communication Elements and Features	701	General	Scope
	702	Alarms [Fire Alarm Systems]	4.28
	703	Signs	4.30
	704	Telephones	4.31
	705	Detectable Warnings	4.29
	706	Assisting Listening Systems	4.3.3.7. Table A2 4.1.3 4.30 10.4
	707	Automatic Teller Machines (ATMs) and Fare Machines	4.34
	708	Two-Way Communication Systems	4.3.11
Special Rooms and Spaces	801	General	Scope
	802	Auditorium and Assembly Areas	4.33
	803	Dressing Fitting and Locker Rooms	4.35
	804	Kitchens	9.2.2(7)
	805	Transportation Facilities	10
	806	Holding Cells and Housing Cells	Not Covered
	807	Courtrooms	Not Covered
Built-in Furnitures & Equipments	901	General	Scope
	902	Dining Surfaces and Work Surfaces	4.32
	903	Benches	4.37
	904	Sales and Service Counters	7.2
	905	Storage Facilities	4.1.3 4.25 9.2
Dwelling and Sleeping Units	1001	General	Scope
	1002	Accessible Units	9.1
	1003	Type A Dwelling Units	Not Covered
	1004	Type B Dwelling Units	Not Covered
	1005	Units with Accessible Communication Features	Not Covered

walkways and the public way. Where only one such route is provided, it may not pass through such areas as storage rooms, kitchens, and so on, except in accessible dwelling units. Connecting routes among buildings and facilities on a site, as well as from arrival points at a site (public streets or sidewalks, accessible parking, or public transportation stops), must be accessible unless the only access is vehicular, with no provision for pedestrian traffic.

Primary circulation routes within employee work areas must be accessible, except where an area is less than three hundred square feet and defined by fixed partitions, casework, or counters. Circulation space specifically required by equipment or located in exterior employee work areas does not need to be accessible.

Where accessible fixed seating is provided in assembly areas, an accessible route must be provided to such seating but does not need to extend to all seating. Mezzanines in one-story buildings need not be accessible. Press boxes in bleachers must be accessible, unless their points of entry are at only one level or they are free-standing and more than twelve feet above grade; in either case, their total area is less than five hundred square feet.

In general, accessible routes must coincide with primary circulation routes; and if the main circulation is interior, the accessible route must also be interior. All accessible levels in multistory buildings, including mezzanines, must be connected by an accessible route unless the aggregate of all floors and mezzanines above and below are less than three thousand square feet. This exception does not apply to multitenant facilities with at least five spaces in the *Mercantile* group, health-care providers in groups *B* or *I*, or passenger transportation facilities and airports in groups *A-3* or *B*.

Floor levels that do not contain accessible areas in groups *A, I, R*, and *S* need not be served by accessible routes unless there are special occupancy requirements (Sections 1107 and 1108). The same applies to one story in a two-story building having no public space and an occupancy load of up to five people. For example, the main control room in an air traffic control tower and the floor below it need not be accessible.

Fifty percent of all public entrances must be accessible, excluding loading areas and services entrances that are separate from the main entrances to tenant spaces, as well as entrances to areas not required to be accessible. If a service entrance is the only entrance, it must be accessible. In secured facilities used by inmates and security personnel, or where there are restricted entrances to a facility, at least one such entrance must be accessible.

Direct pedestrian access from parking structures to facility entrances must be accessible, as must be at least one building entrance leading from a pedestrian tunnel or elevated walkway.

At least one accessible entrance must be provided to tenant spaces, dwelling and sleeping units, unless the units are not required to be *Accessible* units, type *A* or type *B* units (see Section 1107).

Parking and Passenger Loading (1106)

Accessible parking spaces are required in proportion to standard parking as set forth in Table 15-2, as determined by the number of spaces provided for the facility. Excluded from this requirement are facilities for buses, trucks, and other delivery vehicles and other special parking areas where the public parking has an accessible passenger loading zone.

Accessible parking must have the shortest possible route of travel to an accessible building entrance. If a parking facility serves multiple buildings, the designated spaces must be as close as possible to an accessible pedestrian entrance to the parking area or structure. If a building or complex has multiple accessible entrances and/or several different types of parking, the accessible parking must be dispersed among them and located adjacent to each of the entrances. For every six accessible spaces (or fraction thereof), one must be van-accessible, and the van spaces may be concentrated on one level of a multilevel parking structure.

Table 15-2   Accessible Parking Spaces°

Total Parking Spaces Provided	Required Minimum Number of Accessible Spaces
1–25	1
26–50	2
51–75	3
76–100	4
101–150	5
151–200	6
201–300	7
301–400	8
401–500	9
501–1,000	2% of Total
>1,000	20 + 1 for Each 100 > 1,000

Institutional parking facilities, principally in occupancy group *I-2*, have more stringent requirements. Ten percent of the public parking serving hospital outpatient facilities must be accessible, and the requirement is 20 percent (at least one space) for outpatient physical therapy and rehabilitation facilities. Medical and long-term-care facilities where treatment or care is given and stays exceed twenty-four hours must have an accessible loading zone at an accessible entrance.

Buildings in residential groups *R-2* and *R-3* that are required to have *Accessible*, type *A* or type *B* dwelling or sleeping units (1107) must have 2 percent of their parking accessible. If the main parking is within or beneath the building, the accessible spaces must be as well.

Regarding passenger loading zones, the *IBC* refers specifically to *ICC A117.1* as governing their design and construction. Where continuous loading zones are provided, such as at a transportation terminal, one accessible space must be provided every hundred lineal feet. Valet parking services must provide for accessible passenger loading in a manner similar to medical facilities as noted earlier.

Dwelling and Sleeping Units (1107)

In each of the preceding two sections, we made reference to dwelling and sleeping units as being classes as *Accessible*, type *A* or type *B*. Let us now define exactly what the ICC means by these terms:

- *Accessible unit*—A dwelling unit or sleeping unit that complies with this code and chapters 1 through 9 of *ICC A117.1*
- *Type A*—A dwelling unit or sleeping unit designed and constructed for accessibility in accordance with *ICC A117.1*
- *Type B*—A dwelling unit or sleeping unit designed and constructed for accessibility in accordance with *ICC A117.1* and consistent with the design and construction requirements of the federal Fair Housing Act [*Guidelines (FHAG)*]°

Dwelling and sleeping units that are required to be accessible must comply with the requirements in each of the preceding definitions. Spaces available for use by the general public and residents alike, serving units in any of the three categories defined, must be accessible. Examples of such spaces include (shared) living and dining areas, kitchens, and toilet and bathing rooms, as well as exterior spaces such as balconies, patios, and terraces. Recreational facilities as described under Section 1109 are excepted from this provision.

At least one accessible route must be provided from the main entrance of any accessible building or facility to the principal entrance of all units in each of the three categories, and to all interior and exterior shared spaces serving the individual units. Certain exceptions are allowed where grade changes or physical barriers would make complete accessibility impractical or unworkable. In such cases, a vehicular alternative must be provided that complies with parking and passenger loading provisions (1106).

Group *I* Occupancies (1107.5)

Supervised residential facilities (group *I-1*) must have at least one, and not less than 4 percent, of its dwelling and sleeping units in the *Accessible* class. If there are more than four residential units, they must *all* be type *B* units, except as provided under Section 1107.7.

At least half of all nursing home (*I-2*) units must be *Accessible* (one minimum); and if there are more than four living units, they must all be type *B*, again subject to the exceptions in 1107.7.

Group *I-2* hospitals, psychiatric and detoxification facilities, assisted living, and residential care facilities must have at least 10 percent of their dwelling and sleeping units *Accessible*. If there are more than four such units, they must all be type *B*, subject to 1107.7. *All* units must meet the requirements of the *Accessible* class in rehabilitation facilities designed for the treatment of conditions affecting mobility.

A secured (*I-3*) facility must have at least 2 percent of its units *Accessible* (not less than one). At least one of each type of special purpose room or cell must be accessible, such as those used for detention or segregation, detoxification, or medical care. Rooms designed for suicide prevention, without protrusions, need not provide grab bars.

Group *R* Occupancies (1107.6)

Group *R-1* (transient occupancies) must have *Accessible* units dispersed among all classes of residential units provided on a site, in the quantities set forth in Table 15-3. All of the dwelling and sleeping units provided must be type *B* if there are more than four, subject to Section 1107.7 exceptions. Roll-in showers in *Accessible* units must include permanently mounted folding shower seats.

Group *R-2* (permanent), *R-3* (permanent single or duplex), and *R-4* (assisted living) occupancies in quantities greater than four must all be type *B*, again subject to the reductions permitted under Section 1107.7. At least one type *A* or 2 percent of all dwelling and sleeping units over twenty on a site must be provided in all apartment houses, along with convents and monasteries.

General Exceptions (1107.7)

Where certain physical constraints exist in residential occupancies and institutional facilities with sleeping units, the required number of type *A* and type *B* units may be reduced.

If a building has no elevator service, at least one story having an accessible entrance from the exterior must contain type *B* units. The number of type *A* units required remains as previously noted for group *R-2* occupancies. Additional stories containing dwelling or sleeping units where the planned entrance is within fifty feet of a pedestrian or vehicular entrance and the slope of the grade is 10 percent or less must contain type *B* units.

If a building has elevator service only to the lowest story containing dwelling or sleeping units, just the units on that floor that are intended for residential use are required to be type *B*. A multistory dwelling or sleeping unit without an internal elevator does not need to be type *B*. A multistory residential unit having external elevator service to only one floor must comply with the requirements for a type *B* unit and have an accessible toilet on that floor.

If a substantial portion of a site prior to development has grades in excess of 10 percent and multiple nonelevator buildings are planned, the requirement for type *B* units applies only to that percentage of the *entire* site where the grades are less than 10 percent, so long as 20 percent of the units are type *B*. The *IBC* enumerates several other special requirements that must be met in such circumstances, related to the permitted grade slope between pedestrian and vehicular access points: typically 8 to 10 percent.

Requirements for type *A* and type *B* units also do not apply where design floor elevations are a consideration and there are substantial grade differentials between minimum required floor elevations and entrance access points.

Table 15-3 Accessible Dwelling and Sleeping Units°

Total Number of Units Provided	Minimum Required Number of Accessible Units Associated with Roll-In Showers	Total Number of Required Accessible Units
1–25	0	1
26–50	0	2
51–75	1	4
76–100	1	5
101–150	2	7
151–200	2	8
201–300	3	10
301–400	4	12
401–500	4	13
501–1,000	1% of Total	3% of Total
>1,000	10 + 2 for Each 100 > 1,000	30 + 2 for Each 100 > 1,000

Special Occupancies (1108)

International Building Code requirements for special occupancies apply principally to three occupancy types: assembly seating, self-service storage facilities, and judicial facilities.

The requirements of *ICC A117.1* (and the *ADAAG*) are referenced once again here, specifically with respect to accessible wheelchair spaces in all types of assembly seating areas. Table 15-4 shows the required numbers of wheelchair spaces in relation to overall seating capacity. These requirements apply to arenas, bleachers, grandstands, stadiums, and theaters with fixed seating, stating that all facilities and services provided in nonaccessible areas must also be available to those with disabilities. The ICC enumerates general seating and all varieties of club, luxury, and other boxes, as well as the main level and one of every two additional levels in multilevel assembly areas. Every wheelchair space must have a companion seat that meets *A117.1*, and 5 percent of all aisle seating (at least one seat) must be designated as accessible.

The *Code* requires that public address systems and electronic signage, real-time and otherwise, be capable of providing equivalent information in both audible and visual form regarding events and facilities. Where audible communications are integral to the use of the facility, assistive listening systems must be provided.

Where there is a circulation path connecting an assembly seating area with a performance area, that path must be accessible, together with the circulation between performance areas and ancillary facilities used by the performers. Where there is a dining area, all of the spaces allocated to tables and seating must be accessible, unless the seating is on multiple, tiered levels. Mezzanine seating areas need not be accessible when not otherwise required so long as 75 percent of the dining area is accessible and each accessible tier provides the same services. In sports facilities, at least 25 percent of tiered dining areas must provide accessible seating and the same services as other areas. Five percent of all surfaces provided for dining must be accessible and distributed throughout the facility (a minimum of one).

Self-service storage facilities must have accessible units dispersed throughout, although they may be contained in a single building within a multibuilding facility. From one to two hundred units, 5 percent must be accessible. Where there are over two hundred units, ten must be accessible, plus 2 percent of all units over two hundred.

Courtrooms in judicial facilities must be accessible. Where central and court-floor holding cells are separated on a basis of age or sex, one of each type must be accessible. Where the cells are not separated, at least one accessible cell must be provided. Accessible cells may serve more than one courtroom.

Where counters or cubicles are provided in visiting areas, at least 5 percent (minimum of one) must be accessible on both the visitor and detainee sides. Where solid partitions or security glazing are provided, at least one of each counter or cubicle must be accessible. The same requirements generally apply to visitor tables, counters, and work surfaces in group *I-3* occupancies.

Table 15-4   Accessible Wheelchair Spaces°

Seating Capacity in Assembly Areas	Required Minimum Number of Wheelchair Spaces
4–25	1
26–50	2
51–100	4
101–300	5
301–500	6
501–5,000	6 + per 150* > 501–5,000
> 5,000	36 + 1 per 200* > 5,000

*or fraction thereof

Other Features and Facilities (1109)

Toilet rooms and bathing facilities in general must be accessible. If certain areas or floors are not required to be connected by an accessible route, toilet and bathing facilities serving the entire facility may not be placed in those locations. At least one of every type of fixture, control, and dispenser must be accessible.

There are a number of conditions under which toilet rooms and bathing facilities are not required to be accessible, such as those intended for single occupancy and accessed through a private office or through dwelling and sleeping units not required to be accessible (Section 1107). Toilet room fixtures in excess of *International Plumbing Code* requirements in primary school and day-care occupancies and toilet rooms in critical-care patient sleeping rooms are also exempted.

An accessible unisex toilet room is required in assembly and mercantile occupancies where six or more water closets are required for both sexes. In mixed-occupancy buildings, the assembly or mercantile occupancy requirements determine the unisex toilet requirement. Where separate-sex bathing facilities are provided in recreational facilities, accessible unisex toilet and bathing rooms are required. Fixtures located within the unisex toilet and bathing rooms are included when determining the number of fixtures required for an occupancy.

Unisex toilet rooms may include only one water closet and one lavatory, plus one urinal if desired. Unisex bathing rooms may have only one bathtub or shower fixture in addition to one set of toilet fixtures and may be considered as toilet rooms. Where each separate-sex bathing room has only one shower or bathtub fixture, a unisex bathing room is not required.

Unisex toilet rooms must be located on an accessible route and not more than five hundred feet distant nor more than one floor above or below separate-sex toilet rooms. The accessible route between separate-sex and unisex toilet rooms in airports and other transportation facilities may not pass through security checkpoints. Doors to unisex facilities must be able to be secured from within the room.

At least one wheelchair-accessible water closet compartment must be provided in toilet or bathing rooms where such compartments are otherwise provided. If the combined total of water closets and urinals exceeds six, at least one ambulatory-accessible compartment must also be provided that complies with *ICC A117.1*. Where sinks are provided, at least 5 percent (one minimum) must comply with *ICC A117.1*. Sinks used exclusively by children in primary school and day-care occupancies are excepted along with service sinks.

At least 50 percent of drinking fountains (one minimum) must be accessible on floors where drinking fountains are provided. Kitchens and kitchenettes in accessible spaces must comply with *ICC A117.1*.

Service Facilities

Where fixed or built-in cabinets, shelving, closets and drawers, coat hooks, and other such storage components are provided, at least one of each type must comply with *ICC A117.1*. At least 5 percent of lockers provided in accessible spaces (one minimum) must be accessible. Service facilities such as dressing and locker rooms are subject to the same 5 percent rule.

At least 5 percent of seating or standing space at built-in or fixed tables, counters, or work surfaces (minimum of one) must be accessible. Such elements must be dispersed throughout the facility. Food-service lines must be accessible; and where self-service shelves are provided, at least half of each type (one minimum) must be provided. At least one of each type of service counter or point-of-sale counter must be accessible, and all such counters must be dispersed throughout the building.

Accessible checkout aisles must be provided in the proportions shown in Table 15-5, except in selling spaces less than five thousand square feet where only one is required. Where different functions are served, at least one accessible aisle is required for each function; and all accessible aisles must be dispersed throughout the facility. Security and traffic-control devices such as turnstiles located in accessible checkout lanes must be accessible. Queuing areas approaching accessible checkout aisles and service counters must be accessible.

Table 15-5 Accessible Checkout Aisles°

Total Checkout Aisles of Each Function	Minimum Number of Accessible Checkout Aisles of Each Function
1–4	1
5–8	2
9–15	3
> 15	3 + 20% of Additional Aisles

Controls, hardware, and operating mechanisms intended for use by building occupants must generally be accessible, such as switches controlling lighting and ventilation and electrical outlets. This requirement does not apply to electrical or telecommunications receptacles serving a specific purpose, floor receptacles, HVAC diffusers, or operable components intended for use by service personnel only. At least one window in each required accessible room must be accessible, except for bathrooms or kitchens.

Recreational facilities must be provided with accessible features in group *R-2* and *R-3* occupancies where there are type *A* or type *B* units. In single buildings and multiple-building complexes, 25 percent of each type of recreational facility must be accessible (but not less than one).

Changes in Level
Stairways that are integral to an accessible route connecting floor levels where there is no elevator must be designed and constructed in accordance with *ICC A117.1* and the *IBC* requirements for egress.

Passenger elevators on an accessible route must be accessible and in compliance with ICC requirements for elevators and conveying systems (Section 3001.3). Wheelchair platform lifts may be a component of a required accessible route in new construction under specified conditions and must be installed in accordance with *ASME A18.1.*° Such conditions include accessible routes to performing areas and wheelchair spaces that must comply with dispersion requirements in assembly (group *A*) occupancies. Platform lifts are also permitted in dwelling or sleeping units, accessible routes to raised judges' benches, witness and jury boxes in judicial facilities, and situations where the use of a ramp or elevator is infeasible due to site constraints.

Edges of passenger transit platforms where there is a drop-off not protected by guards or screens must have detectable warnings. Such warnings are usually provided by means of a resilient, textured walking surface that visually contrasts with the platform material.° This requirement does not apply to bus stops.

Signage (1110)

Required accessible elements must be identified by the *international symbol of accessibility* (Figure 15-1) at all locations that follow:

- Parking spaces and passenger loading zones (1106)
- Entrances where not all entrances are accessible
- Multiple single-unit toilet or bathing rooms at one location or unisex facilities
- Dressing, fitting, and locker rooms where not all are accessible
- Checkout aisles where not all are accessible
- Accessible areas of refuge (1007)

Directional signage must be provided from nonaccessible elements to the nearest similar accessible elements in locations such as the following:

- Inaccessible building entrances
- Exits and elevators that do not provide an accessible means of egress from a required accessible space

Figure 15-1   International Symbol of Accessibility.°

- Elevators not serving an accessible route
- Public toilets and bathing rooms including directions to unisex facilities where provided (1109)

Other signage requirements include posting doors to egress stairways, passageways, and exit discharges; to areas of refuge; and for assisted rescue. Information signage in assembly occupancies must indicate the availability of assistive listening systems.

CLOSING WORDS

Chapter 15 has covered the principal *IBC* requirements pertaining to *Accessibility*: entrances and routes, parking and passenger loading, dwelling and sleeping units, special occupancies, and signage.

In our final chapter, we turn to the Americans with Disabilities Act of 1990, which formally mandated the right of accessibility to all "places of public accommodation and commercial facilities." In July 2004, the U.S. Access Board administering the ADA guidelines issued a major update to the document, stating a fourfold goal:°

- Updating the guidelines to ensure that the needs of persons with disabilities are adequately met
- Refining the format for easier use in order to promote compliance
- Ensuring consistency among the requirements of the Architectural Barriers Act and the ADA
- "Harmonizing" the guidelines with model building codes and industry standards such as the *IBC*

The final goal specifically refers to *ANSI A117.1*, which is at the heart of the *IBC*'s accessibility provisions as we have seen. The new document contains scoping provisions for ABA and ADA in two parts. The third section contains a common set of technical requirements to which both scope documents refer.

Chapter 16 is directed primarily to the technical requirements of Part 3, consisting largely of interior planning criteria such as sizes and clearances for basic building elements referred to as "building blocks." Dimensional standards for such elements as level changes, obstructions, leg clearances, and reach ranges are shown.

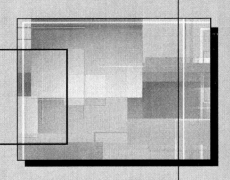

CHAPTER 16

ADA Accessibility Guidelines

*What if physical conditions have built up high walls about us? . . . our world lies
upward; the length and breadth and sweep of the heavens are ours!*
—*Helen Keller (1901)°*

One of the other instructors at the school where I teach remarked recently that we should provide wheelchairs to our students *without disabilities* so that they might experience first-hand some of the impediments of people with disabilities. When William Gibson's play *The Miracle Worker* was first produced in 1959, Anne Bancroft, who portrayed Helen Keller's teacher Anne Sullivan, was reported to have blindfolded herself for a month to prepare for the role. Those of us without serious handicaps tend to take our ease of mobility for granted and sometimes need to be reminded that inaccessibility often creates frustrating and even life-threatening prisons for people with disabilities.

The Americans with Disabilities Act was enacted into law by the U.S. Congress in 1990 with the expressed purpose of setting "guidelines for accessibility to places of public accommodation and commercial facilities by individuals with disabilities."° *Public facilities*, which are "constructed by, on behalf of, or for the use of a public entity" are subject to Title II of the ADA. A *private facility* is a privately owned "place of public accommodation or a commercial facility . . . or a transportation facility subject to Title III of the ADA."°

Predating the ADA by over twenty years was the Architectural Barriers Act (ABA) of 1968, which specifies accessibility requirements for facilities designed, built, or leased using federal funds. The recently issued ADA and ABA guidelines are written much like a performance-oriented building code and are intended to be applied to the alteration of existing facilities as well as the design and construction of new buildings. There are detailed requirements as to the numbers of accessible doors, drinking fountains, and toilets that must be provided, as well as dimensional parameters that determine whether a door, sink, or toilet stall is accessible.

When the ADA first became law, there was some confusion regarding the use of the phrase *to the maximum extent practicable*, which appears some fifteen times (in various forms) throughout the document. The Act prescribes "nondiscrimination" on the basis of disability, which means that its intent is to ensure *equal* accessibility for individuals with disabilities. *Practicable* does not necessarily mean inexpensive or simple, at least with regard to new construction; rather, it means that some terrain-related physical obstacle effectively prevents the incorporation of accessible features. Even so, such exceptions apply only to that portion of the building subject to such limitations. All other portions of the structure are expected to fully comply with ADA guidelines.

One general provision in the ADA that merits particular attention relates to what the standard calls "equivalent facilitation." Although the standard prescribes some very specific dimensional and other requirements for such elements as wheelchairs, guards, grab bars, and plumbing fixtures, the ADA states the following: "Departures from particular technical and scoping requirements of this guideline by the use of other designs and technologies are permitted where the alternative[s] . . . will provide substantially equivalent or greater access to and usability of the facility."° Thus, the ADA standard provides a performance baseline while allowing both for technological innovation and design ingenuity. Equivalent facilitation does not apply to the ABA, as federal agencies have a formal procedure for modifications or waivers of specific provisions when clearly necessary.

TERMINOLOGY AND DOCUMENTATION

As we noted in the previous chapter, a substantial part of the technical requirements set forth in the ADA are based on *ANSI A117.1*. Up until the current version, released on July 23, 2004, the organization of the ADA document was substantially different from *A117.1*. Most of the pertinent scope issues and technical requirements were spelled out in Section 4 of Appendix A of the Act (28 C.F.R. 36), which accounted for nearly two-thirds of the document (see Table 15-1). The accessibility issues specifically addressed by the *IBC* were discussed in Chapter 15, referring repeatedly to *A117.1*.

The current release of the federal accessibility guidelines combines the ADA and ABA requirements and makes them consistent. There are three parts to the new design guidelines. The first two contain scoping provisions for the ADA and ABA, respectively. These sections specify which spaces and elements must comply and include instructions on the application of specific guidelines such as dimensional conventions. Throughout this chapter, we refer to the ADA (or ADAAG) requirements only because they are most pertinent to *places of public accommodation* under Title III. The ABA scoping provisions are substantially similar to those of the ADA.

The third section contains a common set of technical and dimensional criteria referenced by both the ADA and ABA scoping sections. This part is structured to *harmonize* with model building codes and industry standards, notably chapters 3 through 9 of *ICC/ANSI A117.1*. Chapter 10 of *A117.1* deals in detail with dwelling and sleeping units. The first two chapters are devoted to administration and scoping provisions. Chapters 3 through 10 are listed here:

Chapter 3. Building Blocks

Chapter 4. Accessible Routes

Chapter 5. General Site and Building Elements

Chapter 6. Plumbing Elements and Facilities

Chapter 7. Communications Elements and Features

Chapter 8. Special Rooms, Spaces, and Elements

Chapter 9. Built-In Elements

Chapter 10. Recreation Facilities

These subsections will supercede the technical requirements set forth in Appendix A of the 1994 ADA document (29 C.F.R. 36) and the *ADAAG* as amended through September 2002.

Strictly speaking, the *ADA-ABA* guidelines of 2004 are not yet mandatory for the general public. In this respect, they are similar to the current ICC model building codes, which are not obligatory until they have been adopted by an enforcing authority. Formally, the original ADA standards issued by the DOJ and the DOT in 1991 (as amended) remain in effect. This chapter is written with the new guidelines in mind with the expectation that they will ultimately supercede the earlier standards. (As of September, 2006, the ADA-ABA Guidelines Homepage states that the "Guidelines are not enforceable or mandatory in and of themselves, but instead serve as a baseline for updating the standards that are.")°

Given that most commercial projects fall within the ADA's definition of "public use" and, therefore, must comply with the *ADAAG*, we do not attempt to note all the differences between it and *A117.1* here. Where there are significant differences from the earlier *ADAAG* documents, particularly relating to dimensional allowances and clearances, they are highlighted.

Where selected references are made to one or both sets of guidelines, they are given by section number in parentheses in the following format: (*ADA-ABA/ADAAG*).

What Does "Accessible" Mean?

We have already defined *accessible*, which has essentially the same meaning in the ADA as it does in *A117.1* and the *IBC*. In the previous chapter, we discussed *Accessible Means of Egress*. The accessibility guidelines broaden that usage, as follows:

- *Accessible route*—A continuous, unobstructed path connecting all accessible elements and spaces of a building or facility. Interior accessible routes may include corridors, floors, ramps, elevators, lifts, and clear floor space at fixtures. Exterior accessible routes may include parking access aisles, curb ramps, crosswalks at vehicular ways, walks, ramps, and lifts.

The primary concern of the *IBC* and all fire codes is safe egress in the event of emergency. *ADAAG* mandates that all accessible spaces in buildings are reachable by accessible circulation, which in most cases follows the same pathways as those used for egress. In this chapter we focus primarily on *interior* accessibility considerations.

Some other terms that the ADA uses are the following:

- *Addition*—An expansion, extension, or increase in the gross floor area of a building or facility.
- *Alteration*—A change to a building or facility that affects or could affect the usability of the building or facility or parts thereof. Alterations include but are not limited to remodeling, renovation, rehabilitation, reconstruction, historic restoration, resurfacing of circulation paths or vehicular ways, changes or rearrangement of the structural parts or elements, and changes or rearrangement in the plan configuration of walls and full-height partitions. Normal maintenance, re-roofing, painting or wallpapering, or changes to mechanical and electrical systems are not alterations unless they affect the usability of the building or facility.°

Both of these activities require that applicable code provisions be properly implemented and necessary certifications obtained from the appropriate governmental agencies. New construction is expected to be fully compliant, even when it does not involve an entire new building. A completely new tenant buildout, for example, is considered new construction that must fully meet *ADAAG* requirements, even though it is an alteration in the context of the building as a whole. Because the ADA is federal civil law, however, local jurisdictions do not typically review specific projects for compliance.

The ADA provides that alterations must not decrease a building's level of accessibility below that required for new construction at the time the alteration is made. If alterations to a portion of a room or space are such that its usability as a whole is affected, then the entire space must be brought into compliance. On the other hand, no such alteration may compel a greater degree of accessibility than would be required for new construction. In this regard, the Act uses the example of elevators being altered to make them accessible. Such modification does not imply that stairs connecting the same floor levels must, as a matter of course, be upgraded as well. Only if other codes require that the stairs be modified to correct unsafe conditions do they become subject to the ADA. Then, any modifications must comply unless implementing them is technically (meaning structurally) infeasible.

In principle, it is the building owner who determines how the building is operated and how decisions are made affecting accessibility; and it is "that person or entity" that is responsible for making sure the facility is in compliance with ADA. Legally, it is the owner who initiates design and construction of the facility and any additions or alterations and who operates and maintains the facility. Like many other situations in which the courts assign ultimate responsibility, however, the architect, designer, and others involved in the design and construction process may ultimately be held responsible for noncompliance. This is probably as strong an argument as can be made for the professional having a thorough understanding of these requirements and the effective management of facility-planning policy and procedures.

SCOPE

All new construction must be accessible unless to do so is not realistically achievable. Where structural conditions do not permit full compliance, however, the entire building is not automatically exempted, as noted earlier. All portions of the facility where compliance is structurally feasible must comply. Where a facility has more than one use, each portion must meet the *Code* standards for that occupancy classification. Temporary structures such as reviewing stands and bleachers, temporary classrooms or exhibit areas, and pedestrian passageways around construction sites fall within the guidelines. Construction sites themselves and the equipment associated with them, however, do not (201/4.1.1).

Sites and Exterior Facilities: New Construction

The only permissible reason for new construction not meeting the ADA standards is where doing so is "structurally impracticable." The document specifically states that this only applies in "those rare circumstances when the unique characteristics of terrain prevent the incorporation of accessibility features."°

At least one accessible route must be provided from arrival points to a site, such as public transportation stops, accessible parking and loading zones to an accessible building entrance, and between building entrances and other elements on the site. Publicly accessible areas within a sports facility must comply, but accessibility requirements do not pertain to actual areas of sport activity (206/4.1.2).

Some specific types of spaces are *not* required to be accessible. These include areas used only by certain employees as work areas and temporary structures. Spaces used primarily for security observation and nonoccupiable spaces used by service and repair personnel, such as elevator pits, penthouses, and piping or equipment catwalks, are exempted, as are animal-containment areas not for public use (203/4.1.3).

If parking spaces are provided for employees and visitors, a certain proportion must be accessible. These proportions are the same as the *IBC*'s, which are shown in Table 15-2. The ADA formerly required van-accessible spaces in a ratio of 8 to 1.° The *IBC* is less stringent with its requirement of 6 to 1, and the new *ADA-ABA* guidelines follow *IBC*. Parking requirements for outpatient units and facilities rendering services to people with mobility impairments match the *IBC*'s, as do those for valet parking (208/4.1.2).

The remainder of this section refers to other *ADAAG* sections that spell out specific dimensional and other physical requirements, for example: *Toilet Rooms* (603/4.22), *Bathrooms, Bathing Facilities, and Shower Rooms* (607–8/4.23), and *Signage* (703/4.30). If toilet and bathing facilities are provided on a site, they must be accessible. Signage must identify all accessible parking, loading zones, and entrances, as well as toilet and bathing facilities.

Buildings: New Construction

This section enumerates the applicability of specific *ADAAG* performance criteria to elements and spaces within a building. It begins by extending the statement in the last section to include the building's interior: "At least one accessible route complying with 4.3 shall connect accessible building or facility entrances with all accessible spaces and elements within the building or facility" (206/4.1.3(1)(a)). Similar inclusive references are made to sections such as *Protruding Objects* (204.1/4.4), *Ground and Floor Surfaces* (206.2/4.5), *Stairs* (210/4.9), and *Elevators* (206.6/4.10).

It is important to check the latest version of the guidelines for amendments, particularly if you are dealing with an unfamiliar building type. For example, exceptions are made in the newer editions to the effect that requirements dealing with floor surfaces and protruding objects (302 and 307/4.4 and 4.5) do not apply to areas of sport activity. Such exceptions are specific, in this case referring only to "the space needed to play"° the game. Surrounding circulation and related areas such as spectator facilities are not exempted.

At least one accessible route is mandated in new construction by the ADA, with certain exceptions, such as private facilities under three thousand square feet per floor or less than three stories, unless it is a shopping center or mall, a health-care professional office, or a public transportation terminal (206.2.3/4.1.3(5)).

There are other exceptions for certain highly specialized facilities, including such spaces as those related to elevator operation and maintenance, but the *Code* is careful to point out that such exceptions do not "obviate or limit in any way the obligation to comply with . . . other accessibility requirements. . . ."° Once again, the *Code* uses toilet and bathing facilities as an example: if such are provided on a level not served by an elevator, then comparable facilities must be provided on an accessible floor. Where applicable, accessible ramps (405/4.8) or platform (wheelchair) lifts (410/4.11) may be used in lieu of an elevator, the latter only under prescribed conditions.

All of the following types of doors must comply with the accessibility requirements of Section 404/4.13: at least one at each accessible entrance to the building or facility, at least one to each accessible space, each door along an accessible route, and every door required for egress. The new guidelines state that 60 percent of all public entrances (not service entrances or loading areas) must comply with Section 404. This is an increase from 50 percent according to *ADAAG* (206.4.1/4.1.3(8)). *ADAAG* states a preference that *all* entrances be accessible on the basis that they are each potential emergency exits. Nonaccessible entrances must in any case have signage indicating the location of the nearest accessible entrance. Where required exits are not accessible, areas of rescue assistance must be provided. Emergency warning systems must have both audible and visual alarms (702/4.28).

The remainder of this section enumerates several types of devices and amenities, bringing them within the scope of the *Code*. These include drinking fountains (211/4.15), controls and operating mechanisms such as light switches (205.1/4.27), public telephones (217/4.31), and storage and shelving (225/4.25), as well as fixed seating and tables (206/4.32). Toilet facilities are mentioned again. If they are provided, public and common-use toilet rooms must be accessible. Even private toilet rooms must be adaptable to the provisions of Section 213/4.22. The same applies to locker facilities and dressing rooms (222/4.35), as well as to saunas and steam rooms (241/4.36).

Assembly areas must provide specific numbers of accessible wheelchair seating locations compliant with Section 221/4.33. Assistive listening systems are also required where audibility is essential to the function of the space, if more than fifty people can be accommodated or if they have fixed seating or an electronic audio system. These include movie and other types of theaters, concert halls, lecture halls, and meeting facilities. Signage must also be provided to inform patrons that such services are available.

Building Additions and Alterations

Additions are regarded by the ADA as alterations, and the *Code*'s primary concern is that neither reduces the usability or accessibility below the level of performance required for new construction "at the time of the alteration."° In fact, the ADA states that alterations cannot be required to provide a higher level of accessibility than that required for new construction. The *Code* also states, however, that where "alterations of single elements, when considered together, amount to an alteration of a room or space in a building or facility, the entire space shall be made accessible."°

At least one accessible route must connect each story and mezzanine in multistory facilities, as well as accessible entrances and all accessible elements and spaces within the facility, subject to specified exceptions. Although a mezzanine provides a change in level, it is not considered a story, but accessible routes must nonetheless serve all mezzanines. If a stair or escalator is constructed where none existed before, accessible vertical access must then be provided that complies with the applicable provisions of Section 206.2.3 and Section 206.2.4.

If an entrance is altered and the building already has an accessible entrance, the newly altered entrance does not have to be made accessible unless the changes affect access to or the usability of an area containing a *primary function*. If alterations are limited to changes in a building's mechanical, electrical, or plumbing systems; to asbestos abatement; or to the addition of automatic sprinklers and do not materially change any building elements or spaces that must be accessible, the *primary function* rule does not apply.

- *Primary function*—A major activity for which the facility is intended . . . the customer services lobby of a bank, the dining area of a cafeteria, and the meeting rooms in a conference center, as well as offices and all other work areas in which the activities of the public accommodation or other private entities using the facilities are carried out.° *Areas of transportation facilities* include . . . ticket purchase and collection areas, passenger waiting areas, train or bus platforms, baggage checking and return areas, and employment areas (except nonoccupiable service areas).°

- *Technically infeasible*—With respect to an alteration of a building or a facility, means that the alteration has little likelihood of being accomplished because existing structural conditions would require removing or altering a load-bearing member that is an essential part of the structural frame; or because other existing physical or site constraints prohibit modification or addition of elements, spaces, or features that are in full and strict compliance with the minimum requirements for new construction necessary to provide accessibility.°

Beyond structural considerations, if the cost of providing an accessible travel path to an altered primary function area is *disproportionate*, the *Code* allows prioritizing the level of compliance on a basis of cost (202.4). The Department of Justice's guideline is that disproportionate means more than 20 percent of the cost of altering the primary function area, exclusive of the costs directly attributable to accessibility modifications.°

Handrail extensions at stairs are not required where they would be hazardous or not possible due to existing layout constraints (505.10). The 1:12 slope limitation for ramps is relaxed to 1:10 for a six-inch rise and a maximum of 1:8 for a three-inch rise. The inside dimension requirements for existing elevators are sixteen square feet clear floor area, a clear depth of fifty four inches and width of thirty six inches, the final determination being made on a basis of whether standard wheelchair clearances can be accommodated (407.4.1).

Alternate unisex toilet facilities are permitted where it is technically infeasible to modify existing facilities to make them accessible. Where toilets *can* be modified but it is technically infeasible to install a standard accessible stall or where the fixture count cannot be reduced due to other codes, an appropriate unisex toilet as pictured in Figure 16-27 may be used. Unisex toilet rooms have proven themselves useful in addition to standard toilet facilities so that it is considered advantageous to provide them even in new facilities (213.2/A4.22.3).

Similar relaxations of accessibility requirements are made for assembly occupancies and certain other uses when buildings are altered. The essential point is "to ensure that patrons and employees of public accommodations and commercial facilities are able to get to, enter and use the facility."° As far as cost considerations are concerned, the DOJ and DOT guidelines both list the priorities, in order, as follows:

- Entrance
- Route to the altered area
- Restroom facilities
- Telephones
- Drinking fountains
- Parking, storage, alarms, and so on

Historic Preservation

As one may expect, special consideration is given to the requirements of the *ADAAG* when alterations are made to historic buildings. Beyond the question of technical feasibility, the manner in which alterations potentially affect the historic significance of the structure becomes an important concern.

If a building is listed in or eligible for listing in the National Register of Historic Places, Section 106 of the National Historic Preservation Act (16 U.S.C. 470f) provides for a federal authority to consider and guide the effects of proposed alterations. The *Code* also provides for consultation with state officials and interested persons when a building not subject to Section 106 is altered (i.e., local officials, organizations representing persons with disabilities, as well as disabled individuals themselves). The *Code* specifically mentions considerations related to entrances, toilet facilities, and signage; however, such projects are generally handled on a case-by-case basis (202.5/4.1.7).

BUILDING BLOCKS

The new *ADA-ABA Accessibility Guidelines* and *ICC/ANSI A117.1-2003* (first published two months earlier in May of 2004) define "building blocks" of accessibility. This is the title of the first technical chapter in both of the preceding documents.

Including these general principles in one section of the *ADA-ABA* guidelines enables other sections to refer to them repeatedly and eliminates a good deal of the redundancy present in the earlier *ADAAG* document. Each of the *scoping* sections of the new guidelines begins with the following statement:

This document contains scoping and technical requirements for accessibility to *sites, facilities, buildings,* and *elements* by individuals with disabilities. The requirements are to be applied during the design, construction, addition to, alteration (and "lease" in the case of ABA) of *sites, facilities, buildings,* and *elements* to the extent required by regulations. . . .°

It is worth noting in passing that the definitions used by the U.S. Access Board and the ICC are virtually identical, although the definitions of *facility* differ slightly:

- *Site*—A parcel of land bounded by a property line or a designated portion of a public right-of-way.
- *Facility*—All or any portion of buildings, structures, site improvements, elements, and pedestrian routes or vehicular ways located on a site. The ICC definition adds: "wherein specific services are provided or activities are performed."
- *Building*—Any structure used or intended for supporting or sheltering any use or occupancy.
- *Element*—An architectural or mechanical component of a building, facility, space, or site.

The "building blocks" of accessibility, as established by both sets of guidelines, include the following sections:

- Ground and Floor Surfaces (302)
- Changes in Level (303)
- Wheelchair Turning Space (304)
- Clear Floor Space (305)
- Knee and Toe Clearances (306)
- Protruding Objects (307)
- Reach Ranges (308)
- Operable Parts (309).

Each of the basic building block requirements is summarized in this section. The remainder of this chapter follows the structure of the two documents, emphasizing those aspects that are most directly applicable to space planning: accessible routes (corridors, doors, ramps, stairs, and elevators), plumbing elements, and communication systems.

Selected references are made to the new guidelines *and* the older *ADAAG* document to facilitate cross-referencing between them. In such cases, section numbers are shown in parentheses as follows: (*ADA-ABA/ADAAG*). In some of the final sections, specific references are also made to the IBC for clarity. In these cases references are indicated as follows: (ADA-ABA/ADAAG/IBC).

In all of the figures, the small numbers below each dimension line are the equivalent values in millimeters.

Floor or Ground Surfaces (302/4.5)

All horizontal surfaces that are part of an accessible pathway must be firm, stable, and slip resistant. This includes both exterior ground and interior floor surfaces, such as walks, curb ramps, floors, ramps, and stairs. *Firm and stable* means that the surface is resistant to deformation and that its composition is not changed by contaminants or impacts. *Slip resistant* means that the material has sufficient friction to facilitate walking safely.

Carpeting must be securely attached, with firm or no padding, a level pile texture (cut, uncut, or a combination), with a maximum pile thickness of one-half inch (Figure 16-1*a*). Exposed edges must be secured with trim that complies with the level-change requirements that follow.

Gratings in walking surfaces may have openings no greater than one-half inch wide, which must be placed perpendicular to the direction of travel (Figure 16-1*b*).

Changes in Level (303/4.8)

Changes in level, whether vertical or beveled, are subject to the height and proportional limitations shown in Figure 16-2. Gradual level changes greater than one-half inch must be treated as a ramp. Areas of sport activity and animal-containment areas are excluded from this requirement.

Turning Space (304/4.2)

Turning a wheelchair requires sixty inches of unobstructed area, either to turn in a circle or to make a 180-degree turn in a T-shaped space (Figure 16-3). Changes in level are not permitted within a designated turning space, which means that any slope must be less than 1:48.

Figure 16-1 Carpet Tile Height and Elongated Openings (302).

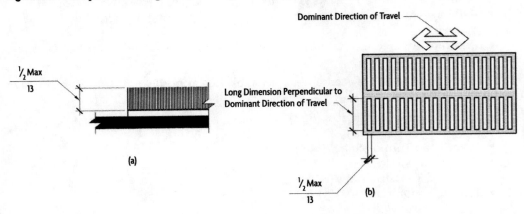

Figure 16-2 Vertical and Beveled Changes in Level (303).

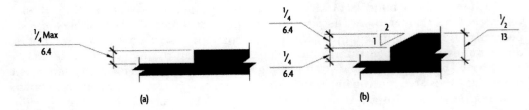

Figure 16-3   Circular and T-Shaped Turning Space (304).

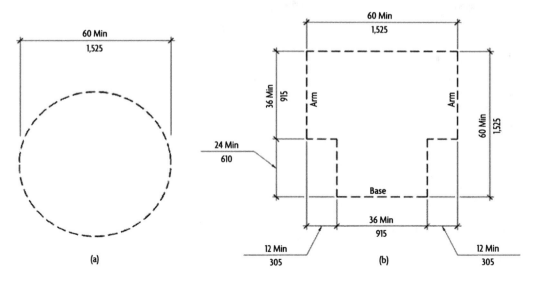

Figure 16-4   Positions in Clear Floor or Ground Space (305).

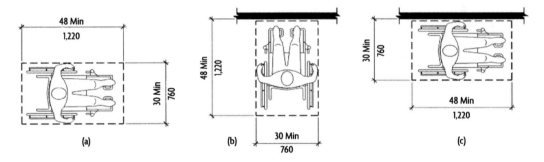

The width of arms and base in a T-shaped turning space must be at least thirty six inches wide. Both arms must be without obstructions for at least twelve inches and the base for twenty four inches. One end of either the base or arm may include knee and toe clearance as specified in Section 306 (Figure 16-3*b*).

Clear Floor or Ground Space (305/4.2)

Surfaces of clear floor or ground space also must comply with Section 302, and any slope must be less than 1:48. Figure 16-4 and Figure 16-5 illustrate a number of different conditions in which a wheelchair requires specified minimum clearances.

Thirty by forty-eight inches is the minimum area needed to accommodate a stationary, occupied wheelchair (Figure 16-4*a*). This same amount of space is also necessary to make either a forward or a side approach to an object (Figure 16-4, *b* and *c*), part of which may be knee space under certain objects such as drinking fountains or tables. The same minimum area is required for a wheelchair in an alcove no deeper than twenty four inches if forward facing or fifteen inches to the side.

If a forward-facing alcove is deeper than twenty four inches, it must be at least thirty six inches wide to provide for maneuvering clearance. If a side alcove is deeper than fifteen inches, it must be at least sixty inches wide (Figure 16-5).

Clear floor or ground space generally must be positioned for approaching an element from the front or the side. One full, unobstructed side of a clear space must directly adjoin an accessible route or another clear floor or ground space.

Figure 16-5 Maneuvering Clearance in Alcoves: Forward and Parallel (305.7).

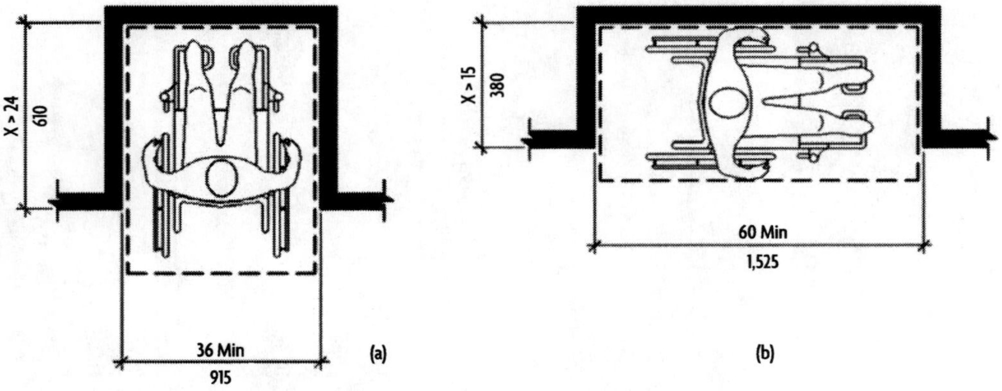

Knee and Toe Clearance (306)

Clearances are not necessarily determined by the vertical support for an element but are a function of the usable clear floor space. Vertical toe clearance includes the zone from the finished floor or ground surface to a height of nine inches as shown in Figure 16-6c. Knee clearance extends from the finished surface to a height of twenty seven inches (Figure 16-6a).

Horizontal knee clearance must extend twenty five inches (Figure 16-6b) and toe clearance from seventeen to twenty five inches under an element (Figure 16-6d). Knee clearance must extend from a minimum of eight inches under the element at the twenty seven-inch height to a minimum of eleven inches at the nine-inch height. The width of both knee and toe clearances must be at least thirty inches.

Protruding Objects (307/4.4)

The *IBC* requirements for protruding objects in an accessible route are the same as those in an egress path (see Chapter 14, 1003). The principal concern is with objects between twenty seven and eighty inches above the floor, whether wall-mounted or free-standing. Typical conditions are pictured in Figure 16-7 and Figure 16-8, and the *ADAAG* provides illustrations of various types of freestanding equipment. Clear headroom must be no less than eighty inches, as shown in Figure 16-7, and no protruding objects may reduce the effective width of the path.

Guardrails or other barriers a maximum of twenty seven inches high must be provided where there is less than eighty inches of vertical clearance above the finished floor or ground surface. Objects having leading edges more than twenty seven inches above the floor or ground cannot protrude more than four inches into the circulation path. Signs or other obstructions mounted between posts or pylons must have their bottom edges located below twenty seven inches or above eighty inches if they are more than one foot wide (see Figure 16-8).

Reach Ranges (308/4.2)

Vertical minimum and maximum reach limits are shown in Figure 16-9 and Figure 16-10:

- Unobstructed low and high forward-reach limits are between fifteen and forty eight inches (Figure 16-9a).
- If the high forward reach is over an obstruction, reach ranges are limited. Referring to Figure 16-9b, and c, the maximum reach depth may not exceed twenty five inches. If the obstruction depth is less than twenty inches, maximum vertical reach may be up to forty eight inches. The maximum reach is limited to forty four inches when the obstruction is deeper than twenty inches.
- Unobstructed low and high side-reach limits are between fifteen and forty eight inches (Figure 16-10a). This is a substantial change from the *ADAAG* limits of between nine and fifty four inches, with exceptions for certain elements such as fuel pumps.

Figure 16-6 Knee and Toe Clearances (306).

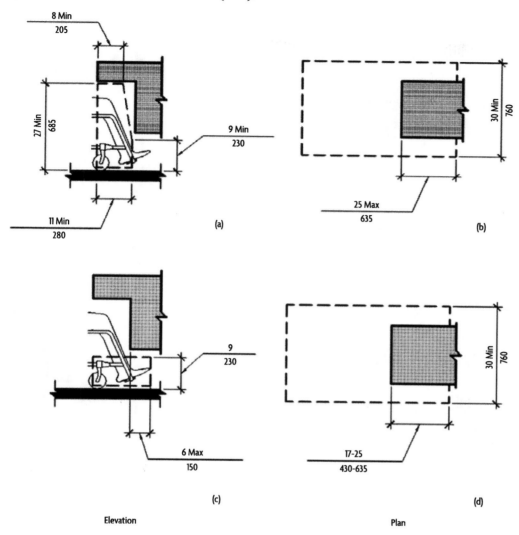

(a)

(b)

(c)

(d)

Elevation   Plan

Figure 16-7 Limits of Protruding Objects and Vertical Clearance (307).

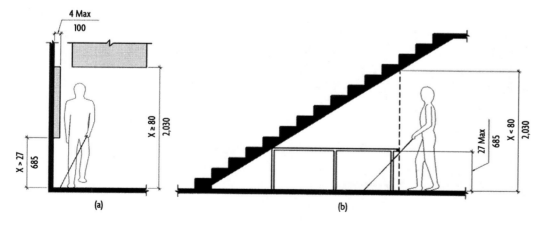

(a)   (b)

- If the high side reach is over an obstruction up to ten inches deep, the reach range may extend up to forty eight inches (Figure 16-10b). If the obstruction is a maximum of twenty four inches deep and thirty four inches high, the maximum vertical reach may not exceed forty six inches (Figure 16-10c).

Figure 16-8 Freestanding Protruding Objects (307.3).

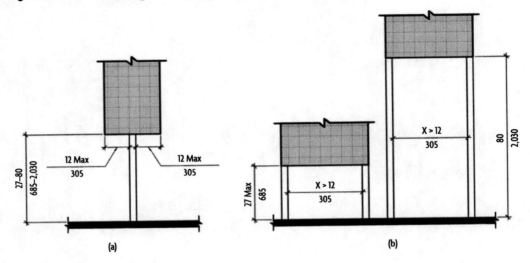

Figure 16-9 Forward-Reach Ranges (308.2).

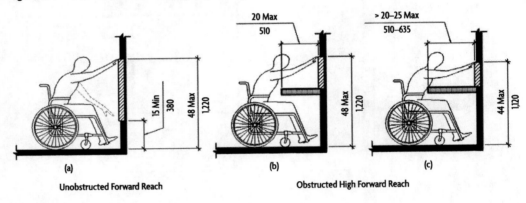

Unobstructed Forward Reach   Obstructed High Forward Reach

Figure 16-10 Side-Reach Ranges (308.3).

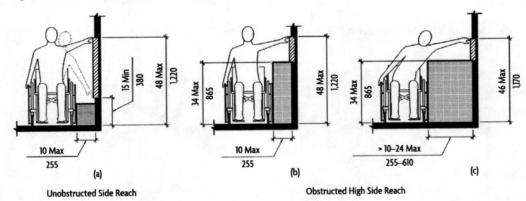

Unobstructed Side Reach   Obstructed High Side Reach

Operable Parts (309/4.27.4)

All controls, dispensers, receptacles, and other equipment intended for use by building occupants must have sufficient clear floor space around them to allow a forward or parallel
approach by a person in a wheelchair; and they must be within one of the reach ranges
specified in Section 308. Operating mechanisms must not require twisting of the wrist or
tight grasping, and the force required to activate operable parts must not exceed five
pounds.

Electrical and telecommunications receptacles must comply with Section 309 except where they are not intended for general use, such as large office equipment, wall-mounted clocks, refrigerators, and so on. Floor outlets need not comply with section 309 (205.1).

ACCESSIBLE ROUTES (400/4.3)

An accessible route from exterior to interior and connecting multiple buildings on the same site includes all walks, skywalks, tunnels, halls, corridors, and aisles, as well as all the connective spaces that provide access to the site, its buildings, and the facilities contained within. It begins with the individual's first arrival at the site, whether by public transportation, automobile, or on foot. Accessible routes should be the same as those provided for the general public. If the building contains dwelling units, the accessible route must extend to all accessible spaces and facilities that serve them.

Minimum acceptable clear widths and passing spaces are determined using the criteria specified in this section. It should be remembered that in many cases these widths will be substantially greater as a consequence of egress requirements, given that the same routes will serve as means of egress in emergencies. Other considerations that are addressed relate to headroom (307.4/4.4), surface textures and changes in level (302–3/4.5), and doors (404/4.13).

Accessible routes must provide *areas of refuge* in the event of emergency. The *ADAAG* uses the term *areas of rescue assistance*, which has substantially the same meaning, but the terminology is now consistent with that of the *IBC* (*IBC* 1007), relative to size, stairway width, communications, and identification (see Chapter 14). The new *ADA-ABA* guidelines directly reference the *IBC-2003* for means of egress and areas of refuge (105-2-4).

Walking Surfaces (403)

Walkways having a running slope (in the direction of travel) not steeper than 1:20, doorways, ramps, elevators, and platform lifts are the basic components of an accessible route. The cross-slope of walkways cannot exceed 1:48.

The *ADAAG* specifies certain critical dimensions for what *ICC/ANSI A117.1* and the new *ADA-ABA* now refer to as *building blocks*, beginning with wheelchair clearances.

As Figure 16-11 shows, a single wheelchair requires thirty six inches clear of continuous passage width, which can occasionally narrow to thirty two inches for a distance of twenty four inches. Such reduced-width segments must be separated by at least forty eight inches.

Figure 16-11   Clear Width of an Accessible Route (403.5.1).

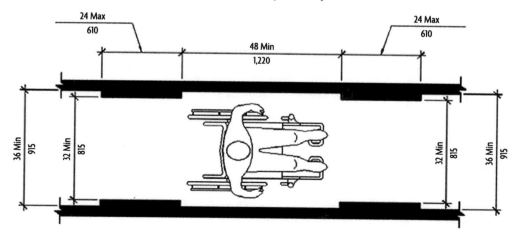

Sixty inches is required for two wheelchairs to pass each other, and an accessible route must provide an area sixty inches square every two hundred feet, or alternatively, a T-shaped space that complies with Section 304 in which the base and arms extend at least forty eight inches beyond the intersection.

When an accessible route makes a 180-degree turn around an element less than forty eight inches wide, at least forty eight inches of turning space must be provided with forty two-inch widths approaching and leaving the turn (Figure 16-12*a*). If at least sixty inches of clear turning space is available for such a turn around an obstacle wider than forty eight inches, the standard thirty six-inch passage width is sufficient (Figure 16-12*b*).

Doors, Doorways, and Gates (404/4.13 and 14)

Required accessible entrances must be part of an accessible route connected to all key points both outside of and within the building (*ADAAG* 4.14).

Doorways must have a minimum clear width of thirty two inches to accommodate a wheelchair (Figure 16-13). This dimension is measured between the face of the door and the opposite stop with the door open ninety degrees. Openings deeper than twenty four inches must be increased to a clear thirty six inches. There may be no projections into the required clear opening below thirty four inches, and such projections between thirty four and eighty inches may not exceed four inches.

Figure 16-12   Clear Width at Turns (403.5.2).

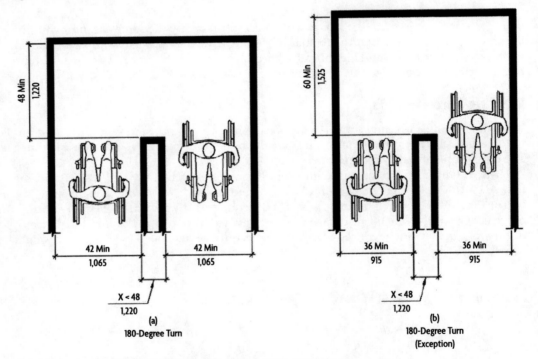

Figure 16-13   Clear Width of Door Openings (404.2.3).

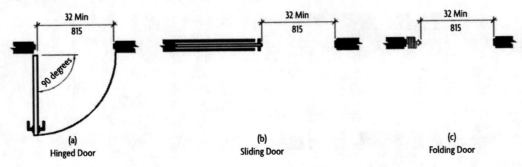

Double-leaf doorways must have one active leaf that meets this specification. Revolving doors, gates, and turnstiles may not be part of an accessible route but must have an adjacent conventional door. These requirements, including two doors in series, are consistent with the *IBC* (Chapter 14, 1008).

Minimum maneuvering clearances at manually operated swinging doors are illustrated in Figure 16-14 and Figure 16-15. In general, approaching a door from the side toward which it opens requires the greatest maneuvering space. A clear and level space sixty inches square, measured from the hinge point, will satisfy most conditions. Requirements pertaining to thresholds and hardware are consistent with the *IBC*.

Interior hinged, sliding, and folding doors may not require a force greater than five pounds to open, notwithstanding the force necessary to disengage latching devices. Door-closer mechanisms, if provided, must be adjusted so that the leading edge of the door takes at least five seconds to close from a ninety-degree open position to twelve degrees from the

Figure 16-14   Maneuvering Clearances at Manual Swinging Doors and Gates (404.2.4).

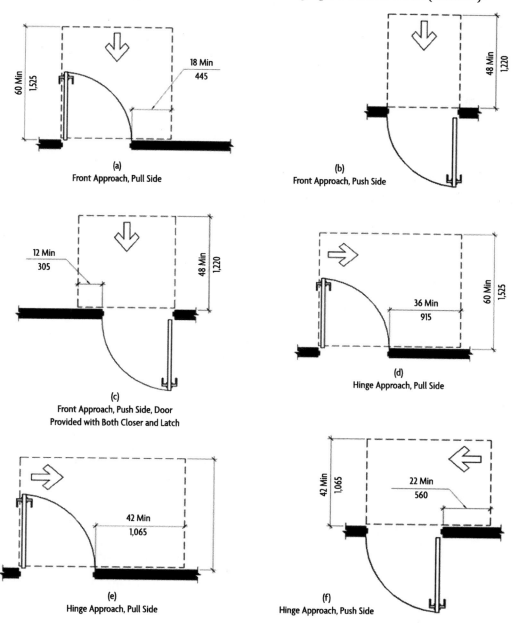

(a) Front Approach, Pull Side

(b) Front Approach, Push Side

(c) Front Approach, Push Side, Door Provided with Both Closer and Latch

(d) Hinge Approach, Pull Side

(e) Hinge Approach, Pull Side

(f) Hinge Approach, Push Side

Figure 16-15 Maneuvering Clearances at Manual Swinging Doors and Gates (404.2.4).

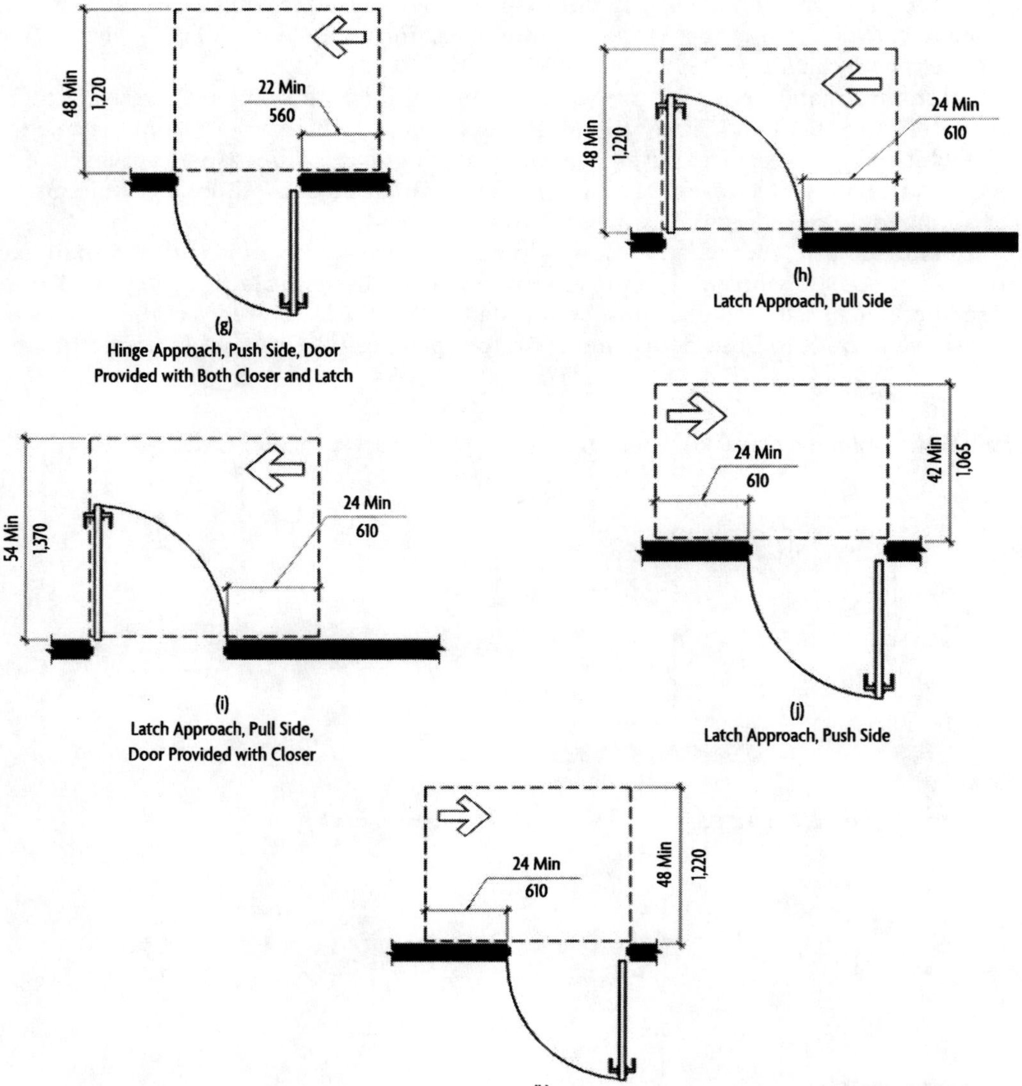

48 Min / 1,220
22 Min / 560

(g)
Hinge Approach, Push Side, Door
Provided with Both Closer and Latch

48 Min / 1,220
24 Min / 610

(h)
Latch Approach, Pull Side

54 Min / 1,370
24 Min / 610

(i)
Latch Approach, Pull Side,
Door Provided with Closer

24 Min / 610
42 Min / 1,065

(j)
Latch Approach, Push Side

24 Min / 610
48 Min / 1,220

(k)
Latch Approach, Push Side
Door Provided with Closer

latch (This is a change from *ADAAG* 4.13.10). Automatic and power-assisted doors are also subject to several ANSI specifications.

Several variations in approach are shown in Figure 16-16 for gates, sliding and folding doors, and openings without doors.

Doors and gates operated by security personnel are exempted from these requirements as they pertain to operating, latching, and closure mechanisms, with the stipulation that such personnel have exclusive control of such doorways. The configuration may not require that persons with disabilities be assisted while others have free access.

This section of the ADA nominally includes windows (*ADAAG* 4.12) as well, but the only specific requirements included are in the scoping sections, which state that mechanically operated windows should require a minimum amount of effort (five pounds) to open or close (309). Casements are also subject to the protrusion limitations of Section 307.

Figure 16-16   Maneuvering Clearances at Other Types of Door Openings (404.2.4).

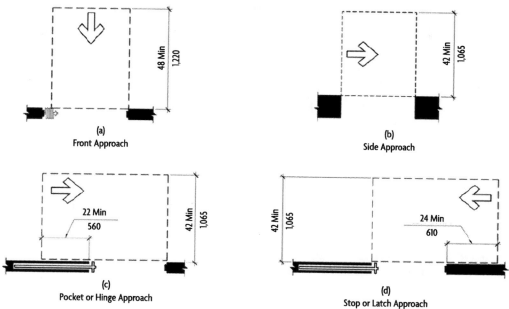

Ramps and Curb Ramps (405–6/4.7 and 8)

The *ADAAG* requirements for ramps, insofar as rise and slope, width, and landings are concerned, are the same as the *IBC*'s (see Chapter 14, 1010). Handrail and edge protection requirements are given in Section 500.

- Handrails are required on both sides if the rise is more than six inches. On switch-back or dogleg ramps, the inside rail must be continuous.
- Handrail height must be thirty four to thirty eight inches above the floor and gripping surfaces must be continuous.
- Clearance between handrail and wall must be 1.5 inches. Noncontiguous rails must extend twelve inches beyond the top and bottom of the ramp segment, parallel with the floor or ground surface. Ends must be rounded or return smoothly to the wall, floor, or post.

The maximum slope of ramps in new construction may not exceed 1:12. In existing buildings where space is limited, ramps may slope up to 1:10 for a maximum rise of up to six inches, or 1:8 for up to three inches. Slopes steeper than 1:8 are not permitted.

The maximum cross-slope of ramp surfaces may not exceed 1:48 (decreased from 1:50 in *ADAAG*), and outdoor ramps must be designed to minimize the accumulation of water. The *ADAAG* requires curbs at least two inches high or railings, or walls must protect ramps and landings with drop-offs.°

Curb Ramps (406/4.7)

The slope of a curb ramp is subject to the same limitation as other ramps: a maximum rise of 1:12. Minimum width is thirty six inches, and surfaces must comply with Section 302 (*ADAAG* 4.5). Curb ramp flares (the sides of the ramp) may not have a slope greater than 1:20. Detectable warnings must be provided over the full curb ramp surface (705/4.29), and the ramps must be located so as to prevent their obstruction by parked vehicles.

Elevators (407/4.10)

The *ADAAG* elevator requirements are instantly observable in almost any modern American high-rise building. Elevator cars must, of course, be wheelchair accessible, with doors having

Table 16-1   Elevator Car Dimensions (407.4.1)

Minimum Dimensions				
Door Location	Door Clear Width	Inside Car, Side to Side	Inside Car, Back Wall to Front Return	Inside Car, Back Wall to Inside Face of Door
Centered	42 Inches	80 Inches	51 Inches	54 Inches
Side (Off-Centered)	36 Inches	68 Inches	51 Inches	54 Inches
Any	36 Inches	54 Inches	80 Inches	80 Inches
Any °	36 Inches	60 Inches	60 Inches	60 Inches

a minimum thirty six-inch clear opening width. The interior depth of the cab must be a minimum of fifty one inches, and the width a minimum of sixty eight to eighty inches depending on whether the doors to the car are side or center opening. Typical dimensions in relation to door placement are given in Table 16-1.

Elevators must be automatic and located on an accessible route. The call buttons must be centered at a height within one of the reach ranges specified in Section 308, with a minimum dimension of three-fourths inch and the *up* button on top. Both visible and audible signals must indicate when a call is being answered. Visual signals must be centered at least seventy two inches above the floor and have a minimum dimension of 2.5 inches. Raised and Braille characters must be provided at elevator entrances centered at sixty inches above the floor, measured to the highest operable part (703/4.30).

Door and signal timing is a function of the interval between call notification and door closing and the distance from the farthest call button to the center of the hoistway door, a minimum of five seconds. Doors must remain fully open in response to a call a minimum of three seconds.

Controls inside the elevator must be placed on the front wall of cabs with center-opening doors or on the front or side wall if the door is side opening. Controls must be designated by Braille and raised alphanumeric characters placed immediately to the left of the button to which they apply. Floor buttons may be no higher than forty eight inches above the finished floor, unless the elevator panel serves more than sixteen openings and a parallel approach is provided. In that case, as well as in existing buildings, a height of fifty four inches is permitted. Emergency stop and alarm controls must be placed at the bottom of the control panel, no less than thirty five inches above the floor.

Car position indicators must be provided above the door or control panel inside the cab to show its position within the hoistway. Illuminated numerals at least one-half inch high, as well as an audible signal, are required. The audible signal may be in the form of an automatic verbal announcement. An emergency communications system must also be provided inside each elevator, connecting to a point outside the hoistway, which may not *require* voice communication. (The system should provide both audio and visual indications that help is on the way.)

Limited-Use and Private-Residence Elevators (408–409)

Separate specifications are now provided for limited-use and residential elevators, which of course must comply with the basic building block sections and doorway clearances. Operation must be automatic, but area requirements are generally reduced, thirty six by forty eight inches being the minimum inside dimensions for residential units. Limited-use cars must be a minimum of forty two by fifty four inches with the doors placed at the narrow end, having a minimum of thirty two inches clear width.

Residential car controls must be located on a side wall at least twelve inches from any adjacent wall. Other requirements refer largely to the provisions of Section 407.

Platform Lifts (410/4.11)

The new *ADA-ABA* guidelines reference a new industry standard for platform or wheelchair lifts: *ASME 18.1-1999, Safety Standard for Platform Lifts and Stairway Chairlifts.*

Where they are used, wheelchair lifts must permit unassisted entry, exit, and operation. Requirements for floor surfaces, clear floor space, and operable parts are similar to those for elevators. Section 4.1.3 of the *ADAAG* provides that lifts may only be used under certain conditions to provide access where site or other constraints make the use of a ramp or elevator infeasible, or for the following:

- Raised judges' benches, witness stands, jury boxes, and so on, or to depressed areas such as the well of a court
- Occupiable rooms and spaces, such as control rooms and projection booths, housing no more than five persons
- An accessible route to a performing area or to comply with viewing position line-of-sight dispersion requirements in an assembly occupancy
- Player seating areas in an area of sports activity

GENERAL SITE AND BUILDING ELEMENTS

Parking Spaces and Passenger-Loading Zones (502–3/4.6)

The *ADAAG* requirements for parking spaces and passenger-loading facilities are extensively covered by the *IBC*, referring repeatedly to *A117.1.* These requirements are discussed in Chapter 15 under Parking and Passenger Loading (1106) and Signage (1110).

Stairways (504/4.9)

Stairs must have uniform tread widths and riser heights with minimum tread widths of eleven inches, the same as *IBC* (1009). Permissible tread nosings are pictured in Figure 16-17. Projections may be no greater than 1.5 inches, the underside having an angle of no more than thirty degrees from vertical. Outdoor stairs must be configured to prevent the accumulation of water. Handrail and handrail extension requirements are identical to those for ramps.

Handrails (505)

Handrails are required on both sides of stairs or ramps, except in assembly areas where there is one handrail at either side or within the width of the aisle. They must be continuous within the full length of every flight of stairs or ramp run including inside rails on switchback or dogleg stairs or ramps *between* flights or runs. There may be breaks in handrails on ramps serving seating in assembly areas.

The top of handrail gripping surfaces must be a consistent height between thirty four and thirty eight inches as shown in Figure 16-18 on stairs (*a*), ramps (*b*), and walking surfaces (*c*) where required. In facilities designed for children, such as elementary schools, a second

Figure 16-17   Stair Nosings (504.5).

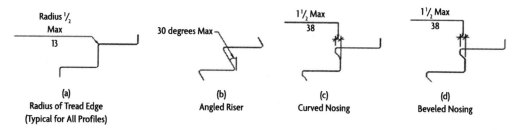

| (a) | (b) | (c) | (d) |
| Radius of Tread Edge (Typical for All Profiles) | Angled Riser | Curved Nosing | Beveled Nosing |

Figure 16-18 Heights of Handrails (505.4).

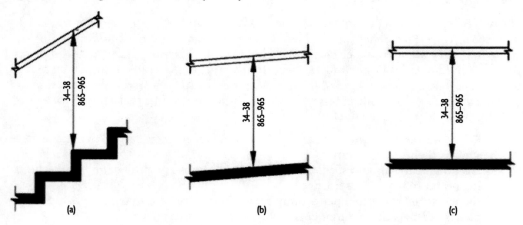

(a) (b) (c)

Figure 16-19 Horizontal and Vertical Handrail Clearances (505.5–6).

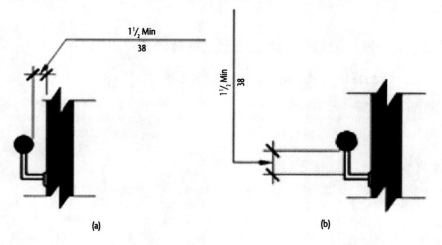

(a) (b)

handrail at a maximum height of twenty eight inches is recommended. In such situations, at least nine inches should be maintained between upper and lower rails to prevent entrapment.

Clearance between gripping surfaces and adjacent elements must be at least 1.5 inches (Figure 16-19a). The new *ADA-ABA* document points out in an advisory that older people and those with disabilities "benefit from continuous gripping surfaces that permit users to reach the fingers outward or downward to grasp the handrail, particularly as the user senses a loss of equilibrium . . ."° Handrail gripping surfaces must therefore be continuous along their length and not obstructed along their tops or sides, nor for more than 20 percent of their length along the bottom (Figure 16-19b).

Gripping surfaces having a circular cross section must have an outside diameter of between 1-¼ and 2 inches. The new *ADA-ABA* guidelines permit a greater variety of handrail shapes including rectangular and elliptical profiles. The range of acceptable perimeter and cross-section dimensions are shown in Figure 16-20.

Handrails at the top and bottom of ramp runs must extend one foot horizontally at the landings and return to the landing surface, adjacent wall, or guard or be continuous with the rail of an adjacent ramp run (Figure 16-21). Extensions are not required in assembly areas where the ramps directly serve seating, at inside turns, or in alterations where they would be potentially hazardous.

Top and bottom handrail extensions at stairs must extend similarly, one foot at the top and at least the depth of one tread at the bottom. Extensions must return to the landing surface, adjacent wall, or guard, or be continuous with the rail of an adjacent flight of stairs (Figure 16-22).

Figure 16-20   Noncircular Handrail Profiles (505.7.2).

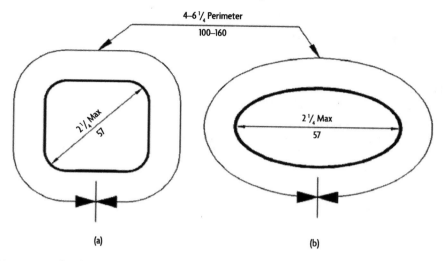

Figure 16-21   Handrail Extensions at Ramps (505.10.1).

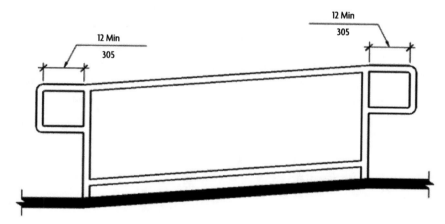

Figure 16-22   Handrail Extensions at Stairs (505.10.2–3).

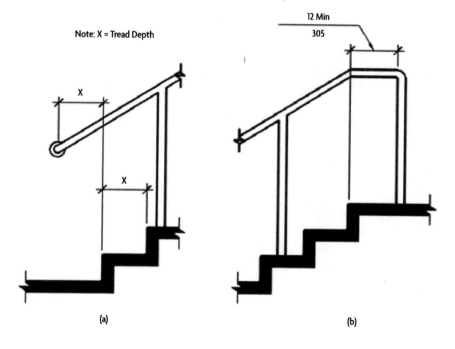

PLUMBING ELEMENTS AND FACILITIES

Drinking Fountains (602/4.15)

The *ADAAG* recommends that two side-by-side drinking fountains be provided to serve both people in wheelchairs and those who find it difficult to bend over. Their placement must satisfy the requirements for clear floor space and knee and toe clearance; and, under the new *ADA-ABA* requirements, a side approach is no longer permitted. The height of wheelchair-accessible drinking spouts must be thirty six inches from the finished floor.

The spout must be placed so the water flow is within five inches of the front of the unit and at least fifteen inches from the vertical support as shown in Figure 16-23. Fountains for children's use may be no more than thirty inches above the floor with the spout a maximum of 3.5 inches from the front of the unit. Fountains for standing persons must be between thirty eight and forty three inches above the finished floor. The flow of water must arc at least four inches high so a container may be used by persons who would otherwise be unable to use the unit.

Bathrooms, Water Closets, and Toilet Compartments (603–4)

Doors into accessible toilet rooms may not swing into the clear floor space required for any fixture. Unobstructed turning space must be provided within the toilet room, but the clear floor spaces required for the turning space, accessible route, and fixtures and controls may overlap.

Mirrors installed above lavatories and countertops must have the bottom edge of their reflecting surface no more than forty inches above the floor. Mirrors placed elsewhere must have their bottom edges thirty five inches above the finished floor. Coat hooks must also be provided within one of the reach ranges specified in Section 308.

Clearances around wheelchair-accessible, wall-hung water closets must be at least sixty inches wide and fifty six inches deep (Figure 16-24*a*). Where there are floor-standing water closets, a fifty nine-inch depth is required. Water closet centerlines must be between sixteen and eighteen inches from an adjacent side wall or partition. If toilet compartments are provided, then at least one must be accessible.

In residential units, a lavatory may be placed adjacent to a water closet so long as it is at least eighteen inches from the water closet's centerline and a sixty six-inch clearance is provided perpendicular to the wall behind the fixtures (Figure 16-24*b*). No overlap is permitted between door swings and fixture clearances.

Clear floor space and dimensional requirements for water closets and grab bars are pictured in Figure 16-25 and are as follows:

- Height, measured to the top of the toilet seat, must be seventeen to nineteen inches. Seats may not have any mechanism to return them to a lifted position.

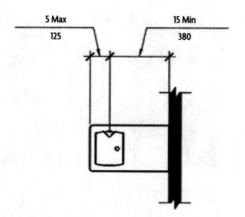

5 Max	15 Min
125	380

Figure 16-23 Drinking Fountain Spout Location (602.5).

Figure 16-24 Clearances and Overlap at Water Closets (604.3).

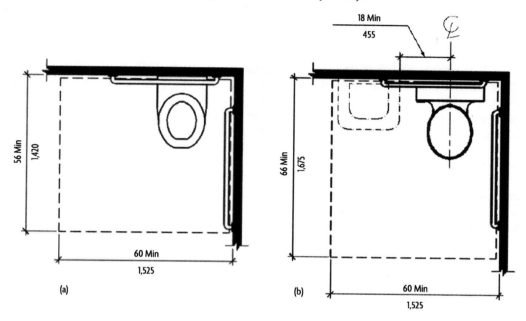

Figure 16-25 Side and Rear Grab Bar Locations at Water Closets (604.5).

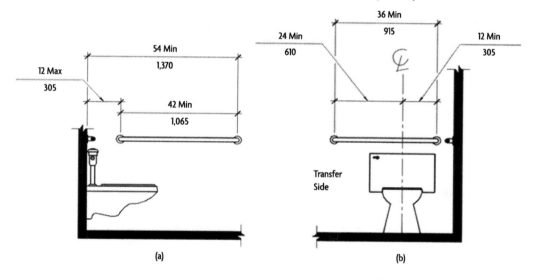

- Grab bars are required on the side and rear walls of toilet rooms intended for public use. They must be mounted between thirty three and thirty six inches above the floor. Side-wall grab bars must be at least forty two inches long and mounted not more than twelve inches from the rear wall. Rear-wall grab bars must be at least thirty six inches long, extending twelve inches from the water closet centerline toward the side wall and twenty four inches in the other direction.
- Flush controls may be automatic or manual but must be located on the side of the water closet away from the side wall. Manual flush controls must comply with requirements for reach ranges and operable parts (308–309).
- Toilet paper dispensers must be within reach and have no mechanism to control paper flow. The dispenser centerline must be between seven and nine inches in front of the water closet, with its outlet located between fifteen and forty five inches above the floor (Figure 16-26).
- Children's water closets (up to age twelve) must be eleven to seventeen inches high and meet the same floor clearance requirements as standard fixtures. Manual flush

controls must be no more than thirty six inches above the floor. Dispensers must be mounted fourteen to nineteen inches high. Grab bars for children's use must be mounted between eighteen and twenty seven inches above the finished floor.

Toilet Compartments (604.8/4.17)

Basic dimensional considerations for toilet stalls are illustrated in Figure 16-27. As with water closets not in stalls, alternate dimensions are given for children ages twelve and younger. Accessible stall doors must be self-closing and swing out. Wheelchair-accessible compartments with wall-hung water closets must be at least sixty inches wide and fifty six inches deep (Figure 16-27*a*). Where there are floor-standing water closets, a fifty nine-inch depth is required (Figure 16-27*b*). Baby-changing tables or other such conveniences may not overlap this space.

Doors on toilet compartments (hardware included) must comply with the requirements for other types of doors (Section 404). If the approach to the compartment door is on the latch side, a minimum clearance of forty two inches between the door side of the partition and any obstruction must be maintained as shown in Figure 16-27*c*. A door may be located in the front of the partition or at the side, but it must be in the corner farthest from (usually diagonally opposite) the water closet. The door opening must be four inches from that nearest corner in either case.

Compartment doors may not swing into the minimum required compartment area. Grab bars must be provided within toilet compartments in the same sizes and locations already discussed.

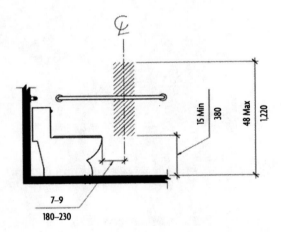

Figure 16-26   Dispenser Outlet Location (604.7).

Figure 16-27   Wheelchair-Accessible Toilet Compartments and Doors (604.8).

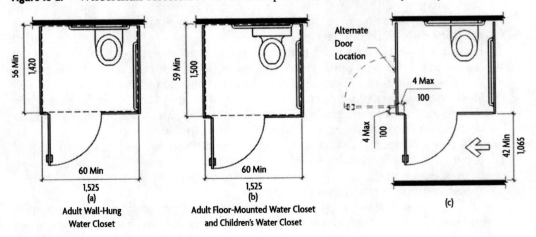

Toe clearances of at least nine inches must be provided in the front and at least one side partition if the stall depth is less than sixty two inches (Figure 16-28a). Children's compartments less than sixty five inches deep must have a vertical toe clearance of at least twelve inches (Figure 16-28b). Toe clearances must extend six inches beyond the compartment-side faces of the partition allowing for its support members, as illustrated in Figure 16-28c.

Urinals (605/4.18)

Urinals may be of the wall-hung or stall type (Figure 16-29). In the former case, the rim must be no higher than seventeen inches above the finished floor. The minimum depth of either type must be 13.5 inches from the outside of the front rim to the back of the fixture.

Placement of urinals must comply with requirements for clear floor space and manual flush controls with requirements for operable parts (305, 309).

Lavatories and Sinks (606/4.19)

Lavatories and sinks must be mounted with the front of their highest element no more than thirty four inches above the finished floor, with at least twenty nine inches clearance to the bottom of the apron. Knee and toe clearances must comply with Section 306 and a clear forward-approach space of thirty by forty eight inches must be allowed in front of each fixture (305). No more than one bowl of a multibowl sink must meet the knee and toe clearance requirement.

Figure 16-28   Wheelchair-Accessible Toilet Compartment Toe Clearance (604.8).

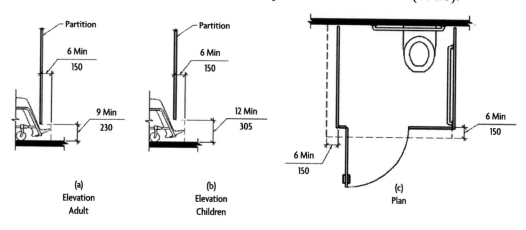

Figure 16-29   Depth and Height of Accessible Urinals (605.2).

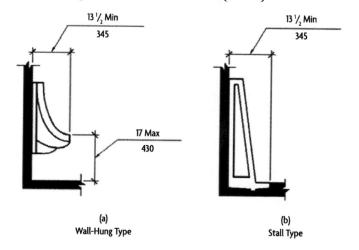

Figure 16-30 Bathtub Clearances (607.2).

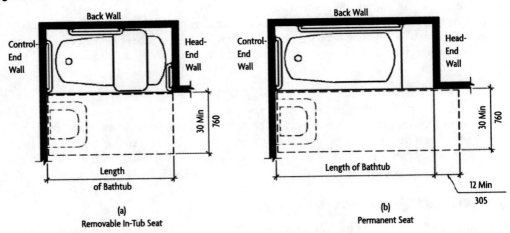

(a)
Removable In-Tub Seat

(b)
Permanent Seat

Lavatories designed for children ages six through twelve may have an apron clearance of twenty four inches if the height of the rim or counter surface is no more than thirty one inches. Lavatories intended for use by children five and younger need not meet these clearance requirements so long as adequate floor space is provided for wheelchair access using a parallel approach (305).

Faucets must comply with the requirements for controls and operating mechanisms (309/4.27), typically lever operated, push type, or electronically controlled. Self-closing valves, if used, must remain open for at least ten seconds. There must be no abrasive surfaces or sharp edges under lavatories, and all pipes must be insulated to protect against contact.

Bathtubs (607/4.20)

Critical dimensions relating to bathtub access are pictured in Figure 16-30. A secure seat must be provided, either within or at the head end of the tub. Clear floor space in front of the tub must be a minimum of thirty inches (a standard tub is nominally sixty inches long). A seat at the head of the tub may be a maximum of fifteen inches deep. A lavatory meeting the requirements of the previous section may be placed adjacent to the control end within the required clearance. Tub enclosures may not have tracks mounted on the tub rim or in any way obstruct transfers from wheelchairs into the tub or onto bathtub seats.

Two grab bars are required on the back wall behind the bathtub; one securely mounted eight to ten inches above the tub rim and the other thirty three to thirty six inches above the floor. Both grab bars must be a minimum of twelve inches from the control end of the tub and not more than fifteen inches from the opposite end if the tub has a fixed seat (Figure 16-31, *a* and *b*). A twenty four inch-long grab bar must also be mounted on the wall above the foot of the tub.

If the tub has a removable seat, grab bars on the rear wall may be no longer than twenty four inches, mounted a maximum of twelve inches from the control-end wall and twenty four inches from the head-end wall. An additional grab bar must be provided at the head wall at the front edge of the tub as pictured in Figure 16-31, *c* and *d*.

Bathtubs located in a private facility accessed only through a private office for the use of one person, or in residential dwelling units, are not required to have grab bars installed as long as suitable reinforcement has been provided to facilitate their installation if needed. Specific requirements for bathtub and other such seats are subsequently discussed in Section 610.

Controls must be located at the foot of the tub on an end wall and comply with Section 309.4 (4.27). They must be mounted between the bathtub and grab bar and between the centerline of the bathtub and its open side (Figure 16-32). A shower unit must be provided having at least a fifty nine-inch flexible hose and a head that can be used either fixed or handheld.

Figure 16-31   Grab Bars for Bathtubs with Seats (607.4).

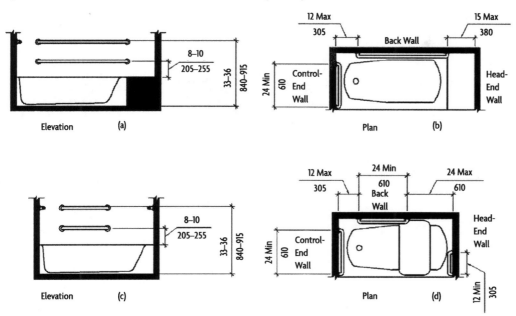

Figure 16-32   Bathtub Control Location (607.5).

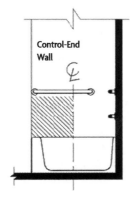

Shower Compartments (608/4.21)

Shower compartments must comply with the dimensions shown in Figure 16-33.

- *Transfer type:* Thirty six by thirty six inches, with thirty six inches of clear floor space in front extending twelve inches past the side on which the seat is mounted (a total of forty eight inches, (Figure 16-33a).
- *Standard roll-in type:* Thirty by sixty inches, with thirty six inches of clear floor space in front extending the full sixty-inch width of the shower opening (a thirty-inch-wide lavatory may be mounted within the required clearance provided it is not on the side adjacent to the clearances for the controls or the seat (Figure 16-33b).
- *Alternate roll-in type:* Thirty six by sixty inches, with a thirty six-inch-wide clear entry at one end of the long side of the compartment (Figure 16-33c).

Grab bars complying with Section 608 (4.26) must be mounted on all enclosure walls that do not have a seat at the standard height of thirty three to thirty six inches above the floor (Section 609). Shower enclosures may not obstruct controls or the transfer from wheelchairs onto shower seats.

- *Transfer type:* Grab bars must be provided across the control and back walls extending eighteen inches from the control wall as shown in Figure 16-34a.

Figure 16-33 Shower Compartment Size and Clearances (608.2).

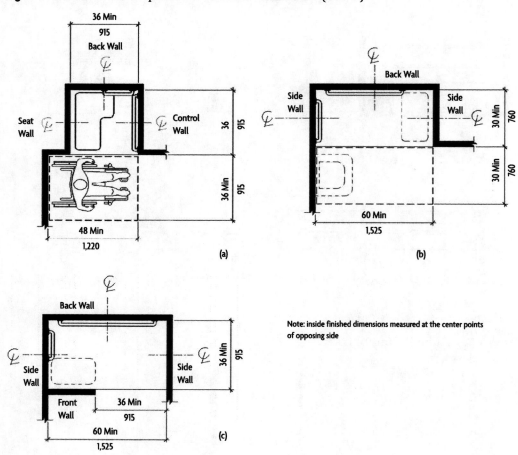

Note: inside finished dimensions measured at the center points of opposing side

- *Standard roll-in type:* Grab bars must be provided on all three walls in compartments without a seat (Figure 16-34b). They may not be located above the seat where a seat is provided but must be provided on the back and opposite wall (Figure 16-34c). Grab bars must extend to a maximum of six inches from adjoining walls.
- *Alternate roll-in type:* Grab bars are required on the back wall and on the side wall farthest from the entry but not above the seat. They must be installed no more than six inches from the adjacent corner (Figure 16-34c).

Controls, faucets, and spray outlets must be located on the side wall opposite the seat in transfer-type shower compartments. The spray unit must be located between thirty eight and forty eight inches above the floor, located no more than fifteen inches from the centerline of the seat toward the opening, as illustrated in Figure 16-35.

Figure 16-36 shows the prescribed arrangements for roll-in-type shower compartments with and without seats. The controls, faucets, and spray units in standard and alternate units must be located above the grab bar but no higher than forty eight inches above the floor (Figure 16-36a).

Where a seat is provided in standard compartments, the control assembly may be located on the back wall up to twenty seven inches from the seat wall (Figure 16-36b).

In alternate roll-in shower compartments where a seat is provided, the control and spray unit assemblies must be located on the wall next to the seat, no more than twenty seven inches from the side wall behind the seat (Figure 16-36c). Alternatively, the assemblies may be located on the back wall opposite the seat within fifteen inches on either side of the seat centerline (Figure 16-36, *d* and *f*). Where there is no seat, the assemblies must be installed on the side farthest from the compartment entry (Figure 16-36e).

Figure 16-34   Grab Bars for Showers (608.3).

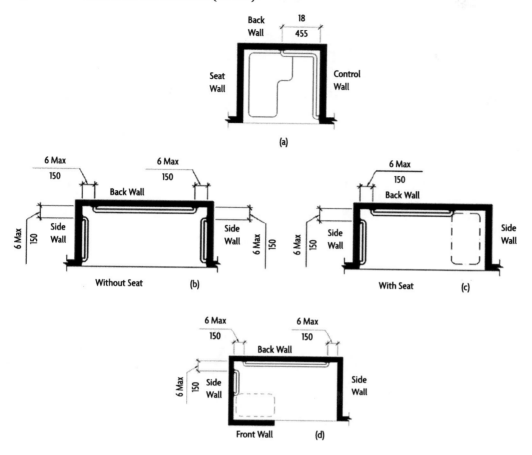

Figure 16-35   Transfer-Type Shower Compartment Control Location (608.5.1).

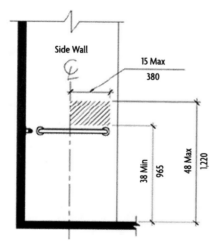

Curbs or thresholds up to one-half inch high may be used in both transfer or roll-in-type showers compliant with Section 303 and may be beveled, rounded, or vertical. In existing facilities where meeting this requirement would disturb the structure of the floor slab, thresholds in transfer-type showers may have a threshold up to two inches high.

The shower unit must provide for both fixed and handheld use and have at least a fifty-nine-inch hose. If a bar-mounted, adjustable-height shower head is used, the assembly may not interfere with the use of grab bars. Shower enclosures may not obstruct controls, outlets, or spray units nor obstruct transfer between wheelchairs and shower seats.

Figure 16-36 Roll-In-Type Shower Compartment Control Location (608.5.2–3).

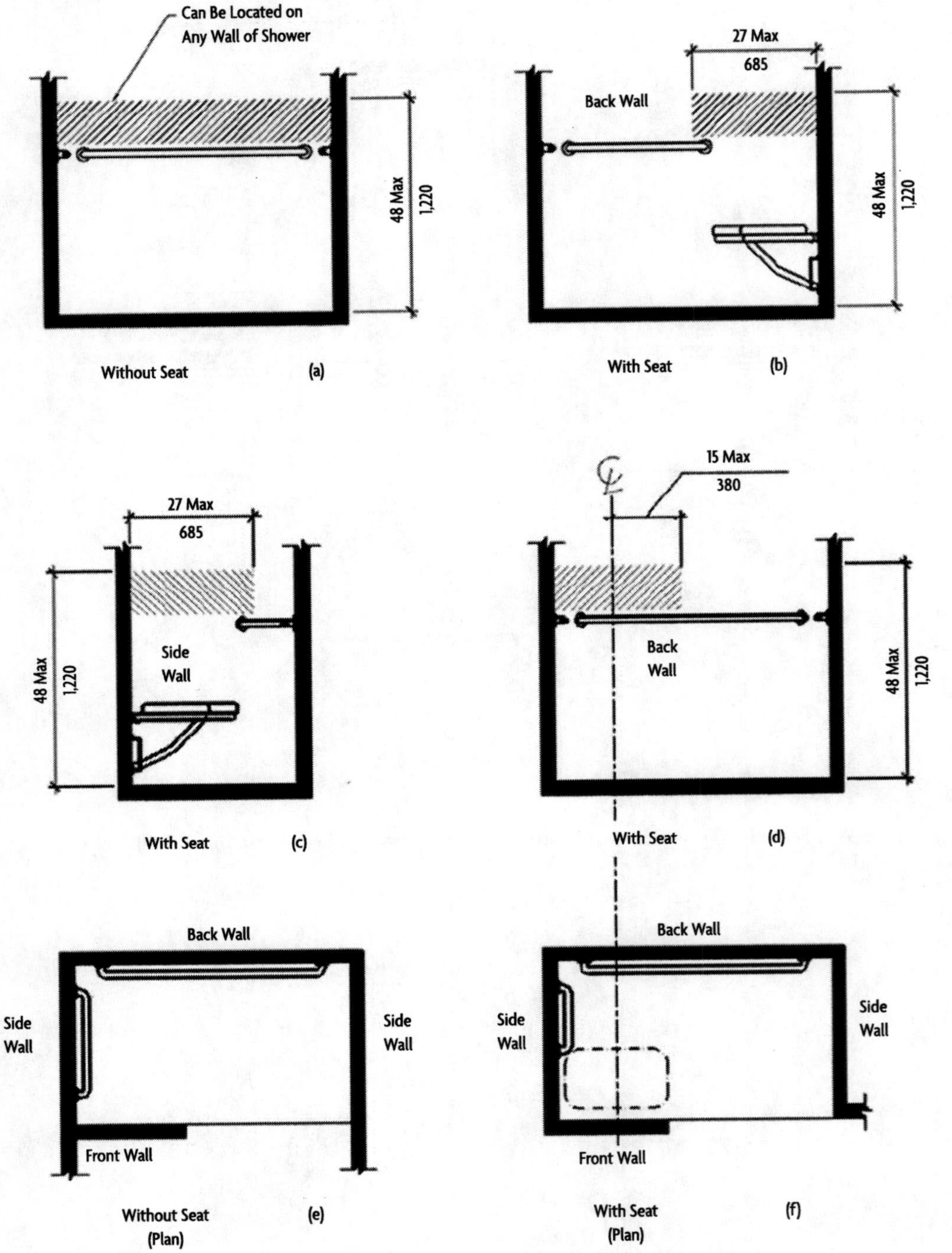

As with bathtubs, showers provided for the use of one person or in residential dwelling units are not required to have grab bars installed as long as suitable reinforcement has been provided.

Grab Bars (609/4.26)

All grab bars and handrails must fit the parameters illustrated in Figure 16-37 and Figure 16-38. Grab bars with a circular cross section must be nominally 1-$\frac{1}{4}$ to 2 inches in diameter (the *ADAAG* upper limit was 1-$\frac{1}{2}$ inches—4.26.2) or have an equivalent

Figure 16-37 Noncircular Grab Bar Profiles (609.2).

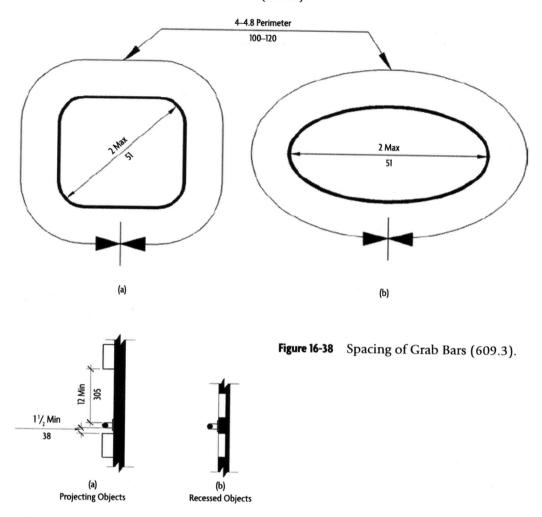

4–4.8 Perimeter
100–120

2 Max
51

2 Max
51

(a)

(b)

Figure 16-38 Spacing of Grab Bars (609.3).

12 Min
305

1½ Min
38

(a)
Projecting Objects

(b)
Recessed Objects

gripping surface. Grab bars with noncircular profiles must have a 2-inch cross section and a perimeter between 4 and 4.8 inches.

Grab bars must be installed in a horizontal position with the top of their gripping surfaces between thirty three and thirty six inches above the finished floor. At water closets for children's use, grab bars must be placed between eighteen and twenty seven inches above the floor.

If grab bars are mounted adjacent to a wall, there must be at least 1.5 inches of clearance between the grab bar and the wall, projecting objects below, and at the ends of the bar. One foot of clearance must be provided between the grab bar and projecting objects above, except for shower control and spray units, which require 1.5 inches of clearance.

Grab bars and seats, together with their mounting devices, must be designed to resist bending, shear, and tensile forces equal to 250 pounds and may not rotate within their fittings. These elements and adjacent surfaces may not have sharp edges or abrasive surfaces, and edges must have a minimum radius of 1/8 inch.

Seats (610/4.27)

Bathtub seats must be placed so that the top of the seat is from seventeen to nineteen inches above the finished floor. Removable seats must be between fifteen and sixteen inches deep and be capable of being securely positioned (Figure 16-39a). Permanent seats must be at least fifteen inches deep, placed at the head of the bathtub and extending from the back wall to the tub's outer edge (Figure 16-39b).

Figure 16-39 Bathtub Seats (610.2).

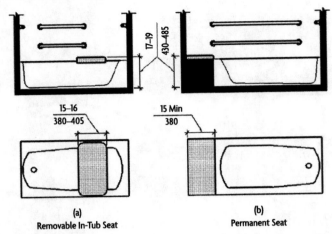

(a)
Removable In-Tub Seat

(b)
Permanent Seat

Figure 16-40 Extent of Shower Seats (610.3).

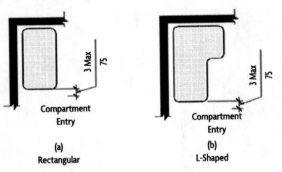

(a)
Rectangular

(b)
L-Shaped

Shower compartment seats (Figure 16-40) must also be placed so that the top is from seventeen to nineteen inches above the finished floor. In both standard and alternate roll-in shower compartments, the seat must be of the folding type. In the standard configuration, the seat must extend from the back wall of the compartment to within three inches of the entry. In the alternate configuration, the seat must be installed on the front wall, extending from the adjacent side wall to within three inches of the entry. In transfer-type showers, the seat must extend from the back wall to within three inches of the compartment entry.

Rectangular seats may extend fifteen to sixteen inches from the seat wall with not more than 2.5 inches of clearance between the rear edge and the wall. The side edge of the seat may be no more than 1.5 inches from the adjacent wall as shown in Figure 16-41*a*. Clearances at the rear and side of L-shaped seats are the same as for rectangular seats (Figure 16-41*c*). The front edge of an L-shaped portion may extend between twenty two and twenty three inches from the seat wall and between fourteen and fifteen inches from the side wall (Figure 16-41*b*).

Seats must be designed to withstand a vertical or horizontal force of 250 pounds applied at any point on the seat, mounting devices, or support structure.

Other Plumbing-Related Elements (611–12)

The last two sections under Plumbing Elements deal with washing machines, clothes dryers, saunas, and steam rooms. Clear floor space and operable parts requirements for laundry facilities reference Sections 305 and 309 under Building Blocks. In addition, top-loading machines must have the door to the laundry compartment thirty-six inches above the finished floor. Front-loading machines must have the bottom of the door opening between fifteen and thirty-six inches above the floor.

More detailed information on dressing areas is given in Sections 803 and 903.

Figure 16-41   Rectangular and L-Shaped Shower Seats (610.3).

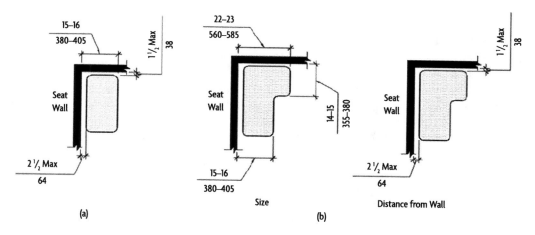

COMMUNICATION ELEMENTS AND FEATURES

Fire Alarm Systems (702/4.28)

Alarm systems must be provided in all areas intended for general use, including hallways, lobbies, toilet rooms, and other such categories listed by the *ADAAG*. The *IBC* elaborates on different types of alarm systems required in Section 907 (see Chapter 13).

Audible alarms must exceed the ambient sound level in a space by at least fifteen dbA, or a maximum sound level of sixty second duration by five dbA but may not be louder than 110 dbA.° The *ADAAG* recommends that clear, intermittent sounds of intermediate frequency—less than ten kHz—be used, as opposed to continuous or reverberating tones.

Visual alarms are also required and must be integrated into the facility's alarm system, but they may be zoned. The *ADAAG* signal requirements for visual alarms are very specific:

- Xenon (halogen) strobe or equivalent, clear or white, with a flash rate between one and three hertz.
- Maximum pulse duration of 0.2 seconds and an intensity of seventy five candela.
- Height must be eighty inches above the highest floor level within a space or six inches below the ceiling, whichever is lower.
- No location in any space where signals are required may be more than fifty feet horizontally from the signal. In large spaces without obstructions, such as auditoriums, they may be spaced one hundred feet apart around the perimeter rather than suspended from the ceiling.

Auxiliary alarms are required in residential units and other occupancies with sleeping accommodations. These may be connected into a central building alarm system or activated through the standard 110-volt electrical wiring. With an uncharacteristic display of (presumably unintentional) levity, the *ADAAG* suggests placing a signal-activated vibrator under the mattress in hotels and other sleeping rooms.°

Signs (703/4.30)

The legibility of text is a function of many factors: viewing distance, color and contrast, the height and style of the letter form, and the width of the stroke comprising different parts of each character. *ICC/ANSI A117.1* and *ADA-ABA* are substantially more detailed in their treatment than the *ADAAG* on this subject.

Tactile characters, with the exception of elevator car controls, must be uppercase, raised at least $\frac{1}{32}$ inch from their contrasting background, and be between $\frac{5}{8}$ and two inches high based on the height of the uppercase letter *I*, and the stroke width-to-height ratio may

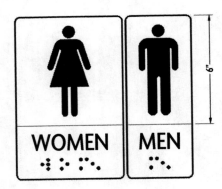

Figure 16-42   Pictogram Layouts (703.6).

be no more than 15 percent. They must be located between forty eight and sixty-inches above the floor or ground. Raised characters must be duplicated in Braille.

The border dimension of pictograms must be at least six inches high, and they must be accompanied by a textual description directly underneath. Neither text nor Braille may appear in the pictogram field (Figure 16-42).

The *ADA-ABA* specifies that both tactile and visual text characters must be selected from fonts where the width of the uppercase *O* is between 55 and 110 percent of the height of the uppercase letter *I*. Spacing between lines must be between 135 and 170 percent of their character height. Characters must be sans serif, neither italic nor having unusual decorative forms.

Braille characters must be placed directly below the entire corresponding text whether single or multilined. A separation of at least 3/8 inch must be provided between Braille and any other tactile characters, raised borders, or decorative elements. Braille dots must have a domed shape using standard dimensions (in inches) as follows:

- Dot height—0.025 to 0.037
- Dot base diameter—0.590 to 0.063
- Distance between dots in the same cell—0.090 to 0.100
- Horizontal distance between dots in adjacent cells—0.241 to 0.300
- Vertical distance between dots in adjacent cells—0.395 to 0.400

Visual text may be a combination of upper- and lowercase and must be placed at least 40 inches above the finished floor or ground. The stroke width-to-height ratio of visual characters must be between 10 and 30 percent using the uppercase *I* as a standard. Spacing between individual characters must be between 20 and 35 percent of their height. Visual characters intended to be viewed at substantial distances are subject to the requirements in Table 16-2.

The international symbol of accessibility must be used to identify all facilities required to be accessible (Figure 15-1).

Text telephones (teletypewriters [TTY], Figure 16-43*a*) and volume-control telephones (Figure 16-43*b*) must be identified by standard international symbols also, with appropriate directional signage adjacent to banks of standard telephones. Assistive-listening systems (Figure 16-43*c*) must be similarly identified where they are available.

Telephones (704/4.31)

Public telephones must have the standard thirty-by-forty-eight inch clear floor or ground space to accommodate a forward or parallel approach (Figure 16-44) and comply with the *ADAAG* requirements as to reach ranges and protruding objects (307–8/4.2 and 4.4). Specific requirements for volume-control and text-control telephones are also enumerated, but one may reasonably expect that public telephones in general will become less common with the continued growth of cellular communications.

Detectable Warnings (705/4.29)

Detectable warnings are most often seen at railway platforms and consist of truncated raised domes of a contrasting surface material. The *ADAAG* specifies that the domes be nominally

Table 16-2   Height of Visual Characters (703.5.5)

Height from Floor or Ground	Horizontal Viewing Distance	Minimum Character Height
40 to 70 Inches	Less Than 72 Inches	⅝ Inch
	72 Inches and Greater	⅝ Inch + ⅛ Inch per Added Foot of Viewing Distance
70 Inches to 10 Feet	Less Than 15 Feet	2 Inches
	15 feet and Greater	2 Inches + ⅛ Inch per Added Foot of Viewing Distance
Greater Than 10 Feet	Less Than 21 Feet	3 Inches
	21 Feet and Greater	3 Inches + ⅛ Inch per Added Foot of Viewing Distance

Figure 16-43   International Symbols (703.7).

(a)   (b)   (c)

Figure 16-44   Approaches to Telephones (704.2).

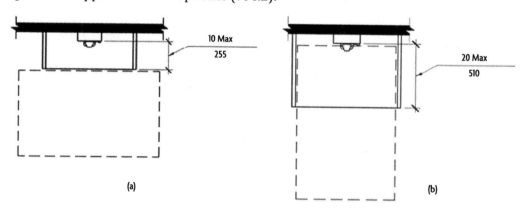

(a)   (b)

0.9 inch diameter, 0.2 inch high, and spaced 2.35 inches center to center. The *ADA-ABA* gives ranges: 0.9 to 1.4 inches for dome size, 1.6 to 2.4 inches on centers, and 0.65 inch measured base to base.

The *ADA-ABA* states that detectable warnings along platform boarding edges are to be twenty four inches wide, extending the full length of the platform used by the public. The *ADAAG* calls for thirty six-inch-wide detectable warnings between hazardous pedestrian and vehicular areas; however, Section 10, Transportation Facilities, specifies a twenty four-inch width along the length of boarding platform drop-offs.

Edges of reflecting pools are to be protected by detectable warnings, curbs, railings, or walls, according to the *ADAAG*. Accessibility notwithstanding, guardrails offer more substantial protection at the perimeter of such features.

Assistive-Listening Systems (706/4.33)

The *ADAAG* describes assistive-listening systems as being "intended to augment standard public address and audio systems by providing signals which can be received directly by persons with special receivers or their own hearing aids and which eliminate or filter background noise."° The new *ADA-ABA*, in keeping with *A117.1*, gives much more detailed requirements.

Criteria for the performance of assistive-listening systems include volume, distortion, and interference. Signal-to-noise (S/N) ratio measures the difference in sound level between a desired signal and background or ambient noise. Systems must be capable of providing volume levels of between 110 and 118 decibels (dB) with a control range of 50 dB. The signal-to-noise ratio must meet or excel 18 dB. Peak clipping (distortion) may not exceed 18 dB relative to the peak levels of speech.

Assistive-listening systems can be hardwired or one of three types of wireless systems: infrared, FM radio, and induction loop. FM generally has the greatest flexibility, especially in open-air assembly spaces, although infrared has the advantage of confidentiality because its transmission is line of sight.

Automated Teller and Fare Machines (707/4.34)

Automatic teller machines (ATM) and fare machines must be approachable by a person in a wheelchair from any direction, have sufficient clear floor space (308/4.2), and meet the requirements for operable parts (309/4.27).

Accessible gates and fare machines should be placed in close proximity to other accessible elements. Token collection devices designed to use perforated tokens allow the visually impaired user to more easily distinguish between tokens and coins. Fare cards that are chamfered at one corner can be more accurately inserted.

Automatic teller machines (ATMs) must provide the same degree of privacy for input and output for all individuals. Machines are required to be speech-enabled for the use of visually impaired people, and mechanisms must be easily used and require no special training. A telephone handset with a volume control may automatically activate speech output when lifted.

Input function keys must contrast visually with background surfaces and be raised to make them tactilely distinct. Numeric input keys must be arranged in a twelve-key ascending or descending keypad layout with the number 5 key identifiable by tactile means. Display screens must be visible from a point forty inches above the center of the clear floor space in front of the machine. Where bins are provided for activities related to ATM transactions, they must be located within the applicable reach ranges.

Two-Way Communication Systems (708)

Emergency two-way communication systems must provide both audible and visual indicators that assistance is on the way. Handset cords, if provided, must be at least twenty nine inches long, but devices not requiring handsets are easier to use by persons with a limited reach. Signs indicating the meaning of visual signals should be provided.

Communication systems provided in residential dwelling units between a dwelling unit and a site, building, or floor entrance must include support for both voice and TTY interfaces.

SPECIAL ROOMS, SPACES, AND ELEMENTS

There are several occupancy types that the *ADA-ABA* and *ADAAG* designate as *special*. Each of them is noted with its corresponding *IBC* classification as follows: (ADA-ABA/ADAAG/IBC). Many of the requirements are not so much unique as they are variations on conditions that the standards treat elsewhere. Without exception, the opening paragraphs of each of these sections states that it is subject to the requirements of Building Block sections such as 302 through 306: floor surfaces and changes in level, clear floor space, and knee and toe clearances. Minimum aisle widths are thirty six inches (although in certain instances forty two inches is preferred), and accessible transaction counters must have a maximum height of thirty four inches.

Wheelchair Spaces, Companion Seats, and Designated Aisle Seats (802/4.33)

The floor or ground at wheelchair locations must be level and in compliance with Section 302/4.5. If forward or rear access is provided, thirty three by forty eight inches of clear floor space must be allocated to each wheelchair. If access is from the side, thirty three by sixty inches is required. Lines of sight must be maintained over the heads of those seated in the rows in front of wheelchair spaces.

This section of the ADA underscores the primary focus of the entire collection of guidelines: assuring equal accommodation for persons with disabilities. In assembly occupancies, wheelchair areas must be incorporated into any fixed-seating layout, so situated as to provide disabled patrons with a choice of location and admission price comparable to those available to the general public. Such areas must be adjacent to an accessible route that is also a means of egress, and adjacent companion seating must also be available. The accessible route must also connect with performing areas such as stages, dressing rooms, and other such spaces used by performers.

Dressing, Fitting, and Locker Rooms (803/4.35–37)

Where seating is provided in dressing facilities, steam rooms, or saunas, there must be at least one accessible bench that complies with Sections 304 (turning space) and 903 under built-in elements. Doors may not swing into the clear floor space required by Section 903. Coat hooks must be provided within one of the reach ranges specified in Section 308.

Kitchens and Kitchenettes (804)

The *ADAAG* only mentions kitchens and kitchenettes in conjunction with accessible transient lodging (9.2.2) and judicial facilities (11.2.2). The new *ADA-ABA* document, in keeping with *A117.1*, distinguishes between U-shaped and pass-through kitchens in residential dwelling units, detailing specific parameters for clearances, work surfaces, sinks, and ovens.

Pass-through kitchens must have two entries with counters and appliances on either side and at least forty inches of clearance between them. Clearances between all base cabinets, countertops, appliances, or walls in U-shaped kitchens must be at least sixty inches to provide sufficient turning space. Knee and toe clearances must meet the requirements of Section 306. Special requirements for sinks, storage, appliances, and ovens have also been added.

Medical Care and Long-Term Facilities (805/6.0/*IBC* I-2)

This classification includes facilities where people receive medical treatment or care for more than twenty four hours and where persons may need assistance in responding to an emergency. The following percentages of patient rooms and toilets and common-use and public spaces must be accessible in the following types of facilities:

- *One hundred percent*—Hospitals and rehabilitation facilities (or units within) that specialize in treating patients with conditions affecting mobility
- *Fifty percent*—Long-term-care facilities and nursing homes
- *Ten percent*—General purpose hospitals, psychiatric, and detoxification facilities°

If any discreet area is altered, whether a department or floor, or patient rooms are added, these proportions must be maintained. When patient bedrooms are added or altered individually, they must be accessible unless the total number of rooms in the department or area already meets the preceding requirements. Every accessible patient room must be on an accessible route and have an accessible toilet/bathroom.

Transient Lodging Guest Rooms (806/9.0/*IBC* R-1)

Transient lodging includes facilities where sleeping accommodations are provided, other than medical care facilities, including hotels and motels, resorts, dormitories, and so on. These must comply unless there are fewer than six rooms for rent within the residence of the proprietor (i.e., bed and breakfast establishment).

The *ADA-ABA* and *ADAAG* provide tables giving the required ratios of accessible rooms and roll-in showers to the total number of sleeping rooms, plus alarm or notification devices for people with impaired hearing. These tables are similar. In general, up to one hundred rooms, one accessible room in twenty five must be provided. From one hundred to four hundred, the requirement is one in fifty, and then approximately 2 percent up to one thousand. The ratios of sleeping accommodations for individuals with hearing impairments are the same. Rooms with roll-in showers are required at the rate of one per one hundred rooms.

Once again, the ADA makes a point of its critical concern with making all classes of accommodations available to patrons with disabilities in terms of room size, cost, and amenities. The accessible sleeping rooms and suites provided must be proportionately distributed among the available rooms. When part of an accessible unit, the following spaces must be on an accessible route:

- Living, dining, and at least one sleeping area
- At least one bathroom (or half bath if only those are provided)
- Terraces, balconies, and car parking facilities

All accessible rooms must have visual alarm systems plus visual and auditory notification devices. Telephones must have volume controls and an accessible electrical outlet within four feet to facilitate the use of a text telephone. Alternatively, sleeping rooms and suites must have electrical and communications wiring to enable hearing-impaired persons to use portable devices supplied by the proprietor.

Certain additional requirements apply to social service establishments such as transient group homes and homeless shelters. In new construction, all public and common-use areas including amenities such as laundry rooms must be accessible to all units and sleeping accommodations. When alterations are made, all doorways including at least one public entrance and doors to sleeping rooms, toilets, and common areas must meet the minimum thirty two-inch clear width standard (404.2.3). Recessed doors must provide at least eighteen-inch maneuvering clearance adjacent to the latch side. At least one accessible route must meet the standards for minimum clear width (thirty six inches) passing and turning space. At least one water closet, one lavatory, and one bathtub or shower must meet the accessibility requirements for plumbing and related fixtures (603/4.16/19–21).

Holding Cells and Housing Cells (807/12.0/*IBC* I-3)

At least 2 percent of the total number of cells, holding cells, and other rooms in a facility (not less than one) must comply with mobility requirements, including those equipped for persons with hearing impairments. Specific requirements include those listed for cells in the previous section, plus the following:

- Fixed or built-in storage facilities
- Controls intended for operation by inmates
- Where audible alarms are provided, visual alarms are required, except where inmates are not allowed independent means of egress
- Permanent telephones installed in accessible cells or rooms must have volume controls

Where special housing or holding cells are provided, at least one serving each purpose must be accessible. Such cells may be multipurpose and include protective custody, administrative or disciplinary segregation, detoxification, and medical isolation. Cells or rooms used exclusively for suicide prevention and without protrusions are not required to have grab bars at water closets. Medical facilities, in addition to special cells where detainees may stay for longer than twenty four hours or for emergency treatment, must comply with the requirements of this section.

Courtrooms (808/11.0/*IBC* A-3)

Where restricted and secured entrances are provided in addition to the regular accessible entrances covered under requirements for new construction, they must be accessible. Restricted entrances are used by judges, public officials, and other authorized personnel. Secured entrances are those used only by detention officers and detainees. Where non-accessible devices are used at security checkpoints, an adjacent accessible route must be provided, as in airports and detention facilities. Where two-way communication systems are used at security checkpoints, they must provide both audible and visual signals.

In courtrooms, areas where there is a level change requiring ramps or platform lifts must provide adequate wheelchair turning space, except where alterations would require extensive reconstruction. Judges' benches and other courtroom stations must comply with Section 902/4.32. Jury boxes and witness stands must be wheelchair accessible, except where ramps or lifts would pose a safety hazard. Spectator and press areas with fixed seating must comply with the requirements for assembly seating. Assistive-listening systems must be available in courtrooms, together with signage indicating their availability. They must be capable of serving at least 4 percent of its occupancy capacity as determined by applicable building codes or have at least two receivers. Jury assembly and deliberation areas must be on an accessible route and include accessible drinking fountains and refreshment areas.

Legislative and regulatory facilities including public meeting and hearing rooms and chambers must comply with essentially the same requirements as courtrooms, as they pertain to raised speaker platforms, spectator and other areas with fixed seating, and assistive-listening systems.

Residential Dwelling Units (809)

In the *ADAAG* document (as amended through 2002), this section was designated as *reserved*, although Chapter 10 of *A117.1-1998* treats the subject extensively, as does the 2003 edition. Consequently, this section of *ADA-ABA* is now more fully detailed as well. *A117.1* distinguishes between two types of residential units meeting different standards, designated as type A and type B. These were discussed in the two preceding chapters covering the *IBC* standards for egress and accessibility (*IBC* 1104–1107 and 1109).

An advisory in the scoping section (233.1) warns that facilities covered by the ADA may also be subject to other federal laws such as the Fair Housing Act and regulations issued by the Department of Housing and Urban Development (HUD) under Section 504 of the Rehabilitation Act of 1973, with their subsequent amendments. The advisory states that "these laws and the appropriate regulations should be consulted before proceeding with the design and construction of residential facilities."°

Section 809 references other sections of the design guidelines, making them specifically applicable to residential units. These include accessible routes; kitchens; toilet and bathing facilities; and communication features including building fire alarm systems, alarm appliances, and activation, as well as site, building, and floor entrances.

Transportation Facilities (810/10.0/*IBC* A-3)

Transportation facilities include bus stops and terminals, fixed facilities and stations for all types of rail service and intercity buses, and rail key stations.° Airports are referenced in the *ADAAG* (Section 10.4) but not in the *ADA-ABA* guidelines except with respect to telephone access.

The *ADAAG* requires, in general, that terminals be planned to minimize the distances that wheelchair users and other people who cannot negotiate steps need to travel compared to the general public. The Act enumerates specific elements whose placement is considered essential to meeting this requirement. These include ramps, elevators and other circulation devices, ticketing, fare vending and collection areas, security checkpoints, and passenger waiting areas. To the maximum extent possible, accessible routes (including entrances) must coincide with the circulation paths used by the general public. Where routes are different, signage compliant with 810.6/4.30 must identify and indicate the location of accessible entrances and routes.

Train and bus terminals must have direct accessible routes to all retail, commercial, or residential facilities from boarding areas and all transportation system elements used by the public, including baggage-handling facilities. Planning should take into account the facilitation of direct future connections on accessible routes as well.

Entrance and direction signage, as well as lists of routes, destinations, and stations served, must comply with the signage requirements of 810.6/4.30. At least one sign identifying the particular station must be provided on each platform or boarding area, and all signs must be placed in uniform locations to the extent practicable. Identification signs on train platforms must be placed at frequent intervals and be visible on both sides from within the vehicle when not obstructed by another train. Lighting along circulation routes and in areas where signage is located must provide uniform illumination. Detectable warnings twenty four inches wide and running the full length of unprotected platform drop-offs must also be provided (810.5.2).

Where escalators are provided, they must be at least thirty two inches in clear width and have at least two contiguous treads at the top and bottom before treads begin to form. Elevators must provide for an unobstructed view into and out of the car and have sufficient turning space for wheelchair traffic (see Figure 16-3).

Ticketing and baggage areas must comply with the requirements for sales, service, and information counters (904/7.2). Terminal information systems in all transportation facilities must include both auditory and visual components including audio paging systems and computer video monitors.

At least one accessible route must be maintained at all security checkpoints having one or more inspection stations. Where x-ray machines, metal detectors, and other such devices are used at security barriers, there must be an accessible path immediately adjacent allowing patrons with disabilities to maintain the same level of visual contact with their personal effects as the general public.

BUILT-IN ELEMENTS

Dining Surfaces and Work Surfaces (902/5.0/*IBC* A-2)

Other than the basic requirements for clear floor space and knee and toe clearance, this section provides that the tops of dining surfaces be between twenty eight and thirty four inches above the finished floor.

The *ADAAG* requirements are somewhat more specific in regard to food-service areas. Where fixed tables or counters are used, at least 5 percent (or at least one) must comply with 4.32. Where food or drink is served at counters higher than thirty four inches, at least a sixty-inch length of counter must be provided at the thirty four-inch height. Alternatively, service must be available at tables in the immediate area.

Access aisles between parallel edges of tables or a table and a wall must be at least thirty six inches. Food-service lines should be forty two inches wide to allow foot traffic to pass a person in a wheelchair, although a thirty five-inch width is permitted.

All dining areas must be accessible in new construction, including raised, sunken, and outdoor seating areas. If no elevator is available, mezzanines constituting less than one-third of the total accessible seating area need not be accessible, so long as the same services and décor are accessible within the same *areas* used by the general public.

Benches (903/4.37)

Benches must be fixed, having seats twenty to twenty four inches deep and at least forty two inches long. Seat height must be seventeen to nineteen inches, and a back support must be provided extending from two to eighteen inches above the seat. Bench seats must be designed to withstand stresses of 250 pounds applied in any direction at any point and positioned to allow a parallel approach to a short end. The surface of the seat must not accumulate water when used in wet locations.

Checkout Aisles, Sales and Service Counters (904/7.0/*IBC* A-3, B, M)

Where sales and service areas have transaction counters and cash registers, at least one of each type must have at least thirty six inches of counter length that is not more than thirty six inches high. Transaction counters that may not have a cash register, such as teller stations or registration counters in transient lodging facilities, must have a portion of the main counter or an auxiliary counter that meets these requirements. If an accessible counter is technically infeasible, equivalent facilitation must be provided by the use of a folding counter or other means that allows handing materials back and forth. Where such facilities have security glazing or solid partitions, at least one of each type must provide for voice communication accessible to persons in wheelchairs, as well as to those who have difficulty bending.

RECREATION FACILITIES (1000/15.0/*IBC* A-3)

Newly designed or altered recreation facilities must comply with applicable requirements of Section 10. These include the following items (a considerably expanded list from Section 4 of the *ADAAG*):

- Amusement Rides (1002)
- Recreational Boating Facilities (1003)
- Exercise Machines and Equipment (1004)
- Fishing Piers and Platforms (1005)
- Golf Facilities (1006)
- Miniature Golf Facilities (1007)
- Play Areas (1008)
- Swimming Pools, Wading Pools, and Spas (1009)
- Shooting Facilities with Firing Positions (1010)

This final section of the *ADA-ABA* enumerates requirements pertaining to heights and clearances, safety and securement devices, load/unload areas, and transfer platforms, all unique to the specific types of facilities covered, in the interest of equivalent facilitation. For example, at least half of the holes on a miniature golf course must be accessible, and they must be consecutive. One break in the sequence of consecutive accessible holes is permitted so long as the final hole in the sequence is the last hole of the course. °

CLOSING WORDS

There are four appendices following this chapter that relate to material covered in earlier chapters. Appendix A contains an outline of possible considerations in developing a project plan. The next contains a space-requirements worksheet for use in developing a facility planning program. Appendix C gives installation instructions for the special AutoCAD routines covered in Chapter 7 as well as instructions on how they may be obtained. Finally, Appendix D contains a detailed outline of the Archibus Space Module.

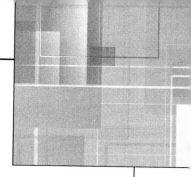

APPENDIX A
Facility-Planning Checklist

This checklist, which is by no means exhaustive, is intended to provide a guideline for identifying issues that may need to be addressed in the course of a detailed facility-requirements analysis or programming study. The nine major categories follow.

1. Space allocation and use
2. Interior systems
3. Building systems
4. Building enclosure
5. Architectural concepts
6. Structural system
7. Site conditions
8. Building services
9. Safety and environmental issues

The general considerations that apply to most if not all of the individual items in this outline are cost determination and allocation along with applicable codes and regulations. It almost goes without saying that for any capital construction project, estimating and budgeting the implementation costs must be integral parts of the options analysis and planning process. Similarly, virtually every item on the checklist falls under the jurisdiction of one or more codes or regulations at the local, state, or federal levels.

Within each category, some considerations are primarily applicable to new construction and many involve evaluating the condition and adequacy of existing buildings and systems. The emphasis given to certain items will vary as a function of project type: new construction, alteration, repair, or some combination. The possible interrelationships among individual items must also be kept in mind.

Finally, the potential need for specialized consultants must be evaluated with respect to many items, such as the design of technically complex spaces, the evaluation of subsurface soil conditions, and fire safety or hazardous materials issues.

1. SPACE ALLOCATION AND USE

Organizational Requirements
Mission Statement

Organization Structure

Personnel Projections

Adjacency Requirements

External Contacts

Workstations and Space Allocation
Planning Module

Space-Allocation Guidelines

Special Furniture and Equipment Needs

Support and Specialized Space Needs

Storage-Space Requirements

Special Needs and Constraints
Operational Considerations

Potential for Systems Changes

Physical Requirements and Limitations

Circulation and Dedicated-Use Corridors

Safety and Handicapped Accessibility Factors

Space-Requirements Summary
Usable or Occupiable Space

Public Circulation and Lobbies

Building Service Areas (Mechanical, Electrical, Etc.)

Enclosing Walls, Structure, Shafts

Total Gross Area

2. INTERIOR SYSTEMS

Partitions
Building Standard Partitions

Special Partition Performance Requirements

Conventionally Partitioned versus Open Office Space

Furniture
New Furniture versus Reuse of Existing

Systems versus Conventional: Functional Considerations

Adequacy and Limitations of Existing Furniture Systems

Flooring
Flooring Materials as Related to Space Utilization

Raised-Flooring Requirements

Under-Floor Duct System Requirements and Availability

Existing Hazardous Materials (Asbestos)

Handicapped Accessibility Requirements

Hardware
Security Requirements

Handicapped Accessibility Requirements

Finishes
Appearance and Durability Considerations

Areas Requiring Special Finishes

Performance Characteristics and Cost

Graphics and Signage
Room Numbering System

Floor Identification and Directories

Building and Site Identification

Prototypical Layouts
Workstation and Furniture System Evaluations

Construction of Full-Scale Prototype Areas

3. BUILDING SYSTEMS

Lighting
Space Types and Recommended Lighting Levels

General Lighting System

Furniture System Task Lighting

Building Automation and Occupancy Sensor Controls

Power
Service Requirements for Present and Future Operations

Special-Use Areas and Equipment

Emergency and Uninterruptable Power System (UPS) Requirements

Availability, Rate Structure, Load Shedding Requirements

Special Metering Requirements

Power-Distribution-System Requirements

Telecommunications
Types of Communication Services and Equipment Required

Telephone Switch Requirements, Line Capacity and Facility Support

Future Expansion Requirements

Performance and HVAC Requirements for Twenty-four-Hour Support

Space Requirements for Wire Closets

Heating, Ventilating, and Air-Conditioning Systems
Special HVAC Requirements

Heating and Cooling Load Estimates

Utility Restrictions and Costs for Connection and Use

System Preferences, Alternatives Evaluation and Costing

Lifecycle Cost Evaluation

Conveyance Systems
Vertical Circulation Requirements

Spaces Requiring Escalators versus Elevators

Effectiveness of Existing versus Optimum Systems for Projected Use

Security
Perimeter and Interior Security Requirements

Appropriate Devices to Accommodate Specific Requirements

Control/Alarm Interface with Door Locks, Elevators, and Building Automation Systems

Special Vault Requirements

Special Requirements for ADP Shielding or Filtering

Acoustics

Background Sound Masking Systems

Recommended Noise Levels to Be Maintained

Sources of Excessive Noise Requiring Special Control

Plumbing

Water and Sewer Requirements

Quantity and Quality Restrictions in Use of Utilities

Age and Condition of Services, Existing Fixtures

Special Plumbing or Accessibility Requirements

Fire Detection and Suppression

Occupancy Types and Hazard Ratings

Building Classification Related to Sprinklers and Interior Standpipes

Size and Location of Water Sources

Emergency Pump and Power Requirements

Special Fire-Suppression Systems

Coordination with Local Fire Department

Building Automation Systems

Systems and Operations to Be Included

Control Strategies and System Architecture

Methodology for Temperature Control, Fire, Security Monitoring, Etc.

Potential Benefits to Operations, Energy Conservation, Etc.

4. BUILDING ENCLOSURE

Walls

Material Use as Related to Image and Appearance

Insulation and Vapor-Barrier Requirements

Historic-Preservation Issues

Windows

Appearance and Performance Priorities

Operable versus Fixed Units

Amount and Type of Glass

Window-Washing Considerations

Historic-Preservation Issues

Roofing

Structural Loads and Service Life

Anticipated Rooftop Activities and Maintenance

Fire Access Requirements

Effects of Corrosive Atmosphere

Historical Considerations

5. ARCHITECTURAL CONCEPTS

Building Location
Setbacks and Zoning Restrictions
Easements and Rights-of-Way
Effects on Surrounding Area, Streets, and Existing Buildings
Subsurface Conditions

Shape and Orientation
Number and Size of Floors in Terms of Building Massing
Floor Area and Shape Related to Planned Occupancy
Site Constraints and Lines of Sight

Appearance and Image
Articulate Desirable Characteristics
Historic-Preservation Issues

6. STRUCTURAL SYSTEM

Foundations and Geotechnical
Geotechnical Data
Subsurface Conditions, Soil Capacity, and Water
Foundation System Alternatives

Superstructure
Proposed Occupancies, Functional Requirements, and Building Module
Applicable Structural Systems
Optimal Column-Free Floor Areas
Optimal Floor-to-Ceiling Heights
Seismic and Live Load Considerations

7. SITE CONDITIONS

Site Data
Plot Plan and Legal Description
Topographic Maps
State and Local or Other Planning Studies
Groundwater Conditions
Neighboring Structures, Lines of Sight, Antenna Effects
Presence of Archaeological Artifacts

Utilities
Water System
Sanitary Sewers
Storm Sewers and Stormwater Management

Pedestrian and Vehicular Access
Expected Variety of Vehicles
Pedestrian, Vehicular Bridges or Tunnels
Loading Docks, Service Vehicle Access

Parking
Parking Space Required or Prohibited
Parking Requirements by Vehicle Type
Structural Parking Requirements

Outdoor-Use Spaces
Types of Spaces Required
Appearance and Image Considerations

Landscaping
Site Grading and Vegetation
Circulation Patterns and Screening Requirements
Historical Considerations
Security Issues

8. BUILDING SERVICES

Guarantees and Warranties
Available Guarantees and Warranties
Components and Systems Requiring Extended Warranties
Performance Characteristics and Costs

Long-Term-Maintenance Contracts
Service-Contracting Alternatives for Building Systems
Contract Cost Considerations

Robotics
Identify Potential Benefit Areas
Special Requirements and Costs

Trash Removal
Type and Quantity of Trash Anticipated
Regulations Regarding Disposal, Separation, and Storage
Alternative Collection, Storage, and Removal Methods

9. SAFETY AND ENVIRONMENTAL ISSUES

Asbestos
Location, Type, Quantity, and Condition of Asbestos-Containing Material (ACM)
Determine Appropriate Abatement Action

Indoor Air Quality
Carbon Dioxide, Carbon Monoxide, Particulates, Volatile Organic Compounds (VOC), and Radon

Monitor HVAC and Other Systems to Reduce Contaminants

Lead in Drinking Water
Quantify Lead Content and Sources of Contamination

Determine Recommended Action to Alleviate Hazards

Storage Tanks
Location, Size, and Condition of Storage Tanks and Piping Systems

Content and Current Utilization

Age, Leakage, and Contamination Issues

Polychlorinated Biphenyl
Location and Condition of Electrical Equipment Containing PCB

Determine Appropriate Course of Action

Hazardous Waste
Type and Amount of Hazardous Waste Involved in Current or Future Operations

Recommended Methods to Minimize Hazard

APPENDIX B
Space-Requirements Worksheet

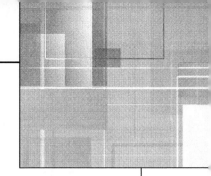

The process of collecting information is a critical part of the data-management process, but it is also one fraught with problems. Unless the right questions are asked of the right people, either bad data or insufficient data are obtained. Even worse, the ability to process huge amounts of statistical data with the computer can lead to information saturation: collecting more data than can be effectively used for space-planning purposes.

The following questionnaire worksheet (Figure B-1 and Figure B-3) is directed to operating managers or a point person in charge of one or several organizational units. The principal criterion for selection is that the person completing a worksheet be responsible for a group small enough so that he or she has direct hands-on knowledge of the day-to-day activities and needs of the group. These individuals are identified during initial orientation meetings and from the organizational data collected.

One or several project kickoff meetings are scheduled for the purpose of distributing the worksheets. These meetings also serve as an opportunity to introduce the project team and client staff. A brief, section-by-section explanation of the questionnaire is

WORKSHEET COVER LETTER

During the next several weeks, we will be conducting an intensive long-range planning study of your future space needs. We are interested in learning about your projected increases in people, departmental adjacency requirements, and specified data with respect to the areas in which you and your co-workers function. The purpose of this program is to assure you that sufficient space will be available to satisfy your future requirements and that it will be well-planned to suit your specific needs. This is only possible if critical decisions can be made carefully, according to the best information available.

Please read the explanation for each of the questions before filling out the worksheet. (A sample answer sheet is also provided to clarify any sections not sufficiently covered by the explanation.) Fill in the spaces that apply to your department, leaving blank all items that do not apply. If you feel that the multiple-choice answers do not adequately cover unique situations that may exist in your area, please feel free to write any comments you may have.

All of your responses will be of the greatest importance to the success of the study and, more importantly, to you and your department. Please give each question your most careful consideration.

made, followed by a question-and-answer session. It is also explained that, after the worksheets have been tabulated and specific questions formulated, personal interviews will be scheduled to follow up on all special requirements and areas.

Figure B-1 Questionnaire Worksheet.

1 Groups	2 Positions	4 Projections			5 Spaces			6 Furnishings		7 Storage			8 Machines	9 Relationships	10 Description	Rating
		61	61	61	Private	Semiprivate	Open	Plans	Seating	Legal	Letter	Shelving				
ADMIN {	Vice Pres.	1	1	1	X		X	DFN CO	2	2 8	2	15		X X X	OPERATIONS & CONTROL ADMINISTRATION RESEARCH & DEVEL.	1
	Secretary	1	1	1					1	8	2	10		X		2
GEN. ACCTG {	Sr. Accountant	3	3	3		X	X	AN B	1		4	5	CRT	X		
	Accountant	4	4	4							2					
FIN. SYST. ANALYSIS {	Sr. Analyst	3	3	3		X	X	AN BQ	1		2	10		X		
	Analyst	3	3	3			X	A			2	2		X X		
	Clerk	2	2	2												
COMM. EXCH. AUDIT {	Sr. Auditor	6	6	6		X	X	AN A	1	2	3			X		
	Auditor	7	7	7			X	C		2	2			X X		
	Secretary	1	1	1			X	AP		4	2	6		X		
	Clerk	1	1	1								8				
TRANS. AUDIT {	Sr. Auditor	2	2	2		X	X	AN B	1	3				X		
	Auditor	2	2	2			X	A		2	2			X X		
	Clerk	1	1	1												

3 Support Facilities		4			5			6						? Comments
Conference					X			TABLE & Q CHAIRS						PRESENT LOCATION TOO FAR FROM OPERATIONS & CONTROL ADMIN.
Extern. Audit	2	3	3	X			AP 1							
Copier														
Files & Stor.														

Name JAMES HARVEY
Title VICE PRESIDENT
Department FINANCIAL ADMIN.

WORKSHEET INSTRUCTIONS

Please fill in the identifying information in the lower right-hand corner of the requirements worksheet.

There are ten sections to this worksheet. Before entering any data, read over the explanations for the first four sections. Columns 1, 2, and 4 are interrelated, and it is important that column 2 be completed first. Understanding columns 1 and 4 will help you to complete column 2 correctly.

1—Groups

Column 1 is to be used to bracket and describe functional subgroupings within your department. For example, if your department has three such sections, the positions should be listed in column two. The people in each section should be grouped together so that they may be bracketed and identified in column 1. (Please refer to the sample worksheet for clarification.)

2—Positions

Beginning with line 1, list yourself and the people with whom you work or who report to you within your own department. Please use descriptive titles rather than names so that similar functions served by more than one person may be grouped together. (A brief look at the explanation for question 4 should clarify the reason for using titles.)

3—Support Facilities

Beginning with line a, list any support spaces required by your department. These may include conference or workrooms, filing and storage areas, interview spaces or reception areas, and so on. Workstations that must be provided to accommodate authorized positions not presently filled should be treated as support facilities in order not to distort the head-count figures.

In general, the upper two-thirds of the worksheet deal with workstation requirements, while the lower third is directed to support space. *Workstations* are defined as spaces assigned to specific individuals on a full-time basis. *Support facilities* are areas that are used to house equipment, shared spaces used by a number of persons, or work space assigned to positions not presently filled or to persons from outside your organization such as auditors.

4—Projections

The amount of work space allocated to your department in future years will be based upon your projections as to the required number of people in job categories that now exist or new positions that may be required. In Section 4, enter on lines 1 through 24 your projections for all the positions listed in column 2.

There are four columns within this section, which are headed by the years 20—, 20—, 20—, and 20—. Your projections should be stated in terms of the total number of people in each position expected at the ends of the given years. The first column should reflect present conditions existing as of 20—, which is our baseline date.

Projected increases in the number of required support facilities should be indicated on lines a through i. You need not total any of your projections as these forms will be extensively analyzed and tabulated by the consultants.

At this point, you should return to question 2 and begin entering the necessary data. If you have completed questions 1 through 4, please proceed to question 5.

5—Spaces

This section defines the kinds of space required for the efficient functioning of each job, whether it be a private space, a semienclosed space, a completely open work area, or some combination in a systems furniture environment. The degree of privacy required should be thought of in relative terms. That is, a private space can be an enclosed office with a door, or it can be a freestanding workstation with sufficient physical and acoustical separation from others to provide the required privacy.

Mark the appropriate box on each line corresponding to the various job categories in your department. A fourth column has been left open should you wish to indicate any special kinds of space required.

Please feel free to use this worksheet creatively. We have designed it to cover the basic information that we need to begin a thorough analysis of your facility requirements. If you have a special concern that you want to discuss in detail, or unique spatial or equipment needs that do not fit the form, please make a notation to that effect and we will make a point of coming to talk with you.

6—Furnishings

When you complete the first of the two columns (Plans) under Section 6, you will need to refer to the drawings of typical office furnishings in Figure B-2.

From the first group titled *Workplace*, you should select the item of furniture that you feel will best serve the needs of each person in your department as a primary workplace. This should be done for each position listed in Section 2 by entering the appropriate letter designation on each line.

The second group, labeled *Additional*, contains illustrations of several pieces of furniture that may be selected in addition to the primary workplace. These should be entered in the same column (Plans) where needed by letter designation. More than one item may be selected if required.

In the second column under this section (Seating), note the number of additional guest chairs required for each workstation (if any) or the amount of seating necessary in any conference rooms, reception areas, or other auxiliary spaces listed in Section 3. For individual workstations, you should assume that the occupant's own chair would be provided and not include it in the guest chair count.

If you feel there is a unique area or set of furnishings that cannot be described in this way, please provide a description or sketch of the particular condition. It may either be attached to or sketched on the reverse side of the worksheet. Assign an unused letter designation to the sketch or description and note it on the worksheet in the appropriate place.

7—Storage

Question 7 deals with existing filing and storage requirements for each position or support space. This is in addition to storage within the basic items of furniture selected in column 7.

Quantities of legal and letter files should be indicated in terms of number of drawers per person, not cabinets. This is so that there will be no confusion in sizes of cabinets (two drawers versus four drawers, for instance) when interpreting quantities. If the existing files are lateral, please note that fact along with their widths.

Shelving should be indicated in total linear feet required.

One column has been left blank for any special needs or other types of storage facilities not listed. Automated systems such as power files or rolling shelving should be noted in the next section under *Machines*.

Notations on lines *a* through *i* pertain to centralized files and storage accommodations. For example, if a file room is required for a particular department, the total number of files or other storage media required should be indicated.

8—Machines

This section is intended to document machines that require additional floor area, such as copiers, large computer terminals, network servers, microfilm units, power files, and so on.

Equipment such as typewriters and calculators, which normally sit atop a piece of furniture, need not be included. You may assume that standard electrical and communications facilities will normally be provided for all personnel.

Figure B-2   Work-Space Furniture Selection Chart.

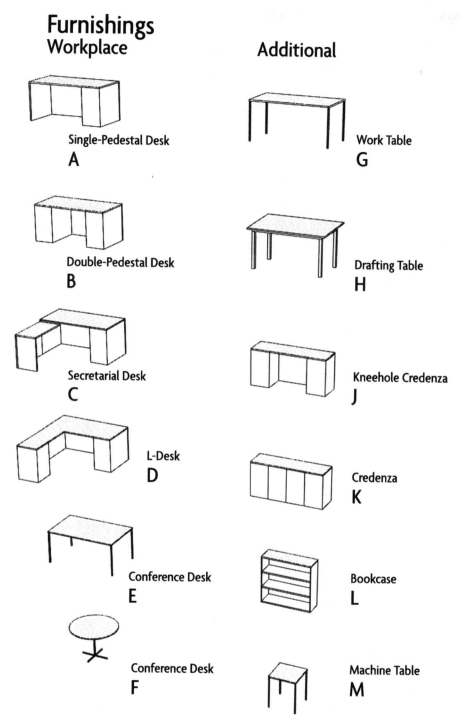

Furnishings
Workplace

Single-Pedestal Desk
A

Double-Pedestal Desk
B

Secretarial Desk
C

L-Desk
D

Conference Desk
E

Conference Desk
F

Additional

Work Table
G

Drafting Table
H

Kneehole Credenza
J

Credenza
K

Bookcase
L

Machine Table
M

If any required furnishings are not represented on this chart, please sketch them and assign them a letter.

9—Relationships

Question 9 deals with desired adjacency relationships that should be established within your department between the people whose requirements are spelled out in the top section of the worksheet and the support spaces listed under Section 3. Lines *a* through *i* are folded vertically to become columns under Section 9, forming a matrix in which you can indicate the people to which each of the support facilities should be closest.

Intradepartmental adjacency requirements should be checked off based upon such considerations as frequency of use (such as a conference room) or the uniqueness of a particular type of space corresponding to a particular job classification (such as a filing and storage room to a file clerk). A look at the sample worksheet (Figure B-1) should clarify these instructions.

10—Departmental Adjacencies

Section 10 should be used to indicate interdepartmental adjacency relationships: the importance of physical adjacencies between your department and other areas within your organization. Such areas may include other departments or physical entities as, for example, an outside entrance, loading dock, service area, and so on.

In the first column (*Description*), print the names of those entities for which an adjacency priority should be specified.

In the second column (*Rating*), please rate the importance of the relationship from 1 to 5. A 1 rating means that immediate physical adjacency is imperative. At the other end of the scale, a 5 means that a remote location is preferred based upon function from your point of view. The other values express the relative degrees of importance between these extremes.

1. Adjacency imperative
2. Adjacency highly desirable
3. Adjacency advantageous
4. Don't care
5. Remote location desirable

The fourth level (don't care) is automatically assigned to those entities not mentioned and does not need to be entered.

11—Comments

Please answer the following questions regarding the information you have entered on the worksheet. Comments may be written in the space provided, on the reverse side of the sheet, or on attached pages as required.

1. What basic assumptions did you make in formulating your projections? That is, did you assume a constant rate of expansion yearly, or are there specific reasons for more or fewer personnel due to work volume changes in existing areas of business? Please describe.

2. Does your department operate on multiple shifts? If so, please indicate which is the largest shift, whether personnel on the other shift(s) use the same workstations as those on the main shift, and whether all shifts have been included in your projections.

3. In Section 7 of the worksheet, you were asked to note existing quantities of filing and other types of storage. Do you foresee any substantial change in future filing methods or systems or quantities of files required? If so, please describe the kind of changes you have in mind, indicating growth either by percentage per year or by referencing specific line entries on the worksheet, as applicable.

Please feel free to add any additional comments that you consider relevant to effective planning of your department's facilities in the future. These may include your expectation of how a required adjacency relationship may change over time or how technological innovation may alter work patterns or clarification of other information you have recorded.

Thank-you for your cooperation.

Figure B-3   Questionnaire Worksheet.

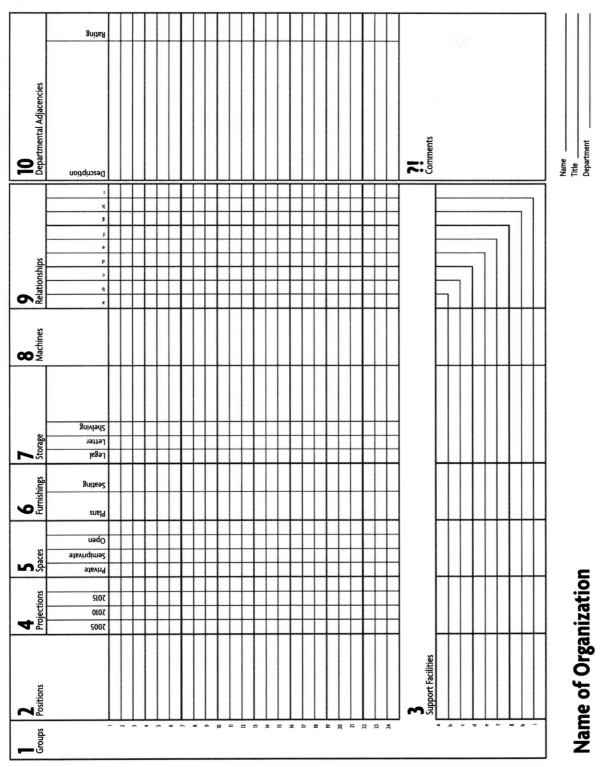

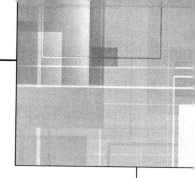

APPENDIX C
Loading AutoCAD Macros

The AutoCAD macros described in this Appendix may be found on the accompanying CD or downloaded from the *Prentice Hall* Web site at http://www.prenhall.

After areas have been defined by polylines in an AutoCAD drawing, they can be exported to a text file, an Excel worksheet, or most any similar application. This can be accomplished by taking advantage of AutoCAD's extensive customization capability. Figure 7-1 shows ways in which data can be moved among different applications, for which we used the term *pipelining*.

Beginning with a facility program prepared using Access, we can export the personnel, area requirements, and other data in order to create blocking and stacking diagrams. These data can also be used to annotate detailed space plans as they are developed. Final as-built plans are then often linked to a CAFM database.

The *Area Allocation Tools* described in Chapter 7 consist of a group of AutoCAD VBA macros that facilitate moving the data among applications (primarily AutoCAD and Excel) and changing them from numeric to graphic form. These macros are accessed using a custom toolbar that must first be added to AutoCAD's menu. The following material describes how to set up and load the VBA macro procedures and their related files. These files are contained in a folder named Allocation_Macros, which must be placed in the root of either the C:\ or D:\ drive. (The root folder of the D:\ drive should only be used if it is a hard-disk and it is not desirable to use drive C:\.)

In order for AutoCAD to locate the Allocation_Macros folder, it must be added to the Support File Search Path. This need only be done once for each AutoCAD user installation. This is accomplished using the Files tab on the Tools → Options dialog box (Figure C-1).

Clicking the Add . . . button creates a new path, after which the Browse . . . button invokes a Browse for Folder subdialog box from which the appropriate folder is selected. Clicking OK adds the selected folder to the search path. Its position in the list of folders may be changed using the Move Up/Down buttons. Figure C-2 shows the Allocation_Macros folder added to the end of the search path.

The next step is to load the VBA procedure that creates the new toolbar. VBA macros are of two types. *Embedded* macros are saved with a drawing and may only be accessed by that drawing. *Global* macros, on the other hand, are more like stand-alone programs in that they are saved in separate files with a DVB extension. They are able to open, close, create, and modify AutoCAD drawings and can function as libraries in which commonly used macros are collected. In this case, we need to search for and open the global VBA project named AreaAllocation.dvb, which is found in the Allocation_Macros folder. (We do not need to worry about the other file, AreaAssignment.dvb—it will automatically be loaded from the toolbar when needed.)

Tools → Macro → Load Project . . . calls the Open VBA Project dialog box as shown in Figure C-3 and Figure C-4. After AreaAllocation.dvb has been selected, clicking the Open button loads the global project.

Figure C-1   Adding a Folder to the Support File Search Path.

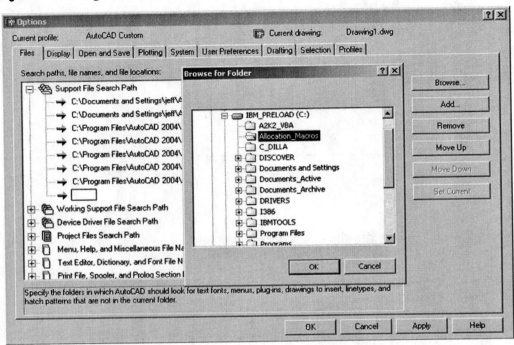

Figure C-2   The Allocation_Macros Folder Added to the Search Path.

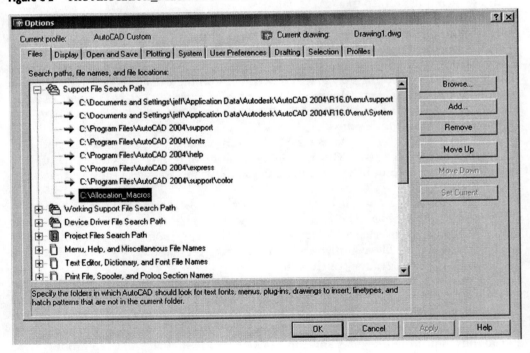

The final step is to execute the procedure that creates the toolbar from which all the other area-allocation tools can be accessed. Returning to the Tools menu, we select Tools → Macro → Macros . . . to call the Macros dialog box (Figure C-5 and Figure C-6). The BlockDiagramAreaToolbar procedure should be selected if it is not already selected when the dialog box appears. (It is the first macro in the list.) Clicking the Run button executes the procedure, and the toolbar should appear on the AutoCAD screen.

Figure C-3   Loading an AutoCAD VBA Project.

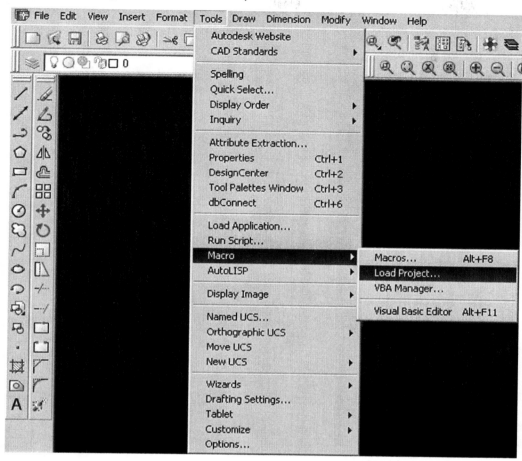

Figure C-4   Opening a VBA Project.

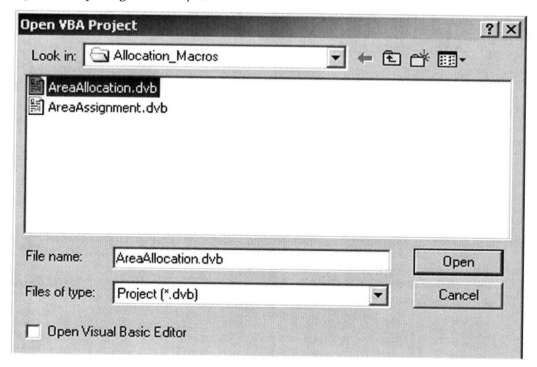

Figure C-5 Executing an AutoCAD VBA Procedure.

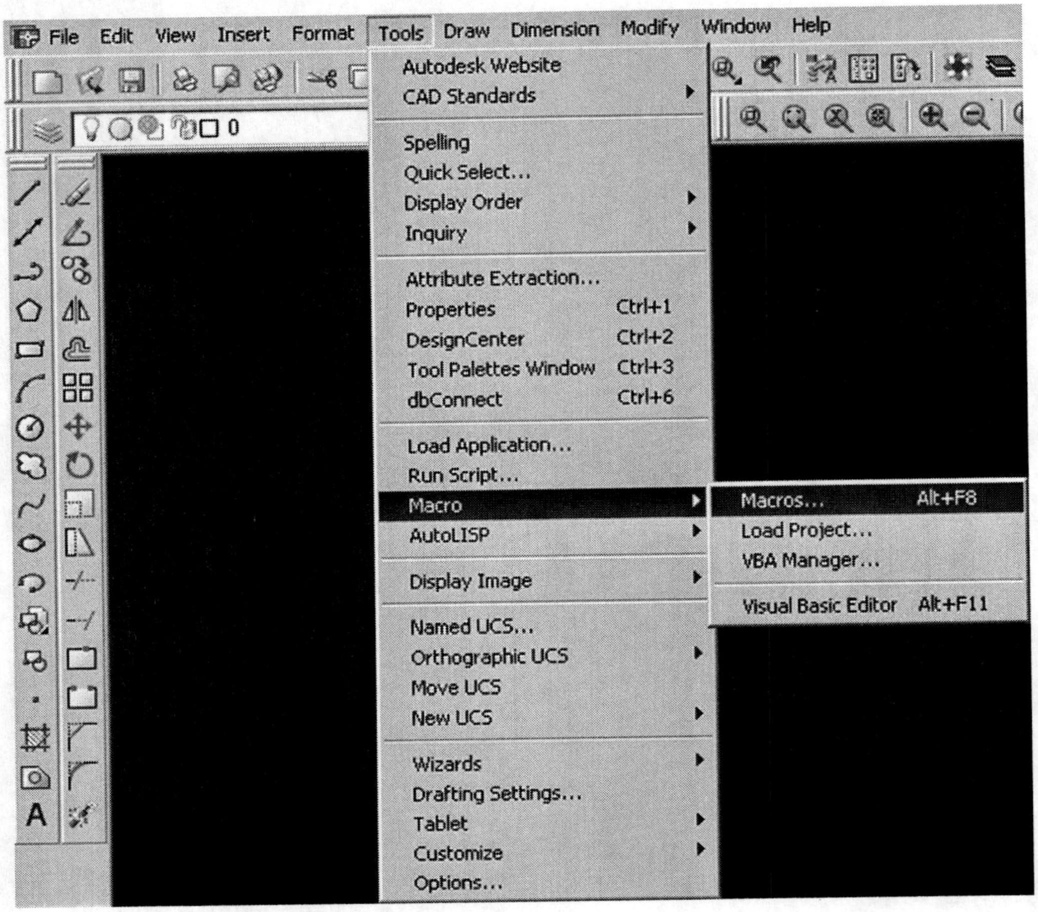

The Area-Allocation toolbar is shown in Figure C-7. The function of each button is described in detail in the "Area-Allocation Tools" section of Chapter 7.

FPDATA.INI

The Allocation_Macros folder contains an initialization file named FPdata.ini. This text file contains five lines of data that initialize several of the macros. FPdata.ini can be edited as required using the Windows Notepad. Its format is as follows:

C:\Documents_Active\2005_Spring\Week07\FacilityData029.mdb
10
64
32
4

- *Line 1*—The full path and file name of the Access database you wish to query to annotate your space layout in AutoCAD. The Program Requirements procedure allows you to drill down to a workstation or support space line entry in the database using the dialog box illustrated in Figure 7-20. The Level/Sequence Codes, Description, SAR Key, and allocated Square Feet can then be added to your drawing as single-line text.

    If the full path is displayed in the Address Bar of Windows Explorer, it can be copied and pasted into FPdata.ini using the Notepad. You must then add a back-

Figure C-6   Running the BlockDiagramAreaToolbar Macro.

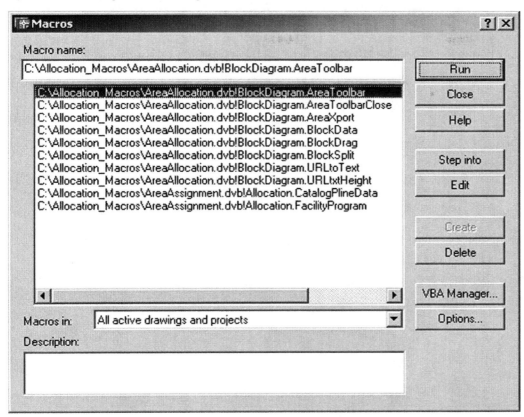

Figure C-7   Area-Allocation Toolbar.

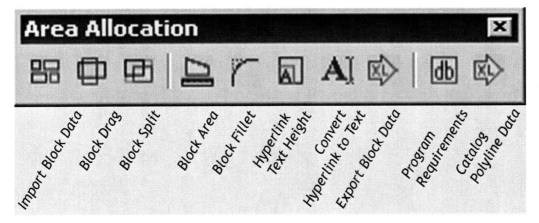

slash "\" followed by the name (with its .mdb extension) of the Access database you wish to query.

- *Line 2*—The height of the annotation text added to your drawing by the Program Requirements text annotation dialog box. Annotations are made using the current text style. The default value is 10 (inches—in scale).

- *Line 3*—Sets the default value of the AutoCAD system variable FILLETRAD, which is the radius of the arc connecting filleted objects. The Block Fillet procedure applies fillets of the specified radius to the four corners of rectangular block objects, if desired, for presentation purposes. The default value is 64, or five feet four inches (5'-4").

- *Line 4*—Sets the default value of the OSMODE system variable, which turns Object Snap on and activates the Intersection Object Snap mode. This setting is necessary for the proper operation of the Block Drag procedure. The default value is 32.
- *Line 5*—Sets the default value of the AutoCAD system variable LUNITS, which sets length units to Architectural. The default value is 4.

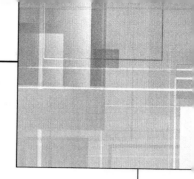

APPENDIX D
Archibus/FM Space-Management Module

All of the standard navigator actions contained in the Archibus space-management module are listed in Appendix D. This outline extends the activity list illustrated in Figure 11-6 (page 242) to include task categories, tasks, and subtasks.

The number of tasks and subtasks shown here is representative of the number of standard tables, reports, actions, queries, and drawing-related functions to be found in each of the other modules. Many of these tasks and subtasks initiate AVW or ABS procedures or a combination of both, and new views (AVW) procedures can be created and saved using the Archibus/FM menus. Customized queries may be written as ABS procedures and added to the navigator.

3.0   Space Management Module

Activity class
Activity
Task Category
Task
SubTask

3. 1	**Composite Inventory**	
3. 1. 1	*Background Data*	
3. 1. 1. 1	**Tables**	
3. 1. 1. 1. 1		*Space Information*
3. 1. 1. 1. 1. 1		Sites
3. 1. 1. 1. 1. 2		Buildings
3. 1. 1. 1. 1. 3		Floors
3. 1. 1. 1. 1. 4		Floors by Buildings
3. 1. 1. 1. 2		*Organization Information*
3. 1. 1. 1. 2. 1		Business Units
3. 1. 1. 1. 2. 2		Divisions
3. 1. 1. 1. 2. 3		Departments
3. 1. 1. 1. 2. 4		Departments by Divisions
3. 1. 1. 2	**Reports**	
3. 1. 1. 2. 1		*Space Information*
3. 1. 1. 2. 1. 1		Sites
3. 1. 1. 2. 1. 2		Buildings
3. 1. 1. 2. 1. 3		Buildings by Site

(*continued*)

3.0   Space Management Module *(continued)*

Activity class
Activity
Task Category
Task
SubTask

3. 1. 1. 2. 1. 4	Floors
3. 1. 1. 2. 1. 5	Floors by Site
3. 1. 1. 2. 1. 6	Floors by Building
3. 1. 1. 2. 1. 7	All Areas by Building
3. 1. 1. 2. 1. 8	All Areas by Floor
3. 1. 1. 2. 2	*Organization Information*
3. 1. 1. 2. 2. 1	Business Units
3. 1. 1. 2. 2. 2	Divisions
3. 1. 1. 2. 2. 3	Divisions by Business Units
3. 1. 1. 2. 2. 4	Departments
3. 1. 1. 2. 2. 5	Departments by Business Units
3. 1. 1. 2. 2. 6	Departments by Divisions
3. 1. 2	*Gross Areas*
3. 1. 2. 1	**Drawings**
3. 1. 2. 1. 1	*Draw Gross Areas*
3. 1. 2. 2	**Actions**
3. 1. 2. 2. 1	*Update Gross Area Totals*
3. 1. 2. 3	**Reports**
3. 1. 2. 3. 1	*Gross Areas*
3. 1. 2. 3. 2	*Gross Area by Site*
3. 1. 2. 3. 3	*Gross Area by Building*
3. 1. 2. 3. 4	*Gross Area by Floor*
3. 1. 3	*Vertical Penetration Areas*
3. 1. 3. 1	**Tables**
3. 1. 3. 1. 1	*Vertical Penetration Standards*
3. 1. 3. 1. 2	*Vertical Penetrations*
3. 1. 3. 2	**Drawings**
3. 1. 3. 2. 1	*Draw Vertical Penetrations*
3. 1. 3. 3	**Actions**
3. 1. 3. 3. 1	*Update Vertical Penetrations Area Totals*
3. 1. 3. 4	**Reports**
3. 1. 3. 4. 1	*Vertical Penetrations*
3. 1. 3. 4. 2	*Vertical Penetrations by . . .*
3. 1. 3. 4. 2. 1	. . . Site
3. 1. 3. 4. 2. 2	. . . Building
3. 1. 3. 4. 2. 3	. . . Floor
3. 1. 3. 4. 2. 4	. . . Standard
3. 1. 3. 4. 3	*Vertical Penetration Standards*
3. 1. 3. 4. 4	*Vertical Penetrations Analysis . . .*
3. 1. 3. 4. 4. 1	. . . Summary
3. 1. 3. 4. 4. 2	. . . by Site
3. 1. 3. 4. 4. 3	. . . by Building
3. 1. 3. 4. 4. 4	. . . by Floor
3. 1. 3. 5	**Queries**
3. 1. 3. 5. 1	*Highlight Vertical Penetrations by Standard*
3. 1. 4	*Service Areas*
3. 1. 4. 1	**Tables**
3. 1. 4. 1. 1	*Service Area Standards*
3. 1. 4. 1. 2	*Service Areas*
3. 1. 4. 1. 3	*Designate Common Service Area*
3. 1. 4. 2	**Drawings**
3. 1. 4. 2. 1	*Draw Service Areas*

3.0   Space Management Module

Activity class
Activity

Task Category
Task

SubTask

3. 1. 4. 3	**Actions**
3. 1. 4. 3. 1	*Update Service Area Totals*
3. 1. 4. 4	**Reports**
3. 1. 4. 4. 1	*Service Areas*
3. 1. 4. 4. 2	*Service Areas by . . .*
3. 1. 4. 4. 2. 1	. . . Site
3. 1. 4. 4. 2. 2	. . . Building
3. 1. 4. 4. 2. 3	. . . Floor
3. 1. 4. 4. 2. 4	. . . Standard
3. 1. 4. 4. 3	*Service Area Standards*
3. 1. 4. 4. 4	*Building Performance*
3. 1. 4. 4. 5	*Service Area Analysis . . .*
3. 1. 4. 4. 5. 1	. . . Summary
3. 1. 4. 4. 5. 2	. . . by Site
3. 1. 4. 4. 5. 3	. . . by Building
3. 1. 4. 4. 5. 4	. . . by Floor
3. 1. 4. 5	**Queries**
3. 1. 4. 5. 1	*Highlight Service Areas by Standard*
3. 1. 4. 5. 2	*Highlight Common Service Areas*
3. 1. 5	*Group Areas*
3. 1. 5. 1	**Tables**
3. 1. 5. 1. 1	*Group Standards*
3. 1. 5. 1. 2	*Groups*
3. 1. 5. 1. 3	*Designate Common Group Area*
3. 1. 5. 2	**Drawings**
3. 1. 5. 2. 1	*Draw Groups*
3. 1. 5. 3	**Actions**
3. 1. 5. 3. 1	*Update Group Area Totals*
3. 1. 5. 3. 2	*Update All Area Totals*
3. 1. 5. 3. 3	*Perform Group Chargeback*
3. 1. 5. 3. 4	*Perform BOMA Chargeback*
3. 1. 5. 3. 5	*Perform Enhanced BOMA Chargeback*
3. 1. 5. 3. 6	*Perform BOMA 96 Chargeback*
3. 1. 5. 4	**Reports**
3. 1. 5. 4. 1	*Groups*
3. 1. 5. 4. 2	*Groups by . . .*
3. 1. 5. 4. 2. 1	. . . Business Unit
3. 1. 5. 4. 2. 2	. . . Division
3. 1. 5. 4. 2. 3	. . . Department
3. 1. 5. 4. 2. 4	. . . Site
3. 1. 5. 4. 2. 5	. . . Building
3. 1. 5. 4. 2. 6	. . . Floor
3. 1. 5. 4. 2. 7	. . . Standard
3. 1. 5. 4. 3	*Group Standards*
3. 1. 5. 4. 4	*Departmental Analysis . . .*
3. 1. 5. 4. 4. 1	. . . Summary
3. 1. 5. 4. 4. 2	. . . by Site
3. 1. 5. 4. 4. 3	. . . by Building
3. 1. 5. 4. 4. 4	. . . by Floor
3. 1. 5. 4. 4. 5	. . . Location Breakdown
3. 1. 5. 4. 5	*Group Standard Analysis . . .*
3. 1. 5. 4. 5. 1	. . . Summary
3. 1. 5. 4. 5. 2	. . . by Site

(continued)

3.0   Space Management Module   (*continued*)

Activity class
    Activity
        Task Category
            Task
                SubTask

3. 1. 5. 4. 5. 3	. . . by Building
3. 1. 5. 4. 5. 4	. . . by Floor
3. 1. 5. 4. 5. 5	. . . by Department
3. 1. 5. 4. 6	*Facility Percentage Analysis . . .*
3. 1. 5. 4. 6. 1	. . . by Site
3. 1. 5. 4. 6. 2	. . . by Building
3. 1. 5. 4. 6. 3	. . . by Floor
3. 1. 5. 4. 7	*Chargeback Analysis . . .*
3. 1. 5. 4. 7. 1	Group Chargeback - Prorate Report
3. 1. 5. 4. 7. 2	Group Chargeback - Detailed Analysis
3. 1. 5. 4. 7. 3	Group Chargeback - Detailed Analysis - Buildings without Sites
3. 1. 5. 4. 7. 4	Group Chargeback - Financial Statement
3. 1. 5. 4. 7. 5	Group Chargeback - Financial Stmt - Divisions w/o Business Units
3. 1. 5. 4. 7. 6	BOMA Chargeback Analysis
3. 1. 5. 4. 8	*Remaining Area*
3. 1. 5. 5	**Queries**
3. 1. 5. 5. 1	*Highlight Groups by Department*
3. 1. 5. 5. 2	*Highlight Groups by Standard*
3. 1. 5. 5. 3	*Highlight Common Area Groups*
3. 1. 5. 5. 4	*Departmental Analysis*
3. 1. 5. 5. 5	*Departmental Stack Plan*
3. 1. 6	***Room Areas***
3. 1. 6. 1	**Tables**
3. 1. 6. 1. 1	*Room Standards*
3. 1. 6. 1. 2	*Room Types by Category*
3. 1. 6. 1. 3	*Room Uses*
3. 1. 6. 1. 4	*Rooms*
3. 1. 6. 1. 5	*Designate Common Area Rooms*
3. 1. 6. 2	**Drawings**
3. 1. 6. 2. 1	*Draw Rooms*
3. 1. 6. 3	**Actions**
3. 1. 6. 3. 1	*Update Room Area Totals*
3. 1. 6. 3. 2	*Update All Area Totals*
3. 1. 6. 3. 3	*Perform Room Chargeback*
3. 1. 6. 3. 4	*Update Room Area from Manual Area*
3. 1. 6. 4	**Reports**
3. 1. 6. 4. 1	*Rooms*
3. 1. 6. 4. 2	*Rooms by . . .*
3. 1. 6. 4. 2. 1	. . . Site
3. 1. 6. 4. 2. 2	. . . Building
3. 1. 6. 4. 2. 3	. . . Floor
3. 1. 6. 4. 2. 4	. . . Floor per Department
3. 1. 6. 4. 2. 5	. . . Business Unit
3. 1. 6. 4. 2. 6	. . . Division
3. 1. 6. 4. 2. 7	. . . Department
3. 1. 6. 4. 2. 8	. . . Standard
3. 1. 6. 4. 2. 9	. . . Type and Category
3. 1. 6. 4. 2. 10	. . . Use
3. 1. 6. 4. 3	*Room Classifications . . .*
3. 1. 6. 4. 3. 1	Room Standards
3. 1. 6. 4. 3. 2	Room Categories
3. 1. 6. 4. 3. 3	Room Types
3. 1. 6. 4. 3. 4	Room Types by Category

3.0   Space Management Module

Activity class
Activity

Task Category
Task

SubTask

3. 1. 6. 4. 3. 5	Room Uses
3. 1. 6. 4. 4	*Departmental Room Analysis . . .*
3. 1. 6. 4. 4. 1	. . . Summary
3. 1. 6. 4. 4. 2	. . . by Building
3. 1. 6. 4. 4. 3	. . . by Floor
3. 1. 6. 4. 4. 4	. . . Location Breakdown
3. 1. 6. 4. 5	*Room Standard Analysis . . .*
3. 1. 6. 4. 5. 1	. . . Summary
3. 1. 6. 4. 5. 2	. . . by Building
3. 1. 6. 4. 5. 3	. . . by Floor
3. 1. 6. 4. 5. 4	. . . by Floor per Department
3. 1. 6. 4. 5. 5	. . . by Department
3. 1. 6. 4. 5. 6	. . . by Department per Floor
3. 1. 6. 4. 5. 7	. . . Area Comparison: Overview
3. 1. 6. 4. 5. 8	. . . Area Comparison: Room-by-Room
3. 1. 6. 4. 6	*Room Type and Category Analysis . . .*
3. 1. 6. 4. 6. 1	. . . Summary
3. 1. 6. 4. 6. 2	. . . by Building
3. 1. 6. 4. 6. 3	. . . by Floor
3. 1. 6. 4. 6. 4	. . . by Floor per Department
3. 1. 6. 4. 6. 5	. . . by Department
3. 1. 6. 4. 6. 6	. . . by Department per Floor
3. 1. 6. 4. 7	*Chargeback Analysis . . .*
3. 1. 6. 4. 7. 1	Prorate Report
3. 1. 6. 4. 7. 2	Detailed Analysis
3. 1. 6. 4. 7. 3	Detailed Analysis - Buildings without Sites
3. 1. 6. 4. 7. 4	Financial Statement
3. 1. 6. 4. 7. 5	Financial Statement - Divisions without Business Units
3. 1. 6. 4. 8	*All Vacant Rooms*
3. 1. 6. 4. 9	*Occupiable Vacant Rooms*
3. 1. 6. 4. 10	*Remaining Area*
3. 1. 6. 5	**Queries**
3. 1. 6. 5. 1	*Highlight Rooms by Department*
3. 1. 6. 5. 2	*Highlight Rooms by Standard*
3. 1. 6. 5. 3	*Highlight Rooms by Category*
3. 1. 6. 5. 4	*Highlight Rooms by Type*
3. 1. 6. 5. 5	*Highlight Common Area Rooms*
3. 1. 6. 5. 6	*Highlight All Vacant Rooms*
3. 1. 6. 5. 7	*Occupancy Plan*
3. 1. 6. 5. 8	*Occupiable vs. Non-Occupiable*
3. 1. 6. 5. 9	*Departmental Stack Plan*
3. 1. 7	*Room Percentages*
3. 1. 7. 1	**Background Tables**
3. 1. 7. 1. 1	*Divisions*
3. 1. 7. 1. 2	*Departments*
3. 1. 7. 1. 3	*Buildings*
3. 1. 7. 1. 4	*Floors*
3. 1. 7. 1. 5	*Rooms*
3. 1. 7. 1. 6	*Room Standards*
3. 1. 7. 1. 7	*Room Types by Category*
3. 1. 7. 2	**Tables**
3. 1. 7. 2. 1	*Allocate Percentages*
3. 1. 7. 3	**Drawings**
3. 1. 7. 3. 1	*Draw Rooms*

(continued)

3.0 **Space Management Module** *(continued)*

Activity class
Activity
Task Category
Task
SubTask

3. 1. 7. 4	**Actions**
3. 1. 7. 4. 1	*Synchronize Room and Percentage Records*
3. 1. 7. 4. 2	*Update Area Totals - Space Percentages*
3. 1. 7. 4. 3	*Update Area Totals - Space & Time Percentages*
3. 1. 7. 4. 4	*Perform Percentage Chargeback*
3. 1. 7. 5	**Reports**
3. 1. 7. 5. 1	*Percentages*
3. 1. 7. 5. 2	*Percentages by . . .*
3. 1. 7. 5. 2. 1	. . . Site
3. 1. 7. 5. 2. 2	. . . Building
3. 1. 7. 5. 2. 3	. . . Floor
3. 1. 7. 5. 2. 4	. . . Floor per Department
3. 1. 7. 5. 2. 5	. . . Room
3. 1. 7. 5. 2. 6	. . . Business Unit
3. 1. 7. 5. 2. 7	. . . Division
3. 1. 7. 5. 2. 8	. . . Department
3. 1. 7. 5. 2. 9	. . . Type and Category
3. 1. 7. 5. 3	*Departmental Analysis . . .*
3. 1. 7. 5. 3. 1	. . . Summary
3. 1. 7. 5. 3. 2	. . . by Building
3. 1. 7. 5. 3. 3	. . . by Floor
3. 1. 7. 5. 3. 4	. . . Location Breakdown
3. 1. 7. 5. 4	*Standard Analysis . . .*
3. 1. 7. 5. 4. 1	. . . Summary
3. 1. 7. 5. 4. 2	. . . by Building
3. 1. 7. 5. 4. 3	. . . by Floor
3. 1. 7. 5. 4. 4	. . . by Floor per Department
3. 1. 7. 5. 4. 5	. . . by Department
3. 1. 7. 5. 4. 6	. . . by Department per Floor
3. 1. 7. 5. 4. 7	. . . Area Comparison: Overview
3. 1. 7. 5. 4. 8	. . . Area Comparison: Room-by-Room
3. 1. 7. 5. 5	*Type and Category Analysis . . .*
3. 1. 7. 5. 5. 1	. . . Summary
3. 1. 7. 5. 5. 2	. . . by Building
3. 1. 7. 5. 5. 3	. . . by Floor
3. 1. 7. 5. 5. 4	. . . by Floor per Department
3. 1. 7. 5. 5. 5	. . . by Department
3. 1. 7. 5. 5. 6	. . . by Department per Floor
3. 1. 7. 5. 6	*Chargeback Analysis . . .*
3. 1. 7. 5. 6. 1	Prorate Report
3. 1. 7. 5. 6. 2	Detailed Analysis
3. 1. 7. 5. 6. 3	Financial Statement
3. 1. 7. 6	**Queries**
3. 1. 7. 6. 1	*Departmental Breakdown*
3. 1. 7. 6. 2	*Room Type Breakdown*
3. 1. 7. 6. 3	*Common Area Breakdown*
3. 2	**All Room Inventory**
3. 2. 1	*Background Data*
3. 2. 1. 1	**Tables**
3. 2. 1. 1. 1	*Space Information*
3. 2. 1. 1. 1. 1	Sites
3. 2. 1. 1. 1. 2	Buildings

3.0   Space Management Module

Activity class
Activity

Task Category
Task

SubTask

3. 2. 1. 1. 1. 3	Floors
3. 2. 1. 1. 1. 4	Floors by Building
3. 2. 1. 1. 1. 5	Room Standards
3. 2. 1. 1. 1. 6	Room Types by Category
3. 2. 1. 1. 2	*Organization Information*
3. 2. 1. 1. 2. 1	Business Units
3. 2. 1. 1. 2. 2	Divisions
3. 2. 1. 1. 2. 3	Departments
3. 2. 1. 1. 2. 4	Departments by Divisions
3. 2. 1. 2	**Actions**
3. 2. 1. 2. 1	*Add Vertical Penetration and Service Area Room Categories*
3. 2. 1. 3	**Reports**
3. 2. 1. 3. 1	*Space Information*
3. 2. 1. 3. 1. 1	Sites
3. 2. 1. 3. 1. 2	Buildings
3. 2. 1. 3. 1. 3	Buildings by Site
3. 2. 1. 3. 1. 4	Floors
3. 2. 1. 3. 1. 5	Floors by Site
3. 2. 1. 3. 1. 6	Floors by Building
3. 2. 1. 3. 1. 7	Room Standards
3. 2. 1. 3. 1. 8	Room Categories
3. 2. 1. 3. 1. 9	Room Types
3. 2. 1. 3. 1. 10	Room Types by Category
3. 2. 1. 3. 2	*Organization Information*
3. 2. 1. 3. 2. 1	Business Units
3. 2. 1. 3. 2. 2	Divisions
3. 2. 1. 3. 2. 3	Divisions by Business Units
3. 2. 1. 3. 2. 4	Departments
3. 2. 1. 3. 2. 5	Departments by Business Units
3. 2. 1. 3. 2. 6	Departments by Divisions
3. 2. 2	*Gross Areas*
3. 2. 2. 1	**Drawings**
3. 2. 2. 1. 1	*Draw Gross Areas*
3. 2. 2. 2	**Actions**
3. 2. 2. 2. 1	*Update Gross Area Totals*
3. 2. 2. 3	**Reports**
3. 2. 2. 3. 1	*Gross Areas*
3. 2. 2. 3. 2	*Gross Areas by Site*
3. 2. 2. 3. 3	*Gross Areas by Building*
3. 2. 2. 3. 4	*Gross Areas by Floor*
3. 2. 3	*Rooms*
3. 2. 3. 1	**Tables**
3. 2. 3. 1. 1	*Room Standards*
3. 2. 3. 1. 2	*Room Types by Category*
3. 2. 3. 1. 3	*Room Uses*
3. 2. 3. 1. 4	*Rooms*
3. 2. 3. 1. 5	*Designate Common Area Rooms*
3. 2. 3. 2	**Drawings**
3. 2. 3. 2. 1	*Draw Rooms*
3. 2. 3. 3	**Actions**
3. 2. 3. 3. 1	*Update Area Totals*
3. 2. 3. 3. 2	*Perform Chargeback*
3. 2. 3. 3. 3	*Update Room Area from Manual Area*

(continued)

3.0   **Space Management Module**   (*continued*)

Activity class
Activity
Task Category
Task
SubTask

---

3. 2. 3. 4	**Reports**
3. 2. 3. 4. 1	*Rooms*
3. 2. 3. 4. 2	*Rooms by . . .*
3. 2. 3. 4. 2. 1	. . . Site
3. 2. 3. 4. 2. 2	. . . Building
3. 2. 3. 4. 2. 3	. . . Floor
3. 2. 3. 4. 2. 4	. . . Floor per Department
3. 2. 3. 4. 2. 5	. . . Business Unit
3. 2. 3. 4. 2. 6	. . . Division
3. 2. 3. 4. 2. 7	. . . Department
3. 2. 3. 4. 2. 8	. . . Standard
3. 2. 3. 4. 2. 9	. . . Type and Category
3. 2. 3. 4. 2. 10	. . . Use
3. 2. 3. 4. 3	*Room Classifications . . .*
3. 2. 3. 4. 3. 1	Room Standards
3. 2. 3. 4. 3. 2	Room Categories
3. 2. 3. 4. 3. 3	Room Types
3. 2. 3. 4. 3. 4	Room Types by Category
3. 2. 3. 4. 3. 5	Room Uses
3. 2. 3. 4. 4	*Non-Occupiable Rooms*
3. 2. 3. 4. 5	*Non-Occupiable Rooms by . . .*
3. 2. 3. 4. 5. 1	. . . Site
3. 2. 3. 4. 5. 2	. . . Building
3. 2. 3. 4. 5. 3	. . . Floor
3. 2. 3. 4. 5. 4	. . . Business Unit
3. 2. 3. 4. 5. 5	. . . Division
3. 2. 3. 4. 5. 6	. . . Department
3. 2. 3. 4. 5. 7	. . . Standard
3. 2. 3. 4. 5. 8	. . . Type and Category
3. 2. 3. 4. 6	*Occupiable Rooms*
3. 2. 3. 4. 7	*Occupiable Rooms by . . .*
3. 2. 3. 4. 7. 1	. . . Site
3. 2. 3. 4. 7. 2	. . . Building
3. 2. 3. 4. 7. 3	. . . Floor
3. 2. 3. 4. 7. 4	. . . Floor per Department
3. 2. 3. 4. 7. 5	. . . Business Unit
3. 2. 3. 4. 7. 6	. . . Division
3. 2. 3. 4. 7. 7	. . . Department
3. 2. 3. 4. 7. 8	. . . Standard
3. 2. 3. 4. 7. 9	. . . Type and Category
3. 2. 3. 4. 8	*Departmental Analysis . . .*
3. 2. 3. 4. 8. 1	. . . Summary
3. 2. 3. 4. 8. 2	. . . by Building
3. 2. 3. 4. 8. 3	. . . by Floor
3. 2. 3. 4. 8. 4	. . . Location Breakdown
3. 2. 3. 4. 9	*Room Standard Analysis . . .*
3. 2. 3. 4. 9. 1	. . . Summary
3. 2. 3. 4. 9. 2	. . . by Building
3. 2. 3. 4. 9. 3	. . . by Floor
3. 2. 3. 4. 9. 4	. . . by Floor per Department
3. 2. 3. 4. 9. 5	. . . by Department
3. 2. 3. 4. 9. 6	. . . by Department per Floor
3. 2. 3. 4. 9. 7	. . . Area Comparison: Overview
3. 2. 3. 4. 9. 8	. . . Area Comparison: Room-by-Room

3.0   Space Management Module

Activity class
Activity

Task Category
Task

SubTask

3. 2. 3. 4. 10	*Room Type and Category Analysis . . .*
3. 2. 3. 4. 10. 1	. . . Summary
3. 2. 3. 4. 10. 2	. . . by Building
3. 2. 3. 4. 10. 3	. . . by Floor
3. 2. 3. 4. 10. 4	. . . by Floor per Department
3. 2. 3. 4. 10. 5	. . . by Department
3. 2. 3. 4. 10. 6	. . . by Department per Floor
3. 2. 3. 4. 11	*Chargeback Analysis . . .*
3. 2. 3. 4. 11. 1	Prorate Report
3. 2. 3. 4. 11. 2	Detailed Analysis
3. 2. 3. 4. 11. 3	Detailed Analysis - Buildings without Sites
3. 2. 3. 4. 11. 4	Financial Statement
3. 2. 3. 4. 11. 5	Financial Statement - Divisions without Business Units
3. 2. 3. 4. 12	*All Vacant Rooms*
3. 2. 3. 4. 13	*Occupiable Vacant Rooms*
3. 2. 3. 4. 14	*Remaining Area*
3. 2. 3. 5	**Queries**
3. 2. 3. 5. 1	*Highlight Rooms by Department*
3. 2. 3. 5. 2	*Highlight Rooms by Standard*
3. 2. 3. 5. 3	*Highlight Rooms by Category*
3. 2. 3. 5. 4	*Highlight Rooms by Type*
3. 2. 3. 5. 5	*Highlight Common Area Rooms*
3. 2. 3. 5. 6	*Occupancy Plan*
3. 2. 3. 5. 7	*Departmental Stack Plan*
3. 2. 4	*Room Percentages*
3. 2. 4. 1	**Tables**
3. 2. 4. 1. 1	*Allocate Percentages*
3. 2. 4. 2	**Drawings**
3. 2. 4. 2. 1	*Draw Rooms*
3. 2. 4. 3	**Actions**
3. 2. 4. 3. 1	*Synchronize Room and Percentage Records*
3. 2. 4. 3. 2	*Update Area Totals - Space Percentage*
3. 2. 4. 3. 3	*Update Area Totals - Space & Time Percentage*
3. 2. 4. 3. 4	*Perform Percentage Chargeback*
3. 2. 4. 4	**Reports**
3. 2. 4. 4. 1	*Percentages*
3. 2. 4. 4. 2	*Percentages by . . .*
3. 2. 4. 4. 2. 1	. . . Site
3. 2. 4. 4. 2. 2	. . . Building
3. 2. 4. 4. 2. 3	. . . Floor
3. 2. 4. 4. 2. 4	. . . Floor per Department
3. 2. 4. 4. 2. 5	. . . Room
3. 2. 4. 4. 2. 6	. . . Business Unit
3. 2. 4. 4. 2. 7	. . . Division
3. 2. 4. 4. 2. 8	. . . Department
3. 2. 4. 4. 2. 9	. . . Type and Category
3. 2. 4. 4. 3	*Non-Occupiable Percentages*
3. 2. 4. 4. 4	*Non-Occupiable Percentages by . . .*
3. 2. 4. 4. 4. 1	. . . Site
3. 2. 4. 4. 4. 2	. . . Building
3. 2. 4. 4. 4. 3	. . . Floor
3. 2. 4. 4. 4. 4	. . . Business Unit

(continued)

3.0   Space Management Module   *(continued)*

Activity class
   Activity
     Task Category
      Task
        SubTask

3. 2. 4. 4. 4. 5		. . . Division
3. 2. 4. 4. 4. 6		. . . Department
3. 2. 4. 4. 4. 7		. . . Type and Category
3. 2. 4. 4. 5		*Occupiable Percentages*
3. 2. 4. 4. 6		*Occupiable Percentages by . . .*
3. 2. 4. 4. 6. 1		. . . Site
3. 2. 4. 4. 6. 2		. . . Building
3. 2. 4. 4. 6. 3		. . . Floor
3. 2. 4. 4. 6. 4		. . . Floor per Department
3. 2. 4. 4. 6. 5		. . . Room
3. 2. 4. 4. 6. 6		. . . Business Units
3. 2. 4. 4. 6. 7		. . . Division
3. 2. 4. 4. 6. 8		. . . Department
3. 2. 4. 4. 6. 9		. . . Type and Category
3. 2. 4. 4. 7		*Departmental Analysis . . .*
3. 2. 4. 4. 7. 1		. . . Summary
3. 2. 4. 4. 7. 2		. . . by Building
3. 2. 4. 4. 7. 3		. . . by Floor
3. 2. 4. 4. 7. 4		. . . by Location
3. 2. 4. 4. 8		*Standard Analysis . . .*
3. 2. 4. 4. 8. 1		. . . Summary
3. 2. 4. 4. 8. 2		. . . by Building
3. 2. 4. 4. 8. 3		. . . by Floor
3. 2. 4. 4. 8. 4		. . . by Floor per Department
3. 2. 4. 4. 8. 5		. . . by Department
3. 2. 4. 4. 8. 6		. . . by Department per Floor
3. 2. 4. 4. 8. 7		. . . Area Comparison: Overview
3. 2. 4. 4. 8. 8		. . . Area Comparison: Room-by-Room
3. 2. 4. 4. 9		*Type and Category Analysis . . .*
3. 2. 4. 4. 9. 1		. . . Summary
3. 2. 4. 4. 9. 2		. . . by Building
3. 2. 4. 4. 9. 3		. . . by Floor
3. 2. 4. 4. 9. 4		. . . by Floor per Department
3. 2. 4. 4. 9. 5		. . . by Department
3. 2. 4. 4. 9. 6		. . . by Department per Floor
3. 2. 4. 4. 10		*Chargeback Analysis . . .*
3. 2. 4. 4. 10. 1		Prorate Report
3. 2. 4. 4. 10. 2		Detailed Analysis
3. 2. 4. 4. 10. 3		Financial Statement
3. 2. 4. 5	*Queries*	
3. 2. 4. 5. 1		*Departmental Breakdown*
3. 2. 4. 5. 2		*Room Type Breakdown*
3. 2. 4. 5. 3		*Common Area Breakdown*
3. 3	**Personnel**	
3. 3. 1	*NONE*	
3. 3. 1. 1		**Tables**
3. 3. 1. 1. 1		*Employee Standards*
3. 3. 1. 1. 2		*Employees*
3. 3. 1. 2		**Drawings**
3. 3. 1. 2. 1		*Insert Employee Designators*
3. 3. 1. 3		**Actions**
3. 3. 1. 3. 1		*Update Employee Headcounts*
3. 3. 1. 3. 2		*Perform Employee Chargeback - Composite*

3.0   Space Management Module

Activity class
Activity

Task Category
Task

SubTask

3. 3. 1. 3. 3	*Perform Employee Chargeback - All Room*
3. 3. 1. 3. 4	*Infer Room Departments from Employees*
3. 3. 1. 4	**Reports**
3. 3. 1. 4. 1	*Employees*
3. 3. 1. 4. 2	*Employees by . . .*
3. 3. 1. 4. 2. 1	. . . Room
3. 3. 1. 4. 2. 2	. . . Floor
3. 3. 1. 4. 2. 3	. . . Building
3. 3. 1. 4. 2. 4	. . . Site
3. 3. 1. 4. 2. 5	. . . Department
3. 3. 1. 4. 2. 6	. . . Division
3. 3. 1. 4. 2. 7	. . . Business Unit
3. 3. 1. 4. 2. 8	. . . Standard
3. 3. 1. 4. 3	*Employee Standards*
3. 3. 1. 4. 4	*Employee Average Area of . . .*
3. 3. 1. 4. 4. 1	. . . Floors, Buildings and Sites
3. 3. 1. 4. 4. 2	. . . Departments, Divisions, Bus. Units
3. 3. 1. 4. 4. 3	. . . Room Types and Categories
3. 3. 1. 4. 4. 4	. . . Room Standards
3. 3. 1. 4. 4. 5	. . . Employee Standards
3. 3. 1. 4. 5	*Departmental Analysis . . .*
3. 3. 1. 4. 5. 1	. . . Summary
3. 3. 1. 4. 5. 2	. . . by Building
3. 3. 1. 4. 5. 3	. . . by Floor
3. 3. 1. 4. 5. 4	. . . by Location
3. 3. 1. 4. 6	*Employee Standard Analysis . . .*
3. 3. 1. 4. 6. 1	. . . Summary
3. 3. 1. 4. 6. 2	. . . by Building
3. 3. 1. 4. 6. 3	. . . by Floor
3. 3. 1. 4. 6. 4	. . . by Floor per Department
3. 3. 1. 4. 6. 5	. . . by Department
3. 3. 1. 4. 6. 6	. . . by Department per Floor
3. 3. 1. 4. 7	*Chargeback Analysis . . .*
3. 3. 1. 4. 7. 1	Detailed Analysis - Composite
3. 3. 1. 4. 7. 2	Detailed Analysis - Composite - Buildings without Sites
3. 3. 1. 4. 7. 3	Financial Statement - Composite
3. 3. 1. 4. 7. 4	Financial Statement - Composite - Divisions w/o Business Units
3. 3. 1. 4. 7. 5	Detailed Analysis - All Room
3. 3. 1. 4. 7. 6	Detailed Analysis - All Room - Buildings without Sites
3. 3. 1. 4. 7. 7	Financial Statement - All Room
3. 3. 1. 4. 7. 8	Financial Statement - All Room - Divisions w/o Business Units
3. 3. 1. 5	**Queries**
3. 3. 1. 5. 1	*Locate Employee*
3. 3. 1. 5. 2	*Occupancy Plan*
3. 4	**Hoteling**
3. 4. 1	*NONE*
3. 4. 1. 1	**Background**
3. 4. 1. 1. 1	*Buildings*
3. 4. 1. 1. 2	*Floors*
3. 4. 1. 1. 3	*Business Units*
3. 4. 1. 1. 4	*Divisions*
3. 4. 1. 1. 5	*Departments*
3. 4. 1. 1. 6	*Employees*

(continued)

3.0   Space Management Module   *(continued)*

Activity class
Activity
Task Category
Task
SubTask

3. 4. 1. 2	**Tables**
3. 4. 1. 2. 1	*Room Standards*
3. 4. 1. 2. 2	*Rooms*
3. 4. 1. 2. 3	*Room Bookings*
3. 4. 1. 3	**Actions**
3. 4. 1. 3. 1	*Infer Room Bookings Dept. Assignments from Employees*
3. 4. 1. 3. 2	*Book Room(s)*
3. 4. 1. 3. 3	*Cancel Room Booking(s)*
3. 4. 1. 3. 4	*Create Employee Move Orders from Bookings*
3. 4. 1. 3. 5	*Create Space Budget from Bookings*
3. 4. 1. 4	**Reports**
3. 4. 1. 4. 1	*Room Bookings for a Date Range*
3. 4. 1. 4. 2	*Room Bookings with Images for a Date Range*
3. 4. 1. 4. 3	*Rooms Without Bookings for a Date Range*
3. 4. 1. 4. 4	*Rooms Without Bookings with Images for a Date Range*
3. 4. 1. 4. 5	*Room Bookings by Employee*
3. 4. 1. 4. 6	*Room Bookings by Department*
3. 4. 1. 4. 7	*Room Bookings by Department for a Date Range*
3. 4. 1. 5	**Queries**
3. 4. 1. 5. 1	*Highlight Room Bookings for a Date Range*
3. 4. 1. 5. 2	*Highlight Rooms Without Bookings for a Date Range*
3. 5	**Room Reservations**
3. 5. 1	*NONE*
3. 5. 1. 1	**Background**
3. 5. 1. 1. 1	*Buildings*
3. 5. 1. 1. 2	*Floors*
3. 5. 1. 1. 3	*Business Units*
3. 5. 1. 1. 4	*Divisions*
3. 5. 1. 1. 5	*Departments*
3. 5. 1. 1. 6	*Employees*
3. 5. 1. 2	**Tables**
3. 5. 1. 2. 1	*Rooms*
3. 5. 1. 2. 2	*Room Amenity Types*
3. 5. 1. 2. 3	*Room Amenities*
3. 5. 1. 2. 4	*Room Amenities by Room*
3. 5. 1. 2. 5	*Room Reservations*
3. 5. 1. 2. 6	*Room Reservations by Room*
3. 5. 1. 3	**Actions**
3. 5. 1. 3. 1	*Reserve a Room*
3. 5. 1. 3. 2	*Confirm a Room Reservation*
3. 5. 1. 3. 3	*Cancel a Room Reservation*
3. 5. 1. 4	**Reports**
3. 5. 1. 4. 1	*Room Amenity Types*
3. 5. 1. 4. 2	*Room Amenities*
3. 5. 1. 4. 3	*Room Amenities by Room*
3. 5. 1. 4. 4	*Room Reservations*
3. 5. 1. 4. 5	*Reservations by Room*
3. 5. 1. 4. 6	*Reservations for a Date Range*
3. 5. 1. 4. 7	*Reservations for a Date Range with Images*
3. 5. 1. 4. 8	*Reservations for Today*
3. 5. 1. 4. 9	*Unconfirmed Reservations for Today*

3.0   Space Management Module

Activity class
Activity

Task Category
Task

SubTask

3. 5. 1. 5		**Queries**
3. 5. 1. 5. 1		*Highlight Rooms Reserved for a Date Range*
3. 5. 1. 5. 2		*Highlight Rooms Reserved for a Date Range with Images*
3. 5. 1. 5. 3		*Highlight Rooms Reserved for Today*
3. 5. 1. 5. 4		*Reserved Rooms Percent Occupancy for Date Range*
3. 6	**Emergency Preparedness**	
3. 6. 1	*NONE*	
3. 6. 1. 1		**Background**
3. 6. 1. 1. 1		*Buildings*
3. 6. 1. 1. 2		*Floors*
3. 6. 1. 1. 3		*Divisions*
3. 6. 1. 1. 4		*System Types*
3. 6. 1. 2		**Tables**
3. 6. 1. 2. 1		*Advisory Bulletin for Employees*
3. 6. 1. 2. 2		*Advisory Bulletin for Managers*
3. 6. 1. 2. 3		*Emergency Contacts*
3. 6. 1. 2. 4		*Contacts by Building*
3. 6. 1. 2. 5		*Escalation Contacts List*
3. 6. 1. 2. 6		*Recovery Team*
3. 6. 1. 2. 7		*Recovery Team Call List*
3. 6. 1. 2. 8		*Employees and Emergency Information*
3. 6. 1. 2. 9		*Systems*
3. 6. 1. 2. 10		*Systems and Dependent Systems*
3. 6. 1. 2. 11		*Zones*
3. 6. 1. 3		**Drawings**
3. 6. 1. 3. 1		*Draw Egress Plans*
3. 6. 1. 3. 2		*Draw Hazardous Material Plans*
3. 6. 1. 3. 3		*Draw Zones*
3. 6. 1. 3. 4		*Update Room Status by Zone*
3. 6. 1. 3. 5		*Update Equipment Status by Zone*
3. 6. 1. 4		**Actions**
3. 6. 1. 4. 1		*Update System Status*
3. 6. 1. 4. 2		*Update Room Status*
3. 6. 1. 4. 3		*Update Equipment Status*
3. 6. 1. 4. 4		*Update Employee Status . . .*
3. 6. 1. 4. 5		*. . . by Employee Name*
3. 6. 1. 4. 6		*. . . by Division*
3. 6. 1. 4. 7		*. . . by Floor*
3. 6. 1. 5		**Reports**
3. 6. 1. 5. 1		*Advisory Bulletin for Employees*
3. 6. 1. 5. 2		*Advisory Bulletin for Managers*
3. 6. 1. 5. 3		*Emergency Contacts*
3. 6. 1. 5. 4		*Contacts by Building*
3. 6. 1. 5. 5		*Escalation Contacts List*
3. 6. 1. 5. 6		*Recovery Team*
3. 6. 1. 5. 7		*Employees and Emergency Information . . .*
3. 6. 1. 5. 7. 1		. . . by Employee Name
3. 6. 1. 5. 7. 2		. . . by Division
3. 6. 1. 5. 7. 3		. . . by Floor
3. 6. 1. 5. 8		*Systems Status*
3. 6. 1. 5. 9		*Systems and Dependent Systems Status*

(continued)

3.0   **Space Management Module**   (*continued*)

Activity class
    Activity

    Task Category
        Task

            SubTask

3. 6. 1. 5. 10	*Room Status*
3. 6. 1. 5. 11	*Employee Status . . .*
3. 6. 1. 5. 11. 1	*. . . by Employee Name*
3. 6. 1. 5. 11. 2	*. . . by Division*
3. 6. 1. 5. 11. 3	*. . . by Floor*
3. 6. 1. 5. 12	*Equipment Status by Floor*
3. 6. 1. 5. 13	*Site Status*
3. 6. 1. 6	**Queries**
3. 6. 1. 6. 1	*Egress Plans*
3. 6. 1. 6. 2	*Hazardous Materials Plans*
3. 6. 1. 6. 3	*Systems and their Dependent Zones*
3. 6. 1. 6. 4	*Systems, Zones, and Rooms*
3. 6. 1. 6. 5	*Systems, Zones, and Equipment*
3. 7	**Quick Access**
3. 7. 1	*NONE*
3. 7. 1. 1	**Actions**
3. 7. 1. 1. 1	*Locate Employee*
3. 7. 1. 1. 2	*Highlight All Vacant Rooms*
3. 7. 1. 1. 3	*Highlight Groups by Department*
3. 7. 1. 1. 4	*Highlight Rooms by Department*
3. 7. 1. 2	**Executive Reports**
3. 7. 1. 2. 1	*Room Standards Report - Exec*
3. 7. 1. 2. 2	*Department Allocation Report - HQ*
3. 7. 1. 2. 3	*Rooms by Department Chargeback Report*
3. 7. 1. 2. 4	*Highlight All Vacant Rooms*
3. 7. 1. 2. 5	*Locate Employee*
3. 7. 1. 2. 6	*Highlight Groups by Department*
3. 7. 1. 2. 7	*Highlight Rooms by Department*
3. 7. 1. 2. 8	*Departmental Stack Plan - Groups*

REFERENCE NOTES

| ii | *Respect for students* | Bertram Cohler, William Rainey Harper Professor in the Social Sciences, The University of Chicago (quoted in *The University of Chicago Chronicle*, May 27, 1999). |

PREFACE

xv	*Bernard of Chartres*	Norman F. Cantor, *Inventing the Middle Ages*, William Morrow, New York, 1991, p. 13.
xv	*Local services*	From the Start Menu, go to PROGRAMS → ADMINISTRATIVE TOOLS → COMPONENT SERVICES
xviii	*Adolescent software*	Jeffrey E. Clark, "An Architect Designs with the PC," *PC Magazine*, December 1983.
xxi	*The ear that hears. . .*	Proverbs, Chapter 15.
xxiii	*Wayfinding*	CIDA: Council for Interior Design Accreditation (formerly FIDER): Strategies used by people to find their way in both new and familiar settings. Incorporates perceptual and cognitive reaction (mental imaging or mapmaking) to architectural, graphic, visual, aural, and tactile elements present in the environment.
xxvi	*Accessibility Guidelines*	U.S. Architectural and Transportation Barriers Compliance Board (Access Board), A Guide to the New ADA-ABA Accessibility Guidelines, July 2004, http://www.access_board.gov.

1 CLASSICAL PRINCIPLES

1	*Architect a generalist*	Vitruvius, *Ten Books of Architecture* (Morris Hicky Morgan trans., orig. pub. Harvard University Press, 1914), Dover Publications, New York, 1960, Book I, Chapter One.
3	*Inflection points*	The *Domesday Book* was the compilation of a royal census conducted during the reign of William the Conqueror.
4	*Three Roman orders*	Francis D. K. Ching, *ARCHITECTURE Form, Space, & Order*, 2nd ed., John Wiley & Sons, 1996.
6	*Four hours of sunlight, shadows*	This has led to the informal designation of the winter solstice by some of his students as Hilberseimer Day.
7	*Umbilicus of the earth*	Manly P. Hall, *The Secret Teachings of All Ages*, The Philosophical Research Society, Inc., Los Angeles, California, 1978, p. 60.
	Mina Purefoy	James Joyce, *Ulysses*, 2nd printing, Random House, 1961.
	Powder magazine	Stuart Gilbert, *James Joyce's* Ulysses, Alfred A. Knopf and Random House, New York, 1952, pp. 51–56.

8	Space imparts an attitude	Howard Dearstyne, "Notes on the Psychology Lectures of Dr. Karlfried Count von Dürckheim" (1930–31), from Hans Maria Wingler, *The Bauhaus*, The Massachusetts Institute of Technology Press, 1969, p. 159. Dearstyne was Professor of Architecture at the Illinois Institute of Technology during the late 1950s and early 1960s when Mies van der Rohe headed the school.
	Center of the sphere, fire	Aristotle, "De Caelo" ("On the Heavens"—trans. J. L. Stocks), Book II, Chapter 13, *The Basic Works of Aristotle*, Random House, New York, 1941.
9	Copernican model	Dennis Richard Danielson, ed., *The Book of the Cosmos (Imagining the Universe from Heraclitus to Hawking)*, Perseus (Helix Books), Cambridge, Mass., 2000.
	Newton/Galileo	Agnes M. Clerke, *The Concise Knowledge Astronomy*, London, 1898 (quoted in Danielson).
	Black hole singularities	Stephen Hawking, *The Illustrated A Brief History of Time*, Bantam Books, New York, 1996.
	Revelation in a lightning flash	William Butler Yeats, *The Variorum Edition of the Poems of W. B. Yeats* (ed. Peter Allt and Russell K. Alspach), The MacMillan Company, New York, 1971. Quotation: opening lines of *The Second Coming*, p. 401.
10	A man alone	J. Bronowski, *William Blake and the Age of Revolution*, Harper & Row, New York, 1965, p. 5.
	Luther's Theses printed	Jacques Barzun, *From Dawn to Decadence*, HarperCollins, New York, 2000, p. 4.

THE RAMESSEUM

11	Ozymandias	A Roman writing in 60 BCE, Diodorus Siculus (*Histories*, I.47), paraphrased a fellow historian in this manner. Quoted in Joyce Tyldesley, *Ramses, Egypt's Greatest Pharaoh*, Penguin Books, 2000, pp. 5–6.
	No circulation	Sir Banister Fletcher, *A History of Architecture on the Comparative Method*, 17th ed., Charles Scribner's Sons, New York, 1961, p. 44.
	Ramses pharaoh	James B. Prichard, *Ancient Near Eastern Texts Relating to the Old Testament*, 3rd ed. with Supplement, Princeton University Press, 1969, pp. 199–201.
12	Karnak labor force	Charles Freeman, *Egypt Greece and Rome, Civilizations of the Ancient Mediterranean*, 2nd ed., Oxford University Press, 2004, pp. 70, 75.
	Ozymandias (Shelley Sonnet)	Also quoted by Tyldesley.

2 GEOMETRY AND PLANNING

| 13 | Nothing has passed away | Sigmund Freud, *Civilization and Its Discontents* (James Strachey trans.), W.W. Norton & Company, New York, London, 1961, p. 18. |
| 14 | 10,002,290 meters | Ken Alder, *The Measure of All Things*, Free Press (Simon & Schuster), New York, 2002. This book is subtitled *The Seven-Year Odyssey and Hidden Error That Transformed the World* and is a highly |

readable account of the origins of the metric system and the survey that determined the length of the meter.

15   *Pole or Perch*   National Institute of Standards and Technology, Weights and Measures Division, *Handbook 44, Appendix C*, March 2003.

17   *Spiritual wisdom*   Some of the theories discussed in this section are described in a fascinating book titled *The Templars' Secret Treasure* by Erling Haagensen and Henry Lincoln (Barnes & Noble, 2002). The "treasure" referred to in the title is found in the precise geometrical relationships among twelve Romanesque churches on the island of Bornholm (plus a thirteenth on a neighboring island) located in the Baltic Sea. The authors make a convincing case that these churches, ostensibly built around the year 1200 by the Knights Templar, demonstrate a sophisticated knowledge of mathematics, geometry, and surveying and that they were intended as a teaching aid. Haagensen and Lincoln theorize that the island was used to pass on knowledge acquired in the Templars' recent travels to the Holy Land, which included measurement and construction techniques used in building the Gothic cathedrals throughout Europe during the next several centuries.

Standards and regulations   Kaoru Isihikawa, *What Is Total Quality Control? The Japanese Way*, Prentice Hall, Inc., Englewood Cliffs, N. J., 1985, p. 56.

The section, golden ratio   Mario Livio, *The Golden Ratio*, Random House, New York, 2002, pp. 5, 62.

18   *Perfect proportions, human figure*   Adolf Zeising, *Neue Lehre von den Proportionen des menschlichen Körpers aus einem bisher unerkannt gebliebenen, die ganze Natur und Kunst durchdringenden morphologischen Grundgesetze entwickelt und mit kieiner vollständigen historischen Übersicht der bisherigen Systeme begleitet* [A new teaching Doctrine of the Proportions of the Human Body, until now unknown, that all Nature and Art possess morphological arrangements, with a brief historical overview from present systems], Weigel, Leipzig, 1854.

19   *Containers/extensions of man*   Le Corbusier (Charles Edouard Jeanneret), *Modulor I and II*, Harvard University Press, Cambridge, Mass., 1980 (*Modulor I*), p. 60.

21   *Metre indifferent to man*   *Modulor I*, pp. 19–20.

His man is a woman . . . brrrh!   *Modulor II*, pp. 20, 52.

22   *Einstein's assessment*   *Modulor I*, p. 58.

23   *Statures of man and woman*   Alvin R. Tilley and Henry Dreyfuss Associates, *The Measure of Man and Woman (Human Factors in Design)*, rev. ed., John Wiley & Sons, New York, 2002.

24   *Three basic forms/diversity of angles*   Johannes Itten, *Design and Form, The Basic Course at the Bauhaus*, trans. John Maass, Reinhold, New York, 1964.

25   *Visualization of man*   Oskar Schlemmer, "Man and Artistic Figure" in "Die Bühne im Bauhaus (The Stage in the Bauhaus)" from Wingler, *The Bauhaus*, p. 119.

	Historical styles of ornament	Owen Jones, *The Grammar of Ornament*, Bernard Quaritch, London, 1910 (Figure 2-12 after diagrams on page 73—see also Plate One).
26	*I needed exercise to straighten up . . .*	Frank Lloyd Wright, *An Autobiography*, Horizon Press, New York, 1943.
27	*To find the true path*	Owen Jones, pp. 1–2.
	This will kill that	Victor Hugo, *Notre-Dame de Paris* (*The Hunchback of Notre Dame*), Paris, 1832. There are three chapters within this novel devoted to architecture: Book Three—Chapters One and Two, Book Five—Chapter Two, as well as significant sections of other chapters.
28	*Stages of development*	Jacques Barzun, *From Dawn to Decadence*, p. xx.
	Principle of tolerance	J. Bronowski, *The Ascent of Man* (television series, 1974), Episode 11, "Knowledge or Certainty." Also Bronowski, *The Ascent of Man*, Little, Brown and Company, Boston/Toronto, 1973, pp. 365, 367.
	Less is more	Sharon M. Kaye and Robert M. Martin, *On Ockham*, Wadsworth/Thomson Learning, Belmont, Calif., 2001. Passage quoted from Ockham's *Opera Theologica*, X, 157. The passage often given as Ockham's razor is: *Entia non sunt multiplicanda praeter necesitatem*, which literally translates as "Beings should not be multiplied beyond necessity."

THE PARTHENON

30	*East and west ends*	Banister Fletcher, p. 121.
	Metopes	Banister Fletcher, p. 123.
	Craftsmen were metics	Freeman, p. 254.
	Metics an economic force	Robert Flaceliere, *Daily Life in Greece at the Time of Pericles* (trans. from French by Peter Green), Phoenix Press, London, 2002, pp. 42–43.
	Annual festival on frieze	Flaceliere, pp. 198–199.
	Council of Five Hundred	Freeman, pp. 252–253.
	Written contracts	Freeman, p. 257.
	Pericles, democratic pride	Freeman, p. 258.
31	*Athena Parthenos*	Banister Fletcher, p. 119.
	Principles of proportion	Banister Fletcher, p. 377.
	Golden sections, squares in plan	One is reminded here of Le Corbusier's "place of the right angle" shown in Figure 1-10.

3 PLACE AND MODERNISM

33	*Preserve the ancient monuments*	Victor Hugo, *Notre-Dame de Paris* (*The Hunchback of Notre Dame*), Paris, 1832.
	Howard Roark laughed	Ayn Rand, *The Fountainhead*, Bobbs-Merrill, New York, 1943.
34	*Books of architecture*	Leon Battista Alberti, *On the Art of Building in Ten Books*, MIT Press, 1991. Andrea Palladio, *The Four Books of Architecture*, Dover—Publications, New York, 1965.
		Sebastino Serlio, *The Five Books of Architecture*, Dover, New York, 1982.

	Descartes' dream(s)	Jacques Barzun, *From Dawn to Decadence*, HarperCollins, New York, p. 201. Barzun says regarding Descartes' experience that "A curious psychological fact is that his first impulse to build a system came to him in a dream, or rather a nightmare. In it he was possessed by a genius and overcome by a dazzling light that suggested to him that he would be given answers to the questions he had been wrestling with . . . The answer Descartes gave to the question in his dream was: unify all knowledge by the use of exact reasoning, the kind used in geometry—the world mathematicized." Barzun goes on to say that there followed three dreams full of disparate images. Sir Anthony Kenny, in his *The Rise of Modern Philosophy* (Oxford, 2006, p. 34) says: "[Descartes'] conviction of [a philosophical] vocation was reinforced when, that night [winter 1619], he had three dreams that he regarded as prophetic."
35	Robert Taylor Homes	Thomas Stearns Eliot, "The Hollow Men," 1925. It should not pass without comment that Le Corbusier is reported to have said that all buildings should be white by law and was critical of the use of ornamentation. Earlier in the poem, Eliot writes, "Shape without form, shade without color, paralyzed force, gesture without motion . . ."
36	Ugliness, incongruity, and incoherence	Le Corbusier, *Quand les cathédrales étaient blanches* (*When the Cathedrals Were White*), 1937, quoted in *Le Corbusier, Architect, Painter, Poet*, Jean Jenger, Harry N. Abrams, New York, 1996, p. 116.
	Principle of tolerance	Oliver Cromwell, Lord Protector of the Commonwealth of England, Scotland, and Ireland, "Letter to the General Assembly of the Church of Scotland, August 3, 1650. Quoted by J. Bronowski, *The Ascent of Man*, Little, Brown and Company, Boston/Toronto, 1973, p. 374.
	Expression of style	The author's recollection of Professor Caldwell's lectures on art history. Alfred Caldwell was a student of Ludwig Hilberseimer, who introduced him to Mies van der Rohe. Caldwell taught several of the basic construction courses during the 1950s and designed the landscaping for a number of Mies van der Rohe's projects. He had also studied with Wright at Taliesin and with Jens Jensen, the noted Danish landscape architect.
37	Two critical factors	Robert Venturi, *Complexity and Contradiction in Architecture*, 2nd ed., Museum of Modern Art, New York, 2002.
	Both-and versus either-or	Cleanth Brooks, *The Well Wrought Urn, Studies in the Structure of Poetry*, Harcourt Brace & Company, 1942.
	More is not less	Robert Venturi, *Complexity and Contradiction*.
38	Modular elements	Le Corbusier, *Modulor I*, p. 92.
	Time had flown	Le Corbusier, *Modulor I*, p. 101.

40	*Beauty brings repose of spirit*	Le Corbusier, *L'Almanach d'architecture moderne* [*The Journal of Modern Architecture*] (1925), quoted in *Le Corbusier, Architect, Painter, Poet*, Jean Jenger, Harry N. Abrams, New York, 1996, p. 129.
	Five points	Jean Jenger, *Le Corbusier*, p. 65.
43	*Mies van der Rohe thought it through*	*Oral History of Alfred Caldwell*, interviewed by Betty J. Blum, Chicago Architects, The Art Institute of Chicago Department of Architecture, 2001.
	A lifetime articulating	These insights are articulated as resulting from four events that occurred roughly ten years apart in the course of Mies van der Rohe's career, as described by Kevin Harrington, professor of architectural history at IIT, in a paper titled "I Gave Myself a Shock: Mies and the Pavilion," given at a symposium at the Politecnico di Milano in 1997.
44	*Quality of serenity*	Maritz Vandenberg, *Farnsworth House*, Phaidon Press Limited, London, 2003, p. 5. Until recently, Lord Peter Palumbo was the owner of the house that he purchased from Edith Farnsworth in 1972. Lord Palumbo wrote the foreword to the book, from which this quote is taken. The *pilotis* upon which the house stands unfortunately did not prevent its being inundated by the waters of the adjacent river on more than one occasion.
46	*Nine principles*	Frank Lloyd Wright, *A Testament*, Horizon Press, New York, 1957.
48	*Dawn seen or sensed*	Lord Peter Palumbo, in *Farnsworth House*.
	A dysfunctional family	Franklin Toker, *Fallingwater Rising*, Alfred A. Knopf, New York, 2003, p. 149.

A RESIDENCE AT DELOS

| 50 | *Second-floor gallery* | Banister Fletcher, pp. 152–153. |
| | *Ad Quadratum: of the square* | Vitruvius, *Ten Books of Architecture* (Morris Hicky Morgan trans. orig. pub. Harvard University Press, 1914), Dover Publications, New York, 1960, Book IX, Introduction, pp. 4–5. |

4 DWELLINGS

| 51 | *In preparing for battle . . .* | Favorite maxim of Dwight D. Eisenhower, thirty-fourth president of the United States (1953–1961), quoted by Richard M. Nixon in *Six Crises* (1962). |

5 MANAGEMENT AND NETWORKS

68	*Working masses*	Peter Drucker, *Management: Tasks, Responsibilities, Practices*, Harper & Row, 1973 (quoted in *Business: The Ultimate Resource*, Perseus Books Group, 2002, p. 1054).
	With a watch in his hand	John Dos Passos, *The Big Money* (third volume of the *USA Trilogy*), 1933.
	Bethlehem Steel	Frederick Winslow Taylor, *The Principles of Scientific Management*, 1911 (quoted in *Business*, Perseus).

	Principal object of management	Taylor, *Principles.*
	Human nature	Henry Laurence Gantt, *Work, Wages, and Profits,* New York Engineering Magazine Co., 1910 (quoted in *Business,* Perseus, p. 990).
	W. Edwards Deming	Kaoru Ishikawa, *What Is Total Quality Control? The Japanese Way,* trans. David J. Lu, Prentice Hall, 1985.
	Manufacturing process	Nancy R. Mann, Ph.D., *The Keys to Excellence, The Story of the Deming Philosophy,* Prestwick Books, Los Angeles, 1989.
69	*Corporate culture*	Edgar Schein, *Organizational Culture and Leadership,* 2nd ed., Jossey-Bass, San Francisco, 1999.
	Intellectual property export	Source: World Trade Organization, World Intellectual Property Organization (patents).
74	*Cofounder of Intel*	Andrew S. Grove, *Only the Paranoid Survive. How to Exploit the Crisis Points That Challenge Every Company and Career,* Doubleday, New York, 1996, pp. 3, 95.
75	*Tim Berners-Lee*	"[A] researcher named Tim Berners-Lee at the CERN atomic research center in Switzerland proposed software and networking protocols in 1989...This effort gained momentum by 1993, culminating in the development of the critical piece of software called a Web browser." Robert H. Reid, *Architects of the Web, 1,000 Days That Built the Future of Business,* John Wiley & Sons, New York, 1997, p. xxiii. See also Tim Berners-Lee, *Weaving the Web, The Original Design and Ultimate Destiny of the World Wide Web by Its Inventor,* HarperCollins, New York, 1999.
77	*Program (programme)*	*The Concise Oxford Dictionary of Current English,* 8th ed., Clarendon Press, Oxford, 1991, p. 953.
	Program definition process	William M. Pena and Steven A. Parshall, *Problem Seeking, An Architectural Programming Primer,* 4th ed., John Wiley & Sons, 2001.
88	*If standards are not revised in six months . . .*	Kaoru Ishikawa, *What Is Total Quality Control? The Japanese Way,* trans. David J. Lu, Prentice Hall, Englewood Cliffs, N.J., 1985, p. 56. The Quality and Facility Management section of this chapter is based on the principles set forth in the section titled "How to Proceed with Control" in Chapter III of Dr. Ishikawa's book (pp. 59–79).

THE PANTHEON

90	*Largest dome for 400 years*	Until Brunelleschi's dome on the Cathedral of Florence, built between 1420 and 1436.
	It has been theorized	Rudolpho Lanciani, *The Ruins and Excavations of Ancient Rome,* 1897.
	Hidden buttressing	Banister Fletcher, pp. 178–179.

6 WORKPLACE PROGRAMMING

93	*Emancipation of humanity*	Simone Weil, *Gravity and Grace,* Routledge, 2004, p. 154.
	Gothic architecture	Max Weber, *The Protestant Ethic and the Spirit of Capitalism* (author's introduction), Routledge

		Classics, 2001 (first published as a two-part article in 1904, trans. Talcott Parsons in 1930).
	The most fateful force	Weber, p. xxxi.
	Rational pursuit of profit	Weber, p. xxxii.
94	*Bureaucracy-based organization*	"Max Weber: The Conceptualization of Bureaucracy," article in *Business: The Ultimate Resource*, Perseus Books Group, 2002, pp. 1060–1061.
	Accelerating technological growth	Gary Hamel and C. K. Prahalad, *Competing for the Future*, Harvard Business School Press, Boston, Mass. 1994.
	Chief information officer	In our love of acronyms, these are often encrypted as CEO, COO, CFO, and CIO.
95	*Vector analysis of adjacencies*	The program was called *CompuGraph* and was based on a mathematical vector analysis technique published in a scientific journal called *Biometrica*. The application was developed by a Boston architectural firm and marketed by Computer Sciences Corporation on a time-sharing computer network in the late 1960s.
	Corporation classifications	*Forbes Magazine*, "The World's 2000 Leading Companies," April 12, 2004.
96	*Level of chaos*	The term *fractal* was coined in 1975 by Benoit B. Mandelbrot, an IBM scientist. It derives from the Latin word *fractus*, which means "to break." Fractals are rough or fragmented geometric shapes that can be subdivided with each part being (if only approximately) a copy of the whole at a smaller scale. The term is also used to describe many structures that do not have simple descriptions or shapes, such as complex matrix organizations.
99	*PEFIC-FICM*	*Postsecondary Education Facilities Inventory and Classification Manual (FICM)*, Working Group on Postsecondary Physical Facilities, National Center for Education Statistics (NCES), July 1992.
100	*ASTM/IFMA standards*	ASTM International (originally the American Society for Testing and Materials) and the International Facility Managers Association.
	Facility usable area	ASTM-E1836, A1.12.1.
101	*Common support area*	ASTM-E1836, A1.18.1.
102	*Primary circulation*	ASTM-E1836, A1.15.1,4.
105	*Data structure representation*	Every advanced AutoCAD user will immediately recognize this format as that of one of its built-in programming languages: Autolisp, the structure of which consists of functions within functions within functions. . . .
	Tree structure representation	After Nicklaus Wirth, *Algorithms + Data Structures = Programs*, Prentice Hall, Englewood Cliffs, N. J., 1976, p. 190.
110	*Hay point evaluation system*	The *Hay Group* was founded in 1943 in Philadelphia and is among the top five worldwide consulting firms whose primary focus is human resources.
	Accountability and problem-solving	It is noteworthy that the Federal Equal Pay Act of 1963 defines the factors upon which equal

work should be based as skill, responsibility, effort, and working conditions.

CATHEDRALS AND THE MEDIEVAL PROGRAM

119	*Well-being of the realm*	Robert A. Scott, *The Gothic Enterprise, A Guide to Understanding the Medieval Cathedral*, University of California Press, 2003, pp. 48–60.
	Economic growth and Christianity	Scott, p. 52.
120	*Suppression of "heresy"*	Joseph R. Strayer, *The Albigensian Crusades*, The University of Michigan Press, Ann Arbor, Mich., 1992, pp. 146–147.
	Templar network	*Atlas of Medieval Europe*, ed. Angus Mackay with David Ditchburn, Routledge, London and New York, 1997: M. C. Barber, *The Templar Network*, p. 92.
	Burned to death in 1314	Malcolm Barber, *The New Knighthood, A History of the Order of the Temple*, Cambridge University Press, 1994, pp. 199–200, 280.
	Satisfying in its harmony	C. S. Lewis, *The Discarded Image*, Cambridge University Press, 1964, p. 99.
	Resonant with music	Lewis, p. 112.
	Four contraries	Lewis, p. 94.
	Dull, sluggish, and pale	Lewis, pp. 169–173.
121	*Card index*	Lewis, p. 10.
	Masters of the orthodoxy	Scott, pp. 74–75. Scott here cites Georges Duby. *The Three Orders*, University of Chicago Press, 1980, and *The Age of the Cathedrals: Art and Society, 980–1420*, University of Chicago Press, 1981.
		[A luminous sphere, radiating light, describes what we now know as the Cosmic Microwave Background, which is the visible horizon of the expanding universe. The CMB was accidentally discovered by Arno Penzias and Robert Wilson, working at Bell Labs in 1964. It provides evidence that the universe began with a hot big bang, approximately 13.7 billion years ago.]
123	*Octagonal chapter house*	The chapter house provided a meeting place for transacting ecclesiastical business and was often adjacent to the cloister. In English cathedrals, as at Salisbury, the chapter house was often octagonal, with its vaulting supported by a central column.
	Profane outer world	Scott, pp. 166–170.
	Form from Muslim art	Banister Fletcher, pp. 343, 358, 361, 371. Possibly the earliest use of the pointed arch in Europe was in the Romanesque Abbey Church at Cluny, France (1089–1131), where it was employed in the nave arcades. Another example can be found in Laach Abbey (German Romanesque, 1093–1156), which shows advances toward the Gothic system in both its plan and the use of the pointed arch in several of its larger windows.
124	*Execution of the idea*	L. R. Shelby, *The Role of the Master Mason in Medieval English Building* (*Speculum*, vol. 39, no. 3, 1964), quoted in Donald Hill, *A History of*

Engineering in Classical and Medieval Times, Barnes & Noble Books, New York, 1997, pp. 112–113.

Increased buttressing

Scott, pp. 125–126, quoting historian Christopher Wilson in *The Gothic Cathedral*, Thames and Hudson, London, 1990.

7 SPACE ALLOCATION

| 127 | Component object model | The component object model, or COM, establishes a standardized means by which one piece of software can call upon another for services. This sharing of objects is accomplished using Microsoft's COM technology through a process known as Automation. AutoCAD's object model is constructed according to the rules of COM, as are Access and Excel, which allows all three applications to communicate directly with one another. |

For more information see Jeffrey E. Clark, *VBA for AutoCAD 2002—Writing AutoCAD Macros*, Prentice Hall PTR, 2002.

SQL script

Virtually all relational databases are constructed using a dialect of *Structured Query Language*, or *SQL. SQL* is used to construct different views of the tables in the database. The language also supports scripting procedures that allow the database tables to be automatically populated with properly formatted data.

Pipeline the information

There are many different relational database products. Microsoft Access is frequently used for small-scale, stand-alone databases although it is capable of being used on a network and supports Web page access. Large-scale database systems include products such as IBM Informix, SQL Server (Microsoft's), Sybase, and Oracle, arguably the market leader among enterprise-level databases.

| 128 | Group table | Many applications have a set of *reserved* words that are used internally by the software and thus cannot be used as table names or as variables. The word *group* is an Access reserved word so the table is named *Groupe* to prevent conflicts (see the window title bar in Figure 7-3). |

Building and floor codes

Relational database tables must be populated starting from the top of the hierarchy and working downward. Conversely, once an entry has been created in a low-level table (such as the *Detail* table) that uses any of the codes in the validating tables above it, you cannot delete those codes from the validating tables. Validated data in relational database tables must always be *deleted* starting from the bottom of the schema and working upward, unless cascading deletes are permitted, which is not generally a good idea (See also Chapter 10).

| 130 | Cutting and pasting | If data are set up correctly, they can be imported to an Access table simply by cutting and pasting them from an Excel worksheet. This is accomplished by highlighting the worksheet (or a |

desired range of cells) and copying them to the Windows clipboard from whence they are pasted to the desired location. The data must be valid and in *exactly* the correct format or errors will result. If some of the data are valid and some are not, Access will create a table named Paste Errors for the invalid records.

134	*Must be done in two steps*	If the report is exported directly to Excel, which the Access File → Export function does permit, all that is obtained is a single row containing the field names used for the *Detail* table. This occurs because Access attempts to export the underlying table data using Excel's *Subtotals* format (see Figure 6-5 and its accompanying explanation). Since the *Block Planning Data* report only contains a tabular list of subtotals, there is no detail for Access to export in this format. [To understand what is really happening here, try exporting one of the other reports from Access to Excel in this manner.]
137	*Snap to intersection*	*Object Snap* is a feature of AutoCAD that allows an object (line, circle, etc.) to be precisely joined to an exact location on another object, such as its endpoint, midpoint, or the intersection of two such objects.
138	*MODIFY → OBJECT → POLYLINE*	*Polylines* are AutoCAD objects consisting of one or more connected line segments or arcs. The PEDIT command allows you to edit polylines (also referred to as *plines*) and to join discrete lines and arcs making them into *plines*.
	Hyperlink text	AutoCAD's EXPLODE command, among other things, transforms a polyline into its constituent *line* and *arc* segments. When a polyline is exploded, it loses any *hyperlinks* that may have been associated with it.
	Circles, ellipses, and polylines	*Polylines*, like other enclosed objects such as *circles* and *ellipses*, have many highly useful properties such as *area*.
139	*AutoCAD layer name*	*Layers*, in AutoCAD, are a means of separating objects so that their properties, such as color, line type, line weight, and visibility, can be independently controlled. All layers have names, and all objects have a layer property.
	Incorrect area results	Areas can be calculated for polylines that are not closed, but the results may be misleading. When an open polyline is selected, AutoCAD calculates the area as if the figure were closed, assuming a straight segment added from the endpoint back to the starting point of the pline.

THREE FACILITY PROGRAMS

151	*An oblique angle*	3DORBIT, DVIEW, and VPOINT, are AutoCAD commands that allow you to create a three-dimensional visualization of your drawing.

8 SPACE PLANNING

153	*Dodecathedral church*	M. C. Escher, *Perspective, Escher on Escher (Exploring the Infinite)*, Harry N. Abrams, 1989, p. 129.

158	*Has just been completed*	The three buildings, coincidentally, are all located on Wacker Drive in Chicago, Illinois. The thirty-story-high Mies van der Rohe building, *One Illinois Center*, was built in 1970 at 111 East Wacker. Figure 8-4*a* shows the forty-eight-floor *Hyatt Center* at 71 South Wacker Drive, designed by Pei Cobb Freed & Partners and completed in 2005. Figure 8-4*b* is *Wacker Plaza*, traditionally known as the *Builders Building*, located at 222 North LaSalle. It was designed in the neoclassical style by Graham, Anderson, Probst, & White and completed in 1927. Wacker Plaza was expanded and extensively renovated in 1986 by the firm of Skidmore, Owings & Merrill. A western addition was made equal in height to the original building, and a four-story tinted glass penthouse was added spanning the length of both structures, resulting in a total of twenty-six floors.
160	*Facility interior gross area*	ASTM-E1836, A1.2.4.2. ASTM-E1836 is the standard used by IFMA for classifying spaces, which we refer to herein as ASTM/IFMA. This standard is discussed in considerably more detail in Chapter 9, "Area Measurement Standards."
164	*Building Owners and Managers Association*	The BOMA Standard is ANSI/BOMA Z65.1-1996, *Standard Method for Measuring Floor Area in Office Buildings* (approved June 7, 1996, by the American National Standards Institute, Inc.).
165	*AIA layer naming*	American Institute of Architects, *CAD Layer Guidelines: Recommended Designations for Architecture, Engineering, and Facility Management Computer-Aided Design*, 1990, and *Computer-Aided Design Management Techniques for Architecture, Engineering, and Facility Management*, 2nd ed., 1997, The American Institute of Architects Press, 1735 New York Avenue, N.W., Washington, DC 20006.

CATHEDRALS OF TODAY

| 176 | *Typological analysis of Frank Lloyd Wright* | *Paul Laseau and James Tice, Frank Lloyd Wright: Between Principle and Form, Van Nostrand Reinhold, 1992.* |

9 AREA-MEASUREMENT STANDARDS

183	*Naming of Cats*	T. S. Eliot, "The Naming of Cats," *Old Possum's Book of Practical Cats*, Harcourt Brace & Company, 1939.
	Tower of Babel	Ibn Ezra, Malbim, *The Torah, A Modern Commentary*, ed. W. Gunther Plaut, Union of American Hebrew Congregations, New York, 1981, p. 85.
184	*New BOMA*	ANSI/BOMA Z65.1-1996, *Standard Method for Measuring Floor Area in Office Buildings* (approved June 7, 1996, by the American National Standards Institute, Inc.).
185	*Floor usable area*	ANSI/BOMA, 16.
186	*Gross building area*	ANSI/BOMA, 10.

186	*Inside finished surface*	ANSI/BOMA, 2.
188	*For record keeping*	ANSI/BOMA, 4.
192	*ASTM-E1836 (IFMA)*	ASTM Designation: E1836-01, *Standard Classification for Building Floor Area Measurements for Facility Management* (ASTM International, Committee E06 on Performance of Buildings, subcommittee E06.25 on Whole Buildings and Facilities).
	Facility area relationships	ASTM-E1836 definitions include the word *facility* in front of all terms relating to interior area. We have omitted them here.
	Bridges and tunnels	ASTM-E1836, Annex A1, Section A1.5.
	Interior gross area	ASTM-E1836, A1.6.1.
193	*Finished interior surface*	ASTM-E1836, A1.7.1.
	Finished surface	ASTM-E1836, A1.3.3.
	Where walls and floors intersect	ASTM-E1836, A1.4.1.1.
	Basis for lease agreements	ASTM-E1836, A1.8.1 and A1.2.4.3.
	Facility rentable	ASTM-E1836, A1.8.3.
	Identical to that of BOMA	ASTM-E1836, A1.9.
	Primary circulation	ANSI/BOMA, 3.
194	*Two ways of measuring*	ASTM-E1836, A1.14.
	Facility usable area	ASTM-E1836, A1.12.1.
195	*Facility assignable*	ASTM-E1836, A1.18.1.
	GSA area measurement	PBS Assignment Drawing Guidance (draft published March 1, 2000, U.S. General Services Administration, Public Buildings Service National CIFM Center, 1800 F Street NW, Room 5026, Washington, DC 20405, Internet http://www.gsa.gov/pbs/cifm).
	PBS CAD standards	*Standard for PBS CAD Deliverables* (October 2001, National CIFM Center, Washington, DC 20405).
197	*Building joint use*	Conference rooms, cafeterias, snack bars, credit unions, etc.
	Zero square feet	Antennas, boat docks, bridges, land, railroad crossings, and wareyards.
	PEFIC-FICM	*Postsecondary Education Facilities Inventory and Classification Manual (FICM)*, Working Group on Postsecondary Physical Facilities, National Center for Education Statistics (NCES), July 1992.
199	*Gross area*	PEFIC, p. 27.
200	*Structural/construction area*	*Federal Construction Council Technical Report No. 50* (Publication 1235), Classification of Building Areas, National Academy of Sciences, Building Research Advisory Board.
	Nonassignable categories	PEFIC, pp. 31–33.
	Mechanical area	PEFIC, pp. 32–33.
	Circulation area	PEFIC, p. 32.
201	*Standardization for comparison*	PEFIC, p. 38.
203	*Room suitability*	PEFIC, pp. 96–105 (Appendixes 2–5).
204	*Pro rata space allocation*	ANSI/BOMA, p. v.

GLOBAL TECHNOLOGY COMPANY

206	*Application integration specialist*	Computerized Facility Integration, L.L.C., of Southfield, Mich., and Chicago, Ill.; http://www.GoCFI.com.
207	*Web interface*	*Facility Asset and Space Tracking* (*FAST*) is an interface developed by CFI that is compatible with many leading facility database systems, including Archibus/FM, TRIRIGA, FAMIS, MAXIMO, Oracle, MS Access, and several others.

10 RELATIONAL DATABASES

209	*The card index*	C. S. Lewis, *The Discarded Image*, Cambridge University Press, 1964, p. 10.
	Relational-database model	*Communications of the ACM*, Vol. 13, No. 6, June 1970.
	Art of Computer Programming	Donald Knuth, *The Art of Computer Programming*, Addison-Wesley, 1973.
210	*Sorting with tape*	Knuth, *The Art of Computer Programming—Volume III, Sorting and Searching*, Section 5.4, "External Sorting," Addison-Wesley, 1973, pp. 247–378.
	Character sets	ASCII, American Standard Code for Information Interchange.
211	*Oracle Corporation*	Source: http://www.oracle.com.
	Codd's ideal database	*Computerworld* articles: "Is Your DBMS Really Relational?," October 14, 1985, and "Does Your DBMS Run By the Rules?," October 21, 1985.
212	*Domains and tuples*	The *Shorter Oxford English Dictionary* does not recognize the word *tuple*, short for *n-tuple*, except as a suffix. It gives the noun form as *tuplet*: "An entity or set with a given number of elements . . .," which is essentially the meaning of *record* as C. J. Date uses it. Date points out that the Codd's formal relational model does not use the term *record* at all because it can be ambiguous. At different times, the word can refer to a record occurrence or a record type; a logical, physical, or virtual record; and so forth. He equates the notion of a *flat record instance* to *tuple* in the same way that *relation* corresponds approximately to the idea of a table. (C. J. Date, *An Introduction to Database Systems*, 6th ed., Addison-Wesley, 1995, p. 57.)
213	*Three-valued logic*	Date, *IDBS*, pp. 123–124 and Chapter 20.
	Abstraction level of SQL	Date, *IDBS*, pp. 58–59.
214	*SQL key words*	C. J. Date with Hugh Darwen, *A Guide to The SQL Standard*, 3rd ed., Addison-Wesley, 1994, pp. 31–32 and Appendix A.
216	*Control of illegal data changes*	A trigger is a stored "procedure that is to be invoked when a specified trigger condition occurs . . . [its purpose] is to carry out a certain compensating action to bring the database back into a state of integrity again." (Date, *IDBS*, p. 453.)
217	*Four properties of all relations*	Date, *IDBS*, pp. 86, 91–93.

219	*Freedom from anomalies*	Date, *IDBS*, p. 334.
	Candidate keys	Date, *IDBS*, Section 5.2., pp. 112–115.
221	*Projection-join dependencies*	Date, *IDBS*, p. 332.
223	*Inner (natural) join*	Date gives the following definition of the *natural join:* It "returns a relation consisting of all possible tuples that are a combination of two tuples, one from each of two specified relations, such that the two tuples contributing to any given combination have a common value for the common attribute(s) of the two relations (and that common value appears just once, not twice, in the result tuple)." *IDBS*, p. 141.
224	*SQL statements in VBA*	A full explanation of this dialog box and its operation can be found in Chapter 19 of *VBA for AutoCAD 2002: Writing AutoCAD Macros*, Jeffrey E. Clark, Prentice Hall PTR, 2002, pp. 581–606. (Note: The procedures function properly with the current version as of this writing: *AutoCAD 2007.*)
	Formalized common sense	Date, *IDBS*, p. 309.
225	*Deadlock*	The concurrency problem was addressed in anecdotal form by Edsger Dijkstra in the early 1970s. His original version involved spaghetti and forks, but since most folks can manage to eat spaghetti with one fork, the story is now often told using Chinese food and chopsticks.
		The story runs thus:
		Five philosophers sit around a table, alternately thinking and eating. In the center of the round table is an unlimited supply of Chinese food. In front of each philosopher is a plate and between each pair of plates is a single chopstick. In order to eat, a philosopher must obtain the two chopsticks on his or her right and left. This calls for a certain amount of cooperation because if each philosopher is greedy and hangs on to one chopstick, all will starve.
		Dijkstra, E. W., *Hierarchical Ordering of Sequential Processes, Acta Informatica*, Volume 1, 1971. The text of Dijkstra's original paper is available at http://www.cs.utexas.edu.
226	*Web page types*	HTML (Hypertext Markup Language) is the standard language in which Web pages are written. The standard is maintained by the World Wide Web Consortium at http://www.w3.org/.

INTERNATIONAL CHEMICAL COMPANY

229	*Application integration specialist*	Computerized Facility Integration, L.L.C., of Southfield, Mich., and Chicago, Ill.; http://www.GoCFI.com.
230	*Market leader survey*	*2005 Market Update: Integrated Workplace Management Systems;* Gartner Inc., Michael A Bell, May 24, 2005, http://www.gartner.com. Products are comparatively evaluated on two-dimensional scales of completeness of vision and ability to execute.

11 COMPUTER-AIDED FACILITY MANAGEMENT

234	TRIRIGA	TRIRIGA, Las Vegas, Nev. Facilities http://www.tririga.com/".
235	*FAMIS*	*Famis Software, Inc.*, Irvine, Calif., http://www.famis.com.
	ODBC	ODBC (Open Database Connectivity) refers to a Microsoft protocol that has become a de facto standard for the data and methods needed to access that data connecting programs to their data sources. ODBC accomplishes this by means of an intermediate procedure (driver) that connects to the data source. *Machine* data sources are stored in the Windows Registry and can only be used by those with authorized access to the machine on which they are defined. *File* data sources are text files that can be used on any machine that has an appropriate ODBC driver.
238	*Archibus/FM*	*Archibus/FM*, Facilities Management Techniques, Inc., Boston, Mass., http://www.archibus.com.
240	*Space budgets*	*Strategic Master Planning*, Archibus/FM 14, Facilities Management Techniques, Inc., Boston, Mass., 2003, p. 77.
244	*Enhanced BOMA method*	SMP, Archibus/FM, p. 59.
246	*Extended entity data*	Extended entity data are *instance-specific* data (one record per object) attached to objects in the AutoCAD drawing database, used by external applications. They provide a powerful tool for utilizing AutoCAD's database capabilities. With XData, it is possible to effectively tie the application's graphic objects to enterprise-level database systems such as Oracle, Microsoft SQL Server.
	Drawing-driven asset data	Entity handle: A unique alphanumeric tag assigned to each object in the AutoCAD drawing database.
251	*Telecommunications infrastructure*	*The Siemon Company*, Watertown, Conn., http://www.siemon.com. (Also, Plate Eight—illustrations courtesy of The Siemon Company, all rights reserved.)
262	*Drawing publishing*	DWF format: Drawing Web Format is a vector-based image format from Autodesk that allows you to view both two- and three-dimensional drawings.

BP AMERICA

264	*British Petroleum and Amoco Corporation*	The information contained herein regarding the leveraging of CAFM and *KPIs* was provided by Ernie Pierz, Midwest Regional Manager of BP's global property portfolio, where he is responsible for real estate, facility planning and construction, medical, safety, and business services for eight BP locations in the Midwest. At the time this CAFM initiative was undertaken, Mr. Pierz was responsible for strategic facility design and tactical management services for BP's divisions in the Western Hemisphere.

Computerized Facility Integration (CFI), L.L.C., was the systems-integration company responsible for the setup and enhancement of the Archibus/FM system, much of the CAD drawing documentation, and the design of the Web interface. The author was CFI's project manager in the Midwest from 1998 to 2001.

12 SUSTAINABLE DESIGN

269   *Control of nature*   Rachel Carson, *Silent Spring*, Houghton Mifflin, New York, 1962, p. 297.

    Subversive view of natural science   Paul Shepard, "Ecology and Man—A View point," introduction to *The Subversive Science: Essays Toward an Ecology of Man*, ed. by Paul Shepard and Daniel McKinley, Houghton Mifflin, Boston, 1969, pp. 1–10.

    Absolute knowledge   J. Bronowski, *The Ascent of Man* (television series, 1974), Episode 11, "Knowledge or Certainty." Also Bronowski, *The Ascent of Man*, Little, Brown and Company, Boston/Toronto, 1973, p. 374.

271   *DDT and malaria*   *Forbes*, February 28, 2005, Vol. 175, No. 4, pp. 17, 22.

273   *ISO 14000 certification*   Source: *The ISO Survey of ISO9001:2000 and ISO14001 Certificates—2003*, ISO Central Secretariat, Geneva, Switzerland.

    Principles of green design   These elements of a sustainable philosophy embody the principles presented by William McDonough at the 1993 AIA convention, later known as the *Hannover Principles* as prepared for the *EXPO2000 World's Fair* in Hannover, Germany.

275   *Efficient use of daylight in offices*   L. E. Abraham, *Daylighting—The Sustainable Building Technical Manual*, IV.7, 1996, http://www.sustainable.doe.gov/pdf/sbt.pdf.

278   *LEED guidelines*   LEED is a registered trademark of the *U.S. Green Building Council*. LEED products and services are licensed through the Council.

    Sustainable performance criteria   LEED Policy Manual: *Foundations of the Leadership in Energy and Environmental Design Environmental Rating System—A Tool for Market Transformation*, Spring 2003, LEED Steering Committee.

279   *LEED CI*   *LEED Green Building Rating System for Commercial Interiors*, Version 2, November 2004, U.S. Green Building Council.

281   *Brownfield redevelopment*   According to the U.S. Environmental Protection Agency, *brownfields* or *brownfield site* refers to "real property, the expansion, redevelopment, or reuse of which may be complicated by the presence or potential presence of a hazardous substance, pollutant, or contaminant," with certain legal exclusions and additions. This definition is found in Public Law 107-118 (H.R. 2869) titled Small Business Liability Relief and Brownfields Revitalization Act signed into law on January 11, 2002. There are various

		kinds of grants and other funding programs to assist interested organizations with the remediation of environmental issues.
	Light pollution	The governing standard is the *Recommended Practice Manual: Lighting for Exterior Environments* of the Illuminating Engineering Society of North America (IESNA).
283	*Minimum energy efficiency*	ASHRAE/IESNA 90.1-2004 or applicable local energy codes if it can be demonstrated that they are more stringent.
	Lighting power	ASHRAE/IESNA 90.1-2004.
	ASHRAE/IESNA	American Society of Heating, Refrigerating, and Air-Conditioning Engineers Illuminating Engineering Society of North America.

Certified Tradable Renewable Certificates

Green-e renewable electricity certification is a nonprofit program that seeks to provide a way for consumers to identify sustainable electricity products in competitive markets. *Tradable Renewable Certificates* (TRCs) are documents representing attributes associated with defined quantities of generated electricity. TRCs represent the attributes of generated renewable energy and may be traded or sold separately from the electricity itself. The TRC's Joint Governing Board lists the following four goals in its mission statement:

1. Bolster consumer confidence in the reliability of retail electrical products reflecting renewable energy generation.

2. Expand the retail market for electricity products incorporating renewable energy, including expanding the demand for new renewable energy generation.

3. Provide consumers with clear information about retail renewable electricity products to enable them to make informed purchasing decisions.

4. Encourage the deployment of electricity products that minimize air pollution and reduce greenhouse gas emissions.

Source: *Code of Conduct, Green-e Renewable Energy Certification Program*, Appendix B, March 15, 2002.

285	*Certified wood products*	According to the Forest Stewardship Council's Principles and Criteria.
	Indoor air quality standards	ASHRAE 62-2004, *Ventilation for Acceptable Air Quality*, and addenda.
	Naturally ventilated spaces	CIBSE (Chartered Institution of Building Services Engineers) *Good Practice Guide 237*, 1998, and recommendations in the CIBSE *Applications Manual 10*, 1997.
	Airborne contaminants and moisture	Meet or exceed the Design Approaches of the *Sheet Metal and Air Conditioning National Contractors Association* (SMACNA) *IAQ Guideline for Occupied Buildings Under Construction*, 1995, Chapter 3.
	Air quality protocols	U.S. Environmental Protection Agency (EPA) *Compendium of Methods for the Determination of Pollutants in Indoor Air*.

286	*Adhesives and sealants*	*Green Seal Standard GS-36* for aerosol adhesives. *South Coast Air Quality Management District* (SCAQMD) Rule #1168, with amendments, for adhesives, sealants, and primers.
	Paints and coatings	*Green Seal Standard GS-11* for topcoat paints, *GC-03* for antirust and anticorrosive paints, and *SCAQMD Rule #1113* for all other architectural coatings and primers.
	Carpet systems	*Green Label Plus* testing standard of the *Carpet and Rug Institute*.
	Temperature and ventilation	ASHRAE 62-2004, *Natural Ventilation.*
	Thermal comfort	ASHRAE 55-2004, *Thermal Comfort Conditions for Human Occupancy.*
287	*LEED-EB*	*Green Building Rating System for Existing Buildings: Upgrades, Operations and Maintenance* (LEED-EB), Version 2, October 2004, U.S. Green Building Council.
290	*LEED-NC*	*LEED Rating System* [NC], Version 2.0, June 2001, U.S. Green Building Council.
291	*Optimize energy performance*	ASHRAE/IESNA 90.1-2004.

13 BUILDING CODES

293	*Code of Hammurabi*	James B. Pritchard, ed., *Ancient Near Eastern Texts*, trans. Theophile J. Meek, Princeton University Press, Princeton, N.J., 1969.
	Performance-related provisions	International Code Council, Inc., *2003 International Building Code*, 4051 West Flossmoor Road, Country Club Hills, Illinois, 60478-5795.
296	*Occupancy separation*	IBC, Table 302.3.2.
304	*Fire-resistance ratings for building elements*	IBC, Table 601.
305	*Separation distance*	IBC, Table 602.
306	*Fire-resistance-rated wall*	IBC, 702.
307	*Building heights and floor areas*	IBC, Table 503.
310	*Exterior wall openings*	IBC, Table 704.8.
311	*Fire wall fire-resistance ratings*	IBC, Table 705.4.
	Roof-covering classification	IBC, Table 1505.1.
	Fire-barrier assemblies	IBC, Table 706.3.7.
314	*Opening protectives*	IBC, Table 715.3.
317	*Halon systems*	Halon systems use halogenated compounds that have been found to erode the ozone layer in the upper atmosphere; therefore, their use is considered hazardous. There are several alternative systems available that should be used for most applications.
321	*Smoke-control systems*	IBC, 909.1.
	Smoke-control implementation	IBC, 909.2.

14 EGRESS

323	*Best-laid schemes*	Robert Burns (1759–1796), *Poems and Songs*, "To a Mouse."
	Accessibility provisions	NMHC/NAA, National Multi-Housing Council/National Apartment Association, *New Model International Code Council Codes*, Suite 540, 1850 M Street NW, Washington, DC 20036-5803, (202)

		974-2300. Memorandum dated January 30, 2000 (updated October 4, 2001), pp. 2–3. http://www.naahq.org and http://www.nmhc.org/.
	Competing fire and life-safety codes	NMHC/NAA (*National Multi-Housing Council and the National Apartment Association*), January 30, 2000, memorandum, p. 4.
324	*Model building codes*	Building Owners and Managers Association (BOMA) International, Position Paper: "*NFPA Building Code*" (NFPA 5000); 1201 New York Avenue NW, Washington, DC 20005, (202) 408-2662, http://www.boma.org, May 23, 2002.
	Fire has an attitude	Statement attributed to Captain Michael Gubricky of the Chicago Fire Department upon reaching the twelfth-floor fire location in the Cook County Administration Building on October 17, 2003, *Chicago Tribune*, October 25, 2003.
	Major architectural firm	*Brunswick Building*, Skidmore Owings and Merrill LLP, Architect, 1964, thirty-five floors.
	Now fully sprinklered	"High-rises Not Alike in Fire Safety Standards, Evacuation Drills Vary," *Chicago Tribune*, October 26, 2003.
	Sprinkler system is present	U.S. Experience with Sprinklers, Kimberly D. Rohr, Fire Analysis and Research Division, National Fire Protection Association, 1 Batterymarch Park, Quincy, MA 02269-9101, September 2001, http://www.nfpa.org.
	Unlocking of stairwell doors	"'Fire Has an Attitude' Questions Unanswered in Official Chain of Events," *Chicago Tribune*, October 25, 2003.
325	*Stayed and waited for help*	"Fire Experts Divided over Survival Strategies; Locked Stairwell Doors, Evacuation Order Stir Debate," *Chicago Tribune*, October 19, 2003.
	People running scared from 9/11	"Few Answers in Deadly Fire; Arson Unlikely. Survivors Say Stairwell Doors Locked," *Chicago Tribune*, October 19, 2003.
	Things going wrong lined up	Chicago Fire Commissioner James Joyce, "'Fire has an Attitude'. . .," *Chicago Tribune*, October 25, 2003.
327	*Means of egress*	IBC, 1002.
	Accessibility	IBC, 1102.
	Americans with Disabilities Act	U.S. Department of Justice, "ADA Standards for Accessible Design," 28 C.F.R. Part 36, Appendix A, *Code of Federal Regulations* (1990), revised as of July 1, 1994, 3.5.
	Continuous unobstructed path	ADA, 3.5.
	Accessible means of egress	IBC, 1002.
329	*Maximum occupancy density*	IBC, Table 1004.1.2.
330	*Egress width per occupant*	IBC, Table 1005.1.
337	*Discernible path of egress travel*	IBC, 1013.2.
338	*A straight line between them*	IBC, 1014.2.1.
	One means of egress	IBC, Table 1014.1.
339	*Natural and unobstructed path*	IBC, 1015.1.
	Exit access travel distance	IBC, Table 1015.1.
340	*Corridor fire-resistance rating*	IBC, Table 1016.1.

341	*Minimum number of exits*	*IBC*, Table 1018.1.
	Buildings with one exit	*IBC*, Table 1018.2.
345	*Steepness of the slope*	*IBC*, Table 1024.6.2, Width of Aisles for Smoke-Protected Assembly.
346	*Proportion to egress capacity*	*IBC*, 1024.9.2.
	Single and dual access	*IBC*, Table 1024.10.1, Smoke-Protected Assembly Aisle Accessways.

15 ACCESSIBILITY

349	*Society's accumulated myths*	William J Brennan, Associate Justice, U.S. Supreme Court, *Majority opinion in 7–2 ruling that people with contagious diseases are covered by law that prohibits discrimination against the handicapped in federally aided programs*, March 3, 1987.
	ADAAG	*ADA Accessibility Guidelines for Buildings and Facilities (ADAAG)*, as amended through September 2002: http://www.access_board.gov/adaag/html/adaag.htm.
350	*ICC/ANSI A117.1*	International Code Council, Inc., *Accessible and Usable Buildings and Facilities*, ICC/ANSI A117.1-2003 (previous edition ICC/ANSI A117.1-1998). ICC, Inc., 5203 Leesburg Pike, Suite 708, Falls Church, VA 22041-3401, ©1998–1999. Names of some sections have changed and new sections added are 409–410, 708, 805–807, and 1002; Section 507 has been deleted.
	ADA technical specification	*ADAAG*, Section 1, Purpose.
	Physically disabled persons	*IBC*, 1101.1.
	Scoping requirements	*IBC*, 1103.1.
353	*Accessible parking spaces*	*IBC*, Table 1106.1.
354	*Type A and B dwelling units*	*IBC*, 1102.
355	*Accessible dwelling and sleeping units*	*IBC*, Table 1107.6.1.1.
356	*Accessible wheelchair spaces*	*IBC*, Table 1108.2.2.1.
358	*Accessible checkout aisles*	*IBC*, Table 1109.12.2.
	ASME A18.1	American Society of Mechanical Engineers, *Safety Standard for Platform Lifts and Stairway Chairlifts*.
	Detectable warnings	*ADAAG*, 4.29.2.
359	*International accessibility symbol*	ICC/ANSI A117.1, Figure 703.7.2.1.
	Major ADA update in 2004	U.S. Architectural and Transportation Barriers Compliance Board (Access Board), *A Guide to the New ADA-ABA Accessibility Guidelines*, July 2004, http://www.access-board.gov.

16 ADA ACCESSIBILITY GUIDELINES

| 361 | *The heavens are ours* | Letter to Dr. Edward Everett Hale, cousin and longtime friend, at the age of 21. Helen Keller, *The Story of My Life* (with selected letters), Penguin Putnam, 2002. |
| | *Individuals with disabilities* | ADA, 28 C.F.R. 36, Appendix A, 1.0, July 1, 1994, edition. |

	Facilities subject to Title III	ADAAG, as amended through September 2002, 3.5 Definitions (all *ADAAG* references are to this edition).
362	Equivalent facilitation	ADAAG, 2.2 Equivalent Facilitation.
	Enforceability of new guidelines	The following Web sites can be checked for updates as to the content of the guidelines and enforceability status: http://www.access_board.gov and http://www.iccsafe.org. Comparison matrices are also available for the new ADA-ABA guidelines, the ADAAG, and the IBC.
363	Alterations and usability	ADA-ABA (2004) 106.5, Definitions, pp. 12–17 (all *ADA-ABA* references are to this edition).
364	Structurally impracticable	ADAAG, 4.1.1 (5)(a).
	Van-accessible spaces	ADA-ABA, 208.2; ADAAG, 4.1.2 (5)(b).
	Space needed to play the game	ADA-ABA, 206.2.2 Advisory, p. 24; ADAAG, 4.1.3 (3).
365	Highly specialized facilities	ADAAG, 4.1.3 (5)(b).
	At the time of the alteration	ADAAG, 4.1.6 (1)(a).
	Entire space shall be made accessible	ADAAG, 4.1.6 (1)(c).
366	Primary function	Department of Justice, *Code of Federal Regulations*, 28 C.F.R. 36.403(b).
	Areas of transportation facilities	Department of Transportation, 49 C.F.R. 37.43(c).
	Technically infeasible	ADA-ABA, 106.5, p. 16; ADAAG, 4.1.6 (1)(j).
	Disproportionate cost	ADA-ABA, 202.4 Advisory, p. 19, 28 C.F.R. 36.403(f)(1).
	Cost priorities	28 C.F.R. 36.403(b), Appendix B, July 26, 1991.
367	Building blocks of accessibility	ADA Chapter 1, 101 Purpose, 101.1 General, p. 5; and *ABA* Chapter 1, F101 Purpose, p. 72.
	Ramps and landings	ADA-ABA does not specifically mention curb height except with regard to fishing piers: Section 1005.3.1.
378	Elevator car dimensions	A tolerance of minus five-eighths inch is permitted for thirty-six-inch-wide doors. Other car configurations that provide turning space compliant with Section 304 with the door closed are permitted.
380	Handrails—loss of equilibrium	ADA-ABA, Advisory 505.6.
393	Fire alarm sound levels	ADAAG, 4.28.2, specifies a maximum of 120 dBA, where ADA-ABA, 702.1, specifies 110 dBA (dBA means A-weighted decibels, in which the weighting compensates to relative loudness as perceived by the human ear).
393	Signal-activated vibrator	ADAAG, Appendix 4.28.4.
396	Filter background noise	ADAAG, 4.33.7.
397	Medical care facilities	ADA-ABA, 223; ADAAG, 6.1.
399	Residential dwelling units	ADA-ABA, Advisory 223.1 General, p. 59.
	Transportation facilities	Key stations serve major activity centers, are transfer or major interchange points with other transportation modes, or are stations where passenger boardings exceed the system average by 15 percent. Department of Transportation, 49 C.F.R. 37.51(b).
401	Last hole of the course	ADA-ABA 239; ADAAG, 15.5.2.

INDEX

A

accessibility
  and wayfinding, 153
  definition, 327
Accessibility (ADA)
  Accessible Routes, 373–379
    Doors, Doorways, and Gates, 374–377
    Elevators, 377–378
    Limited Use and Private Residence
      Elevators, 378
    Platform Lifts, 379
    Ramps and Curb Ramps, 377
    Walking Surfaces, 373–374
  Building Blocks, 367–373
    Changes in Level, 368
    Clear Floor or Ground Space, 369
    Floor or Ground Surfaces, 368
    Knee and Toe Clearance, 370
    Operable Parts, 372
    Protruding Objects, 370
    Reach Ranges, 370–372
    Turning Space, 368–369
  Built-In Elements, 400–401
    Benches, 400
    Check-Out Aisles, Sales and Service
      Counters, 401
    Dining Surfaces and Work Surfaces, 400
  Communication Elements and Features,
    393–396
    Assistive Listening Systems, 396
    Automated Teller and Fare Machines,
      396
    Detectable Warnings, 394–395
    Fire Alarm Systems, 393
    Signs, 393–394
    Telephones, 394
    Two-Way Communication Systems, 396
  General Site and Building
    Elements, 379–381
    Handrails, 379–381
    Parking Spaces and Passenger Loading
      Zones, 379
    Stairways, 379
  Plumbing Elements and Facilities,
    382–393
    Bathrooms, Water Closets, and Toilet
      Compartments, 382–385
    Bathtubs, 386–387
    Drinking Fountains, 382
    Grab Bars, 390–391
    Lavatories and Sinks 385–386
    Other Plumbing-Related Elements, 392
    Seats, 391–393
    Shower Compartments, 387–390

    Urinals, 385
  Recreation Facilities, 401
  Scope, 364–367
    Building Additions and Alterations,
      365–366
    Buildings, New Construction,
      364–365
    Historic Preservation, 366–367
    Sites and Exterior Facilities,
      New Construction, 364
  Special Rooms, Spaces, and Elements,
    396–400
    Courtrooms, 399
    Dressing, Fitting, and Locker
      Rooms, 397
    Holding Cells and Housing Cells, 398
    Kitchens and Kitchenettes, 397
    Medical Care and Long–Term
      Facilities, 397
    Residential Dwelling Units, 399
    Transient Lodging Guest Rooms, 398
    Transportation Facilities, 399–40
    Wheelchair Spaces, Companion Seats
      and Designated Aisle Seats, 397
Accessibility (IBC), 323–324
  Dwelling & Sleeping Units, 354–356
    General Exceptions, 356
    Institutional Occupancies, 354
    Residential Occupancies, 355
  Other Facilities
    Changes in Level, 358
    Service Facilities, 357–358
    Toilet Rooms & Bathing, 357
  Parking & Passenger Loading, 353–354
  Routes & Entrances, 350–353
  Signage, 358–359
  Special Occupancies, 356
accessible means of egress (definition), 327
accessible route (definition, ADA), 363
ad quadratum
  inversion of . . . , 180
  use at Maison de la Colline, 50
  use at Salisbury, 123–124
ADA. *See* Americans with Disabilities Act
ADA Standards for Accessible Design
  (28 CFR Part, 36), 323
ADA-ABA (new)
  harmonized with ICC/ANSI A117.1, 362
  not yet mandatory (2006), 362
ADAAG. *See* Americans with Disability Act
  Accessibility Guidelines
  ICC/ANSI A117.1–1998 (cross reference
    table), 351–352
Addition (definition, ADA), 363

adjacencies
    blocking & stacking, 150–151, 208, 242
    case study—bubble diagram, 143
    case study—bubbles, blocks, 115–118
    case study—ellipse diagram, 148–149
    departmental, 104, 114–115
    matrix matching, 115
    relationships, 114–115
    strategy sessions, 115
    task groupings, 102
adjacency relationship matrix, 54, 56
aesthetic (definition), 34
AIA. *See* American Institute of Architects
Alberti, Leon Battista, 33
Alinsky, Saul (epigram), 67
alteration (definition, ADA), 363
American Hospital Association (AHA), 142
American Institute of Architects (AIA)
    contractual phases of work, 77
    Declaration of Interdependence for a
      Sustainable Future, 273
    layer-naming convention, 164
American Society of Testing Material.
    See ASTM
Americans with Disabilities Act (ADA),
    22, 294, 453 ff.
    concept xxv
    enacted as federal law, 361
    ramps in home, 59
    technical guidelines (basis of), 323
Americans with Disability Act Accessibility
    Guidelines (ADAAG), 323
analysis, strategy, 148
analysis/synthesis. *See* Descartes
annunciator. *See* Fire Alarm & Detection
    Systems, Fire Command Center
ANSI (American National Standards
    Institute), 211
    A177.1-1998 (and IBC), 323, 350 ff.
    ANSI/TIA/EIS-568-B (Telecommunications
      Standard), 250–251
    ICC/ANSI A117.1, harmonize with new
      ADA-ABA, 362
ANSI/BOMA. *See* Building Owners and
    Managers Association (BOMA)
Anthropometry (definition), 22
application service provider, 229
application software (Microsoft Project),
    83–86
arch (development), 90
Archibus/FM, 206–207, 223, 234,
    238–262, 425 ff.
    Building Operations Management,
      253–256
    customization, 264
    Design Management, 245–248
    Furniture and Equipment Management,
      248–250
    Overlay, 238
    Real Property and Lease Management,
      239–240

    Space Management, 242–245
    Space Management, *Appendix D*, 425–433
    Strategic Master Planning, 240–242
    System, 259–262
    task categories, 256–259
    Telecommunications and Cable
      Management, 250–253
    Web Central, 239
architect. *See* Le Corbusier, Mies van der Rohe,
    Wright
    . . . of destruction, 1
    . . . of Record, 1
    dilution of role, 77
    master builder, 1
Architectural Barriers Act (ABA), 361
architectural scales (English), 15
architecture
    as mass media, 27
    Classical Orders, 4–5
    Gothic (and Max Weber), 93
    qualities, 47
    threads in historical fabric, 28
area
    allocation (layer use), 167
    allocation analysis, 162–164
    allocation tools, 135–140
      Block Area, 138
      Block Drag, 137
      Block Fillet, 139
      Block Split, 138
      Catalog Polyline Data, 140, 167,
        170–171
      Convert Hyperlink to Text, 139
      Export Block Data, 139
      Hyperlink Text Height, 139
      Import Block Data, 136
      Program Requirements, 139
    assignable, 100
    assignable (IFMA), 195
    calculations (BOMA), 187–192
    circulation factor, 102
    classification, 100–102
    common support, 101
    common support (IFMA), 194
    gross (BOMA), 186
    gross (IFMA), 192–193
    gross leasable (in malls), 300
    measurement (GSA), 195
    measurement (PEFIC), 197–203
    net (programmable), 100
    polylines. *See* Polylines
    primary circulation, 102
    primary circulation (IFMA), 193
    relationships (IFMA), 192
    rentable (BOMA), 183–185
    rentable (IFMA), 193–194
    secondary circulation (IFMA), 195
    usable, 100
    usable (BOMA), 184
    usable (IFMA), 194–195
areas of refuge. *See* Egress, Means of

Aristotle
  De Caelo (*On the Heavens*), 8
  potency of place, 8
ASP. *See* Application Service Provider
asset management (furnishings and
  equipment), 78
ASTM/IFMA. *See* International Facility
  Managers Association (IFMA)
ASTM-E1836 (IFMA), 192–195
Athena, 30
AutoCAD
  Appendix C (loading macros), 419–424
  polyline object. *See* Polylines

B

Babbage, Charles
  scientific management, 67, 79
Babel, 183
Bach, Johann Sebastian
  Musical Offering (canon), 4–5
Bancroft, Anne (*The Miracle Worker*), 361
Bank of China (Hong Kong), 179–180
Barcelona Pavilion
  separation of structure and
    enclosure, 43
Barzun, Jacques
  . . . on decadence, 28
Bauhaus
  Itten and Schlemmer courses, 24–25
  Mies van der Rohe, director of, 33
Bernard of Chartres (epigram) xv
Berners-Lee, Tim
  World Wide Web, 75
Black Death, 121
block diagrams
  presentation, 150–151
blocking & stacking. *See* adjacencies
Blue Cross & Blue Shield. *See* Case Studies
Blue Cross (Blue Shield) Association, 142
BOCA. *See* Building Officials and Code
  Administrators International, Inc.
BOMA. *See* Building Owners and Managers
  Association
BOMA Method, 183–192
Boyce, Raymond, 209
BP America (case study), 264–268
Braille. *See* tactile characters
Brennan, William J. (epigram), 349
brick construction
  English (Cross) Bond, 41
Bronowski, Jacob
  Industrial Revolution, 269
  principle of tolerance, 28, 269
bubble diagrams. *See* adjacencies
building (definition, ADA), 367
building blocks
  of facility management xvii
building code
  administration, 294
  history, concerns xxv, 291–294
  local regulations, 293

building envelope, 148–150
  limitations in older structures, 158
building evaluation (options analysis), 153
Building Officials and Code Administrators
  International, Inc. (BOCA), 293
Building Owners and Managers Association
  (BOMA), 165, 183–192
  re ICC and NFPA, 323–324
Burnham, Daniel H.
  architecture's grand relapse, 36–37
  Chicago Plan, 15
  Reliance Building (Chicago), 176
Burns, Robert (epigram), 323

C

CABO. *See* Council of Americal Building
  Officials
CAFM. *See* computer-aided facility
  management
Caldwell, Alfred
  quoting Mies, 42–43
  style as an obscenity, 36
capitalism
  emancipation of humanity, 93
  forever renewed profit, 93
Carolingian Dynasty, 119
Carson, Rachel (epigram), 269
case studies
  BP America, 264–268
  global technology company, 206–209
  international chemical company, 229–231
  operations centers, 63–66
  three facility programs
    Blue Cross & Blue Shield, 142–144
    government agency, 148–151
    pharmaceutical company, 144–148
cathedrals
  Beauvais (collapse), 2, 124
  Gothic style, 3
  Salisbury, 121–124
causal factors, 148
Chamberlin, Donald, 209
chargebacks. *See* rooms, categories & types
Charlemagne (Charles the Great), 119
Chaucer, Geoffrey
  four contraries, 120
circulation. *See* Polylines
  and building core, 154
  as closed system, 153
Codd, E.F.: Twelve Rules, 209, 211–217
Codes. *See* building code
  ERP applications, 107
  furniture & equipment, 108
  level and sequence, 102, 107–108,
    125, 140
  room use, 200–203
  space allocation reference (SAR), 105
  TCP/IP, 107
Columbian Exposition, 36
common path of egress travel(definition), 337
component object model (COM), 127

computer applications
  data pipelining, 127
  flow of planning data, 126
  CATIA, 39
computer-aided facility management (CAFM)
  and facility programming, 125
  concept, xxiv
  links to CAD drawings, 126
  major applications, 234
  outsourcing, 229
  real estate portfolio, consolidation, 206
  uploading data, 168
  uploading data (pipelining), 174
Constantine emperor, peace of, 119
Construction Types (IBC), 304–308
  and Use Classifications, 305–306
  Building Height & Area Limitations,
    306–308
contextualism, 37
contracts (in ancient Greece), 30
Cook County Building Fire, Chicago, 2003,
    324–326
Copernicus, Nicolaus
  heliocentric universe, 181
  model of universe, 9
corporate culture
  definition, 69
  influence on design goals, 78
corridor width fire-rated, 340
cost centers, 99–100
Council of American Building Officials
  (CABO), 293
Council of Five Hundred, 30
critical path. *See* project management
Cromwell, Oliver (contemporary of
  Descartes), 36
Crown Hall (architecture building at IIT), 43
Crusades, 120
Cubit, Egyptian Royal (definition), 15

D

Da Vinci, Leonardo
  naval (omphalos), 6–7
  Vitruvian proportions, 33
Dante Alighieri, *Inferno*, 7
database
  administration, 224
  concept xxiv, 205
  hierarchies & networks, 210
  relational model, 209–210
  schema (definition), 213
  user interfaces, 226–228
database, relational, 211
  atomic values, 212
  Codd's Twelve Rules, 211–217
  data independence, 215
  domains, 212
  integrity, 215–216
  metadata, 213
  normalized relations, 217–222
    Boyce-Codd Form, 219–220

fifth form, 221–222
first form, 217–218
fourth form, 220–221
second form, 218
third form, 218–219
  null values, 212–213
  primary keys, 213–221
  relational operations, 214–215
  security, 217
  SQL, 213
  views (queries), 214
databases
  Microsoft Access, 127–133
    exporting block planning data,
      134–135
    management reports, 129–133
    planning/layout reports, 133–134
    populating tables, 128
  schema (definition), 127
Date, C.J., 209
DDT
  and malaria control, 271
  and Rachel Carson, 270
de Borda, John Charles (surveying
  instrument), 13
Dearstyne, Howard: space imparts an
  attitude, 8
Decadence (and Renaissance), 28
deconstructionism, 37
Delambre & Mechain (metric survey), 13–14
Deming, W. Edwards: Total Quality
  Management (TQM), 233, 68, 86–88,
  233, 271
Departmentalization (traditional
  structures), 72
Descartes, René
  Cartesian process, 79
  contemporary of, 36
  coordinate system, 8
  scientific method, 34
detectable warnings (ADA), 394–395
disproportionate cost (ADA), 366
dome (Pantheon), 90
Domesday Book, 3
dominant portion (BOMA), 163, 186
door width & swing, 331
doors
  access control. *See* Egress, Means of
Dos Passos, John (Taylor method)
  See Taylor, Frederick W.
Dreyfuss, Henry and Tilley, Alvin
  measurements of man and woman, 23
drill down, 106
dwelling units,Types A & B, 323

E

Earth Day, 270
École de Beaux-Arts
  concours, esquisse, 36
  method, 36
egress (and wayfinding), 153

egress, means of (definition), 327
Egress, Means of (IBC), 323–324
　Aisles, Corridors & Exit Access, 337–339
　　Corridors, 339–340
　　Doorways, 338–339
　　Travel Distance, 339
　　Travel, Common Path, 337–338
　Areas of Refuge, Accessible, 330–331
　Doors & Access Control, 331–334
　　Access-Controlled, 332
　　Gates & Turnstyles, 333
　　Horizontal Sliding, 332
　　Landings & Thresholds, 332
　　Locks & Latches, 333
　　Revolving, 331
　　Security Grilles, 332
　Egress Width (inches), 328–329
　Exits, 340–344
　　Discharge, 344
　　Exterior Ramps & Stairways, 343
　　Horizontal, 343
　　Number, Continuity, 341
　　Passageways, 342
　　Vertical Enclosures, 341–342
　General Requirements, 327–328
　Illumination & Signage
　　Illumination, 336
　　Signage, 336–337
　Level/Floor Changes
　　Guards, 336
　　Ramps, 335
　　Stairs & Handrails, 334–335
　Occupant Load (area), 328–329
　Special Requirements, 344
　　Assembly, 344–347
　　Emergency Escape & Rescue, 347
Einstein, Albert. *See* Modulor
　General Relativity, 9
Eisenhower, Dwight D. (epigram), 51
Electrical Code, ICC, 294
Element (definition, ADA), 367
elevator lobbies. *See* Fire-Resistance Rated
　Construction, Shaft Enclosures
elevator signals and controls (ADA), 378
Eliot, T.S., *The Hollow Men*, 35, 183
Environmental Quality Council. *See* U.S.
　Environmental Protection Agency
environmental standards. *See* International
　Standards Organization (ISO 14000)
EPA. *See* U.S. Environmental Protection
　Agency
epigrams
　Alinsky, Saul, 67
　Bernard of Chartres xv
　Brennan, William J., 349
　Burns, Robert, 323
　Carson, Rachel, 269
　Deming, W. Edwards, 233
　Eisenhower, Dwight D., 51
　Eliot, Thomas Stearns, 183
　Escher, M.C., 153

　Freud, Sigmund, 13
　Gehry, Frank, 125
　Hammurabi, Code of, 293
　Hugo, Victor, 33
　Keller, Helen, 361
　Lewis, C.S., 209
　Ockham, William of, 1
　Proverbs, Chapter, 15 xxi
　Weil, Simone, 93
equivalent facilitation (ADA), 361–362
Eratosthenes (earth circumference) 8,
　14–15
ergonomics (definition), 22
Escher, M.C. (epigram), 153
Espacio Tower, Madrid, 180
Excel worksheet (Catalog Polyline Data),
　170–171
exit (definition)411. *See* Egress, Means of
exit access (definition), 327
exit discharge (definition), 327

F
facility (definition), xvi
facility (definition, ADA), 367
Facility Center. *See* TRIRIGA
facility data
　developing, 100
Facility Planning Checklist, *Appendix A*,
　403–409
Fair Housing Accessibility Guidelines
　(FHAG), 323
Fallingwater House
　dysfunctional family dynamics, 48–49
　physical environment, 47
FAMIS, 235
Farnsworth House. *See* Mies van der Rohe
Farnsworth House, Plano, Illinois, 44
FHAG. *See* Fair Housing Accessibility
　Guidelines
Fibonacci
　Da Vinci and Zeising (ratios), 18–19
　Kepler, 17
　number series, 17
fire doors (& rated glazing). *See* Fire-
　Resistance Rated Construction, Opening
　Protectives
fire extinguishers. *See* Fire Protection Systems,
　Portable Fire Extinguishers
fire partitions definition, 295
Fire Protection Systems (IBC), 315–322
　Alarm & Detection Systems, 318–320
　Alternative Automatic Systems, 317
　Automatic Sprinkler Systems, 315–317
　Design & Installation Standards, 316
　Emergency Alarm Systems, 321
　Fire Command Center, 322
　Portable Fire Extinguishers, 318
　Smoke & Heat Vents, 321–322
　Smoke Control Systems, 321
　Standpipe Systems, 317–318
　Use/Occupancy Classifications, 316

fire resistance
  definition, 293
  rating (and other related definitions), 309
  testing methodology (ASTM), 309
fire walls
  and area limitations, 306
  definition, 295
Fire-Resistance Rated Construction (IBC),
  308–315
  Ducts & Openings, 314
  Exterior Walls, 310
  Fire Barriers, 311
  Fire Partitions, 312
  Fire Walls, 310–311
  Horizontal Assemblies, 312
  Joint Systems, 313
  Opening Protectives, 313–314
  Other Requirements, 315
  Penetrations, 313
  Shaft Enclosures, 312
  Smoke Barriers & Partitions, 312
  Structural Members, 313
Firmitatis (structural integrity), 35
FIS (Facility Information Systems, Inc. *See* FAMIS
forms (data entry in Access), 130
Fountainhead, Ayn Rand, 33
FPdata.INI (initialization file), 422
Freud, Sigmund (epigram), 13
Froebel, Fredrich
  and Frank Lloyd Wright, 24
  blocks (gifts), 24
  kindergarten and dance forms, 24
function (a moving target), 35
functional requirements
  management questions, 70
  residential project, 53
  site and program, 51
furniture
  3D CAD symbols, 63
  conventional (use of), 148
  system, 112

G

Galileo
  model of universe, 9
  period of pendulum swing, 13
Gantt, Henry
  Gantt Chart, 68, 84
  performance incentives, 68
Gateway Arch, Eero Saarinen, 44
General Motors, 94
General Services Administration
  (GSA), 183
Ghery, Frank, Guggenheim Museum, Bilbao
  xviii, xxi, 39, 125
Gibson, William (*The Miracle Worker*), 361
global technology company (case study),
  206–209
globalization
  intellectual property, 73
  largest public companies, 95

Gödel, Kurt (incompleteness theorem), 37
golden section, 10
  and Adolf Zeising, 18
  and Great Pyramid, 14
  and Parthenon, 31–32
  eurythmic vocabulary, 39
  Mies' brick villa, 42
  phi, Greek letter, 17
  Vitruvius and . . ., 6
  Wright's Fallingwater house, 44–46
golden spiral, 18
Government Agency. *See* Case Studies
grab bars (ADA), 382, 386–391
grades
  General Service (GS), 110,
  Hay points, 110
  job classifications, 102
  use in space allocation, 133
Grammar of Ornament. See Jones, Owen
Great Pyramid (geometry), 14
green design. *See* sustainable design
gross area (BOMA). *See* area, Gross (BOMA)
Grove, Andrew, *Only the Paranoid
  Survive*, 74
GSA. *See* General Services Administration
Guggenheim Museum(s), 38
Gutenberg, Johann (movable type), 95

H

Habitat for Humanity, 28
Hammurabi, Code of (epigram), 293
handrails (ADA), 379–381
Hawthorne Studies. *See* productivity
Hearth (focus of home activities), 58
hierarchies
  management and location, 104
  multiple complexes, 264
Hilberseimer, Ludwig
  sun penetration, 5–6, 177
  urban density, 35
historical connections, xix
Homer, *Odyssey*, 7
horizontal assemblies (definition), 295
hose connections & cabinets. *See* Fire
  Protection Systems, Standpipe Systems
HUD. *See* U.S. Department of Housing and
  Urban Development
Hugo, Victor
  architectural commentaries xix, 27
  epigram, 33
  Paris of plaster, 37
  printing press and architecture, 93

I

IBC. *See* International Building Code
IBM, 209
ICBO. *See* International Conference of
  Building Officials
ICC. *See* International Code Council, Inc.
ICC/ANSI A117.1-1998. *See* ANSI
Ictinius & Callicrates, 30

ideograph: building envelope & principal elements, 52–53

IFMA. *See* International Facility Managers Association

Illinois Institute of Technology (Mies van der Rohe at), 33, 40

Implementation (contract documents), 79

inflection point (engineering term, Grove's use), 75

Insert (command). *See* SQL

Integrated Workplace Management System (IWMS), 233, 263

International Building Code (IBC), 236, 366
    See Use & Occupancy Classification,
    See Special Building Types,
    See Fire-Resistance Rated Construction,
    See Fire Protection Systems,
    See Egress, Means of. *See* Accessibility

international chemical company (case study), 229–231

International Code Council, Inc. (ICC), 293

International Conference of Building Officials (ICBO), 293

International Facility Managers Association (IFMA), 164, 192–195

International Standards Organization (ISO 9000 & ISO 14000, certification), 271–273

Internet
    compared to circulation network, 154
    World Wide Web, 75

interoperability, 233, 237

interviewing, 113–114, 142

ISO. *See* International Standards Organization

Itten, Johannes (creative spirit, classifications), 36

J

JCAHO (Joint Commission on Accreditation of Healthcare Organizations), 99

Jones, Owen
    and Frank Lloyd Wright, 25–27
    and Modulor, 39
    Grammar of Ornament, The, 25–27, 180–181
    Moresque module, 180, 181
    ornamental forms, 36

Joyce, James, *Ulysses*, 7

K

Kaufman House, Bear Run, Pennsylvania, 44. *See Fallingwater*

Keller, Helen (epigram), 361

Kepler, Johannes
    Fibonacci convergence, 17
    Platonic Solids as model of universe, 180
    treasures, 17

key performance indicators (efficiency vs. effectiveness), 265–267

Kissinger, Henry (paranoia), 74

Kitchen triangle, 60

Knights Templar, 7, 120

KPIs. *See* Key Performance Indicators

L

layers (AutoCAD)
    polylining and text annotation, 165–166

Le Corbusier. *See* Modulor
    definition of architecture, 1
    Domino house, 40
    five points, 40
    house as machine for living, 40
    in 1943, 33
    meter vs. foot-and-inch, 19–21
    place of the right angle, 19
    plan for a city, 35
    red and blue series, 19
    regulating lines, 39
    ugliness and incoherence, 36
    Villa Savoye, 40

Leadership in Energy and Environmental Design. *See* LEED

LEED, 278
    CI—Commercial Interiors, 279–287
        categories & point values (table), 287
        energy & atmosphere, 282–283
        indoor environmental quality, 285–287
        innovation & design process, 287
        materials & resources, 283–285
        sustainable sites, 279–282
        water, 282
    EB—Existing Buildings, 287–290
    Green Building Rating System, 278–279
    NB—New Buildings, 290–291
    Organization, 278

Lewis, C.S. (epigram), 120, 209

lifecycle management
    capital assets, 233
    CATIA software, 39

location (evaluation of), 148

locks (egress doors). *See* Egress, Means of

Lucas, Edouard, 17

Luther, Martin (theses), 95

M

management
    agendas, 70
    change, 267
    interest in facility projects, xix
    perspective xxii, 88–89, 110, 114
    scientific, 67
    software, 73
    source for program data, 100
    vision and spatial reality, 76

Mayo, Elton, Hawthorne Studies, 73

McGregor, Douglas (Theory Y), 69

measurement systems. *See* area
    area standards compared, 204–205
    richness of . . ., 15

Mechanical Code, ICC, 294

Medicare & Medicaid, 143
Metcalfe, Robert (Metcalfe's law), 75
Metric System (standard defined), 14
Microsoft Corporation
    Project, 83–86
    SQL Server, 211, 238
Mies van der Rohe, Ludwig, 40–44
    brick construction, 40
    brick villa (1924), 42
    Crown Hall, 43
    Farnsworth House, 44, 48
    four insights, 43–44
    in Chicago, 33
    inaugural address at IIT, 41–42
    less is more, 28
    One Illinois Center (Chicago), 176
Miracle Worker, The, 361
modernism, 37
module
    4-foot grid, 52
    5-foot, 112
    analysis of, 149–150
    and Modulor, 39–40
    building, 112
    symmetry checkpoint, 58
    unit system (Wright), 44
    variations in non-rectilinear buildings, 159
Modulor, 1, 6, 90, 181
    and Golden Section, 6
    as game, 39
    block planning, use in, 56–58
    Einstein's assessment, 22
    red and blue series, 39–40
    scaleable system of proportion, 21
monastic orders, 119
Monticello, 91
Moore, Gordon (Moore's law), 74
mortuary temple. *See* Ramesseum

N

National Apartment Association. *See*
    NMHC/NAA
National Fire Protection Association (NFPA)
    Life Safety Code (NFPA, 101), 298, 326
National Multi-Housing Council.
    See NMHC/NAA
network
    anarchical, 96
    containers, 106
    drill down, 106
    open, 105
network model
    closed. *See* Project Management-Critical
        Path
    web—structure in tension, 76
Newton, Sir Isaac
    earthly mechanics, 35
    gravitation, 180
    planetary mechanics, 9
NFPA, 101, *See* National Fire Protection
    Association

Nixon, Richard M. *See* U.S. Environmental
    Protection Agency
NMHC/NAA, 323
normal forms. *See* database

O

Occupancy Classification. *See* Use & Occu-
    pancy Classification
occupant load. *See* Egress, Means of
Ockham, William of, 37
    epigram, 1
    principle of parsimony, 28
    Razor, xx
office landscape, Bürolandschaft, 75
office planning (hot desk & hotelling), 97
omphalos, 6–7
    center of home, 53, 58
    center of identity, 35
OPEC, 270
operations center
    details, 60
    kitchen as, 59
    modular equipment, 63
Oracle, Delphic, 6–7
Oracle Corporation, 211, 236, 238, 262
order (cosmic & earthly), 121
organization, types of
    advertising & public relations, 97
    consulting & accounting, 97
    cultural institutions, 98
    educational, 99
    financial, 97
    government, 98
    health care, 99
    high-technology, 96
    industrial, 96
    legal, 98
organization structure
    analysis of, 100
    bureaucracy-based, 93–95
    differences, forces driving change, 94
    downsizing and outsourcing, 72
    matrix, 108
owner profile, 54
Ozymandias, 11

P

Palladio, Andrea, 33
Pantheon
    original and second structures, Hadrian's,
        90–91
parapets. *See* Fire Resistance Rated
    Construction, Exterior Walls
parking, accessible (ratios, ADA), 364
Parthenon, 30–32
    as powder magazine, 8
parti pris (option taken), 44
Peace of God, 119
PEFIC, 99, *See* Postsecondary Educational
    Facilities Inventory and Classification
    Manual

PERT (Programmed Evaluation and Review Technique). *See* project management-Critical Path
Petronas Towers (Kuala Lumpur), 180
pharmaceutical company. *See* Case Studies
physical requirements (site and program), 51
PI, Archimedean (22/7), 15–16
pilotis (Le Corbusier and Mies), 43
Planck, Max (quanta), 9
Plato (*Meno*, ad Quadratum), 50
Plumbing Code, ICC, 294
polylines
    BPoly command (AutoCAD), 162
    circulation (primary), 164
    circulation (secondary), 164
    common (service areas), 164
    drawing area, 162–164
    floor, 163
    usable area, 164
    vertical penetrations, 163
post-modernism, 37
Postsecondary Education Facilities Inventory and Classification Manual (PEFIC)
    measurement guidelines, 197–203
    room use codes, 200–203
practicable, maximum extent (ADA), 361
pre-modernism, 36–37
primary function (definition, ADA), 366
private facility (definition, ADA), 361
processions
    at Salisbury, 122–123
    Panathenaic, 30
    Sarum Use rules, 123
product lifecycle (four stages), 71
productivity (Hawthorne Studies), 69, 73
profit centers, 99–100
programming, facility
    and CAFM, 126
    concept development, 78
    definition, 77
    differential growth, 76
    process, 105, 142
    residential xxi
    spatial continuity, 181
    workplace xxii, 132 ff.
project management
    budgetary targets, 78, 83
    budgeting points of view, 77
    Critical Path, 34, 76–80, 211
    different from project administration, 86
    drawing conversion, 207
    internal and external views, 80–81
    milestones, 83
    phased development of CAFM system, 206
    project plan, 82
    resource allocation, 83–86
    task dependencies, 83–85
projections
    personnel, 103
Proverbs, Chapter, 15 (epigram) xxi

Ptolemy, Claudius (model of universe), 8, 120, 180–181
public accommodation, places of (ADA), 362
Pythagorus, 17

Q
quality control
    entry point into TQM cycle, 87, 125
    TQM, 68, 271
Quantum. *See* Planck, Schrödinger
queries
    in Access, 130
    virtual tables, views, 214

R
Rameses II, 11
Ramesseum, 11–12
Rand, Ayn, 33
regulating lines
    boundaries and edges, 31
    points of alignment, 53
regulating masses (anchors to plan), 57
regulating planes (at Fallingwater), 45
relation (definition, essential properties), 217
relational database. *See* database, relational
relational model (origins), 209–210
Reliance Building. *See* Burnham, Daniel H.
Renaissance, 3, 18, 27–28, 33, 36–37, 180–181
    and printing, 93
    Classical Orders, 36
rentable area. *See* area, rentable
revolutions
    American, French, 10
    Industrial, 3, 10, 34
Roark, Howard, 33
Robert Taylor Homes, 35
    and environmentalism, 269
    destroys lifestyle, 37
rod (definition), 15
rooms
    categories and types, 171–173, 196–199
    inventory (Archibus/FM)
        all room, 243
        composite, 244
    names (GSA), 197
    numbering, 170–171
Roosevelt, Franklin (New Deal), 270
Roosevelt, Theodore (National Park System), 270
rotunda (Pantheon), 90
Royal Society (yard based on one second), 13

S
Saarinen, Eero (Gateway Arch, St. Louis), 44
Salisbury Cathedral. *See* Cathedrals
Sarbanes-Oxley Act, 233
SBC. *See* Standard Building Code
SBCCI. *See* Southern Building Code Congress International, Inc.

Schein, Edgar. *See* corporate culture
schema. *See* databases
Schlemmer, Oskar. *See* Bauhaus
Schrödinger, Erwin (quantum), 9
scientific method, 35–36
Sears Tower (Chicago), 178–179
Select (command). *See* SQL
Serlio, Sebastiano, 34,
Serralta and Maissier (Divina Proportione), 21
service-oriented architecture xv, 237–238
Shelley, Percy Bysshe, 12
Shephard, Paul, 269
sightlines (to computer monitors), 63
signage (wayfinding), 336
site (definition, ADA), 367
smoke barriers (definition), 295
Soil Conservation Service, 270
Southern Building Code Congress
    International, Inc. (SBCCI), 293
space
    allocation plan (text annotations),
        166–167
    allocation plan (with building
        core), 167
    as a liability, 73
    orientation within, 154
    to be occupied, 78
space allocation
    bubble diagram (residence), 56–57
    by space type, 112
    data validation, 170
    guidelines, 109–113, 148
    reference (SAR) key, 108–112
    toolbar. *See* Area Allocation Tools
space layout
    development of circulation, 153–159
    stages of development, 160–162
space type
    categories & types (GSA), 196–197
    sorting by, 110
Space-Requirements Worksheet (*Appendix B*),
    411–417
Span FM. *See* TRIRIGA
Special Building Types (IBC)
    Atriums, 299
    Covered Mall Buildings, 300
    High-Rise, 299
    Institutional
        Correctional, 301–302
        Medical, 300–301
    Motor-Vehicle Related, 303–304
    Performing Arts
        Motion Picture, 302
        Stages & Platforms, 302–303
spirituality (in principles of form), 41
SQL xxiv, 127, 174, 209, 211, 213–217,
    222–227, 238, 262
St. Denis (Abbey Church), 119
stacking & blocking. *See* adjacencies
stairways (width & headroom). *See* Egress,
    Means of

Standard Building Code (SBC), 293
Strategic Advisory Group on the Environment
    (SAGE), 271
Structured English Query Language. *See* SQL
Structured Query Language. *See* SQL
style
    as an obscenity, 36
    designer's personal, 34
Suger (Abbot), 119
Sullivan, Anne (Helen Keller's teacher), 361
Sullivan, Louis
    cotyledon & ornament, 181
    rejection of Beaux-Arts method, 36
suspended ceilings. *See* Fire-Resistance Rated
    Construction, Horizontal Assemblies
sustainable design
    asset management, 233
    philosophy, 273
    six major areas of awareness, 274–278
        energy management, 275
        interior environment, 276–277
        materials utilization, 276
        site considerations, 274
        waste management, 277–278
        water supply, 274
Sybase, Inc., 211, 223–224, 238
systems integration (defined), 73

T
tactile characters (signage, ADA), 394
Talleyrand, Charles Maurice de
    meter based on unit of time, 13
Taylor, Frederick Winslow
    and Dos Passos, 93
    work measurement, 68, 73, 79
technically infeasible (definition, ADA), 366
technology
    construction, 34
    convergence, 73–74, 79, 176, 269
    process reengineering, 72
    twelfth field of study, 34
    uses and expression, 37
Tennessee Valley Authority, 270
Theory X and Theory Y, 69–70
Thoreau, Henry David, *Walden*, 270
three facility programs (case studies),
    142–148
Thucydides (quoting Pericles), 30
TIA. *See* ANSI (American National Standards
    Institute)
toilets and toilet compartments (ADA),
    382–385
top-down xv
Total Quality Management (TQM). *See*
    quality control
TRIRIGA, 229–230, 234–235
types of construction. *See* construction types

U
U.S. Department of Housing and Urban
    Development (HUD), 323

U.S. Environmental Protection Agency (Richard Nixon and . . .), 270
U.S. Green Building Council (USGBC). *See* LEED
UBC. *See* Uniform Building Code
Uniform Building Code (UBC), 293
Universe, models of. *See* Copernicus, Galileo, Newton, Ptolemy
usable area. *See* area, usable
Use & Occupancy Classification (IBC), 294–304
    Assembly, 295–297
    Business, 297
    Construction Types, 305–306
    Educational, 297
    Factory & Industrial, 297
    Fire Alarm Systems. *See* Fire Protection Systems, Alarm & Detection Systems
    High-Hazard, 297–298
    Institutional, 298
    Mercantile, 298
    Residential, 298
    Special Building Types. *See* Special Building Types
    Storage, 299
    Utility & Miscellaneous, 299
utilitatis (function), 35
utilization rate (plan efficiency), 157

V
Venturi, Robert
    less is a bore, 37
    quoting Cleanth Brooks, 37
venustatis (delight, beauty), 35
Vertical Penetrations (BOMA), 185
Villa Savoye. *See* Le Corbusier
virtual private network (VPN), 230
visible (fire) alarms, 320
visual alarms (ADA), 393
Vitruvian Man, 6–7
Vitruvius
    ad Quadratum. *See* ad Quadratum
    Classical Orders xx, 4–5, 34
    job description, 1
    project management, 77
    qualifications, 1
    six fundamental principles xx, 3–6, 21, 28–29, 33
    triad, 47

W
wayfinding
    code requirements (IBC), 336–347
    definition xxiii
    geometric organization, 31
    primary circulation, 141, 153–159
    signage (accessibility—IBC), 358–359
    signage (egress—IBC), 336
Web services, 237
Weber, Max
    capitalism (definition), 93
    Gothic architecture, 93
    symbolism, 176–177
Weil, Simone (epigram), 93
wheelchair lifts (safety standard), 379
wheelchair space
    area of refuge, 331
    assembly facilities, 356
worksheet, space requirements, 100. *See Appendix B*
World Trade Center (New York), 177
World War II (post-war era), 270
World Wide Web. *See* Internet
Wright, Frank Lloyd
    and Froebel blocks, 24
    and Owen Jones, 25–27
    as character prototype, 33
    early work, 43
    Guggenheim Museum, New York, 38
    inherent human factor, 8
    Mile-High skyscraper (Chicago), 176
    nine principles, 46–47
    quoting Daniel Burnham, 36–37
    quoting Victor Hugo in Testament, 27
    unit system, 44

X
X—in space allocation reference (SAR) key, 111

Y
Yeats, William Butler (qualities of life), 9

Z
Zeising, Adolf (design methodology), 18